The Accelerating Expansion of The Universe

" The Result of Dark Energy ? "

Edited by Paul F. Kisak

Contents

Chapter 1

Accelerating expansion of the universe

The **accelerating expansion of the universe** is the observation that the universe appears to be expanding at an increasing rate,[1] so that the velocity at which a distant galaxy is receding from the observer is continuously increasing with time.[2]

The accelerated expansion was discovered in 1998, by two independent projects, the Supernova Cosmology Project and the High-Z Supernova Search Team, which both used distant type Ia supernovae as standard candles to measure the acceleration.[3][4][5] The discovery was unexpected, cosmologists at the time expecting that the expansion would be decelerating due to the gravitational attraction of the matter in the universe. Three members of these two groups have subsequently been awarded Nobel Prizes for their discovery.[6] Confirmatory evidence has been found in baryon acoustic oscillations and in analyses of the clustering of galaxies.

The expansion of the universe is thought to have been accelerating since the universe entered its dark-energy-dominated era roughly 5 billion years ago.[7][notes 1] Within the framework of general relativity, an accelerating expansion can be accounted for by a positive value of the cosmological constant Λ, equivalent to the presence of a positive vacuum energy, dubbed "dark energy". While there are alternative possible explanations, the description assuming dark energy (positive Λ) is used in the current standard model of cosmology, which also includes cold dark matter (CDM) and is known as the Lambda-CDM model.

In the decades since the detection of cosmic microwave background (CMB) in 1965,[8] the Big Bang model has become the most accepted model explaining the evolution of our universe. The Friedmann equation defines how the energy in the universe drives its expansion.

$$H^2 = \left(\frac{\dot{a}}{a}\right)^2 = \frac{8\pi G}{3}\rho - \frac{\mathrm{K}c^2}{R^2a^2}$$

where K represents the curvature of the universe, $a(t)$ is the scale factor, ρ is the total energy density of the universe, and H is the Hubble parameter.[9]

We define a critical density

$$\rho_c = \frac{3H^2}{8\pi G}$$

and the density parameter

$$\Omega = \frac{\rho}{\rho_c}$$

We can then rewrite the Hubble parameter as

$$H(a) = H_0\sqrt{\Omega_k a^{-2} + \Omega_m a^{-3} + \Omega_r a^{-4} + \Omega_{\mathrm{DE}} a^{-3(1+w)}}$$

where the four currently hypothesized contributors to the energy density of the universe are curvature, matter, radiation and dark energy.[10] Each of the components decreases with the expansion of the universe (increasing scale factor), except perhaps the dark energy term. It is the values of these cosmological parameters which physicists use to determine the acceleration of the universe.

The acceleration equation describes the evolution of the scale factor with time

$$\frac{\ddot{a}}{a} = -\frac{4\pi G}{3}\left(\rho + \frac{3P}{c^2}\right)$$

where the pressure P is defined by the cosmological model chosen. (see explanatory models below)

Physicists at one time were so assured of the deceleration of the universe's expansion that they introduced a so-called deceleration parameter q_0.[11] Current observations point towards this deceleration parameter being negative.

1.2 Evidence for acceleration

To learn about the rate of expansion of the universe we look at the magnitude-redshift relationship of astronomical objects using standard candles, or their distance-redshift relationship using standard rulers. We can also look at the growth of large-scale structure, and find that the observed values of the cosmological parameters are best described by models which include an accelerating expansion.

1.2.1 Supernova observation

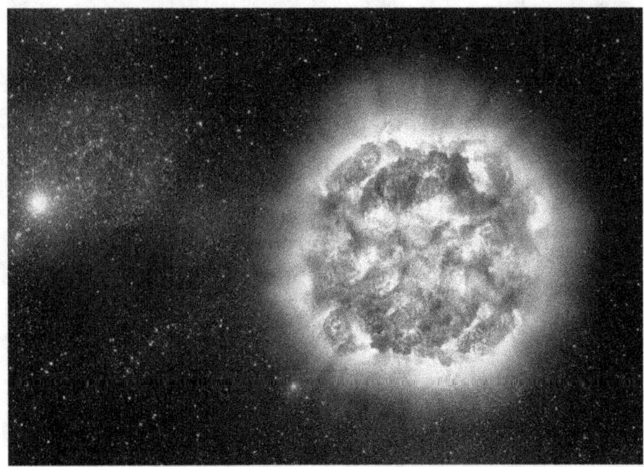

Artist's impression of a Type Ia supernova, as revealed by spectro-polarimetry observations

The first evidence for acceleration came from the observation of Type Ia supernovae, which are exploding white dwarfs that have exceeded their stability limit. Because they all have similar masses, their intrinsic luminosity is standardizable. Repeated imaging of selected areas of sky is used to discover the supernovae, then follow-up observations give their peak brightness, which is converted into a quantity known as luminosity distance (see distance measures in cosmology for details).[12] Spectral lines of their light can be used to determine their redshift.

For supernovae at redshift less than around 0.1, or light travel time less than 10 percent of the age of the universe, this gives a nearly linear distance–redshift relation due to Hubble's law. At larger distances, since the expansion rate of the universe has changed over time, the distance-redshift relation deviates from linearity, and this deviation depends on how the expansion rate has changed over time. The full calculation requires integration of the Friedmann equation, but a simple derivation can be given as follows: the redshift z directly gives the cosmic scale factor at the time the supernova exploded.

$$a(t) = \frac{1}{1+z}$$

So a supernova with a measured redshift $z = 0.5$ implies the universe was 1/1 + 0.5 = 2/3 of its present size when the supernova exploded. In an accelerating universe, the universe was expanding more slowly in the past than it is today, which means it took a longer time to expand from two thirds its present size to its present size compared to a non-accelerating universe. This results in a larger light-travel time, larger distance and fainter supernovae, which corresponds to the actual observations. Adam Riess found that "the distances of the high-redshift SNe Ia were, on average, 10% to 15% farther than expected in a low mass density ΩM = 0.2 universe without a cosmological constant".[13] This means that the measured high-redshift distances were too large, compared to nearby ones, for a decelerating universe.[14]

1.2.2 Baryon acoustic oscillations

Main article: Baryon acoustic oscillations

In the early universe before recombination and decoupling took place, photons and matter existed in a primordial plasma. Points of higher density in the photon-baryon plasma would contract, being compressed by gravity until the pressure became too large and they expanded again.[11] This contraction and expansion created vibrations in the plasma analogous to sound waves. Since dark matter only interacts gravitationally it stayed at the centre of the sound wave, the origin of the original overdensity. When decoupling occurred, approximately 380,000 years after the Big Bang,[15] photons separated from matter and were able to stream freely through the universe, creating the cosmic microwave background as we know it. This left shells of baryonic matter at a fixed radius from the overdensities of dark matter, a distance known as the sound horizon. As time passed and the universe expanded, it was at these anisotropies of matter density where galaxies started to form. So by looking at the distances at which galaxies at different redshifts tend to cluster, it is possible to determine a standard angular diameter distance and use that to compare to the distances predicted by different cosmological models.

Peaks have been found in the correlation function (the probability that two galaxies will be a certain distance apart) at 100 h^{-1} Mpc,[10] indicating that this is the size of the sound horizon today, and by comparing this to the sound horizon at the time of decoupling (using the CMB), we can confirm that the expansion of the universe is accelerating.[16]

1.2.3 Clusters of galaxies

Measuring the mass functions of galaxy clusters, which describe the number density of the clusters above a threshold mass, also provides evidence for dark energy.[17] By comparing these mass functions at high and low redshifts to those predicted by different cosmological models, values for w and Ω_m are obtained which confirm a low matter density and a non zero amount of dark energy.[14]

1.2.4 Age of the universe

See also: Age of the universe

Given a cosmological model with certain values of the cosmological density parameters, it is possible to integrate the Friedmann equations and derive the age of the universe.

$$t_0 = \int_0^1 \frac{da}{\dot{a}}$$

By comparing this to actual measured values of the cosmological parameters, we can confirm the validity of a model which is accelerating now, and had a slower expansion in the past.[14]

1.3 Explanatory models

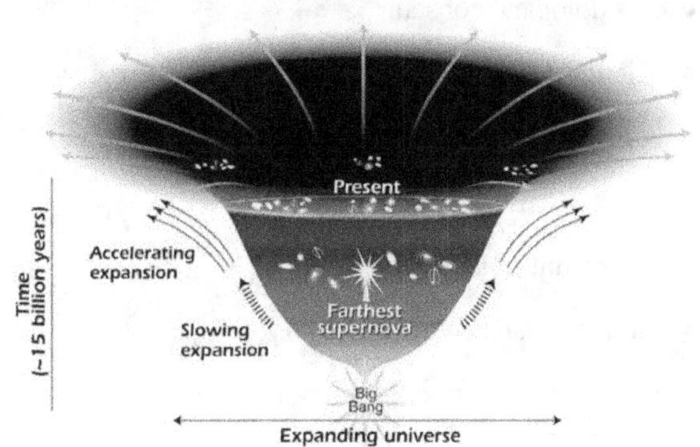

This diagram reveals changes in the rate of expansion since the universe's birth 15 billion years ago. The more shallow the curve, the faster the rate of expansion. The curve changes noticeably about 7.5 billion years ago, when objects in the universe began flying apart at a faster rate. Astronomers theorize that the faster expansion rate is due to a mysterious, dark force that is pushing galaxies apart.

The expansion of the Universe accelerating. Time flows from bottom to top

1.3.1 Dark energy

Main article: Dark energy

The most important property of dark energy is that it has negative pressure which is distributed relatively homogeneously in space.

$$P = wc^2\rho$$

where c is the speed of light and ρ is the energy density. Different theories of dark energy suggest different values of w, with $w < -1/3$ for cosmic acceleration (this leads to a positive value of ä in the acceleration equation above).

The simplest explanation for dark energy is that it is a cosmological constant or vacuum energy; in this case $w = -1$. This leads to the Lambda-CDM model, which has generally been known as the Standard Model of Cosmology from 2003 through the present, since it is the simplest model in good agreement with a variety of recent observations. Riess found that their results from supernovae observations favoured expanding models with positive cosmological constant ($\Omega\lambda > 0$) and a current acceleration of the expansion ($q_0 < 0$).[13]

1.3.2 Phantom energy

Main article: Phantom energy

Current observations allow the possibility of a cosmological model containing a dark energy component with equation of state $w < -1$. This phantom energy density would become infinite in finite time, causing such a huge gravitational repulsion that the universe would lose all structure and end in a Big Rip.[18] For example, for $w = -3/2$ and H_0 = 70 km·s^{-1}·Mpc^{-1}, the time remaining before the universe ends in this "Big Rip" is 22 billion years.[19]

1.3.3 Alternative theories

Other explanations for the accelerating universe include quintessence, a proposed form of dark energy with a non-constant state equation, whose density decreases with time. Dark fluid is an alternative explanation for accelerating expansion which attempts to unite dark matter and dark energy into a single framework.[20] Alternatively, some authors have argued that the universe expansion acceleration could be due to a repulsive gravitational interaction of antimatter.[21][22][23]

Another type of model, the backreaction conjecture,[24][25] was proposed by cosmologist Syksy Räsänen:[26] the rate of expansion is not homogenous, but we are in a region where expansion is faster than the background. Inhomogeneities in the early universe cause the formation of walls and bubbles, where the inside of a bubble has less matter than on average. According to general relativity, space is less curved than on the walls, and thus appears to have more volume and a higher expansion rate. In the denser regions, the expansion is retarded by a higher gravitational attraction. Therefore, the inward collapse of the denser regions looks the same as an accelerating expansion of the bubbles, leading us to conclude that the universe is expanding at an accelerating rate.[27] The benefit is that it does not require any new physics such as dark energy. Räsänen does not consider the model likely, but without any falsification, it must remain a possibility. It would require rather large density fluctuations (20%) to work.[26]

1.4 Theories for the consequences to the universe

See also: Future of an expanding universe

As the universe expands, the density of radiation and ordinary dark matter declines more quickly than the density of dark energy (see equation of state) and, eventually, dark energy dominates. Specifically, when the scale of the universe doubles, the density of matter is reduced by a factor of 8, but the density of dark energy is nearly unchanged (it is exactly constant if the dark energy is a cosmological constant).[11]

In models where dark energy is a cosmological constant, the universe will expand exponentially with time from now on, coming closer and closer to a de Sitter spacetime. This will eventually lead to all evidence for the Big Bang disappearing, as the cosmic microwave background is redshifted to lower intensities and longer wavelengths. Eventually its frequency will be low enough that it will be absorbed by the interstellar medium, and so be screened from any observer within the galaxy. This will occur when the universe is less than 50 times its current age, leading to the end of cosmology as we know it as the distant universe turns dark.[28]

A constantly expanding universe with non-zero cosmological constant has mass density decreasing over time, to an undetermined point when zero matter density is reached. All matter (electrons, protons and neutrons) would ionize and disintegrate, with objects dissipating away.[29]

Alternatives for the ultimate fate of the universe include the Big Rip mentioned above, a Big Bounce, Big Freeze, Big Crunch or possible proton decay.

1.5 Evidence against accelerating expansion

A 2016 report from Oxford University's Department of Physics and the Niels Bohr Institute in Copenhagen working with a much larger data set has cast doubt upon the arguments for accelerated expansion.[30] Sarkar comments:[31]

> The discovery of the accelerating expansion of the universe won the Nobel Prize, the Gruber Cosmology Prize, and the Breakthrough Prize in Fundamental Physics. It led to the widespread acceptance of the idea that the universe is dominated by "dark energy" that behaves like a cosmological constant – this is now the "standard model" of cosmology... However, there now exists a much bigger database of supernovae on which to perform rigorous and detailed statistical analyses. We analysed the latest catalogue of 740 Type Ia supernovae – over ten times bigger than the original samples on which the discovery claim was based – and found that the evidence for accelerated expansion is, at most, what physicists call "3 sigma". This is far short of the 5 sigma standard required to claim a discovery of fundamental significance.

1.6 See also

- Cosmological constant
- Friedmann–Lemaître–Robertson–Walker metric
- High-z Supernova Search Team
- Lambda-CDM model
- List of multiple discoveries
- Metric expansion of space
- Scale factor (cosmology)
- Supernova Cosmology Project

1.7 Notes

[1] [7]Frieman, Turner & Huterer (2008) p. 6: "The Universe has gone through three distinct eras: radiation-dominated, $z \gtrsim 3000$; matter-dominated, $3000 \gtrsim z \gtrsim 0.5$; and dark-energy-dominated, $z \lesssim 0.5$. The evolution of the scale factor is controlled by the dominant energy form: $a(t) \propto t^{2/3}(1 + w)$ (for constant w). During the radiation-dominated era, $a(t) \propto t^{1/2}$; during the matter-dominated era, $a(t) \propto t^{2/3}$; and for the dark energy-dominated era, assuming $w = -1$, asymptotically $a(t) \propto \exp(Ht)$."
p. 44: "Taken together, all the current data provide strong evidence for the existence of dark energy; they constrain the fraction of critical density contributed by dark energy, 0.76 ± 0.02, and the equation-of-state parameter, $w \approx -1 \pm 0.1$ (stat) ± 0.1 (sys), assuming that w is constant. This implies that the Universe began accelerating at redshift $z \sim 0.4$ and age $t \sim 10$ Gyr. These results are robust – data from any one method can be removed without compromising the constraints – and they are not substantially weakened by dropping the assumption of spatial flatness."

1.8 References

[1] Overbye, Dennis (20 February 2017). "Cosmos Controversy: The Universe Is Expanding, but How Fast?". *New York Times*. Retrieved 21 February 2017.

[2] "Is the universe expanding faster than the speed of light?".

[3] "Nobel physics prize honours accelerating universe find". BBC News. 2011-10-04.

[4] "The Nobel Prize in Physics 2011". Nobelprize.org. Retrieved 2011-10-06.

[5] Peebles, P. J. E.; Ratra, Bharat (2003). "The cosmological constant and dark energy". *Reviews of Modern Physics*. **75** (2): 559–606. Bibcode:2003RvMP...75..559P. arXiv:astro-ph/0207347. doi:10.1103/RevModPhys.75.559.

[6] Weinberg, Steven (2008). *Cosmology*. Oxford University Press. ISBN 9780198526827.

[7] Frieman, Joshua A.; Turner, Michael S.; Huterer, Dragan (2008-01-01). "Dark Energy and the Accelerating Universe". *Annual Review of Astronomy and Astrophysics*. **46** (1): 385–432. Bibcode:2008ARA&A..46..385F. arXiv:0803.0982. doi:10.1146/annurev.astro.46.060407.145243.

[8] Penzias, A. A.; Wilson, R. W. (1965). "A Measurement of Excess Antenna Temperature at 4080 Mc/s". *The Astrophysical Journal*. **142** (1): 419–421. Bibcode:1965ApJ...142..419P. doi:10.1086/148307.

[9] Nemiroff, Robert J.; Patla, Bijunath. "Adventures in Friedmann cosmology: A detailed expansion of the cosmological Friedmann equations". *American Journal of Physics*. **76** (3): 265. Bibcode:2008AmJPh..76..265N. arXiv:astro-ph/0703739. doi:10.1119/1.2830536.

[10] Lapuente, P. (2010). "Baryon Acoustic Oscillations". *Dark Energy: Observational and Theoretical Approaches*. Cambridge, UK: Cambridge University Press. ISBN 978-0521518888.

[11] Ryden, Barbara (2003). *Introduction to Cosmology*. San Francisco, CA: Addison Wesley. ISBN 978-0-8053-8912-8.

[12] Albrecht, Andreas; Bernstein, Gary; Cahn, Robert; Freedman, Wendy L.; Hewitt, Jacqueline; Hu, Wayne; Huth, John; Kamionkowski, Marc; Kolb, Edward W.; Knox, Lloyd; Mather, John C.; Staggs, Suzanne; Suntzeff, Nicholas B. (2006-09-20). "Report of the Dark Energy Task Force". arXiv:astro-ph/0609591.

[13] Riess, Adam G.; Filippenko, Alexei V.; Challis, Peter; Clocchiatti, Alejandro; Diercks, Alan; Garnavich, Peter M.; Gilliland, Ron L.; Hogan, Craig J.; Jha, Saurabh; Kirshner, Robert P.; Leibundgut, B.; Phillips, M. M.; Reiss, David; Schmidt, Brian P.; Schommer, Robert A.; Smith, R. Chris; Spyromilio, J.; Stubbs, Christopher; Suntzeff, Nicholas B.; Tonry, John. "Observational Evidence from Supernovae for an Accelerating Universe and a Cosmological Constant". *The Astronomical Journal*. **116** (3): 1009–1038. Bibcode:1998AJ....116.1009R. arXiv:astro-ph/9805201. doi:10.1086/300499.

[14] Pain, Reynald; Astier, Pierre (2012). "Observational evidence of the accelerated expansion of the Universe". *Comptes Rendus Physique*. **13** (6): 521–538. arXiv:1204.5493. doi:10.1016/j.crhy.2012.04.009.

[15] Hinshaw, G. (2014). "Five-Year Wilkinson Microwave Anisotropy Probe (WMAP) Observations: Data Processing, Sky Maps, and Basic Results". *Astrophysical Journal Supplement*. **180**: 225–245. Bibcode:2009ApJS..180..225H. arXiv:0803.0732. doi:10.1088/0067-0049/180/2/225.

[16] Eisenstein, Daniel J.; Zehavi, Idit; Hogg, David W.; Scoccimarro, Roman; Blanton, Michael R.; Nichol, Robert C.; Scranton, Ryan; Seo, Hee-Jong; Tegmark, Max; Zheng, Zheng; Anderson, Scott F.; Annis, Jim; Bahcall, Neta; Brinkmann, Jon; Burles, Scott; Castander, Francisco J.; Connolly, Andrew; Csabai, Istvan; Doi, Mamoru; Fukugita, Masataka; Frieman, Joshua A.; Glazebrook, Karl; Gunn, James E.; Hendry, John S.; Hennessy, Gregory; Ivezić, Zeljko; Kent, Stephen; Knapp, Gillian R.; Lin, Huan; Loh, Yeong-Shang; Lupton, Robert H.; Margon, Bruce; McKay, Timothy A.; Meiksin, Avery; Munn, Jeffery A.; Pope, Adrian; Richmond, Michael W.; Schlegel, David; Schneider, Donald P.; Shimasaku, Kazuhiro; Stoughton, Christopher; Strauss, Michael A.; SubbaRao, Mark; Szalay, Alexander S.; Szapudi, Istvan; Tucker, Douglas L.; Yanny, Brian; York, Donald G. (2005-11-10). "Detection of the Baryon Acoustic Peak in the Large-Scale Correlation Function of SDSS Luminous Red Galaxies". *The Astrophysical Journal*. **633** (2): 560–574. Bibcode:2005ApJ...633..560E. arXiv:astro-ph/0501171. doi:10.1086/466512.

[17] Dekel, Avishai (1999). *Formation of Structure in the Universe*. New York, NY: Cambridge University Press. ISBN 9780521586320.

[18] Caldwell, Robert; Kamionkowski, Marc; Weinberg, Nevin (August 2003). "Phantom Energy: Dark Energy with $w < -1$ Causes a Cosmic Doomsday". *Physical Review Letters*. **91** (7): 071301. Bibcode:2003PhRvL..91g1301C. PMID 12935004. arXiv:astro-ph/0302506. doi:10.1103/PhysRevLett.91.071301.

[19] Caldwell, R. R. (2002). "A phantom menace? Cosmological consequences of a dark energy component with super-negative equation of state". *Physics Letters B*. **545** (1–2): 23–29. Bibcode:2002PhLB..545...23C. arXiv:astro-ph/9908168. doi:10.1016/S0370-2693(02)02589-3.

[20] Halle, Anaelle; Zhao, Hongsheng; Li, Baojiu (2008). "=Perturbations in a non-uniform dark energy fluid: equations reveal effects of modified gravity and dark matter". *Astrophys-*

ical Journal Supplement Series. **177** (1). arXiv:0711.0958 . doi:10.1086/587744.

[21] Benoit-Lévy, A.; Chardin, G. (2012). “Introducing the Dirac–Milne universe”. *Astronomy and Astrophysics.* **537**: A78. doi:10.1051/0004-6361/201016103.

[22] Hajduković, D. S. (2012). “Quantum vacuum and virtual gravitational dipoles: the solution to the dark energy problem?". *Astrophysics and Space Science.* **339** (1): 1–5. doi:10.1007/s10509-012-0992-y.

[23] Villata, M. (2013). “On the nature of dark energy: the lattice Universe”. *Astrophysics and Space Science.* **345**: 1. arXiv:1302.3515 . doi:10.1007/s10509-013-1388-3.

[24] “Backreaction: directions of progress”. *Classical and Quantum Gravity.* **28**: 164008. Bibcode:2011CQGra..28p4008R. arXiv:1102.0408 . doi:10.1088/0264-9381/28/16/164008.

[25] “Backreaction in Late-Time Cosmology”. *Annual Review of Nuclear and Particle Science.* **62**: 57–79. Bibcode:2012ARNPS..62...57B. arXiv:1112.5335 . doi:10.1146/annurev.nucl.012809.104435.

[26] “Is dark energy an illusion?". *New Scientist.* 2007.

[27] “A Cosmic 'Tardis': What the Universe Has In Common with 'Doctor Who'". *Space.com.*

[28] Krauss, Lawrence M.; Scherrer, Robert J. (2007-06-28). “The return of a static universe and the end of cosmology”. *General Relativity and Gravitation.* **39** (10): 1545–1550. Bibcode:2007GReGr..39.1545K. arXiv:0704.0221 . doi:10.1007/s10714-007-0472-9.

[29] John Baez, “The End of the Universe”, 7 February 2016. http://math.ucr.edu/home/baez/end.html

[30] J. T. Nielsen; A. Guffanti; S. Sarkar (21 October 2016). “Marginal evidence for cosmic acceleration from Type Ia supernovae”. *Nature Scientific Reports.* **6**.

[31] Stuart Gillespie (21 October 2016). “The universe is expanding at an accelerating rate – or is it?". *University of Oxford - News & Events - Science Blog (WP:NEWSBLOG).*

Chapter 2

Supernova Cosmology Project

The **Supernova Cosmology Project** is one of two research teams that determined the likelihood of an accelerating universe and therefore a positive cosmological constant, using data from the redshift of Type Ia supernovae.[1] The project is headed by Saul Perlmutter at Lawrence Berkeley National Laboratory, with members from Australia, Chile, France, Portugal, Spain, Sweden, United Kingdom and United States.

This discovery was named “Breakthrough of the Year for 1998” by Science Magazine[2] and, along with the High-z Supernova Search Team, the project team won the 2007 Gruber Prize in Cosmology[3] and the 2015 Breakthrough Prize in Fundamental Physics.[4] In 2011, Perlmutter was awarded the Nobel Prize in Physics for this work, alongside Adam Riess and Brian P. Schmidt from the High-z team.[5]

2.1 Project Members

The team members are:[4][6]

- Saul Perlmutter, Lawrence Berkeley National Laboratory
- Gregory Aldering, Lawrence Berkeley National Laboratory
- Brian J. Boyle, Australia Telescope National Facility
- Patricia G. Castro, Instituto Superior Técnico, Lisbon
- Warrick Couch, Swinburne University of Technology
- Susana Deustua, American Astronomical Society
- Richard Ellis, California Institute of Technology
- Sebastien Fabbro, Instituto Superior Técnico, Lisbon
- Alexei Filippenko, University of California, Berkeley (later a member of the High-z Supernova Search Team)
- Andrew Fruchter, Space Telescope Science Institute
- Gerson Goldhaber, Lawrence Berkeley National Laboratory
- Ariel Goobar, University of Stockholm
- Donald Groom, Lawrence Berkeley National Laboratory
- Isobel Hook, University of Oxford
- Mike Irwin, University of Cambridge
- Alex Kim, Lawrence Berkeley National Laboratory
- Matthew Kim
- Robert Knop, Vanderbilt University
- Julia C. Lee, Harvard University
- Chris Lidman, European Southern Observatory
- Thomas Matheson, NOAO Gemini Science Center
- Richard McMahon, University of Cambridge
- Richard Muller, University of California, Berkeley
- Heidi Newberg, Rensselaer Polytechnic Institute
- Peter Nugent, Lawrence Berkeley National Laboratory
- Nelson Nunes, University of Cambridge
- Reynald Pain, CNRS-IN2P3, Paris
- Nino Panagia, Space Telescope Science Institute
- Carl Pennypacker, University of California, Berkeley
- Robert Quimby, The University of Texas
- Pilar Ruiz-Lapuente, University of Barcelona
- Bradley E. Schaefer, Louisiana State University
- Nicholas Walton, University of Cambridge

2.2 References

[1] Goldhaber, Gerson (2009). "The Acceleration of the Expansion of the Universe: A Brief Early History of the Supernova Cosmology Project (SCP)". *AIP Conference Proceedings*. **1166**: 53. Bibcode:2009AIPC.1166...53G. arXiv:0907.3526. doi:10.1063/1.3232196.

[2] Cosmic Motion Revealed *Science* **282**(5397), 2156-2157

[3] Gruber Foundation Prize in Cosmology Press Release

[4] Recipients Of The 2015 Breakthrough Prizes In Fundamental Physics And Life Sciences Announced

[5] "Nobel physics prize honours accelerating Universe find". BBC News. 2011-10-04.

[6] Gruber Foundation: Saul Perlmutter & the Supernova Cosmology Project

2.3 External links

- Supernova Cosmology Project Mainsite

Chapter 3

Shape of the universe

"Edge of the Universe" redirects here. For the Bee Gees song, see Edge of the Universe (song).

The **shape of the universe** is the local and global geometry of the universe. The local features of the geometry of the universe are primarily described by its curvature, whereas the topology of the universe describes general global properties of its shape as of a continuous object. The shape of the universe is related to general relativity which describes how spacetime is curved and bent by mass and energy.

Cosmologists distinguish between the observable universe and the global universe. The observable universe consists of the part of the universe that can, in principle, be observed by light reaching Earth within the age of the universe. It encompasses a region of space which currently forms a ball centered at Earth of estimated radius 46 billion light-years (4.4×10^{26} m) (however we observe distant areas only in their past, being closer at the time). Assuming an isotropic nature, the observable universe is similar for all contemporary vantage points.

According to the book *Our Mathematical Universe*, the shape of the global universe can be explained with three categories:[1]

1. Finite or infinite
2. Flat (no curvature), open (negative curvature) or closed (positive curvature)
3. Connectivity, how the universe is put together, i.e., simply connected space or multiply connected.

There are certain logical connections among these properties. For example, a universe with positive curvature is necessarily finite.[2] Although it is usually assumed in the literature that a flat or negatively curved universe is infinite, this need not be the case if the topology is not the trivial one.[2]

The exact shape is still a matter of debate in physical cosmology, but experimental data from various, independent sources (WMAP, BOOMERanG and Planck for example) confirm that the observable universe is flat with only a 0.4% margin of error.[3][4][5] Theorists have been trying to construct a formal mathematical model of the shape of the universe. In formal terms, this is a 3-manifold model corresponding to the spatial section (in comoving coordinates) of the 4-dimensional space-time of the universe. The model most theorists currently use is the Friedmann–Lemaître–Robertson–Walker (FLRW) model. Arguments have been put forward that the observational data best fit with the conclusion that the shape of the global universe is infinite and flat,[6] but the data are also consistent with other possible shapes, such as the so-called Poincaré dodecahedral space[7][8] and the Sokolov-Starobinskii space (quotient of the upper half-space model of hyperbolic space by 2-dimensional lattice).[9]

3.1 Shape of the observable universe

Main article: Observable universe
See also: Distance measures (cosmology)

As stated in the introduction, there are two aspects to consider:

1. its *local* geometry, which predominantly concerns the curvature of the universe, particularly the observable universe, and
2. its *global* geometry, which concerns the topology of the universe as a whole.

The observable universe can be thought of as a sphere that extends outwards from any observation point for 46.5 billion light years, going farther back in time and more redshifted the more distant away one looks. Ideally, one can continue to look back all the way to the Big Bang; in practice, however, the farthest away one can look is the cosmic microwave background (CMB), as anything past that was opaque. Experimental investigations show that the observable universe is very close to isotropic and homogeneous.

If the observable universe encompasses the entire universe, we may be able to determine the global structure of the entire universe by observation. However, if the observable universe is smaller than the entire universe, our observations will be limited to only a part of the whole, and we may not be able to determine its global geometry through measurement. From experiments, it is possible to construct different mathematical models of the global geometry of the entire universe all of which are consistent with current observational data and so it is currently unknown whether the observable universe is identical to the global universe or it is instead many orders of magnitude smaller than it. The universe may be small in some dimensions and not in others (analogous to the way a cuboid is longer in the dimension of length than it is in the dimensions of width and depth). To test whether a given mathematical model describes the universe accurately, scientists look for the model's novel implications—what are some phenomena in the universe that we have not yet observed, but that must exist if the model is correct—and they devise experiments to test whether those phenomena occur or not. For example, if the universe is a small closed loop, one would expect to see multiple images of an object in the sky, although not necessarily images of the same age.

Cosmologists normally work with a given space-like slice of spacetime called the comoving coordinates, the existence of a preferred set of which is possible and widely accepted in present-day physical cosmology. The section of spacetime that can be observed is the backward light cone (all points within the cosmic light horizon, given time to reach a given observer), while the related term Hubble volume can be used to describe either the past light cone or comoving space up to the surface of last scattering. To speak of "the shape of the universe (at a point in time)" is ontologically naive from the point of view of special relativity alone: due to the relativity of simultaneity we cannot speak of different points in space as being "at the same point in time" nor, therefore, of "the shape of the universe at a point in time". However, the comoving coordinates (if well-defined) provide a strict sense to those by using the time since the Big Bang (measured in the reference of CMB) as a distinguished universal time.

3.2 Curvature of the universe

Main article: spatial curvature

The curvature is a quantity describing how the geometry of a space differs locally from the one of the flat space. The curvature of any isotropic space (and hence of an isotropic universe) falls into one of the following three cases:

1. Zero curvature (flat); a drawn triangle's angles add up to 180° and the Pythagorean theorem holds
2. Positive curvature; a drawn triangle's angles add up to more than 180°
3. Negative curvature; a drawn triangle's angles add up to less than 180°

An example of a flat space would be any subset of a Euclidean space, e.g., a triangle in a plane.

Curved geometries are in the domain of Non-Euclidean geometry. An example of a positively curved space would be the surface of a sphere such as the Earth. A triangle drawn from the equator to a pole will have at least two angles equal 90°, which makes the sum of the 3 angles greater than 180°. An example of a negatively curved surface would be the shape of a saddle or mountain pass. A triangle drawn on a saddle surface will have the sum of the angles adding up to less than 180°.

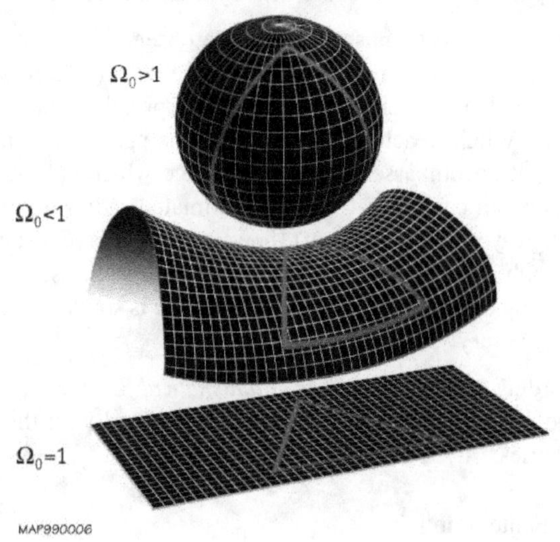

The local geometry of the universe is determined by whether the density parameter Ω is greater than, less than, or equal to 1. From top to bottom: a spherical universe with Ω > 1, a hyperbolic universe with Ω < 1, and a flat universe with Ω = 1. Note that these depictions of two-dimensional surfaces are merely easily visualizable analogs to the 3-dimensional structure of (local) space.

General relativity explains that mass and energy bend the curvature of spacetime and is used to determine what curvature the universe has by using a value called the density parameter, represented with Omega (Ω). The density parameter is the average density of the universe divided by the critical energy density, that is, the mass energy needed for a universe to be flat. Put another way

- If $\Omega = 1$, the universe is flat
- If $\Omega > 1$, there is positive curvature
- if $\Omega < 1$ there is negative curvature

The geometry of the universe is usually represented in the system of comoving coordinates, according to which the expansion of the universe can be ignored. Comoving coordinates form a single frame of reference according to which the universe has a static geometry of three spatial dimensions.

Under the assumption that the universe is homogeneous and isotropic, the curvature of the observable universe, or the local geometry, is described by one of the three "primitive" geometries (in mathematics these are called the model geometries):

- 3-dimensional Flat Euclidean geometry, generally notated as E^3
- 3-dimensional spherical geometry with a small curvature, often notated as S^3
- 3-dimensional hyperbolic geometry with a small curvature

One can experimentally calculate this Ω to determine the curvature two ways. One is to count up all the mass-energy in the universe and take its average density then divide that average by the critical energy density. Data from Wilkinson Microwave Anisotropy Probe (WMAP) as well as the Planck spacecraft give values for the three constituents of all the mass-energy in the universe - normal mass (baryonic matter and dark matter), relativistic particles (photon and neutrinos) and dark energy or the cosmological constant:[10][11]

$\Omega_{mass} \approx 0.315 \pm 0.018$

$\Omega_{relativistic} \approx 9.24 \times 10^{-5}$

$\Omega\Lambda \approx 0.6817 \pm 0.0018$

$\Omega_{total} = \Omega_{mass} + \Omega_{relativistic} + \Omega\Lambda = 1.00 \pm 0.02$

The actual value for critical density value is measured as $\rho_{critical} = 9.47 \times 10^{-27}$ kg m^{-3}. From these values, it seems that within experimental error, the universe seems to be flat.

Another way to measure Ω is to do so geometrically by measuring an angle across the observable universe. We can do this by using the CMB and measuring the power spectrum and temperature anisotropy. For an intuition, one can imagine finding a gas cloud that is not in thermal equilibrium due to being so large that light speed cannot propagate the thermal information. Knowing this propagation speed, we then know the size of the gas cloud as well as the distance to the gas cloud, we then have two sides of a triangle and can then determine the angles. Using a method similar to this, the BOOMERanG experiment has determined that the sum of the angles to 180° within experimental error, corresponding to an $\Omega_{total} \approx 1.00 \pm 0.12$.[12]

These and other astronomical measurements constrain the spatial curvature to be very close to zero, although they do not constrain its sign. This means that although the local geometries of spacetime are generated by the theory of relativity based on spacetime intervals, we can approximate *3-space* by the familiar Euclidean geometry.

The Friedmann–Lemaître–Robertson–Walker (FLRW) model using Friedmann equations is commonly used to model the universe. The FLRW model provides a curvature of the universe based on the mathematics of fluid dynamics, that is, modeling the matter within the universe as a perfect fluid. Although stars and structures of mass can be introduced into an "almost FLRW" model, a strictly FLRW model is used to approximate the local geometry of the observable universe. Another way of saying this is that if all forms of dark energy are ignored, then the curvature of the universe can be determined by measuring the average density of matter within it, assuming that all matter is evenly distributed (rather than the distortions caused by 'dense' objects such as galaxies). This assumption is justified by the observations that, while the universe is "weakly" inhomogeneous and anisotropic (see the large-scale structure of the cosmos), it is on average homogeneous and isotropic.

3.3 Global universe structure

Global structure covers the geometry and the topology of the whole universe—both the observable universe and beyond. While the local geometry does not determine the global geometry completely, it does limit the possibilities, particularly a geometry of a constant curvature. The universe is often taken to be a geodesic manifold, free of topological defects; relaxing either of these complicates the analysis considerably. A global geometry is a local geometry plus a topology. It follows that a topology alone does not give a global geometry: for instance, Euclidean 3-space and hyperbolic 3-space have the same topology but different global geometries.

As stated in the introduction, investigations within the study of the global structure of the universe include:

- Whether the universe is infinite or finite in extent
- Whether the geometry of the global universe is flat, positively curved, or negatively curved
- Whether the topology is simply connected like a sphere or multiply connected, like a torus[13]

3.3.1 Infinite or finite

One of the presently unanswered questions about the universe is whether it is infinite or finite in extent. For intuition, it can be understood that a finite universe has a finite volume that, for example, could be in theory filled up with a finite amount of material, while an infinite universe is unbounded and no numerical volume could possibly fill it. Mathematically, the question of whether the universe is infinite or finite is referred to as boundedness. An infinite universe (unbounded metric space) means that there are points *arbitrarily* far apart: for any distance d, there are points that are of a distance at least d apart. A finite universe is a bounded metric space, where there is some distance d such that all points are within distance d of each other. The smallest such d is called the diameter of the universe, in which case the universe has a well-defined "volume" or "scale."

Bounded or unbounded

Assuming a finite universe, the universe can either have an edge or no edge. Many finite mathematical spaces, e.g., a disc, have an edge or boundary. Spaces that have an edge are difficult to treat, both conceptually and mathematically. Namely, it is very difficult to state what would happen at the edge of such a universe. For this reason, spaces that have an edge are typically excluded from consideration.

However, there exist many finite spaces, such as the 3-sphere and 3-torus, which have no edges. Mathematically, these spaces are referred to as being compact without boundary. The term compact basically means that it is finite in extent ("bounded") and is a closed set. The term "without boundary" means that the space has no edges. Moreover, so that calculus can be applied, the universe is typically assumed to be a differentiable manifold. A mathematical object that possesses all these properties, compact without boundary and differentiable, is termed a closed manifold. The 3-sphere and 3-torus are both closed manifolds.

An infinite universe (or infinite in a specific spatial direction) must be unbounded in that direction.

3.3.2 Curvature

The curvature of the universe places constraints on the topology. If the spatial geometry is spherical, i.e., possess positive curvature, the topology is compact. For a flat (zero curvature) or a hyperbolic (negative curvature) spatial geometry, the topology can be either compact or infinite.[14] It's very important to note that many textbooks erroneously state that a flat universe implies an infinite universe; however, the correct statement is that a flat universe that is also simply connected implies an infinite universe.[14] For example, Euclidean space is flat, simply connected and infinite, but the torus is flat, multiply connected, finite and compact.

In general, local to global theorems in Riemannian geometry relate the local geometry to the global geometry. If the local geometry has constant curvature, the global geometry is very constrained, as described in Thurston geometries.

The latest research shows that even the most powerful future experiments (like SKA, Planck..) will not be able to distinguish between flat, open and closed universe if the true value of cosmological curvature parameter is smaller than 10^{-4}. If the true value of the cosmological curvature parameter is larger than 10^{-3} we will be able to distinguish between these three models even now.[15]

Results of the *Planck* mission released in 2015 show the cosmological curvature parameter, ΩK, to be 0.000±0.005, consistent with a flat universe.[16]

Universe with zero curvature

In a universe with zero curvature, the local geometry is flat. The most obvious global structure is that of Euclidean space, which is infinite in extent. Flat universes that are finite in extent include the torus and Klein bottle. Moreover, in three dimensions, there are 10 finite closed flat 3-manifolds, of which 6 are orientable and 4 are non-orientable. These are the Bieberbach manifolds. The most familiar is the aforementioned 3-Torus universe.

In the absence of dark energy, a flat universe expands forever but at a continually decelerating rate, with expansion asymptotically approaching zero. With dark energy, the expansion rate of the universe initially slows down, due to the effect of gravity, but eventually increases. The ultimate fate of the universe is the same as that of an open universe.

A flat universe can have zero total energy.

Universe with positive curvature

A positively curved universe is described by elliptic geometry, and can be thought of as a three-dimensional hypersphere, or some other spherical 3-manifold (such as the Poincaré dodecahedral space), all of which are quotients of the 3-sphere.

Poincaré dodecahedral space, a positively curved space, colloquially described as "soccerball-shaped", as it is the quotient of the 3-sphere by the binary icosahedral group, which is very close to icosahedral symmetry, the symmetry of a soccer ball. This was proposed by Jean-Pierre Luminet and colleagues in 2003[7][17] and an optimal orientation on the sky for the model was estimated in 2008.[8]

Universe with negative curvature

A hyperbolic universe, one of a negative spatial curvature, is described by hyperbolic geometry, and can be thought of locally as a three-dimensional analog of an infinitely extended saddle shape. There are a great variety of hyperbolic 3-manifolds, and their classification is not completely understood. Those of finite volume can be understood via the Mostow rigidity theorem. For hyperbolic local geometry, many of the possible three-dimensional spaces are informally called **horn topologies**, so called because of the

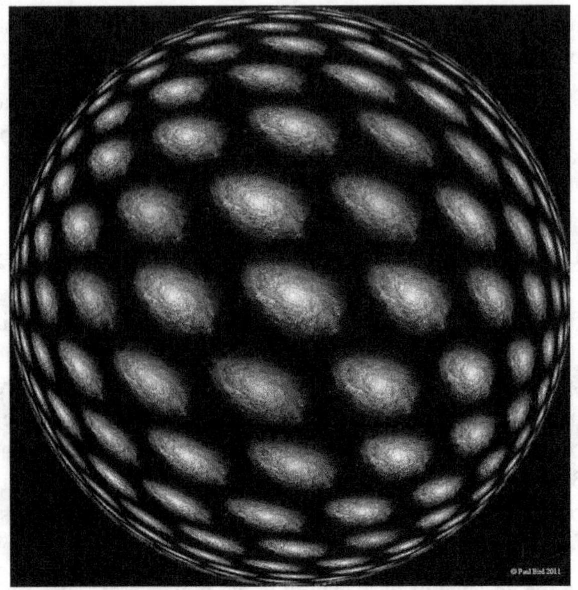

Universe in an expanding sphere. *The galaxies farthest away are moving fastest and hence experience length contraction and so become smaller to an observer in the centre.*

shape of the pseudosphere, a canonical model of hyperbolic geometry. An example is the Picard horn, a negatively curved space, colloquially described as "funnel-shaped".[9]

Curvature: open or closed

When cosmologists speak of the universe as being "open" or "closed", they most commonly are referring to whether the curvature is negative or positive. These meanings of open and closed are different from the mathematical meaning of open and closed used for sets in topological spaces and for the mathematical meaning of open and closed manifolds, which gives rise to ambiguity and confusion. In mathematics, there are definitions for a closed manifold (i.e., compact without boundary) and open manifold (i.e., one that is not compact and without boundary). A "closed universe" is necessarily a closed manifold. An "open universe" can be either a closed or open manifold. For example, in the Friedmann–Lemaître–Robertson–Walker (FLRW) model the universe is considered to be without boundaries, in which case "compact universe" could describe a universe that is a closed manifold.

Milne model ("spherical" expanding)

Main article: Milne model

If one applies Minkowski space-based Special Relativity to expansion of the universe, without resorting to the concept of a curved spacetime, then one obtains the Milne model. Any spatial section of the universe of a constant age (the proper time elapsed from the Big Bang) will have a negative curvature; this is merely a pseudo-Euclidean geometric fact analogous to one that concentric spheres in the *flat* Euclidean space are nevertheless curved. Spatial geometry of this model is an unbounded hyperbolic space. The entire universe is contained within a light cone, namely the future cone of the Big Bang. For any given moment $t > 0$ of coordinate time (assuming the Big Bang has $t = 0$), the entire universe is bounded by a sphere of radius exactly $c\ t$. The apparent paradox of an infinite universe contained within a sphere is explained with length contraction: the galaxies farther away, which are travelling away from the observer the fastest, will appear thinner.

This model is essentially a degenerate FLRW for $\Omega = 0$. It is incompatible with observations that definitely rule out such a large negative spatial curvature. However, as a background in which gravitational fields (or gravitons) can operate, due to diffeomorphism invariance, the space on the macroscopic scale, is equivalent to any other (open) solution of Einstein's field equations.

3.4 See also

- De Sitter space
- Ekpyrotic universe – a String theory-related model depicting a five-dimensional, membrane-shaped universe; an alternative to the Hot Big Bang Model, whereby the universe is described to have originated when two membranes collided at the fifth dimension
- Extra dimensions in String Theory for 6 or 7 extra space-like dimensions all with a *compact* topology.
- History of the Center of the Universe
- Holographic Universe
- List of Cosmology paradoxes
- Theorema Egregium – The "remarkable theorem" discovered by Gauss which showed there is an intrinsic notion of curvature for surfaces. This is used by Riemann to generalize the (intrinsic) notion of curvature to higher-dimensional spaces
- Three-torus model of the universe
- Zero-energy universe

3.5 References

[1] Tegmark, Max (2014). *Our Mathematical Universe: My Quest for the Ultimate Nature of Reality* (1 ed.). Knopf. ISBN 978-0307599803.

[2] G. F. R. Ellis; H. van Elst (1999). "Cosmological models (Cargèse lectures 1998)". In Marc Lachièze-Rey. *Theoretical and Observational Cosmology*. NATO Science Series C. **541**. p. 22. Bibcode:1999toc..conf....1E. ISBN 978-0792359463. arXiv:gr-qc/9812046.

[3] "Will the Universe expand forever?". NASA. 24 January 2014. Retrieved 16 March 2015.

[4] "Our universe is Flat". FermiLab/SLAC. 7 April 2015.

[5] Marcus Y. Yoo (2011). "Unexpected connections". *Engineering & Science*. Caltech. LXXIV1: 30.

[6] Demianski, Marek; Sánchez, Norma; Parijskij, Yuri N. (2003). "Topology of the universe and the cosmic microwave background radiation". *The Early Universe and the Cosmic Microwave Background: Theory and Observations. Proceedings of the NATO Advanced Study Institute*. The early universe and the cosmic microwave background: theory and observations. Springer. **130**: 161. Bibcode:2003eucm.book..159D. ISBN 1-4020-1800-2.

[7] Luminet, Jean-Pierre; Weeks, Jeff; Riazuelo, Alain; Lehoucq, Roland; Uzan, Jean-Phillipe (2003-10-09). "Dodecahedral space topology as an explanation for weak wide-angle temperature correlations in the cosmic microwave background". *Nature*. **425** (6958): 593–5. Bibcode:2003Natur.425..593L. PMID 14534579. arXiv:astro-ph/0310253. doi:10.1038/nature01944.

[8] Roukema, Boudewijn; Zbigniew Buliński; Agnieszka Szaniewska; Nicolas E. Gaudin (2008). "A test of the Poincare dodecahedral space topology hypothesis with the WMAP CMB data". *Astronomy and Astrophysics*. **482** (3): 747. Bibcode:2008A&A...482..747L. arXiv:0801.0006. doi:10.1051/0004-6361:20078777.

[9] Aurich, Ralf; Lustig, S.; Steiner, F.; Then, H. (2004). "Hyperbolic Universes with a Horned Topology and the CMB Anisotropy". *Classical and Quantum Gravity*. **21** (21): 4901–4926. Bibcode:2004CQGra..21.4901A. arXiv:astro-ph/0403597. doi:10.1088/0264-9381/21/21/010.

[10] "Density Parameter, Omega". *hyperphysics.phy-astr.gsu.edu*. Retrieved 2015-06-01.

[11] Ade, P. A. R.; Aghanim, N.; Armitage-Caplan, C.; Arnaud, M.; Ashdown, M.; Atrio-Barandela, F.; Aumont, J.; Baccigalupi, C.; Banday, A. J.; Barreiro, R. B.; Bartlett, J. G.; Battaner, E.; Benabed, K.; Benoît, A.; Benoit-Lévy, A.; Bernard, J.-P.; Bersanelli, M.; Bielewicz, P.; Bobin, J.; Bock, J. J.; Bonaldi, A.; Bond, J. R.; Borrill, J.; Bouchet, F. R.; Bridges, M.; Bucher, M.; Burigana, C.; Butler, R. C.; Calabrese, E.; et al. (2014). "Planck2013 results. XVI. Cosmological parameters". *Astronomy & Astrophysics*. **571**: A16. Bibcode:2014A&A...571A..16P. arXiv:1303.5076. doi:10.1051/0004-6361/201321591.

[12] De Bernardis, P.; Ade, P. A. R.; Bock, J. J.; Bond, J. R.; Borrill, J.; Boscaleri, A.; Coble, K.; Crill, B. P.; De Gasperis, G.; Farese, P. C.; Ferreira, P. G.; Ganga, K.; Giacometti, M.; Hivon, E.; Hristov, V. V.; Iacoangeli, A.; Jaffe, A. H.; Lange, A. E.; Martinis, L.; Masi, S.; Mason, P. V.; Mauskopf, P. D.; Melchiorri, A.; Miglio, L.; Montroy, T.; Netterfield, C. B.; Pascale, E.; Piacentini, F.; Pogosyan, D.; et al. (2000). "A flat Universe from high-resolution maps of the cosmic microwave background radiation". *Nature*. **404** (6781): 955–9. Bibcode:2000Natur.404..955D. PMID 10801117. arXiv:astro-ph/0004404. doi:10.1038/35010035.

[13] P.C.W.Davies (1977). *Space and time in the modern universe*. cambridge university press. ISBN 0-521-29151-8.

[14] Luminet, Jean-Pierre; Lachièze-Rey, Marc (1995). "Cosmic Topology". *Physics Reports*. **254** (3): 135–214. Bibcode:1995PhR...254..135L. arXiv:gr-qc/9605010. doi:10.1016/0370-1573(94)00085-h.

[15] Vardanyan, Mihran; Trotta, Roberto; Silk, Joseph (2009). "How flat can you get? A model comparison perspective on the curvature of the Universe". *Monthly Notices of the Royal Astronomical Society*. **397**: 431. Bibcode:2009MNRAS.397..431V. arXiv:0901.3354. doi:10.1111/j.1365-2966.2009.14938.x.

[16] Planck Collaboration; Ade, P. A. R.; Aghanim, N.; Arnaud, M.; Ashdown, M.; Aumont, J.; Baccigalupi, C.; Banday, A. J.; Barreiro, R. B.; Bartlett, J. G.; Bartolo, N.; Battaner, E.; Battye, R.; Benabed, K.; Benoit, A.; Benoit-Levy, A.; Bernard, J.-P.; Bersanelli, M.; Bielewicz, P.; Bonaldi, A.; Bonavera, L.; Bond, J. R.; Borrill, J.; Bouchet, F. R.; Boulanger, F.; Bucher, M.; Burigana, C.; Butler, R. C.; Calabrese, E.; et al. (2015). "Planck 2015 results. XIII. Cosmological parameters". arXiv:1502.01589 [astro-ph.CO].

[17] "Is the universe a dodecahedron?", article at PhysicsWeb.

3.6 External links

- How do we know that the universe is flat A video explains how astrophysicists measure the geometry of the universe at Physicsworld.com
- Geometry of the Universe at icosmos.co.uk
- Janna Levin, Evan Scannapieco & Joseph Silk (1998). "The topology of the universe: the biggest manifold of them all". *Classical and Quantum Gravity*. **15** (9). Bibcode:1998CQGra..15.2689L. arXiv:gr-qc/9803026. doi:10.1088/0264-9381/15/9/015.

- Lachièze-Rey, M., Luminet, J.P. (1995). “Cosmic Topology”. *Physics Reports*. **254** (3): 135–214. Bibcode:1995PhR...254..135L. arXiv:gr-qc/9605010 🔓. doi:10.1016/0370-1573(94)00085-H.
- Universe is Finite, “Soccer Ball"-Shaped, Study Hints. Possible wrap-around dodecahedral shape of the universe
- Classification of possible universes in the Lambda-CDM model.
- Fagundes, Helio V. (2002). “Exploring the global topology of the universe”. *Brazilian Journal of Physics*. **32** (4). doi:10.1590/S0103-97332002000500012.
- Grime, James. "π_{39} (Pi and the size of the Universe)". *Numberphile*. Brady Haran.
- What do you mean the universe is flat? Scientific American Blog explanation of a flat universe and the curved spacetime in the universe.

Chapter 4

High-Z Supernova Search Team

The **High-Z Supernova Search Team** was an international cosmology collaboration which used Type Ia supernovae to chart the expansion of the universe. The team was formed in 1994 by Brian P. Schmidt, then a post-doctoral research associate at Harvard University, and Nicholas B. Suntzeff, a staff astronomer at the Cerro Tololo Inter-American Observatory (CTIO) in Chile. The original team first proposed for the research on September 29, 1994 in a proposal called *A Pilot Project to Search for Distant Type Ia Supernova* to the CTIO Inter-American Observatory. The original team as co-listed on the first observing proposal was: Nicholas Suntzeff (PI); Brian Schmidt (Co-I); (other Co-Is) R. Chris Smith, Robert Schommer, Mark M. Phillips, Mario Hamuy, Roberto Aviles, Jose Maza, Adam Riess, Robert Kirshner, Jason Spiromilio, and Bruno Leibundgut. The original project was awarded four nights of telescope time on the CTIO Victor M. Blanco Telescope on the nights of February 25, 1995, and March 6, 24, and 29, 1995. The pilot project led to the discovery of supernova SN1995Y. In 1995, the HZT elected Brian P. Schmidt of the Mount Stromlo Observatory which is part of the Australian National University to manage the team.

The team expanded to roughly 20 astronomers located in the United States, Europe, Australia, and Chile. They used the Victor M. Blanco telescope to discover Type Ia supernovae out to redshifts of $z = 0.9$. The discoveries were verified with spectra taken mostly from the telescopes of the Keck Observatory, and the European Southern Observatory.

In a 1998 study led by Adam Riess, the High-Z Team became the first to publish evidence that the expansion of the Universe is accelerating (Riess et al. 1998, AJ, 116, 1009, submitted March 13, 1998, accepted May 1998). The team later spawned Project ESSENCE led by Christopher Stubbs of Harvard University and the Higher-Z Team led by Adam Riess of Johns Hopkins University and Space Telescope Science Institute.

In 2011, Riess and Schmidt, along with Saul Perlmutter of the Supernova Cosmology Project, were awarded the Nobel Prize in Physics for this work.[1]

4.1 Major International Awards to High-Z Team and Team Members

- 1998: Breakthrough of the Year, Science Magazine
- 2006: Shaw Prize
- 2007: Gruber Prize in Cosmology
- 2011: Nobel Prize in Physics
- 2011: Albert Einstein Medal
- 2015: Breakthrough Prize in Fundamental Physics
- 2015: Wolf Prize in Physics

4.2 Members

- Mount Stromlo Observatory and the Australian National University
  - Brian P. Schmidt
- CTIO
  - Nicholas Suntzeff
  - Robert Schommer
  - R. Chris Smith
  - Mario Hamuy (1994–1997)
- Las Campanas Observatory
  - Mark M. Phillips (1994–2000)
- Pontificia Universidad Católica de Chile

Observing Proposal
Cerro Tololo Inter-American Observatory

Date: September 29, 1994 *Proposal number:*

TITLE: A Pilot Project to Search for Distant Type Ia Supernovae

PI: N. Suntzeff Grad student? N nsuntzeff@ctio.noao.edu
CTIO, Casilla 603, La Serena Chile 56-51-225415

CoI: B. Schmidt Grad student? N brian@cfanewton.harvard.edu
CfA/MSSSO, 60 Garden St., Cambridge, MA 02138 617 495 7390

Other CoIs: C. Smith, R. Schommer, M. Phillips, M. Hamuy, R. Aviles (CTIO); J. Maza (UChile); A. Riess, R. Kirshner (Harvard); J. Spyromilio, B. Leibundgut (ESO)

Abstract of Scientific Justification:

We propose to initiate a search for Type Ia supernovae at redshifts to $z \sim 0.3 - 0.5$ in equatorial fields using the CTIO 4m telescope. This program is the next step in the Calán/Tololo SN survey, where we have found ~ 30 Type Ia supernovae out to $z \sim 0.1$. The proposed program is a pilot project to discover fainter SN Ia's using multiple-epoch CCD images from the 4m telescope. We will follow up these discoveries with CCD photometry and spectroscopy both at CTIO and at several observatories in both hemispheres. With the spectral classification and light curve shapes, we can use our calibrations of the absolute magnitudes of SN Ia's from the Calán/Tololo survey to place stringent limits (Figure 2) on q_0 in a reasonable time-frame. Based on the statistics of discovery from the Calán/Tololo SN survey, we can expect to find about 3 SNe Ia per month.

The goal of this pilot project is to obtain enough imaging data to allow us to verify that we can discover faint supernovae at the expected rates. The success rate of discovery for this program and the related (but independent) program proposed for the 0.9m rests primarily in the software used to search for the supernovae in digital images. The data set from this pilot project will allow us to tune the software to provide the most efficient discovery techniques.

- *Is this proposal part of a PhD thesis? If 'Y', you must send a letter; see instructions.* N
- *Are you requesting long-term status? If 'Y', please give details on the line below.* N

Summary of observing runs requested for this project

Run	Telescope	Instrument, detectors, gratings, filters, camera optics, etc.
1	4m	PFCCD Tek2048K, our set of reshifted *BV* filters
2		
3		

Run	No. nights	Moon age (d)	Optimal dates	Acceptable dates
1	4	2× -5, 2 × +3 (see proposal)	March and April	February-April
2				
3				

- *List dates you cannot use for non-astronomical reasons on the next line.*

The original telescope time proposal in 1994 to the Cerro Tololo Inter-American Observatory which began the High-Z Team.

  - Alejandro Clocchiatti (starting in 1996)
- University of Chile
  - Jose Maza (1994–1997)
- European Southern Observatory
  - Bruno Leibundgut
  - Jason Spyromilio
- University of Hawaii
  - John Tonry (starting in 1996)
- University of California, Berkeley
  - Alexei Filippenko (starting in 1996)
  - Weidong Li (starting in 1999)
- Space Telescope Science Institute
  - Adam Riess
  - Ron Gilliland (1996–2000)
- University of Washington
  - Christopher Stubbs (starting in 1995)
  - Craig Hogan (starting in 1995)
  - David Reiss (1995–1999)
  - Alan Diercks (1995–1999)
- Harvard University
  - Christopher Stubbs (starting in 2003)
  - Robert Kirshner
  - Thomas Matheson (starting 1999)
  - Saurabh Jha (starting 1997)
  - Peter Challis
- University of Notre Dame
  - Peter Garnavich
  - Stephen Holland (starting 2000)

4.3 References

[1] "Nobel physics prize honours accelerating Universe find". BBC News. 2011-10-04.

4.4 External links

- High-Z Supernova Search Team Mainsite

Chapter 5

Type Ia supernova

A **type Ia supernova** (type one-a) is a type of supernova that occurs in binary systems (two stars orbiting one another) in which one of the stars is a white dwarf. The other star can be anything from a giant star to an even smaller white dwarf.[1]

Physically, carbon–oxygen white dwarfs with a low rate of rotation are limited to below 1.44 solar masses ($M_\odot$).[2][3] Beyond this, they re-ignite and in some cases trigger a supernova explosion. Somewhat confusingly, this limit is often referred to as the Chandrasekhar mass, despite being marginally different from the absolute Chandrasekhar limit where electron degeneracy pressure is unable to prevent catastrophic collapse. If a white dwarf gradually accretes mass from a binary companion, the general hypothesis is that its core will reach the ignition temperature for carbon fusion as it approaches the limit. If the white dwarf merges with another white dwarf (a very rare event), it will momentarily exceed the limit and begin to collapse, again raising its temperature past the nuclear fusion ignition point. Within a few seconds of initiation of nuclear fusion, a substantial fraction of the matter in the white dwarf undergoes a runaway reaction, releasing enough energy (1–2×10^{44} J)[4] to unbind the star in a supernova explosion.[5]

This type Ia category of supernovae produces consistent peak luminosity because of the uniform mass of white dwarfs that explode via the accretion mechanism. The stability of this value allows these explosions to be used as standard candles to measure the distance to their host galaxies because the visual magnitude of the supernovae depends primarily on the distance.

In May 2015, NASA reported that the *Kepler* space observatory observed KSN 2011b, a Type Ia supernova in the process of exploding. Details of the pre-nova moments may help scientists better judge the quality of Type Ia supernovae as standard candles, which is an important link in the argument for dark energy.[6]

5.1 Consensus model

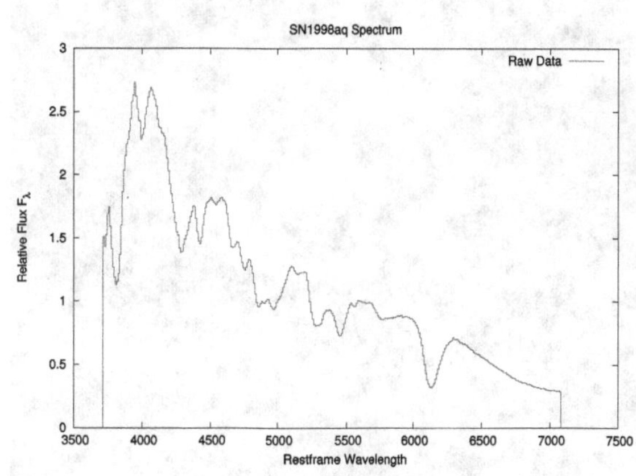

Spectrum of SN1998aq, a Type Ia supernova, one day after maximum light in the B band[7]

The Type Ia supernova is a sub-category in the Minkowski-Zwicky supernova classification scheme, which was devised by German-American astronomer Rudolph Minkowski and Swiss astronomer Fritz Zwicky.[8] There are several means by which a supernova of this type can form, but they share a common underlying mechanism. Theoretical astronomers long believed the progenitor star for this type of supernova is a white dwarf and empirical evidence for this was found in 2014 when a Type Ia supernova was observed in the galaxy Messier 82.[9] When a slowly-rotating[2] carbon-oxygen white dwarf accretes matter from a companion, it can exceed the Chandrasekhar limit of about 1.44 $M_\odot$, beyond which it can no longer support its weight with electron degeneracy pressure.[10] In the absence of a countervailing process, the white dwarf would collapse to form a neutron star, in an accretion-induced non-ejective process,[11] as normally occurs in the case of a white dwarf that is primarily composed of magnesium, neon, and oxygen.[12]

The current view among astronomers who model Type Ia supernova explosions, however, is that this limit is never actually attained and collapse is never initiated. Instead, the increase in pressure and density due to the increasing weight raises the temperature of the core,[3] and as the white dwarf approaches about 99% of the limit,[13] a period of convection ensues, lasting approximately 1,000 years.[14] At some point in this simmering phase, a deflagration flame front is born, powered by carbon fusion. The details of the ignition are still unknown, including the location and number of points where the flame begins.[15] Oxygen fusion is initiated shortly thereafter, but this fuel is not consumed as completely as carbon.[16]

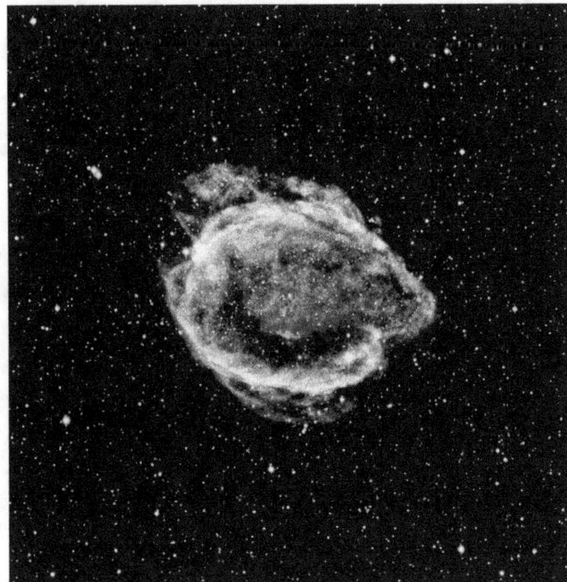

G299 Type Ia supernova remnant.

Once fusion has begun, the temperature of the white dwarf starts to rise. A main sequence star supported by thermal pressure would expand and cool which automatically counterbalances an increase in thermal energy. However, degeneracy pressure is independent of temperature; the white dwarf is unable to regulate the fusion process in the manner of normal stars, so it is vulnerable to a runaway fusion reaction. The flame accelerates dramatically, in part due to the Rayleigh–Taylor instability and interactions with turbulence. It is still a matter of considerable debate whether this flame transforms into a supersonic detonation from a subsonic deflagration.[14][17]

Regardless of the exact details of this nuclear fusion, it is generally accepted that a substantial fraction of the carbon and oxygen in the white dwarf is converted into heavier elements within a period of only a few seconds,[16] raising the internal temperature to billions of degrees. This energy release from thermonuclear fusion (1–2×10^{44} J[4]) is more than enough to unbind the star; that is, the individual particles making up the white dwarf gain enough kinetic energy to fly apart from each other. The star explodes violently and releases a shock wave in which matter is typically ejected at speeds on the order of 5,000–20000 km/s, roughly 6% of the speed of light. The energy released in the explosion also causes an extreme increase in luminosity. The typical visual absolute magnitude of Type Ia supernovae is M_v = −19.3 (about 5 billion times brighter than the Sun), with little variation.[14]

The theory of this type of supernovae is similar to that of novae, in which a white dwarf accretes matter more slowly and does not approach the Chandrasekhar limit. In the case of a nova, the in-falling matter causes a hydrogen fusion surface explosion that does not disrupt the star.[14] This type of supernova differs from a core-collapse supernova, which is caused by the cataclysmic explosion of the outer layers of a massive star as its core implodes.[18]

5.2 Formation

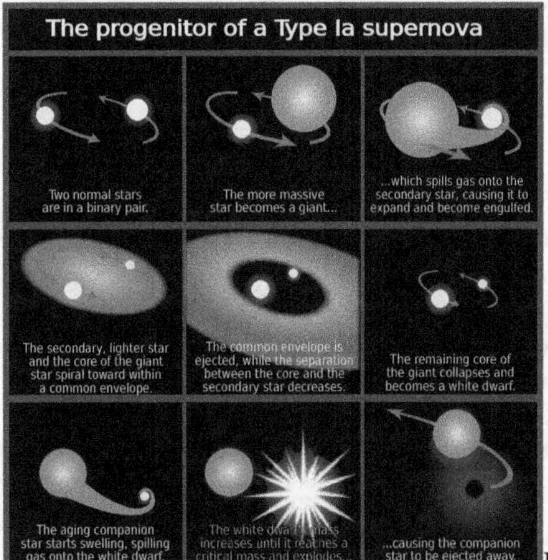

Formation process

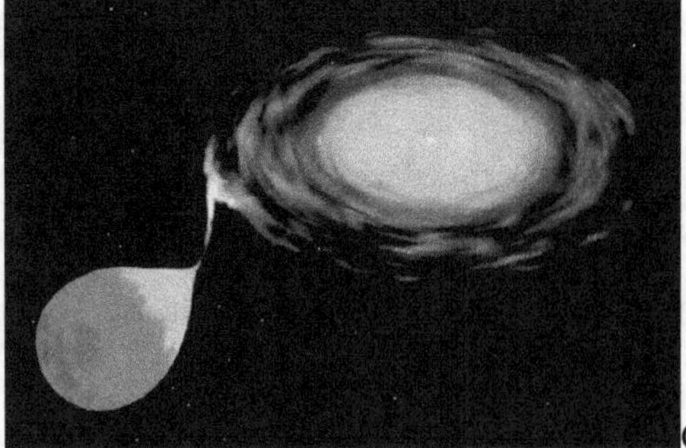

Gas is being stripped from a giant star to form an accretion disc around a compact companion (such as a white dwarf star). *NASA image*

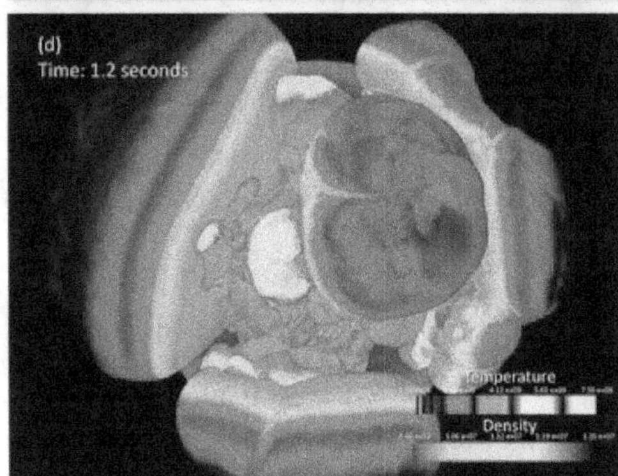

Simulation of the explosion phase of the deflagration-to-detonation model of supernovae formation, run on scientific super-computer.

5.2.1 Single degenerate progenitors

One model for the formation of this category of supernova is a close binary star system. The progenitor binary system consists of main sequence stars, with the primary possessing more mass than the secondary. Being greater in mass, the primary is the first of the pair to evolve onto the asymptotic giant branch, where the star's envelope expands considerably. If the two stars share a common envelope then the system can lose significant amounts of mass, reducing the angular momentum, orbital radius and period. After the primary has degenerated into a white dwarf, the secondary star later evolves into a red giant and the stage is set for mass accretion onto the primary. During this final shared-envelope phase, the two stars spiral in closer together as angular momentum is lost. The resulting orbit can have a period as brief as a few hours.[19][20] If the accretion continues long enough, the white dwarf may eventually approach the Chandrasekhar limit.

The white dwarf companion could also accrete matter from other types of companions, including a subgiant or (if the orbit is sufficiently close) even a main sequence star. The actual evolutionary process during this accretion stage remains uncertain, as it can depend both on the rate of accretion and the transfer of angular momentum to the white dwarf companion.[21]

It has been estimated that single degenerate progenitors account for no more than 20% of all Type Ia supernovae.[22]

5.2.2 Double degenerate progenitors

A second possible mechanism for triggering a Type Ia supernova is the merger of two white dwarfs whose combined mass exceeds the Chandrasekhar limit. The resulting merger is called a super-Chandrasekhar mass white dwarf.[23][24] In such a case, the total mass would not be constrained by the Chandrasekhar limit.

Collisions of solitary stars within the Milky Way occur only once every $10^7 - 10^{13}$ years; far less frequently than the appearance of novae.[25] Collisions occur with greater frequency in the dense core regions of globular clusters[26] (*cf.* blue stragglers). A likely scenario is a collision with a binary star system, or between two binary systems containing white dwarfs. This collision can leave behind a close binary system of two white dwarfs. Their orbit decays and they merge through their shared envelope.[27] However, a study based on SDSS spectra found 15 double systems of the 4,000 white dwarfs tested, implying a double white dwarf merger every 100 years in the Milky Way. Conveniently, this rate matches the number of Type Ia supernovae detected in our neighborhood.[28]

A double degenerate scenario is one of several explanations

proposed for the anomalously massive (2 $M_\odot$) progenitor of the SN 2003fg.[29][30] It is the only possible explanation for SNR 0509-67.5, as all possible models with only one white dwarf have been ruled out.[31] It has also been strongly suggested for SN 1006, given that no companion star remnant has been found there.[22] Observations made with NASA's Swift space telescope ruled out existing supergiant or giant companion stars of every Type Ia supernovae studied. The supergiant companion's blown out outer shell should emit X-rays, but this glow was not detected by Swift's *XRT (X-Ray telescope)* in the 53 closest supernova remnants. For 12 Type Ia supernovae observed within 10 days of the explosion, the satellite's *UVOT (Ultraviolet/Optical Telescope)* showed no ultraviolet radiation originating from the heated companion star's surface hit by the supernova shock wave, meaning there were no red giants or larger stars orbiting those supernova progenitors. In the case of SN 2011fe, the companion star must have been smaller than the Sun, if it existed.[32] The Chandra X-ray Observatory revealed that the X-ray radiation of five elliptical galaxies and the bulge of the Andromeda galaxy is 30-50 times fainter than expected. X-ray radiation should be emitted by the accretion discs of Type Ia supernova progenitors. The missing radiation indicates that few white dwarfs possess accretion discs, ruling out the common, accretion-based model of Ia supernovae.[33] Inward spiraling white dwarf pairs are strong sources of gravitational waves.

Double degenerate scenarios raise questions about the applicability of Type Ia supernovae as standard candles, since total mass of the two merging white dwarfs varies significantly, meaning luminosity also varies.

5.2.3 Type Iax

It has been proposed that a group of sub-luminous supernovae that occur when helium accretes onto a white dwarf should be classified as **Type Iax**.[34][35] This type of supernova may not always completely destroy the white dwarf progenitor.[36]

5.3 Observation

Unlike the other types of supernovae, Type Ia supernovae generally occur in all types of galaxies, including ellipticals. They show no preference for regions of current stellar formation.[38] As white dwarf stars form at the end of a star's main sequence evolutionary period, such a long-lived star system may have wandered far from the region where it originally formed. Thereafter a close binary system may spend another million years in the mass transfer stage (possibly forming persistent nova outbursts) before the conditions are ripe for a Type Ia supernova to occur.[39]

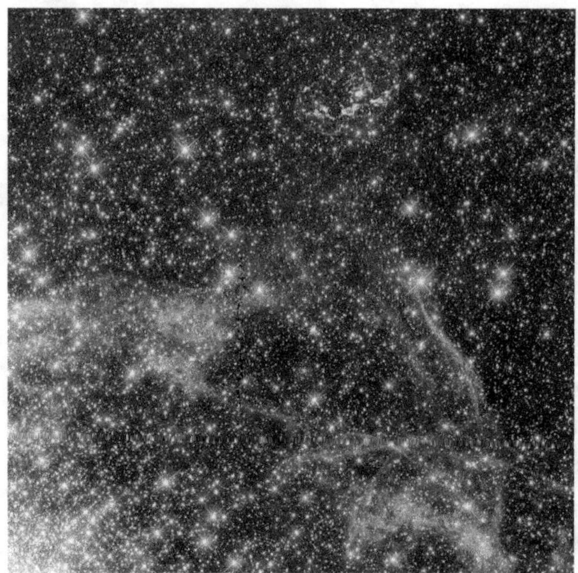

Supernova remnant N103B taken by the Hubble Space Telescope.[37]

A long-standing problem in astronomy has been the identification of supernova progenitors. Direct observation of a progenitor would provide useful constraints on supernova models. As of 2006, the search for such a progenitor had been ongoing for longer than a century.[40] Observation of the supernova SN 2011fe has provided useful constraints. Previous observations with the Hubble Space Telescope did not show a star at the position of the event, thereby excluding a red giant as the source. The expanding plasma from the explosion was found to contain carbon and oxygen, making it likely the progenitor was a white dwarf primarily composed of these elements.[41] Similarly, observations of the nearby SN PTF 11kx,[42] discovered January 16, 2011 (UT) by the Palomar Transient Factory (PTF), lead to the conclusion that this explosion arises from single-degenerate progenitor, with a red giant companion, thus suggesting there is no single progenitor path to SN Ia. Direct observations of the progenitor of PTF11kx were reported in the August 24 edition of Science and support this conclusion, and also show that the progenitor star experienced periodic nova eruptions before the supernova - another surprising discovery. [43][44] However, later analysis revealed that the circumstellar material (CSM) is too massive for the single-degenerate scenario, and fits better the core-degenerate scenario.[45]

5.3.1 Light curve

Type Ia supernovae have a characteristic light curve, their graph of luminosity as a function of time after the explo-

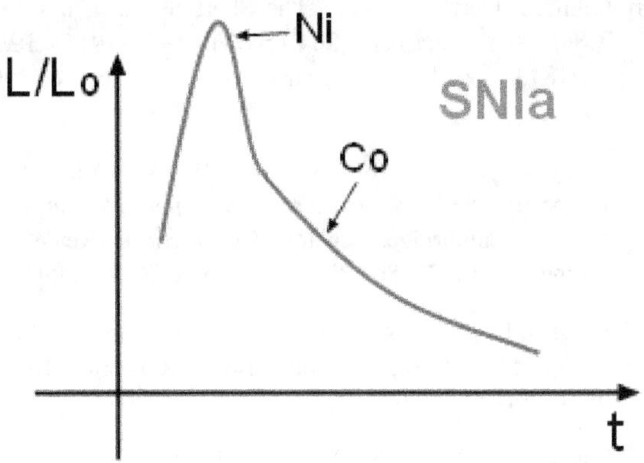

This plot of luminosity (relative to the Sun, L_0) versus time shows the characteristic light curve for a Type Ia supernova. The peak is primarily due to the decay of nickel (Ni), while the later stage is powered by cobalt (Co).

sion. Near the time of maximal luminosity, the spectrum contains lines of intermediate-mass elements from oxygen to calcium; these are the main constituents of the outer layers of the star. Months after the explosion, when the outer layers have expanded to the point of transparency, the spectrum is dominated by light emitted by material near the core of the star, heavy elements synthesized during the explosion; most prominently isotopes close to the mass of iron (iron-peak elements). The radioactive decay of nickel−56 through cobalt−56 to iron−56 produces high-energy photons, which dominate the energy output of the ejecta at intermediate to late times.[14]

The use of Type Ia supernovae to measure precise distances was pioneered by a collaboration of Chilean and US astronomers, the Calán/Tololo Supernova Survey.[46] In a series of papers in the 1990s the survey showed that while Type Ia supernovae do not all reach the same peak luminosity, a single parameter measured from the light curve can be used to correct unreddened Type Ia supernovae to standard candle values. The original correction to standard candle value is known as the Phillips relationship[47] and was shown by this group to be able to measure relative distances to 7% accuracy.[48] The cause of this uniformity in peak brightness is related to the amount of nickel-56 produced in white dwarfs presumably exploding near the Chandrasekhar limit.[49]

The similarity in the absolute luminosity profiles of nearly all known Type Ia supernovae has led to their use as a secondary standard candle in extragalactic astronomy.[50] Improved calibrations of the Cepheid variable distance scale[51] and direct geometric distance measurements to NGC 4258 from the dynamics of maser emission[52] when combined with the Hubble diagram of the Type Ia supernova distances have led to an improved value of the Hubble constant.

In 1998, observations of distant Type Ia supernovae indicated the unexpected result that the Universe seems to undergo an accelerating expansion.[53][54] Three members from two teams were subsequently awarded Nobel Prizes for this discovery.[55]

5.4 Types

It has been discovered that Type Ia supernovae that were considered the same are in fact different; moreover, a form of the Type Ia supernova that is relatively infrequent today was far more common earlier in the history of the universe. This could have far reaching cosmological significance and could lead to revision of estimation of the rate of expansion of the universe and the prevalence of dark energy.[56][57]

5.5 See also

- Carbon detonation
- Cosmic distance ladder
- History of supernova observation
- Supernova remnant

5.6 References

[1] HubbleSite - Dark Energy - Type Ia Supernovae

[2] Yoon, S.-C.; Langer, L. (2004). "Presupernova Evolution of Accreting White Dwarfs with Rotation". *Astronomy and Astrophysics*. **419** (2): 623–644. Bibcode:2004A&A...419..623Y. arXiv:astro-ph/0402287. doi:10.1051/0004-6361:20035822. Retrieved 2007-05-30.

[3] Mazzali, P. A.; Röpke, F. K.; Benetti, S.; Hillebrandt, W. (2007). "A Common Explosion Mechanism for Type Ia Supernovae". *Science*. **315** (5813): 825–828. Bibcode:2007Sci...315..825M. PMID 17289993. arXiv:astro-ph/0702351. doi:10.1126/science.1136259.

[4] Khokhlov, A.; Müller, E.; Höflich, P. (1993). "Light curves of Type IA supernova models with different explosion mechanisms". *Astronomy and Astrophysics*. **270** (1–2): 223–248. Bibcode:1993A&A...270..223K.

[5] Staff (2006-09-07). "Introduction to Supernova Remnants". NASA Goddard/SAO. Retrieved 2007-05-01.

[6] Johnson, Michele; Chandler, Lynn (May 20, 2015). "NASA Spacecraft Capture Rare, Early Moments of Baby Supernovae". *NASA*. Retrieved May 21, 2015.

[7] Matheson, Thomas; Kirshner, Robert; Challis, Pete; Jha, Saurabh; et al. (2008). "Optical Spectroscopy of Type Ia Supernovae". *Astronomical Journal*. **135** (4): 1598–1615. Bibcode:2008AJ....135.1598M. arXiv:0803.1705. doi:10.1088/0004-6256/135/4/1598.

[8] da Silva, L. A. L. (1993). "The Classification of Supernovae". *Astrophysics and Space Science*. **202** (2): 215–236. Bibcode:1993Ap&SS.202..215D. doi:10.1007/BF00626878.

[9] Type 1a Supernovae: Why Our Standard Candle Isn't Really Standard

[10] Lieb, E. H.; Yau, H.-T. (1987). "A rigorous examination of the Chandrasekhar theory of stellar collapse". *Astrophysical Journal*. **323** (1): 140–144. Bibcode:1987ApJ...323..140L. doi:10.1086/165813.

[11] Canal, R.; Gutiérrez, J. (1997). "The possible white dwarf-neutron star connection". *Astrophysics and Space Science Library*. Astrophysics and Space Science Library. **214**: 49–55. Bibcode:1997astro.ph..1225C. ISBN 978-0-7923-4585-5. arXiv:astro-ph/9701225. doi:10.1007/978-94-011-5542-7_7.

[12] Fryer, C. L.; New, K. C. B. (2006-01-24). "2.1 Collapse scenario". *Gravitational Waves from Gravitational Collapse*. Max-Planck-Gesellschaft. Retrieved 2007-06-07.

[13] Wheeler, J. Craig (2000-01-15). *Cosmic Catastrophes: Supernovae, Gamma-Ray Bursts, and Adventures in Hyperspace*. Cambridge, UK: Cambridge University Press. p. 96. ISBN 0-521-65195-6.

[14] Hillebrandt, W.; Niemeyer, J. C. (2000). "Type IA Supernova Explosion Models". *Annual Review of Astronomy and Astrophysics*. **38** (1): 191–230. Bibcode:2000ARA&A..38..191H. arXiv:astro-ph/0006305. doi:10.1146/annurev.astro.38.1.191.

[15] "Science Summary". ASC / Alliances Center for Astrophysical Thermonuclear Flashes. 2004. Retrieved 2017-04-25.

[16] Röpke, F. K.; Hillebrandt, W. (2004). "The case against the progenitor's carbon-to-oxygen ratio as a source of peak luminosity variations in Type Ia supernovae". *Astronomy and Astrophysics*. **420** (1): L1–L4. Bibcode:2004A&A...420L...1R. arXiv:astro-ph/0403509. doi:10.1051/0004-6361:20040135.

[17] Gamezo, V. N.; Khokhlov, A. M.; Oran, E. S.; Chtchelkanova, A. Y.; Rosenberg, R. O. (2003-01-03). "Thermonuclear Supernovae: Simulations of the Deflagration Stage and Their Implications". *Science*. **299** (5603): 77–81. PMID 12446871. doi:10.1126/science.1078129. Retrieved 2006-11-28.

[18] Gilmore, Gerry (2004). "The Short Spectacular Life of a Superstar". *Science*. **304** (5697): 1915–1916. PMID 15218132. doi:10.1126/science.1100370. Retrieved 2007-05-01.

[19] Paczynski, B. (July 28 – August 1, 1975). "Common Envelope Binaries". *Structure and Evolution of Close Binary Systems*. Cambridge, England: Dordrecht, D. Reidel Publishing Co. pp. 75–80. Bibcode:1976IAUS...73...75P.

[20] Postnov, K. A.; Yungelson, L. R. (2006). "The Evolution of Compact Binary Star Systems". Living Reviews in Relativity. Retrieved 2007-01-08.

[21] Langer, N.; Yoon, S.-C.; Wellstein, S.; Scheithauer, S. (2002). "On the evolution of interacting binaries which contain a white dwarf". In Gänsicke, B. T.; Beuermann, K.; Rein, K. *The Physics of Cataclysmic Variables and Related Objects, ASP Conference Proceedings*. San Francisco, California: Astronomical Society of the Pacific. p. 252. Bibcode:2002ASPC..261..252L.

[22] González Hernández, J. I.; Ruiz-Lapuente, P.; Tabernero, H. M.; Montes, D.; Canal, R.; Méndez, J.; Bedin, L. R. (2012). "No surviving evolved companions of the progenitor of SN 1006". *Nature*. **489** (7417): 533–536. Bibcode:2012Natur.489..533G. PMID 23018963. arXiv:1210.1948. doi:10.1038/nature11447. See also lay reference: John Matson (December 2012). "No Star Left Behind". *Scientific American*. **307** (6). p. 16

[23] Staff. "Type Ia Supernova Progenitors". Swinburne University. Retrieved 2007-05-20.

[24] "Brightest supernova discovery hints at stellar collision". New Scientist. 2007-01-03. Retrieved 2007-01-06.

[25] Whipple, Fred L. (1939). "Supernovae and Stellar Collisions". *Proceedings of the National Academy of Sciences of the United States of America*. **25** (3): 118–125. Bibcode:1939PNAS...25..118W. doi:10.1073/pnas.25.3.118.

[26] Rubin, V. C.; Ford, W. K. J. (1999). "A Thousand Blazing Suns: The Inner Life of Globular Clusters". *Mercury*. **28**: 26. Bibcode:1999Mercu..28d..26M. Retrieved 2006-06-02.

[27] Middleditch, J. (2004). "A White Dwarf Merger Paradigm for Supernovae and Gamma-Ray Bursts". *The Astrophysical Journal*. **601** (2): L167–L170. Bibcode:2003astro.ph.11484M. arXiv:astro-ph/0311484. doi:10.1086/382074.

[28] "Important Clue Uncovered for the Origins of a Type of Supernovae Explosion, Thanks to a Research Team at the University of Pittsburgh". University of Pittsburgh. Retrieved 23 March 2012.

[29] "The Weirdest Type Ia Supernova Yet". Lawrence Berkeley National Laboratory. 2006-09-20. Retrieved 2006-11-02.

[30] "Bizarre Supernova Breaks All The Rules". New Scientist. 2006-09-20. Retrieved 2007-01-08.

[31] Schaefer, Bradley E.; Pagnotta, Ashley (2012). "An absence of ex-companion stars in the type Ia supernova remnant SNR 0509-67.5". *Nature*. **481** (7380): 164–166. Bibcode:2012Natur.481..164S. PMID 22237107. doi:10.1038/nature10692.

[32] "NASA'S Swift Narrows Down Origin of Important Supernova Class". NASA. Retrieved 24 March 2012.

[33] "NASA's Chandra Reveals Origin of Key Cosmic Explosions". Chandra X-ray Observatory website. Retrieved 28 March 2012.

[34] Bo Wang; Stephen Justham; Zhanwen Han (2013). "Double-detonation explosions as progenitors of Type Iax supernovae". arXiv:1301.1047v1 [astro-ph.SR].

[35] Ryan J. Foley; P. J. Challis; R. Chornock; M. Ganeshalingam; W. Li; G. H. Marion; N. I. Morrell; G. Pignata; M. D. Stritzinger; J. M. Silverman; X. Wang; J. P. Anderson; A. V. Filippenko; W. L. Freedman; M. Hamuy; S. W. Jha; R. P. Kirshner; C. McCully; S. E. Persson; M. M. Phillips; D. E. Reichart; A. M. Soderberg (2012). "Type Iax Supernovae: A New Class of Stellar Explosion". arXiv:1212.2209v2 [astro-ph.SR].

[36] "Hubble finds supernova star system linked to potential 'zombie star'". SpaceDaily. 6 August 2014.

[37] "Search for stellar survivor of a supernova explosion". *www.spacetelescope.org*. Retrieved 30 March 2017.

[38] van Dyk, Schuyler D. (1992). "Association of supernovae with recent star formation regions in late type galaxies". *Astronomical Journal*. **103** (6): 1788–1803. Bibcode:1992AJ....103.1788V. doi:10.1086/116195.

[39] Hoeflich, N.; Deutschmann, A.; Wellstein, S.; Höflich, P. (1999). "The evolution of main sequence star + white dwarf binary systems towards Type Ia supernovae". *Astronomy and Astrophysics*. **362**: 1046–1064. Bibcode:2000A&A...362.1046L. arXiv:astro-ph/0008444.

[40] Kotak, R. (December 2008). "Progenitors of Type Ia Supernovae". Written at Keele University, Keele, United Kingdom. In Evans, A.; Bode, M. F.; O'Brien, T. J.; Darnley, M. J. *RS Ophiuchi (2006) and the Recurrent Nova Phenomenon, proceedings of the conference held 12–14 June 2007*. ASP Conference Series. **401**. San Francisco: Astronomical Society of the Pacific, 2008. p. 150. Bibcode:2008ASPC..401..150K.

[41] Nugent, Peter E.; Sullivan, Mark; Cenko, S. Bradley; Thomas, Rollin C.; Kasen, Daniel; Howell, D. Andrew; Bersier, David; Bloom, Joshua S.; Kulkarni, S. R.; Kandrashoff, Michael T.; Filippenko, Alexei V.; Silverman, Jeffrey M.; Marcy, Geoffrey W.; Howard, Andrew W.; Isaacson, Howard T.; Maguire, Kate; Suzuki, Nao; Tarlton, James E.; Pan, Yen-Chen; Bildsten, Lars; Fulton, Benjamin J.; Parrent, Jerod T.; Sand, David; Podsiadlowski, Philipp; Bianco, Federica B.; Dilday, Benjamin; Graham, Melissa L.; Lyman, Joe; James, Phil; et al. (December 2011). "Supernova 2011fe from an Exploding Carbon-Oxygen White Dwarf Star". *Nature*. **480** (7377): 344–347. Bibcode:2011Natur.480..344N. PMID 22170680. arXiv:1110.6201. doi:10.1038/nature10644

[42] Dilday, B.; Howell, DA; Cenko, SB; Silverman, JM; Nugent, PE; Sullivan, M; Ben-Ami, S; Bildsten, L; Bolte, M; Endl, M; Filippenko, A. V.; Gnat, O; Horesh, A; Hsiao, E; Kasliwal, MM; Kirkman, D; Maguire, K; Marcy, GW; Moore, K; Pan, Y; Parrent, J. T.; Podsiadlowski, P; Quimby, RM; Sternberg, A; Suzuki, N; Tytler, DR; Xu, D; Bloom, JS; Gal-Yam, A; et al. (2012). "PTF11kx: A Type-Ia Supernova with a Symbiotic Nova Progenitor". *Science*. **337** (6097): 942–5. Bibcode:2012Sci...337..942D. PMID 22923575. arXiv:1207.1306. doi:10.1126/science.1219164.

[43] Dilday, B.; Howell, DA; Cenko, SB; Silverman, JM; Nugent, PE; Sullivan, M; Ben-Ami, S; Bildsten, L; Bolte, M; Endl, M; Filippenko, A. V.; Gnat, O; Horesh, A; Hsiao, E; Kasliwal, MM; Kirkman, D; Maguire, K; Marcy, GW; Moore, K; Pan, Y; Parrent, J. T.; Podsiadlowski, P; Quimby, RM; Sternberg, A; Suzuki, N; Tytler, DR; Xu, D; Bloom, JS; Gal-Yam, A; et al. (24 August 2012). "PTF 11kx: A Type Ia Supernova with a Symbiotic Nova Progenitor". *Science*. **337** (6097): 942–945. Bibcode:2012Sci...337..942D. PMID 22923575. arXiv:1207.1306. doi:10.1126/science.1219164.

[44] "The First-Ever Direct Observations of a Type 1a Supernova Progenitor System". *Scitech daily*. - popular account of the discovery

[45] Soker, Noam; Kashi, Amit; García-Berro, Enrique; Torres, Santiago; Camacho, Judit (2013). "Explaining the Type Ia supernova PTF 11kx with a violent prompt merger scenario". *Monthly Notices of the Royal Astronomical Society*. **431**: 1541–1546. Bibcode:2013MNRAS.431.1541S. arXiv:1207.5770. doi:10.1093/mnras/stt271.

[46] Hamuy, M.; et al. (1993). "The 1990 Calan/Tololo Supernova Search". *Astronomical Journal*. **106** (6): 2392. Bibcode:1993AJ....106.2392H. doi:10.1086/116811.

[47] Phillips, M. M. (1993). "The absolute magnitudes of Type IA supernovae". *Astrophysical Journal Letters*. **413** (2): L105. Bibcode:1993ApJ...413L.105P. doi:10.1086/186970.

[48] Hamuy, M.; et al. (1996). "The Absolute Luminosities of the Calan/Tololo Type IA Supernovae". *Astronomical Journal*. **112**: 2391. Bibcode:1996AJ....112.2391H. arXiv:astro-ph/9609059. doi:10.1086/118190.

[49] Colgate, S. A. (1979). "Supernovae as a standard candle for cosmology". *Astrophysical Journal*. **232** (1): 404–408. Bibcode:1979ApJ...232..404C. doi:10.1086/157300.

[50] Hamuy, M.; et al. (1996). "A Hubble diagram of distant type IA supernovae". *Astronomical Journal*. **109**: 1. Bibcode:1995AJ....109....1H. doi:10.1086/117251.

[51] Freedman, W.; et al. "Final Results from the Hubble Space Telescope Key Project to Measure the Hubble Constant". *Astrophysical Journal*. **553** (1): 47–72. Bibcode:2001ApJ...553...47F. arXiv:astro-ph/0012376. doi:10.1086/320638.

[52] Macri, L. M.; Stanek, K. Z.; Bersier, D.; Greenhill, L. J.; Reid, M. J. (2006). "A New Cepheid Distance to the Maser-Host Galaxy NGC 4258 and Its Implications for the Hubble Constant". *Astrophysical Journal*. **652** (2): 1133–1149. Bibcode:2006ApJ...652.1133M. arXiv:astro-ph/0608211. doi:10.1086/508530.

[53] Perlmutter S, Supernova Cosmology Project, Goldhaber G, Knop RA, Nugent P, Castro PG, Deustua S, Fabbro S, Goobar A, Groom DE, Hook IM, Kim AG, Kim MY, Lee JC, Nunes NJ, Pain R, Pennypacker CR, Quimby R, Lidman C, Ellis RS, Irwin M, McMahon RG, Ruiz-Lapuente P, Walton N, Schaefer B, Boyle BJ, Filippenko AV, Matheson T, Fruchter AS, et al. (1999). "Measurements of Omega and Lambda from 42 high redshift supernovae". *Astrophysical Journal*. **517** (2): 565–86. Bibcode:1999ApJ...517..565P. arXiv:astro-ph/9812133. doi:10.1086/307221.

[54] Riess AG, et al. (1998). "Observational evidence from supernovae for an accelerating Universe and a cosmological constant". *Astronomical Journal*. **116** (3): 1009–38. Bibcode:1998AJ....116.1009R. arXiv:astro-ph/9805201. doi:10.1086/300499.

[55] *Cosmology*, Steven Weinberg, Oxford University Press, 2008.

[56] Accelerating universe? Not so fast

[57] Nielsen, J. T.; Guffanti, A.; Sarkar, S. (2015). "Marginal evidence for cosmic acceleration from Type Ia supernovae". *Scientific Reports*. **6**: 35596. Bibcode:2016NatSR...635596N. arXiv:1506.01354. doi:10.1038/srep35596.

5.7 External links

- List of all known Type Ia supernovae at The Open Supernova Catalog.
- Falck, Bridget (2006). "Type Ia Supernova Cosmology with ADEPT". Johns Hopkins University. Retrieved 2007-05-20.
- Staff (February 27, 2007). "Sloan Supernova Survey". Sloan Digital Sky Survey. Retrieved 2007-05-25.
- "Novae and Supernovae". peripatus.gen.nz. Retrieved 2007-05-25.
- "Source for major type of supernova". Pole Star Publications Ltd. August 6, 2003. Retrieved 2007-11-25. (A Type Ia progenitor found)
- "Novae and Supernovae explosions found". peripatus.gen.nz. Retrieved 2007-05-25.
- SNFactory Shows Type Ia 'Standard Candles' Have Many Masses (March 4, 2014)

Chapter 6

Cosmic distance ladder

For various definitions of distance in cosmology, see Distance measures (cosmology).

The **cosmic distance ladder** (also known as the **extragalactic distance scale**) is the succession of methods by which astronomers determine the distances to celestial objects. A real *direct* distance measurement of an astronomical object is possible only for those objects that are "close enough" (within about a thousand parsecs) to Earth. The techniques for determining distances to more distant objects are all based on various measured correlations between methods that work at close distances and methods that work at larger distances. Several methods rely on a **standard candle**, which is an astronomical object that has a known luminosity.

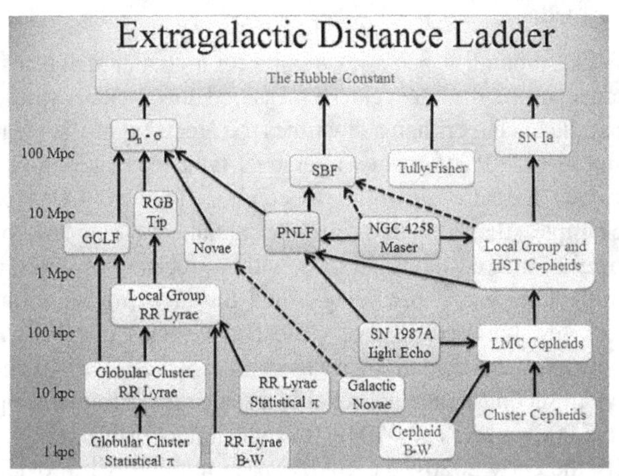

- *Light green boxes: Technique applicable to star-forming galaxies.*
- *Light blue boxes: Technique applicable to Population II galaxies.*
- *Light Purple boxes: Geometric distance technique.*
- *Light Red box: The planetary nebula luminosity function technique is applicable to all populations of the Virgo Supercluster.*
- *Solid black lines: Well calibrated ladder step.*
- *Dashed black lines: Uncertain calibration ladder step.*

The ladder analogy arises because no single technique can measure distances at all ranges encountered in astronomy. Instead, one method can be used to measure nearby distances, a second can be used to measure nearby to intermediate distances, and so on. Each rung of the ladder provides information that can be used to determine the distances at the next higher rung.

6.1 Direct measurement

At the base of the ladder are *fundamental* distance measurements, in which distances are determined directly, with no physical assumptions about the nature of the object in question. The precise measurement of stellar positions is part of the discipline of astrometry.

6.1.1 Astronomical unit

Main article: Astronomical Unit

Direct distance measurements are based upon the astronomical unit (AU), which is the distance between the Earth and the Sun. Historically, observations of transits of Venus were crucial in determining the AU; in the first half of the 20th century, observations of asteroids were also important. Presently the orbit of Earth is determined with high precision using radar measurements of distances to Venus and other nearby planets and asteroids,[1] and by tracking interplanetary spacecraft in their orbits around the Sun through the Solar System. Kepler's laws provide precise ratios of the sizes of the orbits of objects orbiting the Sun, but provides no measurement of the overall scale of the orbit system. Radar is used to measure the distance between the orbits of the Earth and of a second body. From that measurement and the ratio of the two orbit sizes, the size of Earth's orbit is calculated. The Earth's orbit is known with a precision of a few meters.

Statue of an astronomer and the concept of the cosmic distance ladder by the parallax method, made from the azimuth ring and other parts of the Yale–Columbia Refractor (telescope) (c 1925) wrecked by the 2003 Canberra bushfires which burned out the Mount Stromlo Observatory; at Questacon, Canberra, Australian Capital Territory

6.1.2 Parallax

Main article: Parallax
See also: Stellar Parallax and Parsec

The most important fundamental distance measurements come from trigonometric parallax. As the Earth orbits the Sun, the position of nearby stars will appear to shift slightly against the more distant background. These shifts are angles in an isosceles triangle, with 2 AU (the distance between the extreme positions of Earth's orbit around the Sun) making the base leg of the triangle and the distance to the star being the long equal length legs. The amount of shift is quite small, measuring 1 arcsecond for an object at the 1 parsec (3.26 light-years) distance of the nearest stars, and thereafter decreasing in angular amount as the distance increases. Astronomers usually express distances in units of parsecs (parallax arcseconds); light-years are used in popular media.

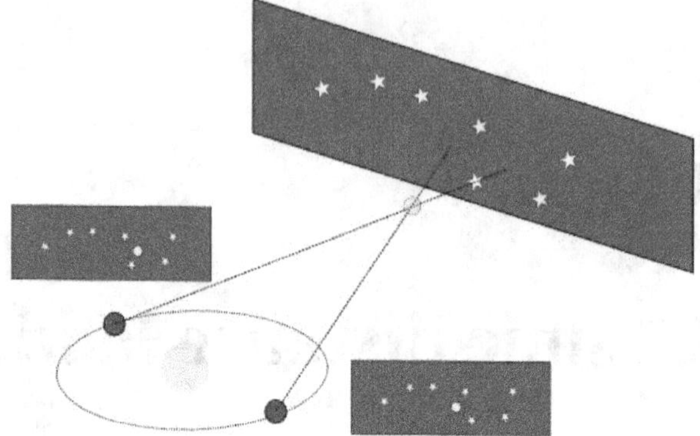

Stellar parallax motion from annual parallax. Half the apex angle is the parallax angle

Because parallax becomes smaller for a greater stellar distance, useful distances can be measured only for stars whose parallax is larger than a few times the precision of the measurement. Parallax measurements typically have an accuracy measured in milliarcseconds.[2] In the 1990s, for example, the Hipparcos mission obtained parallaxes for over a hundred thousand stars with a precision of about a milliarcsecond,[3] providing useful distances for stars out to a few hundred parsecs. The Hubble telescope WFC3 now has the potential to provide a precision of 20 to 40 *micro*arcseconds, enabling reliable distance measurements up to 5,000 parsecs (20,000 ly) for small numbers of stars.[4][5] By the early 2020s, the GAIA space mission will provide similarly accurate distances to all moderately bright stars.

Stars have a velocity relative to the Sun that causes proper motion (transverse across the sky) and radial velocity (motion toward or away from the Sun). The former is determined by plotting the changing position of the stars over many years, while the latter comes from measuring the Doppler shift of the star's spectrum caused by motion along the line of sight. For a group of stars with the same spectral class and a similar magnitude range, a mean parallax can be derived from statistical analysis of the proper motions relative to their radial velocities. This statistical parallax method is useful for measuring the distances of bright stars beyond 50 parsecs and giant variable stars, including Cepheids and the RR Lyrae variables.[6]

The motion of the Sun through space provides a longer baseline that will increase the accuracy of parallax measurements, known as secular parallax. For stars in the Milky Way disk, this corresponds to a mean baseline of 4 AU per year, while for halo stars the baseline is 40 AU per year. After several decades, the baseline can be orders of magnitude greater than the Earth–Sun baseline used for traditional parallax. However, secular parallax introduces a higher level of uncertainty because the relative velocity of observed stars

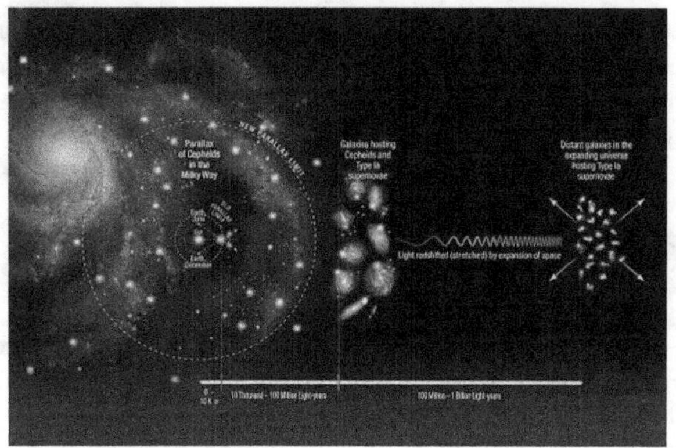

Parallax measurements may be an important clue to understanding three of the Universe's most elusive components: dark matter, dark energy and neutrinos.[7]

Hubble precision stellar distance measurement has been extended 10 times further into the Milky Way.[8]

is an additional unknown. When applied to samples of multiple stars, the uncertainty can be reduced; the uncertainty is inversely proportional to the square root of the sample size.[9]

Moving cluster parallax is a technique where the motions of individual stars in a nearby star cluster can be used to find the distance to the cluster. Only open clusters are near enough for this technique to be useful. In particular the distance obtained for the Hyades has historically been an important step in the distance ladder.

Other individual objects can have fundamental distance estimates made for them under special circumstances. If the expansion of a gas cloud, like a supernova remnant or planetary nebula, can be observed over time, then an *expansion parallax* distance to that cloud can be estimated. Those measurements however suffer from uncertainties in the deviation of the object from sphericity. Binary stars which are both visual and spectroscopic binaries also can have their distance estimated by similar means, and don't suffer from the above geometric uncertainty. The common characteristic to these methods is that a measurement of angular motion is combined with a measurement of the absolute velocity (usually obtained via the Doppler effect). The distance estimate comes from computing how far the object must be to make its observed absolute velocity appear with the observed angular motion.

Expansion parallaxes in particular can give fundamental distance estimates for objects that are very far, because supernova ejecta have large expansion velocities and large sizes (compared to stars). Further, they can be observed with radio interferometers which can measure very small angular motions. These combine to provide fundamental distance estimates to supernovae in other galaxies.[10] Though valuable, such cases are quite rare, so they serve as important consistency checks on the distance ladder rather than workhorse steps by themselves.

6.2 Standard candles

Almost all astronomical objects used as physical distance indicators belong to a class that has a known brightness. By comparing this known luminosity to an object's observed brightness, the distance to the object can be computed using the inverse square law. These objects of known brightness are termed *standard candles*.

The brightness of an object can be expressed in terms of its absolute magnitude. This quantity is derived from the logarithm of its luminosity as seen from a distance of 10 parsecs. The apparent magnitude, the magnitude as seen by the observer (an instrument called a bolometer is used), can be measured and used with the absolute magnitude to calculate the distance D to the object in kiloparsecs (where 1 kpc equals 1000 parsecs) as follows:

$$5 \cdot \log_{10} D = m - M - 10,$$

or

$$D = 0.2 \cdot 10^{m-M-10},$$

where m the apparent magnitude and M the absolute magnitude. For this to be accurate, both magnitudes must be in the same frequency band and there can be no relative motion in the radial direction.

Some means of correcting for interstellar extinction, which also makes objects appear fainter and more red, is needed, especially if the object lies within a dusty or gaseous region.[11] The difference between an object's absolute and apparent magnitudes is called its distance modulus, and astronomical distances, especially intergalactic ones, are sometimes tabulated in this way.

6.2.1 Problems

Two problems exist for any class of standard candle. The principal one is calibration, that is the determination of exactly what the absolute magnitude of the candle is. This includes defining the class well enough that members can be recognized, and finding enough members of that class with well-known distances to allow their true absolute magnitude to be determined with enough accuracy. The second problem lies in recognizing members of the class, and not mistakenly using a standard candle calibration on an object which does not belong to the class. At extreme distances, which is where one most wishes to use a distance indicator, this recognition problem can be quite serious.

A significant issue with standard candles is the recurring question of how standard they are. For example, all observations seem to indicate that Type Ia supernovae that are of known distance have the same brightness (corrected by the shape of the light curve). The basis for this closeness in brightness is discussed below; however, the possibility exists that the distant Type Ia supernovae have different properties than nearby Type Ia supernovae. The use of Type Ia supernovae is crucial in determining the correct cosmological model. If indeed the properties of Type Ia supernovae are different at large distances, i.e. if the extrapolation of their calibration to arbitrary distances is not valid, ignoring this variation can dangerously bias the reconstruction of the cosmological parameters, in particular the reconstruction of the matter density parameter.[12]

That this is not merely a philosophical issue can be seen from the history of distance measurements using Cepheid variables. In the 1950s, Walter Baade discovered that the nearby Cepheid variables used to calibrate the standard candle were of a different type than the ones used to measure distances to nearby galaxies. The nearby Cepheid variables were population I stars with much higher metal content than the distant population II stars. As a result, the population II stars were actually much brighter than believed, and when corrected, this had the effect of doubling the distances to the globular clusters, the nearby galaxies, and the diameter of the Milky Way.

6.3 Standard ruler

Another class of physical distance indicator is the standard ruler. In 2008, galaxy diameters have been proposed as a possible standard ruler for cosmological parameter determination.[13] The method exploits regularity in baryon acoustic oscillations (BAO) in the early universe: while small structures occur at all scales, acoustic oscillations in the primordial plasma did not exceed a certain length. The inhomogeneities in the plasma were then organized by gravity into galaxy filaments and voids. Consequently, cosmic voids are not expected to exceed a maximum size determined by the size of the BAOs that gave rise to them. The method requires an extensive galaxy survey in order to make this scale visible. The WiggleZ galaxy survey was able to resolve this scale for detecting the expansion effect of dark energy.

Light echos can be also used as standard rulers.

6.4 Galactic distance indicators

See also: distance measures (cosmology)

With few exceptions, distances based on direct measurements are available only out to about a thousand parsecs, which is a modest portion of our own Galaxy. For distances beyond that, measures depend upon physical assumptions, that is, the assertion that one recognizes the object in question, and the class of objects is homogeneous enough that its members can be used for meaningful estimation of distance.

Physical distance indicators, used on progressively larger distance scales, include:

- Dynamical parallax, uses orbital parameters of visual binaries to measure the mass of the system, and hence use the mass-luminosity relation to determine the luminosity
    - Eclipsing binaries — In the last decade, measurement of eclipsing binaries' fundamental parameters has become possible with 8—meter class telescopes. This makes it feasible to use them as indicators of distance. Recently, they have been used to give direct distance estimates

to the Large Magellanic Cloud (LMC), Small Magellanic Cloud (SMC), Andromeda Galaxy and Triangulum Galaxy. Eclipsing binaries offer a direct method to gauge the distance to galaxies to a new improved 5% level of accuracy which is feasible with current technology to a distance of around 3 Mpc (3 million parsecs).[14]

- RR Lyrae variables — used for measuring distances within the galaxy and in nearby globular clusters.
- The following four indicators all use stars in the old stellar populations (Population II):[15]
  - Tip of the red giant branch (TRGB) distance indicator.
  - Planetary nebula luminosity function (PNLF)
  - Globular cluster luminosity function (GCLF)
  - Surface brightness fluctuation (SBF)
- In galactic astronomy, X-ray bursts (thermonuclear flashes on the surface of a neutron star) are used as standard candles. Observations of X-ray burst sometimes show X-ray spectra indicating radius expansion. Therefore, the X-ray flux at the peak of the burst should correspond to Eddington luminosity, which can be calculated once the mass of the neutron star is known (1.5 solar masses is a commonly used assumption). This method allows distance determination of some low-mass X-ray binaries. Low-mass X-ray binaries are very faint in the optical, making their distances extremely difficult to determine.
- Interstellar masers can be used to derive distances to galactic and some extragalactic objects that have maser emission.
- Cepheids and novae
- The Tully–Fisher relation
- The Faber–Jackson relation
- Type Ia supernovae that have a very well-determined maximum absolute magnitude as a function of the shape of their light curve and are useful in determining extragalactic distances up to a few hundred Mpc.[16] A notable exception is SN 2003fg, the "Champagne Supernova", a Type Ia supernova of unusual nature.
- Redshifts and Hubble's law

6.4.1 Main sequence fitting

When the absolute magnitude for a group of stars is plotted against the spectral classification of the star, in a Hertzsprung–Russell diagram, evolutionary patterns are found that relate to the mass, age and composition of the star. In particular, during their hydrogen burning period, stars lie along a curve in the diagram called the main sequence. By measuring these properties from a star's spectrum, the position of a main sequence star on the H–R diagram can be determined, and thereby the star's absolute magnitude estimated. A comparison of this value with the apparent magnitude allows the approximate distance to be determined, after correcting for interstellar extinction of the luminosity because of gas and dust.

In a gravitationally-bound star cluster such as the Hyades, the stars formed at approximately the same age and lie at the same distance. This allows relatively accurate main sequence fitting, providing both age and distance determination.

6.5 Extragalactic distance scale

The extragalactic distance scale is a series of techniques used today by astronomers to determine the distance of cosmological bodies beyond our own galaxy, which are not easily obtained with traditional methods. Some procedures utilize properties of these objects, such as stars, globular clusters, nebulae, and galaxies as a whole. Other methods are based more on the statistics and probabilities of things such as entire galaxy clusters.

6.5.1 Wilson–Bappu effect

Main article: Wilson–Bappu effect

Discovered in 1956 by Olin Wilson and M.K. Vainu Bappu, **The Wilson–Bappu effect** utilizes the effect known as spectroscopic parallax. Certain stars have features in their emission/absorption spectra allowing relatively easy absolute magnitude calculation. Certain spectral lines are directly related to an object's magnitude, such as the K absorption line of calcium. Distance to the star can be calculated from magnitude by the distance modulus:

$$M - m = -2.5 \log_{10}(F_1/F_2)\,.$$

Though in theory this method has the ability to provide reliable distance calculations to stars roughly at 7 megaparsecs (Mpc), it is generally only used for stars at hundreds kiloparsecs (kpc).

This method is only valid for stars over 15 magnitudes.

6.5.2 Classical Cepheids

Beyond the reach of the Wilson–Bappu effect, the next method relies on the period-luminosity relation of classical Cepheid variable stars. The following relation can be used to calculate the distance to Galactic and extragalactic classical Cepheids:

$$5\log_{10} d = V + (3.34)\log_{10} P - (2.45)(V - I) + 7.52\,.$$ [18]

$$5\log_{10} d = V + (3.37)\log_{10} P - (2.55)(V - I) + 7.48\,.$$ [19]

Several problems complicate the use of Cepheids as standard candles and are actively debated, chief among them are: the nature and linearity of the period-luminosity relation in various passbands and the impact of metallicity on both the zero-point and slope of those relations, and the effects of photometric contamination (blending) and a changing (typically unknown) extinction law on Cepheid distances.[20][21][22][23][24][25][26][27][28]

These unresolved matters have resulted in cited values for the Hubble Constant ranging between 60 km/s/Mpc and 80 km/s/Mpc. Resolving this discrepancy is one of the foremost problems in astronomy since the cosmological parameters of the Universe may be constrained by supplying a precise value of the Hubble constant.[29] [30]

Cepheid variable stars were the key instrument in Edwin Hubble's 1923 conclusion that M31 (Andromeda) was an external galaxy, as opposed to a smaller nebula within the Milky Way. He was able to calculate the distance of M31 to 285 Kpc, today's value being 770 Kpc.

As detected thus far, NGC 3370, a spiral galaxy in the constellation Leo, contains the farthest Cepheids yet found at a distance of 29 Mpc. Cepheid variable stars are in no way perfect distance markers: at nearby galaxies they have an error of about 7% and up to a 15% error for the most distant.

6.5.3 Supernovae

There are several different methods for which supernovae can be used to measure extragalactic distances.

Measuring a supernova's photosphere

We can assume that a supernova expands in a spherically symmetric manner. If the supernova is close enough such that we can measure the angular extent, $\theta(t)$, of its photosphere, we can use the equation

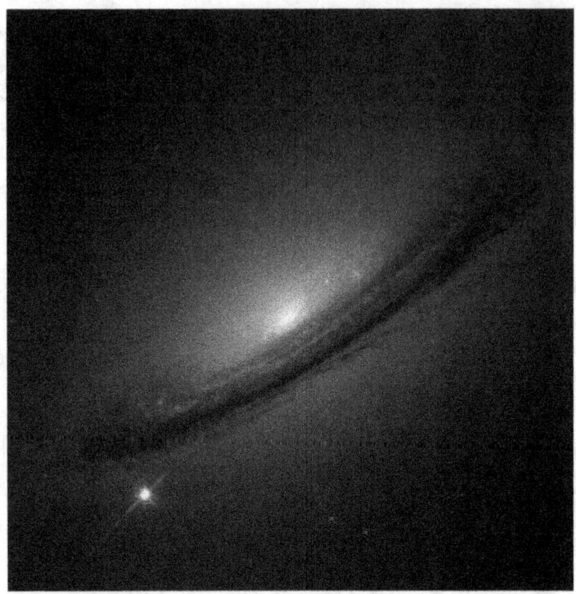

SN 1994D (bright spot on the lower left) in the NGC 4526 galaxy. Image by NASA, ESA, The Hubble Key Project Team, and The High-Z Supernova Search Team

$$\omega = \frac{\Delta\theta}{\Delta t}\,,$$

where ω is angular velocity, θ is angular extent. In order to get an accurate measurement, it is necessary to make two observations separated by time Δt. Subsequently, we can use

$$d = \frac{V_{ej}}{\omega}\,,$$

where d is the distance to the supernova, *Vej* is the supernova's ejecta's radial velocity (it can be assumed that *Vej* equals $V\theta$ if spherically symmetric).

This method works only if the supernova is close enough to be able to measure accurately the photosphere. Similarly, the expanding shell of gas is in fact not perfectly spherical nor a perfect blackbody. Also interstellar extinction can hinder the accurate measurements of the photosphere. This problem is further exacerbated by core-collapse supernova. All of these factors contribute to the distance error of up to 25%.

Type Ia light curves

Type Ia supernovae are some of the best ways to determine extragalactic distances. Ia's occur when a binary white dwarf star begins to accrete matter from its companion star.

As the white dwarf gains matter, eventually it reaches its Chandrasekhar Limit of $1.4 M_{\odot}$.

Once reached, the star becomes unstable and undergoes a runaway nuclear fusion reaction. Because all Type Ia supernovae explode at about the same mass, their absolute magnitudes are all the same. This makes them very useful as standard candles. All Type Ia supernovae have a standard blue and visual magnitude of

$$M_B \approx M_V \approx -19.3 \pm 0.3\,.$$

Therefore, when observing a Type Ia supernova, if it is possible to determine what its peak magnitude was, then its distance can be calculated. It is not intrinsically necessary to capture the supernova directly at its peak magnitude; using the **multicolor light curve shape** method (**MLCS**), the shape of the light curve (taken at any reasonable time after the initial explosion) is compared to a family of parameterized curves that will determine the absolute magnitude at the maximum brightness. This method also takes into effect interstellar extinction/dimming from dust and gas.

Similarly, the **stretch method** fits the particular supernovae magnitude light curves to a template light curve. This template, as opposed to being several light curves at different wavelengths (MLCS) is just a single light curve that has been stretched (or compressed) in time. By using this *Stretch Factor*, the peak magnitude can be determined.

Using Type Ia supernovae is one of the most accurate methods, particularly since supernova explosions can be visible at great distances (their luminosities rival that of the galaxy in which they are situated), much farther than Cepheid Variables (500 times farther). Much time has been devoted to the refining of this method. The current uncertainty approaches a mere 5%, corresponding to an uncertainty of just 0.1 magnitudes.

Novae in distance determinations

Novae can be used in much the same way as supernovae to derive extragalactic distances. There is a direct relation between a nova's max magnitude and the time for its visible light to decline by two magnitudes. This relation is shown to be:

$$M_V^{\max} = -9.96 - 2.31 \log_{10} \dot{x}\,.$$

Where $\dot{x}$ is the time derivative of the nova's mag, describing the average rate of decline over the first 2 magnitudes.

After novae fade, they are about as bright as the most luminous Cepheid Variable stars, therefore both these techniques have about the same max distance: ~ 20 Mpc. The error in this method produces an uncertainty in magnitude of about ±0.4

6.5.4 Globular cluster luminosity function

Based on the method of comparing the luminosities of globular clusters (located in galactic halos) from distant galaxies to that of the Virgo cluster, the globular cluster luminosity function carries an uncertainty of distance of about 20% (or 0.4 magnitudes).

US astronomer William Alvin Baum first attempted to use globular clusters to measure distant elliptical galaxies. He compared the brightest globular clusters in Virgo A galaxy with those in Andromeda, assuming the luminosities of the clusters were the same in both. Knowing the distance to Andromeda, Baum has assumed a direct correlation and estimated Virgo A's distance.

Baum used just a single globular cluster, but individual formations are often poor standard candles. Canadian astronomer René Racine assumed the use of the globular cluster luminosity function (GCLF) would lead to a better approximation. The number of globular clusters as a function of magnitude is given by:

$$\Phi(m) = A e^{(m-m_0)^2/2\sigma^2}$$

where m_0 is the turnover magnitude, M_0 is the magnitude of the Virgo cluster, and sigma is the dispersion ~ 1.4 mag.

It is important to remember that it is assumed that globular clusters all have roughly the same luminosities within the universe. There is no universal globular cluster luminosity function that applies to all galaxies.

6.5.5 Planetary nebula luminosity function

Like the GCLF method, a similar numerical analysis can be used for planetary nebulae (note the use of more than one!) within far off galaxies. The planetary nebula luminosity function (PNLF) was first proposed in the late 1970s by Holland Cole and David Jenner. They suggested that all planetary nebulae might all have similar maximum intrinsic brightness, now calculated to be M = −4.53. This would therefore make them potential standard candles for determining extragalactic distances.

Astronomer George Howard Jacoby and his colleagues later proposed that the PNLF function equaled:

$$N(M) \propto e^{0.307M}(1 - e^{3(M^*-M)})\,.$$

Where N(M) is number of planetary nebula, having absolute magnitude M. M* is equal to the nebula with the brightest magnitude.

6.5.6 Surface brightness fluctuation method

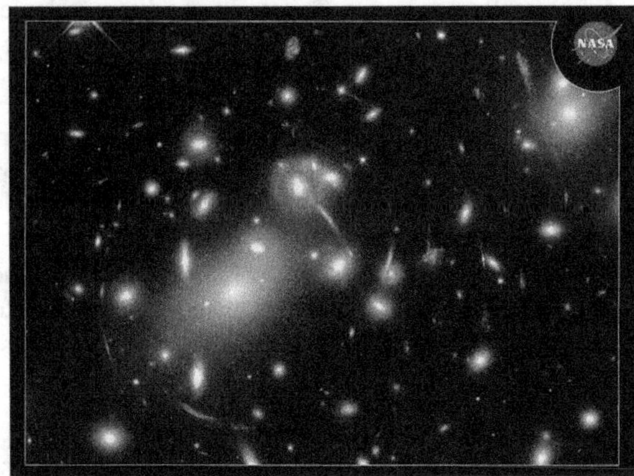

Galaxy cluster

The following method deals with the overall inherent properties of galaxies. These methods, though with varying error percentages, have the ability to make distance estimates beyond 100 Mpc, though it is usually applied more locally.

The surface brightness fluctuation (SBF) method takes advantage of the use of CCD cameras on telescopes. Because of spatial fluctuations in a galaxy's surface brightness, some pixels on these cameras will pick up more stars than others. However, as distance increases the picture will become increasingly smoother. Analysis of this describes a magnitude of the pixel-to-pixel variation, which is directly related to a galaxy's distance.

6.5.7 D–σ relation

The D–σ relation, used in elliptical galaxies, relates the angular diameter (D) of the galaxy to its velocity dispersion. It is important to describe exactly what D represents, in order to understand this method. It is, more precisely, the galaxy's angular diameter out to the surface brightness level of 20.75 B-mag arcsec^{-2}. This surface brightness is independent of the galaxy's actual distance from us. Instead, D is inversely proportional to the galaxy's distance, represented as d. Thus, this relation does not employ standard candles. Rather, D provides a standard ruler. This relation between D and σ is

$$\log_{10}(D) = 1.333\log(\sigma) + C\,.$$

Where C is a constant which depends on the distance to the galaxy clusters.

This method has the potential to become one of the strongest methods of galactic distance calculators, perhaps exceeding the range of even the Tully–Fisher method. As of today, however, elliptical galaxies aren't bright enough to provide a calibration for this method through the use of techniques such as Cepheids. Instead, calibration is done using more crude methods.

6.6 Overlap and scaling

A succession of distance indicators, which is the distance ladder, is needed for determining distances to other galaxies. The reason is that objects bright enough to be recognized and measured at such distances are so rare that few or none are present nearby, so there are too few examples close enough with reliable trigonometric parallax to calibrate the indicator. For example, Cepheid variables, one of the best indicators for nearby spiral galaxies, cannot yet be satisfactorily calibrated by parallax alone. The situation is further complicated by the fact that different stellar populations generally do not have all types of stars in them. Cepheids in particular are massive stars, with short lifetimes, so they will only be found in places where stars have very recently been formed. Consequently, because elliptical galaxies usually have long ceased to have large-scale star formation, they will not have Cepheids. Instead, distance indicators whose origins are in an older stellar population (like novae and RR Lyrae variables) must be used. However, RR Lyrae variables are less luminous than Cepheids, and novae are unpredictable and an intensive monitoring program – and luck during that program – is needed to gather enough novae in the target galaxy for a good distance estimate.

Because the more distant steps of the cosmic distance ladder depend upon the nearer ones, the more distant steps include the effects of errors in the nearer steps, both systematic and statistical ones. The result of these propagating errors means that distances in astronomy are rarely known to the same level of precision as measurements in the other sciences, and that the precision necessarily is poorer for more distant types of object.

Another concern, especially for the very brightest standard candles, is their "standardness": how homogeneous the objects are in their true absolute magnitude. For some of these different standard candles, the homogeneity is based on theories about the formation and evolution of stars and galaxies, and is thus also subject to uncertainties in those aspects. For the most luminous of distance indicators, the Type Ia supernovae, this homogeneity is known to be poor[31]; however, no other class of object is bright enough to be detected at such large distances, so the class is useful simply because there is no real alternative.

The observational result of Hubble's Law, the proportional relationship between distance and the speed with which a galaxy is moving away from us (usually referred to as redshift) is a product of the cosmic distance ladder. Hubble observed that fainter galaxies are more redshifted. Finding the value of the Hubble constant was the result of decades of work by many astronomers, both in amassing the measurements of galaxy redshifts and in calibrating the steps of the distance ladder. Hubble's Law is the primary means we have for estimating the distances of quasars and distant galaxies in which individual distance indicators cannot be seen.

6.7 See also

- Distance measures (cosmology)
- Orders of magnitude (length)#Astronomical
- Standard ruler

6.8 References

[1] Ash, M. E.; Shapiro, I. I.; Smith, W. B. (1967). "Astronomical constants and planetary ephemerides deduced from radar and optical observations". *The Astronomical Journal*. **72**: 338. Bibcode:1967AJ.....72..338A. doi:10.1086/110230.

[2] Staff. "Trigonometric Parallax". *The SAO Encyclopedia of Astronomy*. Swinburne Centre for Astrophysics and Supercomputing. Retrieved 2008-10-18.

[3] Perryman, M. A. C.; et al. (1999). "The HIPPARCOS Catalogue". *Astronomy and Astrophysics*. **323**: L49–L52. Bibcode:1997A&A...323L..49P.

[4] Harrington, J. D.; Villard, R. (10 April 2014). "NASA's Hubble Extends Stellar Tape Measure 10 Times Farther Into Space". NASA. Retrieved 17 October 2014.

[5] Riess, A. G.; Casertano, S.; Anderson, J.; MacKenty, J.; Filippenko, A. V. (2014). "Parallax Beyond a Kiloparsec from Spatially Scanning the Wide Field Camera 3 on the Hubble Space Telescope". *The Astrophysical Journal*. **785** (2): 161. Bibcode:2014ApJ...785..161R. arXiv:1401.0484. doi:10.1088/0004-637X/785/2/161.

[6] B., Baidyanath (2003). *An Introduction to Astrophysics*. PHI Learning Private Limited. ISBN 81-203-1121-3.

[7] "Hubble finds Universe may be expanding faster than expected". Retrieved 3 June 2016.

[8] "Hubble stretches the stellar tape measure ten times further". *ESA/Hubble Images*. Retrieved April 12, 2014.

[9] Popowski, P.; Gould, A. (1998). "Mathematics of Statistical Parallax and the Local Distance Scale". arXiv:astro-ph/9703140.

[10] Bartel, N.; et al. (1994). "The shape, expansion rate and distance of supernova 1993J from VLBI measurements". *Nature*. **368** (6472): 610–613. Bibcode:1994Natur.368..610B. doi:10.1038/368610a0.

[11] "Type Ia Supernova". *Weekly Topic*. Caglow. Retrieved 30 January 2012.

[12] Linden, S.; Virey, J.-M.; Tilquin, A. (2009). "Cosmological parameter extraction and biases from type Ia supernova magnitude evolution". *Astronomy and Astrophysics*. **506** (3): 1095–1105. Bibcode:2009A&A...506.1095L. arXiv:0907.4495. doi:10.1051/0004-6361/200912811. (And references therein.)

[13] Marinoni, C.; et al. (2008). "Geometrical tests of cosmological models. I. Probing dark energy using the kinematics of high redshift galaxies". *Astronomy and Astrophysics*. **478** (1): 43–55. Bibcode:2008A&A...478...43M. arXiv:0710.0759. doi:10.1051/0004-6361:20077116.

[14] Bonanos, A. Z. (2006). "Eclipsing Binaries: Tools for Calibrating the Extragalactic Distance Scale". *Proceedings of IAU Symposium*. **240**: 79–87. Bibcode:2007IAUS..240...79B. arXiv:astro-ph/0610923. doi:10.1017/S1743921307003845.

[15] Ferrarese, L; et al. (2000). "A Database of Cepheid Distance Moduli and Tip of the Red Giant Branch, Globular Cluster Luminosity Function, Planetary Nebula Luminosity Function, and Surface Brightness Fluctuation Data Useful for Distance Determinations". *The Astrophysical Journal Supplement Series*. **128** (2): 431–459. Bibcode:2000ApJS..128..431F. arXiv:astro-ph/9910501. doi:10.1086/313391.

[16] Colgate, S. A. (1979). "Supernovae as a standard candle for cosmology". *Astrophysical Journal*. **232** (1): 404–408. Bibcode:1979ApJ...232..404C. doi:10.1086/157300.

[17] Adapted from George H. Jacoby; David Branch; Robin Ciardullo; Roger L. Davies; William E. Harris; Michael J. Pierce; Christopher J. Pritchet; John L. Tonry; Douglas L. Welch (1992). "A critical review of selected techniques for measuring extragalactic distances". *Publications of the Astronomical Society of the Pacific*. **104** (678): 599–662. Bibcode:1992PASP..104..599J. JSTOR 40679907. doi:10.1086/133035.

[18] Benedict, G. Fritz et al. "Hubble Space Telescope Fine Guidance Sensor Parallaxes of Galactic Cepheid Variable Stars: Period-Luminosity Relations", *The Astronomical Journal*, Volume 133, Issue 4, pp. 1810–1827 (2007)

[19] Majaess, Daniel; Turner, David; Moni Bidin, Christian; Mauro, Francesco; Geisler, Douglas; Gieren, Wolfgang; Minniti, Dante; Chené, André-Nicolas; Lucas, Philip; Borissova, Jura; Kurtev, Radostn; Dékány, Istvan; Saito, Roberto K. "New Evidence Supporting Membership for TW Nor in Lyngå 6 and the Centaurus Spiral Arm", *ApJ Letters*, Volume 741, Issue 2, article id. L2 (2011)

[20] Stanek, K. Z.; Udalski, A. (1999). "The Optical Gravitational Lensing Experiment. Investigating the Influence of Blending on the Cepheid Distance Scale with Cepheids in the Large Magellanic Cloud". arXiv:astro-ph/9909346.

[21] Udalski, A.; Wyrzykowski, L.; Pietrzynski, G.; Szewczyk, O.; Szymanski, M.; Kubiak, M.; Soszynski, I.; Zebrun, K. (2001). "The Optical Gravitational Lensing Experiment. Cepheids in the Galaxy IC1613: No Dependence of the Period-Luminosity Relation on Metallicity". *Acta Astronomica*. **51**: 221. Bibcode:2001AcA....51..221U. arXiv:astro-ph/0109446.

[22] Ngeow, C.; Kanbur, S. M. (2006). "The Hubble Constant from Type Ia Supernovae Calibrated with the Linear and Nonlinear Cepheid Period-Luminosity Relations". *The Astrophysical Journal*. **642**: L29. Bibcode:2006ApJ...642L..29N. arXiv:astro-ph/0603643. doi:10.1086/504478.

[23] Macri, L. M.; Stanek, K. Z.; Bersier, D.; Greenhill, L. J.; Reid, M. J. (2006). "A New Cepheid Distance to the Maser–Host Galaxy NGC 4258 and Its Implications for the Hubble Constant". *The Astrophysical Journal*. **652** (2): 1133. Bibcode:2006ApJ...652.1133M. arXiv:astro-ph/0608211. doi:10.1086/508530.

[24] Bono, G.; Caputo, F.; Fiorentino, G.; Marconi, M.; Musella, I. (2008). "Cepheids in External Galaxies. I. The Maser–Host Galaxy NGC 4258 and the Metallicity Dependence of Period–Luminosity and Period–Wesenheit Relations". *The Astrophysical Journal*. **684**: 102. Bibcode:2008ApJ...684..102B. arXiv:0805.1592. doi:10.1086/589965.

[25] Majaess, D.; Turner, D.; Lane, D. (2009). "Type II Cepheids as Extragalactic Distance Candles". *Acta Astronomica*. **59**: 403. Bibcode:2009AcA....59..403M. arXiv:0909.0181.

[26] Madore, Barry F.; Freedman, Wendy L. (2009). "Concerning the Slope of the Cepheid Period–Luminosity Relation". *The Astrophysical Journal*. **696** (2): 1498. Bibcode:2009ApJ...696.1498M. arXiv:0902.3747. doi:10.1088/0004-637X/696/2/1498.

[27] Scowcroft, V.; Bersier, D.; Mould, J. R.; Wood, P. R. (2009). "The effect of metallicity on Cepheid magnitudes and the distance to M33". *Monthly Notices of the Royal Astronomical Society*. **396** (3): 1287. Bibcode:2009MNRAS.396.1287S. doi:10.1111/j.1365-2966.2009.14822.x.

[28] Majaess, D. (2010). "The Cepheids of Centaurus A (NGC 5128) and Implications for H0". *Acta Astronomica*. **60**: 121. Bibcode:2010AcA....60..121M. arXiv:1006.2458.

[29] Tammann, G. A.; Sandage, A.; Reindl, B. (2008). "The expansion field: The value of H 0". *Annual Review of Astronomy and Astrophysics*. **15** (4): 289. Bibcode:2008A&ARv..15..289T. arXiv:0806.3018. doi:10.1007/s00159-008-0012-y.

[30] Freedman, Wendy L.; Madore, Barry F. (2010). "The Hubble Constant". *Annual Review of Astronomy and Astrophysics*. **48**: 673. Bibcode:2010ARA&A..48..673F. arXiv:1004.1856. doi:10.1146/annurev-astro-082708-101829.

[31] Gilfanov, Marat; Bogdán, Ákos (2010). "An upper limit on the contribution of accreting white dwarfs to the type Ia supernova rate". *Nature*. **463** (3): 924–925. Bibcode:2010Natur.463..924G. arXiv:1002.3359. doi:10.1038/nature08685.

6.9 Bibliography

- *An Introduction to Modern Astrophysics*, Carroll and Ostlie, copyright 2007.
- *Measuring the Universe The Cosmological Distance Ladder*, Stephen Webb, copyright 2001.
- Pasachoff, JM & Filippenko, AV, *The Cosmos: Astronomy in the New Millennium*, Cambridge: Cambridge University Press, 4th edition, 2013 ISBN 9781107687561.
- *The Astrophysical Journal, The Globular Cluster Luminosity Function as a Distance Indicator: Dynamical Effects*, Ostriker and Gnedin, May 5, 1997.
- *An Introduction to Distance Measurement in Astronomy*, Richard de Grijs, Chichester: John Wiley & Sons, 2011, ISBN 978-0-470-51180-0.

6.10 External links

- The ABC's of distances (UCLA)
- The Extragalactic Distance Scale by Bill Keel
- The Hubble Space Telescope Key Project on the Extragalactic Distance Scale
- The Hubble Constant, a historical discussion
- NASA Cosmic Distance Scale

- PNLF information database
- The Astrophysical Journal

Chapter 7

Scale factor (cosmology)

The relative expansion of the universe is parametrized by a dimensionless **scale factor** a . Also known as the **cosmic scale factor** or sometimes the **Robertson-Walker scale factor**,[1] this is a key parameter of the Friedmann equations.

In the early stages of the big bang, most of the energy was in the form of radiation, and that radiation was the dominant influence on the expansion of the universe. Later, with cooling from the expansion the roles of mass and radiation changed and the universe entered a mass-dominated era. Recently results suggest that we have already entered an era dominated by dark energy, but examination of the roles of mass and radiation are most important for understanding the early universe.

Using the dimensionless scale factor to characterize the expansion of the universe, the effective energy densities of radiation and mass scale differently. This leads to a **radiation-dominated era** in the very early universe but a transition to a **matter-dominated era** at a later time and, since about 5 billion years ago, a subsequent **dark energy-dominated era**.[2][notes 1]

7.1 Detail

Some insight into the expansion can be obtained from a Newtonian expansion model which leads to a simplified version of the Friedman equation. It relates the proper distance (which can change over time, unlike the comoving distance which is constant) between a pair of objects, e.g. two galaxy clusters, moving with the Hubble flow in an expanding or contracting FLRW universe at any arbitrary time t to their distance at some reference time t_0 . The formula for this is:

$$d(t) = a(t)d_0,$$

where $d(t)$ is the proper distance at epoch t , d_0 is the distance at the reference time t_0 and $a(t)$ is the scale factor.[3] Thus, by definition, $a(t_0) = 1$.

The scale factor is dimensionless, with t counted from the birth of the universe and t_0 set to the present age of the universe: 13.799 ± 0.021 Gyr [4] giving the current value of a as $a(t_0)$ or 1 .

The evolution of the scale factor is a dynamical question, determined by the equations of general relativity, which are presented in the case of a locally isotropic, locally homogeneous universe by the Friedmann equations.

The Hubble parameter is defined:

$$H \equiv \frac{\dot{a}(t)}{a(t)}$$

where the dot represents a time derivative. From the previous equation $d(t) = d_0 a(t)$ one can see that $\dot{d}(t) = d_0 \dot{a}(t)$, and also that $d_0 = \frac{d(t)}{a(t)}$, so combining these gives $\dot{d}(t) = \frac{d(t)\dot{a}(t)}{a(t)}$, and substituting the above definition of the Hubble parameter gives $\dot{d}(t) = Hd(t)$ which is just Hubble's law.

Current evidence suggests that the expansion rate of the universe is accelerating, which means that the second derivative of the scale factor $\ddot{a}(t)$ is positive, or equivalently that the first derivative $\dot{a}(t)$ is increasing over time.[5] This also implies that any given galaxy recedes from us with increasing speed over time, i.e. for that galaxy $\dot{d}(t)$ is increasing with time. In contrast, the Hubble parameter seems to be decreasing with time, meaning that if we were to look at some fixed distance d and watch a series of different galaxies pass that distance, later galaxies would pass that distance at a smaller velocity than earlier ones.[6]

According to the Friedmann–Lemaître–Robertson–Walker metric which is used to model the expanding universe, if at the present time we receive light from a distant object with a redshift of z, then the scale factor at the time the object originally emitted that light is $a(t) = \frac{1}{1+z}$.[7][8]

7.2 Chronology

Further information: Chronology of the universe

7.2.1 Radiation-dominated era

After Inflation, and until about 47,000 years after the Big Bang, the dynamics of the early universe were set by radiation (referring generally to the constituents of the universe which moved relativistically, principally photons and neutrinos).[9]

For a radiation-dominated universe the evolution of the scale factor in the Friedmann–Lemaître–Robertson–Walker metric is obtained solving the Friedmann equations:

$$a(t) \propto t^{1/2}.$$ [10]

7.2.2 Matter-dominated era

Between about 47,000 years and 9.8 billion years after the Big Bang,[11] the energy density of matter exceeded both the energy density of radiation and the vacuum energy density.[12]

When the early universe was about 47,000 years old (redshift 3600), mass–energy density surpassed the radiation energy, although the universe remained optically thick to radiation until the universe was about 378,000 years old (redshift 1100). This second moment in time (close to the time of recombination) at which point the photons which compose the cosmic microwave background radiation were last scattered, is often mistaken as marking the end of the radiation era.

For a matter dominated universe the evolution of the scale factor in the Friedmann–Lemaître–Robertson–Walker metric is easily obtained solving the Friedmann equations:

$$a(t) \propto t^{2/3}$$

7.2.3 Dark energy-dominated era

In physical cosmology, the **dark-energy-dominated era** is proposed as the last of the three phases of the known universe, the other two being the matter-dominated era and the radiation-dominated era. The dark-energy-dominated era began after the matter-dominated era, i.e. when the Universe was about 9.8 billion years old.[13]

The cosmological constant is given the symbol Λ, and, considered as a source term in the Einstein field equation, can be viewed as equivalent to a "mass" of empty space, or dark energy. Since this increases with the volume of the universe, the effective expansion pressure is effectively constant, independent of the scale of the universe, while the other terms decrease with time. Thus, as the density of other forms of matter – dust and radiation – drops to very low concentrations, the cosmological constant (or "dark energy") term will eventually dominate the energy density of the Universe. Recent measurements of the change in Hubble constant with time, based on observations of distant supernovae, show this acceleration in expansion rate,[14] indicating the presence of such dark energy.

For a dark-energy-dominated universe, the evolution of the scale factor in the Friedmann–Lemaître–Robertson–Walker metric is easily obtained solving the Friedmann equations:

$$a(t) \propto \exp(Ht)$$

Here, the coefficient *H* in the exponential, the Hubble constant, is

$$H = \sqrt{8\pi G\rho_{\text{full}}/3} = \sqrt{\Lambda/3}.$$

This exponential dependence on time makes the spacetime geometry identical to the de Sitter Universe, and only holds for a positive sign of the cosmological constant, the sign that was observed to be realized in Nature anyway. The current density of the observable universe is of the order of 9.44 x 10^{-27}kg m^{-3} and the age of the universe is of the order of 13.8 billion years, or 4.358 x 10^{17}s. The Hubble parameter, H, is ~70.88 km $s^{-1}Mpc^{-1}$. (The Hubble time is 13.79 billion years.) The value of the cosmological constant, Λ, is ~2 x $10^{-35}s^{-2}$.

7.3 See also

- Cosmological principle
- Lambda-CDM model
- Redshift

7.4 Notes

[1] [2]p. 6: "The Universe has gone through three distinct eras: radiation-dominated, z≳3000; matter-dominated,

3000≳z≳0.5; and dark-energy dominated, z≲0.5. The evolution of the scale factor is controlled by the dominant energy form: a(t) ∝ t2/3(1+w) (for constant w). During the radiation-dominated era, a(t) ∝ t1/2; during the matter-dominated era, a(t) ∝ t2/3; and for the dark energy-dominated era, assuming w = −1, asymptotically a(t) ∝ exp(Ht)."
p. 44: "Taken together, all the current data provide strong evidence for the existence of dark energy; they constrain the fraction of critical density contributed by dark energy, 0.76 ± 0.02, and the equation-of-state parameter, w ≈ −1 ± 0.1 (stat) ±0.1 (sys), assuming that w is constant. This implies that the Universe began accelerating at redshift z ∼ 0.4 and age t ∼ 10 Gyr. These results are robust – data from any one method can be removed without compromising the constraints – and they are not substantially weakened by dropping the assumption of spatial flatness."

7.5 References

[1] Steven Weinberg (2008). *Cosmology*. Oxford University Press. p. 3. ISBN 978-0-19-852682-7.

[2] Frieman, Joshua A.; Turner, Michael S.; Huterer, Dragan (2008-01-01). "Dark Energy and the Accelerating Universe". *Annual Review of Astronomy and Astrophysics*. **46** (1): 385–432. Bibcode:2008ARA&A..46..385F. arXiv:0803.0982. doi:10.1146/annurev.astro.46.060407.145243.

[3] Schutz, Bernard (2003). *Gravity from the Ground Up: An Introductory Guide to Gravity and General Relativity*. Cambridge University Press. p. 363. ISBN 978-0-521-45506-0.

[4] Planck Collaboration (2015). "Planck 2015 results. XIII. Cosmological parameters (See Table 4 on page 31 of pfd).". *Astronomy & Astrophysics*. **594**: A13. Bibcode:2016A&A...594A..13P. arXiv:1502.01589. doi:10.1051/0004-6361/201525830.

[5] Jones, Mark H.; Robert J. Lambourne (2004). *An Introduction to Galaxies and Cosmology*. Cambridge University Press. p. 244. ISBN 978-0-521-83738-5.

[6] Is the universe expanding faster than the speed of light? (see final paragraph) Archived November 28, 2010, at the Wayback Machine.

[7] Davies, Paul (1992), *The New Physics*, p. 187.

[8] Mukhanov, V. F. (2005), *Physical Foundations of Cosmology*, p. 58.

[9] Ryden, Barbara, "Introduction to Cosmology", 2006, eqn. 5.25, 6.41

[10] Padmanabhan (1993), p. 64.

[11] Ryden, Barbara, "Introduction to Cosmology", 2006, eqn. 6.33, 6.41

[12] Zelik, M and Gregory, S: "Introductory Astronomy & Astrophysics", page 497. Thompson Learning, Inc. 1998

[13] Ryden, Barbara, "Introduction to Cosmology", 2006, eqn. 6.33

[14] The Nobel Prize in Physics 2011. Retrieved 18 May 2017.

- Padmanabhan, Thanu (1993). *Structure formation in the universe*. Cambridge: Cambridge University Press. ISBN 0-521-42486-0.

- Spergel, D. N.; et al. (2003). "First-Year Wilkinson Microwave Anisotropy Probe (WMAP) Observations: Determination of Cosmological Parameters". *Astrophysical Journal Supplement*. **148** (1): 175–194. Bibcode:2003ApJS..148..175S. arXiv:astro-ph/0302209. doi:10.1086/377226.

7.6 External links

- Relation of the scale factor with the cosmological constant and the Hubble constant

Chapter 8

Vacuum energy

For articles related to vacuum energy, see Quantum vacuum (disambiguation).

Vacuum energy is an underlying background energy that exists in space throughout the entire Universe. One contribution to the vacuum energy may be from virtual particles which are thought to be particle pairs that blink into existence and then annihilate in a timespan too short to observe. Their behavior is codified in Heisenberg's energy–time uncertainty principle. Still, the exact effect of such fleeting bits of energy is difficult to quantify. The vacuum energy is a special case of zero-point energy that relates to the quantum vacuum.[1]

The effects of vacuum energy can be experimentally observed in various phenomena such as spontaneous emission, the Casimir effect and the Lamb shift, and are thought to influence the behavior of the Universe on cosmological scales. Using the upper limit of the cosmological constant, the vacuum energy of free space has been estimated to be 10^{-9} joules (10^{-2} ergs) per cubic meter.[2] However, in both quantum electrodynamics (QED) and stochastic electrodynamics (SED), consistency with the principle of Lorentz covariance and with the magnitude of the Planck constant requires it to have a much larger value of 10^{113} joules per cubic meter.[3][4] This huge discrepancy is known as the vacuum catastrophe.

8.1 Origin

Quantum field theory states that all fundamental fields, such as the electromagnetic field, must be quantized at each and every point in space. A field in physics may be envisioned as if space were filled with interconnected vibrating balls and springs, and the strength of the field were like the displacement of a ball from its rest position. The theory requires "vibrations" in, or more accurately changes in the strength of, such a field to propagate as per the appropriate wave equation for the particular field in question. The second quantization of quantum field theory requires that each such ball-spring combination be quantized, that is, that the strength of the field be quantized at each point in space. Canonically, if the field at each point in space is a simple harmonic oscillator, its quantization places a quantum harmonic oscillator at each point. Excitations of the field correspond to the elementary particles of particle physics. Thus, according to the theory, even the vacuum has a vastly complex structure and all calculations of quantum field theory must be made in relation to this model of the vacuum.

The theory considers vacuum to implicitly have the same properties as a particle, such as spin or polarization in the case of light, energy, and so on. According to the theory, most of these properties cancel out on average leaving the vacuum empty in the literal sense of the word. One important exception, however, is the vacuum energy or the vacuum expectation value of the energy. The quantization of a simple harmonic oscillator requires the lowest possible energy, or zero-point energy of such an oscillator to be:

$E = \frac{1}{2}h\nu.$

Summing over all possible oscillators at all points in space gives an infinite quantity. To remove this infinity, one may argue that only differences in energy are physically measurable, much as the concept of potential energy has been treated in classical mechanics for centuries. This argument is the underpinning of the theory of renormalization. In all practical calculations, this is how the infinity is handled.

Vacuum energy can also be thought of in terms of virtual particles (also known as vacuum fluctuations) which are created and destroyed out of the vacuum. These particles are always created out of the vacuum in particle-antiparticle pairs, which in most cases shortly annihilate each other and disappear. However, these particles and antiparticles may interact with others before disappearing, a process which can be mapped using Feynman diagrams. Note that this method of computing vacuum energy is mathematically equivalent to having a quantum harmonic oscillator at each point and, therefore, suffers the same renormalization problems.

Additional contributions to the vacuum energy come from spontaneous symmetry breaking in quantum field theory.

8.2 Implications

Vacuum energy has a number of consequences. In 1948, Dutch physicists Hendrik B. G. Casimir and Dirk Polder predicted the existence of a tiny attractive force between closely placed metal plates due to resonances in the vacuum energy in the space between them. This is now known as the Casimir effect and has since been extensively experimentally verified. It is therefore believed that the vacuum energy is "real" in the same sense that more familiar conceptual objects such as electrons, magnetic fields, etc., are real. However, alternative explanations for the Casimir effect have since been proposed.[5]

Other predictions are harder to verify. Vacuum fluctuations are always created as particle–antiparticle pairs. The creation of these virtual particles near the event horizon of a black hole has been hypothesized by physicist Stephen Hawking to be a mechanism for the eventual "evaporation" of black holes.[6] If one of the pair is pulled into the black hole before this, then the other particle becomes "real" and energy/mass is essentially radiated into space from the black hole. This loss is cumulative and could result in the black hole's disappearance over time. The time required is dependent on the mass of the black hole (the equations indicate that the smaller the black hole, the more rapidly it evaporates) but could be on the order of 10^{100} years for large solar-mass black holes.[6]

The vacuum energy also has important consequences for physical cosmology. General relativity predicts that energy is equivalent to mass, and therefore, if the vacuum energy is "really there", it should exert a gravitational force. Essentially, a non-zero vacuum energy is expected to contribute to the cosmological constant, which affects the expansion of the universe. In the special case of vacuum energy, general relativity stipulates that the gravitational field is proportional to $\rho+3p$ (where ρ is the mass-energy density, and p is the pressure). Quantum theory of the vacuum further stipulates that the pressure of the zero-state vacuum energy is always negative and equal in magnitude to ρ. Thus, the total is $\rho+3p = \rho-3\rho = -2\rho$, a negative value. If indeed the vacuum ground state has non-zero energy, the calculation implies a repulsive gravitational field, giving rise to acceleration of the expansion of the universe,. However, the vacuum energy is mathematically infinite without renormalization, which is based on the assumption that we can only measure energy in a relative sense, which is not true if we can observe it indirectly via the cosmological constant.

The existence of vacuum energy is also sometimes used as theoretical justification for the possibility of free-energy machines. It has been argued that due to the broken symmetry (in QED), free energy does not violate conservation of energy, since the laws of thermodynamics only apply to equilibrium systems. However, consensus amongst physicists is that this is unknown as the nature of vacuum energy remains an unsolved problem.[7] In particular, the second law of thermodynamics is unaffected by the existence of vacuum energy. However, in Stochastic Electrodynamics, the energy density is taken to be a classical random noise wave field which consists of real electromagnetic noise waves propagating isotropically in all directions. The energy in such a wave field would seem to be accessible, e.g., with nothing more complicated than a directional coupler. The most obvious difficulty appears to be the spectral distribution of the energy, which compatibility with Lorentz invariance requires to take the form Kf^3, where K is a constant and f denotes frequency.[3][8] It follows that the energy and momentum flux in this wave field only becomes significant at extremely short wavelengths where directional coupler technology is currently lacking.

8.3 History

In 1934, Georges Lemaître used an unusual perfect-fluid equation of state to interpret the cosmological constant as due to vacuum energy. In 1948, the Casimir effect was provided an experimental method for a verification of the existence of vacuum energy, however, in 1955, Evgeny Lifshitz offered a different origin for the Casimir effect. In 1957, Lee and Yang proved the concepts of broken symmetry and parity violation, for which they won the Nobel prize. In 1973, Edward Tryon proposed the zero-energy universe hypothesis: that the Universe may be a large-scale quantum-mechanical vacuum fluctuation where positive mass-energy is balanced by negative gravitational potential energy. During the 1980s, there were many attempts to relate the fields that generate the vacuum energy to specific fields that were predicted by attempts at a Grand unification theory and to use observations of the Universe to confirm one or another version. However, the exact nature of the particles (or fields) that generate vacuum energy, with a density such as that required by inflation theory, remains a mystery.

8.4 See also

- Casimir effect
- Cosmological constant
- Dark energy

- False vacuum
- Heisenberg's uncertainty principle
- Lambdavacuum solution
- Quantum electrodynamics
- Stochastic electrodynamics
- Vacuum state
- Virtual particles
- Zero-point energy
- Zero-point field
- Zero-energy universe
- Normal ordering

8.5 External articles and references

- Free pdf copy of The Structured Vacuum - thinking about nothing by Johann Rafelski and Berndt Muller (1985) ISBN 3-87144-889-3.
- Saunders, S., & Brown, H. R. (1991). *The Philosophy of Vacuum.* Oxford [England]: Clarendon Press.
- Poincaré Seminar, Duplantier, B., & Rivasseau, V. (2003). "Poincaré Seminar 2002: vacuum energy-renormalization". *Progress in mathematical physics*, v. 30. Basel: Birkhäuser Verlag.
- Futamase & Yoshida *Possible measurement of vacuum energy*
- YAN Kun. Vacuum energy and superluminal velocity(2006), Zero-point energy step equation(2011).
- Study of Vacuum Energy Physics for Breakthrough Propulsion 2004, NASA Glenn Technical Reports Server, (pdf, 57 pages, Retrieved 2013-09-18).

8.6 Notes

[1] Scientific American. 1997. FOLLOW-UP: What is the 'zero-point energy' (or 'vacuum energy') in quantum physics? Is it really possible that we could harness this energy? - Scientific American. [ONLINE] Available at: http://www.scientificamerican.com/article/follow-up-what-is-the-zer/. [Accessed 27 September 2016].

[2] Sean Carroll, Sr Research Associate - Physics, California Institute of Technology, June 22, 2006C-SPAN broadcast of Cosmology at Yearly Kos Science Panel, Part 1

[3] Peter W. Milonni - "The Quantum Vacuum"

[4] de la Pena and Cetto "The Quantum Dice: An Introduction to Stochastic Electrodynamics"

[5] R. L. Jaffe: *The Casimir Effect and the Quantum Vacuum.* In: *Physical Review D.* Band 72, 2005

[6] Page, Don N. (1976). "Particle emission rates from a black hole: Massless particles from an uncharged, nonrotating hole". *Physical Review D.* **13** (2): 198–206. Bibcode:1976PhRvD..13..198P. doi:10.1103/PhysRevD.13.198.

[7] IEEE Trans. Ed., 1996, p.7

[8] de la Pena and Cetto "The Quantum Dice: An Introduction to Stochastic Electrodynamics"

Chapter 9

Dark energy

Not to be confused with dark fluid, dark flow, or dark matter.

In physical cosmology and astronomy, **dark energy** is an unknown form of energy which is hypothesized to permeate all of space, tending to accelerate the expansion of the universe.[1][2] Dark energy is the most accepted hypothesis to explain the observations since the 1990s indicating that the universe is expanding at an accelerating rate.

Assuming that the standard model of cosmology is correct, the best current measurements indicate that dark energy contributes 68.3% of the total energy in the present-day observable universe. The mass–energy of dark matter and ordinary (baryonic) matter contribute 26.8% and 4.9%, respectively, and other components such as neutrinos and photons contribute a very small amount.[3][4][5][6] The density of dark energy (~ 7×10^{-30} g/cm^3) is very low, much less than the density of ordinary matter or dark matter within galaxies. However, it comes to dominate the mass–energy of the universe because it is uniform across space.[7][8][9]

Two proposed forms for dark energy are the cosmological constant,[10][11] representing a constant energy density filling space homogeneously, and scalar fields such as quintessence or moduli, dynamic quantities whose energy density can vary in time and space. Contributions from scalar fields that are constant in space are usually also included in the cosmological constant. The cosmological constant can be formulated to be equivalent to the zero-point radiation of space i.e. the vacuum energy.[12] Scalar fields that change in space can be difficult to distinguish from a cosmological constant because the change may be extremely slow.

9.1 History of discovery and previous speculation

9.1.1 Einstein's Cosmological Constant

The "cosmological constant" is a constant term that can be added to Einstein's field equation of General Relativity. If considered as a "source term" in the field equation, it can be viewed as equivalent to the mass of empty space (which conceptually could be either positive or negative), or "vacuum energy".

The cosmological constant was first proposed by Einstein as a mechanism to obtain a solution of the gravitational field equation that would lead to a static universe, effectively using dark energy to balance gravity.[13] Einstein gave the cosmological constant the symbol Λ (capital lambda).

The mechanism was an example of fine-tuning, and it was later realized that Einstein's static universe would not be stable: local inhomogeneities would ultimately lead to either the runaway expansion or contraction of the universe. The equilibrium is unstable: if the universe expands slightly, then the expansion releases vacuum energy, which causes yet more expansion. Likewise, a universe which contracts slightly will continue contracting. These sorts of disturbances are inevitable, due to the uneven distribution of matter throughout the universe. Further, observations made by Edwin Hubble in 1929 showed that the universe appears to be expanding and not static at all. Einstein reportedly referred to his failure to predict the idea of a dynamic universe, in contrast to a static universe, as his greatest blunder.[14]

9.1.2 Inflationary Dark Energy

Alan Guth and Alexei Starobinsky proposed in 1980 that a negative pressure field, similar in concept to dark energy, could drive cosmic inflation in the very early universe. Inflation postulates that some repulsive force, qualitatively similar to dark energy, resulted in an enormous and exponential expansion of the universe slightly after the Big Bang. Such expansion is an essential feature of most current models of the Big Bang. However, inflation must have occurred at a much higher energy density than the dark energy we observe today and is thought to have completely ended when the universe was just a fraction of a second old. It is unclear what relation, if any, exists between dark energy and inflation. Even after inflationary models became accepted, the cosmological constant was thought to be irrelevant to the current universe.

Nearly all inflation models predict that the total (matter+energy) density of the universe should be very close to the critical density. During the 1980s, most cosmological research focused on models with critical density in matter only, usually 95% cold dark matter and 5% ordinary matter (baryons). These models were found to be successful at forming realistic galaxies and clusters, but some problems appeared in the late 1980s: in particular, the model required a value for the Hubble constant lower than preferred by observations, and the model under-predicted observations of large-scale galaxy clustering. These difficulties became stronger after the discovery of anisotropy in the cosmic microwave background by the COBE spacecraft in 1992, and several modified CDM models came under active study through the mid-1990s: these included the Lambda-CDM model and a mixed cold/hot dark matter model. The first direct evidence for dark energy came from supernova observations in 1998 of accelerated expansion in Riess *et al.*[15] and in Perlmutter *et al.*,[16] and the Lambda-CDM model then became the leading model. Soon after, dark energy was supported by independent observations: in 2000, the BOOMERanG and Maxima cosmic microwave background experiments observed the first acoustic peak in the CMB, showing that the total (matter+energy) density is close to 100% of critical density. Then in 2001, the 2dF Galaxy Redshift Survey gave strong evidence that the matter density is around 30% of critical. The large difference between these two supports a smooth component of dark energy making up the difference. Much more precise measurements from WMAP in 2003–2010 have continued to support the standard model and give more accurate measurements of the key parameters.

The term “dark energy”, echoing Fritz Zwicky's “dark matter” from the 1930s, was coined by Michael Turner in 1998.[17]

9.1.3 Change in expansion over time

High-precision measurements of the expansion of the universe are required to understand how the expansion rate changes over time and space. In general relativity, the evolution of the expansion rate is estimated from the curvature of the universe and the cosmological equation of state (the relationship between temperature, pressure, and combined matter, energy, and vacuum energy density for any region of space). Measuring the equation of state for dark energy is one of the biggest efforts in observational cosmology today. Adding the cosmological constant to cosmology's standard FLRW metric leads to the Lambda-CDM model, which has been referred to as the "*standard model of cosmology*" because of its precise agreement with observations.

As of 2013, the Lambda-CDM model is consistent with a series of increasingly rigorous cosmological observations, including the Planck spacecraft and the Supernova Legacy Survey. First results from the SNLS reveal that the average behavior (i.e., equation of state) of dark energy behaves like Einstein's cosmological constant to a precision of 10%.[18] Recent results from the Hubble Space Telescope Higher-Z Team indicate that dark energy has been present for at least 9 billion years and during the period preceding cosmic acceleration.

9.2 Nature

1

The nature of dark energy is more hypothetical than that of dark matter, and many things about the nature of dark energy remain matters of speculation.[19] Dark energy is thought to be very homogeneous, not very dense and is not known to interact through any of the fundamental forces other than gravity. Since it is quite rarefied — roughly 10^{-27} kg/m^3 — it is unlikely to be detectable in laboratory experiments. The reason dark energy can have such a profound effect on the universe, making up 68% of universal density, in spite of being so rarefied is because it uniformly fills otherwise empty space.

Independently of its actual nature, dark energy would need to have a strong negative pressure (acting repulsively) like radiation pressure in a metamaterial[20] to explain the observed acceleration of the expansion of the universe. According to general relativity, the pressure within a substance contributes to its gravitational attraction for other things just as its mass density does. This happens because the physical quantity that causes matter to generate gravitational effects is the stress–energy tensor, which contains both the energy (or matter) density of a substance and its pressure and viscosity. In the Friedmann–Lemaître–Robertson–Walker metric, it can be shown that a strong constant negative pressure in all the universe causes an acceleration in universe expansion if the universe is already expanding, or a deceleration in universe contraction if the universe is already contracting. This accelerating expansion effect is sometimes labeled "gravitational repulsion".

9.2.1 Technical definition

See also: Friedmann equations

In standard cosmology, there are three components of the universe: matter, radiation and dark energy. Matter is anything whose energy density scales with the inverse cube of the scale factor, i.e. $\rho \propto a^{-3}$, while radiation is anything which scales to the inverse fourth power of the scale factor $\rho \propto a^{-4}$. This can be understood intuitively: for an ordinary particle in a square box, doubling the length of a side of the box decreases the density (and hence energy density) by a factor of eight (2^3). For radiation, the decrease in energy density is greater, because an increase in spatial distance also causes a redshift.[21]

The final component, dark energy, is an intrinsic property of space, and so has a constant energy density regardless of the volume under consideration ($\rho \propto a^0$).

9.3 Evidence of existence

The evidence for dark energy is indirect but comes from three independent sources:

- Distance measurements and their relation to redshift, which suggest the universe has expanded more in the last half of its life.[22]
- The theoretical need for a type of additional energy that is not matter or dark matter to form the observationally flat universe (absence of any detectable global curvature).
- It can be inferred from measures of large scale wave-patterns of mass density in the universe.

9.3.1 Supernovae

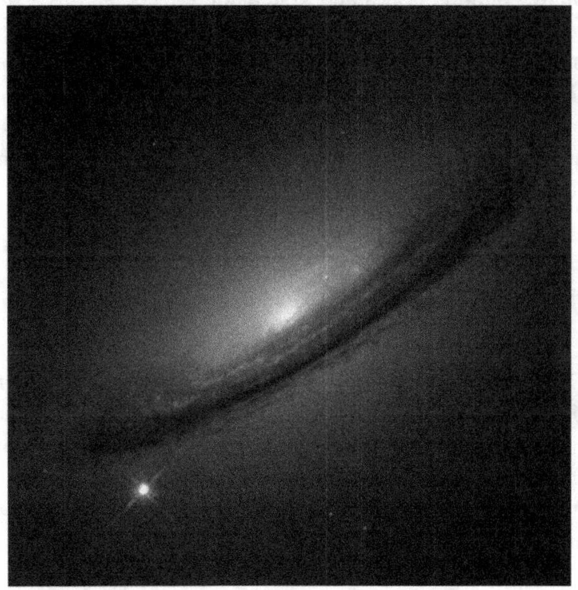

A Type Ia supernova (bright spot on the bottom-left) near a galaxy

In 1998, the High-Z Supernova Search Team[15] published observations of Type Ia ("one-A") supernovae. In 1999, the Supernova Cosmology Project[16] followed by suggesting that the expansion of the universe is accelerating.[23] The 2011 Nobel Prize in Physics was awarded to Saul Perlmutter, Brian P. Schmidt and Adam G. Riess for their leadership in the discovery.[24][25]

Since then, these observations have been corroborated by several independent sources. Measurements of the cosmic microwave background, gravitational lensing, and the large-scale structure of the cosmos as well as improved measurements of supernovae have been consistent with the Lambda-CDM model.[26] Some people argue that the only indication for the existence of dark energy is observations of distance measurements and the associated redshifts. Cosmic microwave background anisotropies and baryon acoustic oscillations only serve to demonstrate that distances to a given redshift are larger than would be expected from a "dusty" Friedmann–Lemaître universe and the local measured Hubble constant.[27]

Supernovae are useful for cosmology because they are excellent standard candles across cosmological distances. They allow the expansion history of the universe to be measured by looking at the relationship between the distance to an object and its redshift, which gives how fast it is receding from us. The relationship is roughly linear, according to Hubble's law. It is relatively easy to measure redshift, but finding the distance to an object is more difficult. Usually, astronomers use standard candles: objects for which the intrinsic brightness, the absolute magnitude, is known. This allows the object's distance to be measured from its actual observed brightness, or apparent magnitude. Type Ia supernovae are the best-known standard candles across cosmological distances because of their extreme and consistent luminosity.

Recent observations of supernovae are consistent with a universe made up 71.3% of dark energy and 27.4% of a combination of dark matter and baryonic matter.[28]

9.3.2 Cosmic microwave background

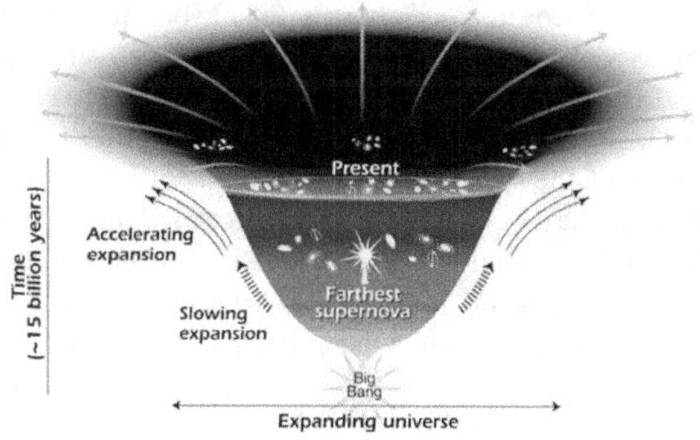
Present
Time
(~15 billion years)
Accelerating expansion
Farthest supernova
Slowing expansion
Big Bang
Expanding universe
This diagram reveals changes in the rate of expansion since the universe's birth 15 billion years ago. The more shallow the curve, the faster the rate of expansion. The curve changes noticeably about 7.5 billion years ago, when objects in the universe began flying apart at a faster rate. Astronomers theorize that the faster expansion rate is due to a mysterious, dark force that is pushing galaxies apart

Diagram representing the accelerated expansion of the universe due to dark energy.

The existence of dark energy, in whatever form, is needed to reconcile the measured geometry of space with the total amount of matter in the universe. Measurements of cosmic microwave background (CMB) anisotropies indicate that the universe is close to flat. For the shape of the universe to be flat, the mass/energy density of the universe must be equal to the critical density. The total amount of matter in the universe (including baryons and dark matter), as measured from the CMB spectrum, accounts for only about 30% of the critical density. This implies the existence of an additional form of energy to account for the remaining 70%.[26] The Wilkinson Microwave Anisotropy Probe (WMAP) spacecraft seven-year analysis estimated a universe made up of 72.8% dark energy, 22.7% dark matter and 4.5% ordinary matter.[5] Work done in 2013 based on the Planck spacecraft observations of the CMB gave a more accurate estimate of 68.3% of dark energy, 26.8% of dark matter and 4.9% of ordinary matter.[29]

9.3.3 Large-scale structure

The theory of large-scale structure, which governs the formation of structures in the universe (stars, quasars, galaxies and galaxy groups and clusters), also suggests that the density of matter in the universe is only 30% of the critical density.

A 2011 survey, the WiggleZ galaxy survey of more than 200,000 galaxies, provided further evidence towards the existence of dark energy, although the exact physics behind it remains unknown.[30][31] The WiggleZ survey from the Australian Astronomical Observatory scanned the galaxies to determine their redshift. Then, by exploiting the fact that baryon acoustic oscillations have left voids regularly of ~150 Mpc diameter, surrounded by the galaxies, the voids were used as standard rulers to estimate distances to galaxies as far as 2,000 Mpc (redshift 0.6), allowing for accurate estimate of the speeds of galaxies from their redshift and distance. The data confirmed cosmic acceleration up to half of the age of the universe (7 billion years) and constrain its inhomogeneity to 1 part in 10.[31] This provides a confirmation to cosmic acceleration independent of supernovae.

9.3.4 Late-time integrated Sachs-Wolfe effect

Accelerated cosmic expansion causes gravitational potential wells and hills to flatten as photons pass through them, producing cold spots and hot spots on the CMB aligned with vast supervoids and superclusters. This so-called late-time Integrated Sachs–Wolfe effect (ISW) is a direct signal of dark energy in a flat universe.[32] It was reported at high significance in 2008 by Ho *et al.*[33] and Giannantonio *et al.*[34]

9.3.5 Observational Hubble constant data

A new approach to test evidence of dark energy through observational Hubble constant data (OHD) has gained significant attention in recent years.[35][36][37][38] The Hubble constant, $H(z)$, is measured as a function of cosmological redshift. OHD directly tracks the expansion history of the universe by taking passively evolving early-type galaxies as “cosmic chronometers”.[39] From this point, this approach provides standard clocks in the universe. The core of this idea is the measurement of the differential age evolution as a function of redshift of these cosmic chronometers. Thus, it provides a direct estimate of the Hubble parameter

$$H(z) = -\frac{1}{1+z}\frac{dz}{dt} \approx -\frac{1}{1+z}\frac{\Delta z}{\Delta t}.$$

The reliance on a differential quantity, $\Delta z/\Delta t$, can minimize many common issues and systematic effects; and as a direct measurement of the Hubble parameter instead of its integral, like supernovae and baryon acoustic oscillations (BAO), it brings more information and is appealing in computation. For these reasons, it has been widely used to examine the accelerated cosmic expansion and study properties of dark energy.

9.4 Theories of dark energy

Dark energy’s status as a hypothetical force with unknown properties makes it a very active target of research. The problem is attacked from a great variety of angles, such as modifying the prevailing theory of gravity (general relativity), attempting to pin down the properties of dark energy, and finding alternative ways to explain the observational data.

9.4.1 Cosmological constant

Main article: Cosmological constant
For more details on this topic, see Equation of state (cosmology).

The simplest explanation for dark energy is that it is an intrinsic, fundamental energy of space. This is the cosmological constant, usually represented by the Greek letter Λ (Lambda, hence Lambda-CDM model). Since energy and mass are related according to the equation $E = mc^2$, Einstein’s theory of general relativity predicts that this energy will have a gravitational effect. It is sometimes called a vacuum energy because it is the energy density of empty vacuum.

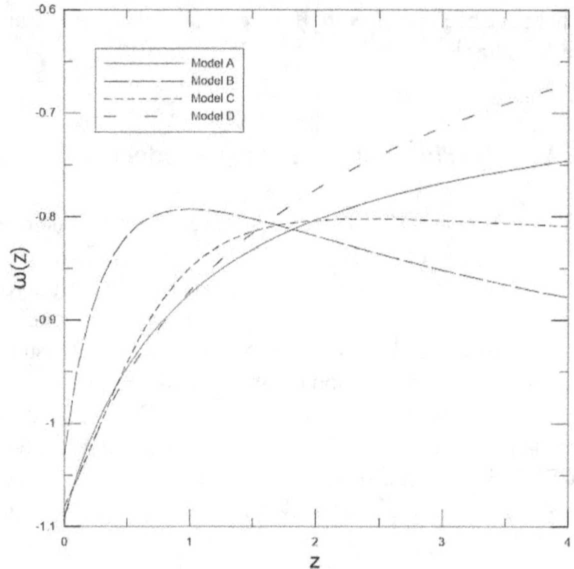

The equation of state of Dark Energy for 4 common models by Redshift.[40]
A: CPL Model,
B: Jassal Model,
C: Barboza & Alcaniz Model,
D: Wetterich Model

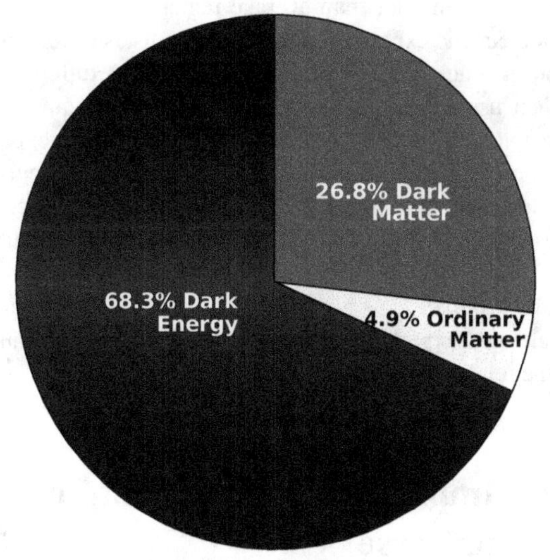

Estimated distribution of matter and energy in the universe[41]

The cosmological constant has negative pressure equal to its energy density and so causes the expansion of the universe to accelerate. The reason a cosmological constant has negative pressure can be seen from classical thermodynamics. In general, energy must be lost from inside a container (the container must do work on its environment) in order for the volume to increase. Specifically, a change in volume dV requires work done equal to a change of energy $-P$ dV, where P is the pressure. But the amount of energy in a container full of vacuum actually increases when the volume increases, because the energy is equal to ϱV, where ϱ is the energy density of the cosmological constant. Therefore, P is negative and, in fact, $P = -\varrho$.

There are two major advantages for the cosmological constant. The first is that it is simple. Einstein had in fact introduced this term in his original formulation of general relativity such as to get a static universe. Although he later discarded the term after Hubble found that the universe is expanding, a nonzero cosmological constant can act as dark energy, without otherwise changing the Einstein field equations. The other advantage is that there is a natural explanation for its origin. Most quantum field theories predict vacuum fluctuations that would give the vacuum this sort of energy. This is related to the Casimir effect, in which there is a small suction into regions where virtual particles are geometrically inhibited from forming (e.g. between plates with tiny separation).

A major outstanding problem is that the same quantum field theories predict a huge cosmological constant, more than 100 orders of magnitude too large.[11] This would need to be almost, but not exactly, cancelled by an equally large term of the opposite sign. Some supersymmetric theories require a cosmological constant that is exactly zero,[42] which does not help because supersymmetry must be broken.

Nonetheless, the cosmological constant is the most economical solution to the problem of cosmic acceleration. Thus, the current standard model of cosmology, the Lambda-CDM model, includes the cosmological constant as an essential feature.

9.4.2 Modified gravity

The evidence for dark energy is heavily dependent on the theory of general relativity. Therefore, it is conceivable that a modification to general relativity also eliminates the need for dark energy. There are very many such theories, and research is ongoing.[43][44]

Quintessence

Main article: Quintessence (physics)

In quintessence models of dark energy, the observed acceleration of the scale factor is caused by the potential energy of a dynamical field, referred to as quintessence field. Quintessence differs from the cosmological constant in that it can vary in space and time. In order for it not to clump and form structure like matter, the field must be very light so that it has a large Compton wavelength.

No evidence of quintessence is yet available, but it has not been ruled out either. It generally predicts a slightly slower acceleration of the expansion of the universe than the cosmological constant. Some scientists think that the best evidence for quintessence would come from violations of Einstein's equivalence principle and variation of the fundamental constants in space or time.[45] Scalar fields are predicted by the Standard Model of particle physics and string theory, but an analogous problem to the cosmological constant problem (or the problem of constructing models of cosmological inflation) occurs: renormalization theory predicts that scalar fields should acquire large masses.

The coincidence problem asks why the acceleration of the Universe began when it did. If acceleration began earlier in the universe, structures such as galaxies would never have had time to form and life, at least as we know it, would never have had a chance to exist. Proponents of the anthropic principle view this as support for their arguments. However, many models of quintessence have a so-called "tracker" behavior, which solves this problem. In these models, the quintessence field has a density which closely tracks (but is less than) the radiation density until matter-radiation equality, which triggers quintessence to start behaving as dark energy, eventually dominating the universe. This naturally sets the low energy scale of the dark energy.[46][47]

In 2004, when scientists fit the evolution of dark energy with the cosmological data, they found that the equation of state had possibly crossed the cosmological constant boundary ($w = -1$) from above to below. A No-Go theorem has been proved that gives this scenario at least two degrees of freedom as required for dark energy models. This scenario is so-called Quintom scenario.

Some special cases of quintessence are phantom energy, in which the energy density of quintessence actually increases with time, and k-essence (short for kinetic quintessence) which has a non-standard form of kinetic energy such as a negative kinetic energy.[48] They can have unusual properties: phantom energy, for example, can cause a Big Rip.

Interacting dark energy

This class of theories attempts to come up with an all-encompassing theory of both dark matter and dark energy as a single phenomenon that modifies the laws of gravity at various scales. This could for example treat dark energy and dark matter as different facets of the same unknown substance,[49] or postulate that cold dark matter decays into dark energy.[50] Another class of theories that unifies dark matter and dark energy are suggested to be covariant theories of modified gravities. These theories alter the dynamics of the space-time such that the modified dynamic stems what have been assigned to the presence of dark energy and dark matter.[51]

9.4.3 Variable dark energy models

The density of dark energy might have varied in time over the history of the universe. Modern observational data allow for estimates of the present density. Using baryon acoustic oscillations, it is possible to investigate the effect of dark energy in the history of the Universe, and constrain parameters of the equation of state of dark energy. To that end, several models have been proposed. One of the most popular models is the Chevallier–Polarski–Linder model (CPL).[52][53] Some other common models are, (Barboza & Alcaniz. 2008),[54] (Jassal et al. 2005),[55] (Wetterich. 2004).[56]

9.4.4 Observational skepticism

Some alternatives to dark energy aim to explain the observational data by a more refined use of established theories. In this scenario, dark energy doesn't actually exist, and is merely a measurement artifact. For example, if we are located in an emptier-than-average region of space, the observed cosmic expansion rate could be mistaken for a variation in time, or acceleration.[57][58][59][60] A different approach uses a cosmological extension of the equivalence principle to show how space might appear to be expanding more rapidly in the voids surrounding our local cluster. While weak, such effects considered cumulatively over billions of years could become significant, creating the illusion of cosmic acceleration, and making it appear as if we live in a Hubble bubble.[61][62][63] Yet another possibility is that the accelerated expansion of the universe is an illusion caused by the relative motion of us to the rest of the universe.[64][65]

9.5 Implications for the fate of the universe

Cosmologists estimate that the acceleration began roughly 5 billion years ago. Before that, it is thought that the expansion was decelerating, due to the attractive influence of dark matter and baryons. The density of dark matter in an expanding universe decreases more quickly than dark energy, and eventually the dark energy dominates. Specifically, when the volume of the universe doubles, the density of dark matter is halved, but the density of dark energy is nearly unchanged (it is exactly constant in the case of a cosmological constant).

Projections into the future can differ radically for different models of dark energy. For a cosmological constant, or any other model that predicts that the acceleration will continue indefinitely, the ultimate result will be that galaxies outside the Local Group will have a line-of-sight velocity that continually increases with time, eventually far exceeding the speed of light.[66] This is not a violation of special relativity because the notion of "velocity" used here is different from that of velocity in a local inertial frame of reference, which is still constrained to be less than the speed of light for any massive object (see Uses of the proper distance for a discussion of the subtleties of defining any notion of relative velocity in cosmology). Because the Hubble parameter is decreasing with time, there can actually be cases where a galaxy that is receding from us faster than light does manage to emit a signal which reaches us eventually.[67][68] However, because of the accelerating expansion, it is projected that most galaxies will eventually cross a type of cosmological event horizon where any light they emit past that point will never be able to reach us at any time in the infinite future[69] because the light never reaches a point where its "peculiar velocity" toward us exceeds the expansion velocity away from us (these two notions of velocity are also discussed in Uses of the proper distance). Assuming the dark energy is constant (a cosmological constant), the current distance to this cosmological event horizon is about 16 billion light years, meaning that a signal from an event happening *at present* would eventually be able to reach us in the future if the event were less than 16 billion light years away, but the signal would never reach us if the event were more than 16 billion light years away.[68]

As galaxies approach the point of crossing this cosmological event horizon, the light from them will become more and more redshifted, to the point where the wavelength becomes too large to detect in practice and the galaxies appear to vanish completely[70][71] (*see* Future of an expanding universe). The Earth, the Milky Way, and the Local Group of which the Milky way is a part, would all remain virtually undisturbed as the rest of the universe recedes and disappears from view. In this scenario, the Local Group would ultimately suffer heat death, just as was hypothesized for the flat, matter-dominated universe before measurements of cosmic acceleration.

There are other, more speculative ideas about the future of the universe. The phantom energy model of dark energy results in *divergent* expansion, which would imply that the effective force of dark energy continues growing until it dominates all other forces in the universe. Under this scenario, dark energy would ultimately tear apart all gravitationally bound structures, including galaxies and solar systems, and eventually overcome the electrical and nuclear forces to tear apart atoms themselves, ending the universe in a "Big Rip". It is also possible the universe may never have an end and continue in its present state forever (see The Second Law as a law of disorder). On the other hand, dark energy might dissipate with time or even become attractive. Such uncertainties leave open the possibility that gravity might yet rule the day and lead to a universe that contracts in on itself in a "Big Crunch",[72] or that there may even be a dark energy cycle, which implies a cyclic model of the universe in which every iteration (Big Bang then eventually a Big Crunch) takes about a trillion (10^{12}) years.[73][74] While none of these are supported by observations, they are not ruled out.

9.6 In philosophy of science

In philosophy of science, dark energy is an example of an "auxiliary hypothesis", an ad hoc postulate that is added to a theory in response to observations that falsify it. It has been argued that the dark energy hypothesis is a conventionalist hypothesis, that is, a hypothesis that adds no empirical content and hence is unfalsifiable in the sense defined by Karl Popper.[75]

9.7 See also

- Conformal gravity
- Dark Energy Spectroscopic Instrument
- De Sitter relativity
- Illustris project
- *The Dark Energy Survey*
- *Quintessence: The Search for Missing Mass in the Universe*
- Vacuum state
- Negative mass

9.8 References

[1] Overbye, Dennis (20 February 2017). "Cosmos Controversy: The Universe Is Expanding, but How Fast?". *New York Times.* Retrieved 21 February 2017.

[2] Peebles, P. J. E.; Ratra, Bharat (2003). "The cosmological constant and dark energy". *Reviews of Modern Physics.* **75** (2): 559–606. Bibcode:2003RvMP...75..559P. arXiv:astro-ph/0207347. doi:10.1103/RevModPhys.75.559.

[3] Ade, P. A. R.; Aghanim, N.; Armitage-Caplan, C.; et al. (Planck Collaboration), C.; Arnaud, M.; Ashdown, M.; Atrio-Barandela, F.; Aumont, J.; Aussel, H.; Baccigalupi, C.; Banday, A. J.; Barreiro, R. B.; Barrena, R.; Bartelmann, M.; Bartlett, J. G.; Bartolo, N.; Basak, S.; Battaner, E.; Battye, R.; Benabed, K.; Benoît, A.; Benoit-Lévy, A.; Bernard, J.-P.; Bersanelli, M.; Bertincourt, B.; Bethermin, M.; Bielewicz, P.; Bikmaev, I.; Blanchard, A.; et al. (22 March 2013). "Planck 2013 results. I. Overview of products and scientific results – Table 9". *Astronomy and Astrophysics*. **571**: A1. Bibcode:2014A&A...571A...1P. arXiv:1303.5062. doi:10.1051/0004-6361/201321529.

[4] Ade, P. A. R.; Aghanim, N.; Armitage-Caplan, C.; et al. (Planck Collaboration), C.; Arnaud, M.; Ashdown, M.; Atrio-Barandela, F.; Aumont, J.; Aussel, H.; Baccigalupi, C.; Banday, A. J.; Barreiro, R. B.; Barrena, R.; Bartelmann, M.; Bartlett, J. G.; Bartolo, N.; Basak, S.; Battaner, E.; Battye, R.; Benabed, K.; Benoît, A.; Benoit-Lévy, A.; Bernard, J.-P.; Bersanelli, M.; Bertincourt, B.; Bethermin, M.; Bielewicz, P.; Bikmaev, I.; Blanchard, A.; et al. (31 March 2013). "Planck 2013 Results Papers". *Astronomy and Astrophysics*. **571**: A1. Bibcode:2014A&A...571A...1P. arXiv:1303.5062. doi:10.1051/0004-6361/201321529.

[5] "First Planck results: the Universe is still weird and interesting".

[6] Sean Carroll, Ph.D., Caltech, 2007, The Teaching Company, *Dark Matter, Dark Energy: The Dark Side of the Universe*, Guidebook Part 2 page 46. Retrieved Oct. 7, 2013, "...dark energy: A smooth, persistent component of invisible energy, thought to make up about 70 percent of the current energy density of the universe. Dark energy is known to be smooth because it doesn't accumulate preferentially in galaxies and clusters..."

[7] Paul J. Steinhardt, Neil Turok (2006). "Why the cosmological constant is small and positive". *Science*. **312** (5777): 1180–1183. Bibcode:2006Sci...312.1180S. PMID 16675662. arXiv:astro-ph/0605173. doi:10.1126/science.1126231.

[8] "Dark Energy". *Hyperphysics*. Retrieved January 4, 2014.

[9] Ferris, Timothy. "Dark Matter(Dark Energy)". Retrieved 2015-06-10.

[10] "Moon findings muddy the water".

[11] Carroll, Sean (2001). "The cosmological constant". *Living Reviews in Relativity*. **4**. Bibcode:2001LRR.....4....1C. arXiv:astro-ph/0004075. doi:10.12942/lrr-2001-1. Retrieved 2006-09-28.

[12] Kragh, H. 2012. Preludes to dark energy: zero-point energy and vacuum speculations. Archive for History of Exact Sciences. Volume 66, Issue 3, pp 199–240

[13] Harvey, Alex (2012). "How Einstein Discovered Dark Energy". arXiv:1211.6338.

[14] Gamow, George (1970) *My World Line: An Informal Autobiography*. p. 44: "Much later, when I was discussing cosmological problems with Einstein, he remarked that the introduction of the cosmological term was the biggest blunder he ever made in his life." – Here the "cosmological term" refers to the cosmological constant in the equations of general relativity, whose value Einstein initially picked to ensure that his model of the universe would neither expand nor contract; if he hadn't done this he might have theoretically predicted the universal expansion that was first observed by Edwin Hubble.

[15] Riess, Adam G.; Filippenko; Challis; Clocchiatti; Diercks; Garnavich; Gilliland; Hogan; Jha; Kirshner; Leibundgut; Phillips; Reiss; Schmidt; Schommer; Smith; Spyromilio; Stubbs; Suntzeff; Tonry (1998). "Observational evidence from supernovae for an accelerating universe and a cosmological constant". *Astronomical Journal*. **116** (3): 1009–38. Bibcode:1998AJ....116.1009R. arXiv:astro-ph/9805201. doi:10.1086/300499.

[16] Perlmutter, S.; Aldering; Goldhaber; Knop; Nugent; Castro; Deustua; Fabbro; Goobar; Groom; Hook; Kim; Kim; Lee; Nunes; Pain; Pennypacker; Quimby; Lidman; Ellis; Irwin; McMahon; Ruiz-Lapuente; Walton; Schaefer; Boyle; Filippenko; Matheson; Fruchter; et al. (1999). "Measurements of Omega and Lambda from 42 high redshift supernovae". *Astrophysical Journal*. **517** (2): 565–86. Bibcode:1999ApJ...517..565P. arXiv:astro-ph/9812133. doi:10.1086/307221.

[17] The first appearance of the term "dark energy" is in the article with another cosmologist and Turner's student at the time, Dragan Huterer, "Prospects for Probing the Dark Energy via Supernova Distance Measurements", which was posted to the ArXiv.org e-print archive in August 1998 and published in Huterer, D.; Turner, M. (1999). "Prospects for probing the dark energy via supernova distance measurements". *Physical Review D*. **60** (8). Bibcode:1999PhRvD..60h1301H. arXiv:astro-ph/9808133. doi:10.1103/PhysRevD.60.081301., although the manner in which the term is treated there suggests it was already in general use. Cosmologist Saul Perlmutter has credited Turner with coining the term in an article they wrote together with Martin White, where it is introduced in quotation marks as if it were a neologism. Perlmutter, S.; Turner, M.; White, M. (1999). "Constraining Dark Energy with Type Ia Supernovae and Large-Scale Structure". *Physical Review Letters*. **83** (4): 670. Bibcode:1999PhRvL..83..670P. arXiv:astro-ph/9901052. doi:10.1103/PhysRevLett.83.670.

[18] Astier, Pierre (Supernova Legacy Survey); Guy; Regnault; Pain; Aubourg; Balam; Basa; Carlberg; Fabbro; Fouchez; Hook; Howell; Lafoux; Neill; Palanque-Delabrouille; Perrett; Pritchet; Rich; Sullivan; Taillet; Aldering; Antilogus; Arsenijevic; Balland; Baumont; Bronder; Courtois; Ellis; Filiol; et al. (2006). "The Supernova legacy survey: Measurement of ΩM, ΩΛ and W from the first year data set". *Astronomy and Astrophysics*.

447: 31–48. Bibcode:2006A&A...447...31A. arXiv:astro-ph/0510447 . doi:10.1051/0004-6361:20054185.

[19] Overbye, Dennis. “Astronomers Report Evidence of 'Dark Energy' Splitting the Universe”. The New York Times. Retrieved August 5, 2015.

[20] Zhong-Yue Wang (2016). “Modern Theory for Electromagnetic Metamaterials”. *Plasmonics*. **11** (2): 503–508. doi:10.1007/s11468-015-0071-7.

[21] Daniel Baumann. “Cosmology: Part III Mathematical Tripos, Cambridge University” (PDF). p. 21–22.

[22] Durrer, R. (2011). “What do we really know about Dark Energy?". *Philosophical Transactions of the Royal Society A: Mathematical, Physical and Engineering Sciences*. **369** (1957): 5102. Bibcode:2011RSPTA.369.5102D. arXiv:1103.5331 . doi:10.1098/rsta.2011.0285.

[23] The first paper, using observed data, which claimed a positive Lambda term was Paál, G.; et al. (1992). “Inflation and compactification from galaxy redshifts?". *Astrophysics and Space Science*. **191**: 107–24. Bibcode:1992Ap&SS.191..107P. doi:10.1007/BF00644200.

[24] “The Nobel Prize in Physics 2011”. Nobel Foundation. Retrieved 2011-10-04.

[25] The Nobel Prize in Physics 2011. Perlmutter got half the prize, and the other half was shared between Schmidt and Riess.

[26] Spergel, D. N. (WMAP collaboration); et al. (March 2006). “Wilkinson Microwave Anisotropy Probe (WMAP) three year results: implications for cosmology”.

[27] Durrer, R. (2011). “What do we really know about dark energy?". *Philosophical Transactions of the Royal Society A*. **369** (1957): 5102–5114. Bibcode:2011RSPTA.369.5102D. arXiv:1103.5331 . doi:10.1098/rsta.2011.0285.

[28] Kowalski, Marek; Rubin, David; Aldering, G.; Agostinho, R. J.; Amadon, A.; Amanullah, R.; Balland, C.; Barbary, K.; Blanc, G.; Challis, P. J.; Conley, A.; Connolly, N. V.; Covarrubias, R.; Dawson, K. S.; Deustua, S. E.; Ellis, R.; Fabbro, S.; Fadeyev, V.; Fan, X.; Farris, B.; Folatelli, G.; Frye, B. L.; Garavini, G.; Gates, E. L.; Germany, L.; Goldhaber, G.; Goldman, B.; Goobar, A.; Groom, D. E.; et al. (October 27, 2008). “Improved Cosmological Constraints from New, Old and Combined Supernova Datasets”. *The Astrophysical Journal*. Chicago: University of Chicago Press. **686** (2): 749–778. Bibcode:2008ApJ...686..749K. arXiv:0804.4142 . doi:10.1086/589937.. They find a best fit value of the dark energy density, ΩΛ of 0.713+0.027–0.029(stat)+0.036–0.039(sys), of the total matter density, ΩM, of 0.274+0.016–0.016(stat)+0.013–0.012(sys) with an equation of state parameter w of –0.969+0.059–0.063(stat)+0.063–0.066(sys).

[29] “Big Bang's afterglow shows universe is 80 million years older than scientists first thought”. *The Washington Post*. Archived from the original on 22 March 2013. Retrieved 22 March 2013.

[30] “New method 'confirms dark energy'". BBC News. 2011-05-19.

[31] Dark energy is real, Swinburne University of Technology, 19 May 2011

[32] Crittenden; Neil Turok (1995). “Looking for Λ with the Rees-Sciama Effect”. *Physical Review Letters*. **76** (4): 575–578. Bibcode:1996PhRvL..76..575C. PMID 10061494. arXiv:astro-ph/9510072 . doi:10.1103/PhysRevLett.76.575.

[33] Shirley Ho; Hirata; Nikhil Padmanabhan; Uros Seljak; Neta Bahcall (2008). “Correlation of CMB with large-scale structure: I. ISW Tomography and Cosmological Implications”. *Physical Review D*. **78** (4): 043519. Bibcode:2008PhRvD..78d3519H. arXiv:0801.0642 . doi:10.1103/PhysRevD.78.043519.

[34] Tommaso Giannantonio; Ryan Scranton; Crittenden; Nichol; Boughn; Myers; Richards (2008). “Combined analysis of the integrated Sachs-Wolfe effect and cosmological implications”. *Physical Review D*. **77** (12): 123520. Bibcode:2008PhRvD..77l3520G. arXiv:0801.4380 . doi:10.1103/PhysRevD.77.123520.

[35] Zelong Yi; Tongjie Zhang (2007). “Constraints on holographic dark energy models using the differential ages of passively evolving galaxies”. *Modern Physics Letters A*. **22** (1): 41. Bibcode:2007MPLA...22...41Y. arXiv:astro-ph/0605596 . doi:10.1142/S0217732307020889.

[36] Haoyi Wan; Zelong Yi; Tongjie Zhang; Jie Zhou (2007). “Constraints on the DGP Universe Using Observational Hubble parameter”. *Physics Letters B*. **651** (5): 352. Bibcode:2007PhLB..651..352W. arXiv:0706.2723 . doi:10.1016/j.physletb.2007.06.053.

[37] Cong Ma; Tongjie Zhang (2010). “Power of Observational Hubble Parameter Data: a Figure of Merit Exploration”. *Astrophysical Journal*. **730** (2): 74. Bibcode:2011ApJ...730...74M. arXiv:1007.3787 . doi:10.1088/0004-637X/730/2/74.

[38] Tongjie Zhang; Cong Ma; Tian Lan (2010). “Constraints on the Dark Side of the Universe and Observational Hubble Parameter Data”. *Advances in Astronomy*. **2010** (1): 1. Bibcode:2010AdAst2010E..81Z. arXiv:1010.1307 . doi:10.1155/2010/184284.

[39] Joan Simon; Licia Verde; Raul Jimenez (2005). “Constraints on the redshift dependence of the dark energy potential”. *Physical Review D*. **71** (12): 123001. Bibcode:2005PhRvD..71l3001S. arXiv:astro-ph/0412269 . doi:10.1103/PhysRevD.71.123001.

[40] by Ehsan Sadri Astrophysics MSc, Azad University, Tehran

[41] "Planck reveals an almost perfect universe". *Planck*. ESA. 2013-03-21. Retrieved 2013-03-21.

[42] Wess, Julius; Bagger, Jonathan. *Supersymmetry and Supergravity*. ISBN 978-0691025308.

[43] See M. Sami; R. Myrzakulov (2015). "Late time cosmic acceleration: ABCD of dark energy and modified theories of gravity". *International Journal of Modern Physics D*. **25** (12). arXiv:1309.4188. doi:10.1142/S0218271816300317. for a recent review

[44] Austin Joyce; Lucas Lombriser; Fabian Schmidt (2016). "Dark Energy vs. Modified Gravity". *Annual Review of Nuclear and Particle Science*. **66**. arXiv:1601.06133. doi:10.1146/annurev-nucl-102115-044553.

[45] Carroll, Sean M. (1998). "Quintessence and the Rest of the World: Suppressing Long-Range Interactions". *Physical Review Letters*. **81** (15): 3067–3070. Bibcode:1998PhRvL..81.3067C. ISSN 0031-9007. arXiv:astro-ph/9806099. doi:10.1103/PhysRevLett.81.3067.

[46] Ratra, Bharat; Peebles, P.J.E. "Cosmological consequences of a rolling homogeneous scalar field". *Phys. Rev*. **D37**: 3406. Bibcode:1988PhRvD..37.3406R. doi:10.1103/PhysRevD.37.3406.

[47] Steinhardt, Paul J.; Wang, Li-Min; Zlatev, Ivaylo. "Cosmological tracking solutions". *Phys. Rev*. **D59**: 123504. Bibcode:1999PhRvD..59l3504S. arXiv:astro-ph/9812313. doi:10.1103/PhysRevD.59.123504.

[48] R.R.Caldwell (2002). "A phantom menace? Cosmological consequences of a dark energy component with super-negative equation of state". *Physics Letters B*. **545** (1–2): 23–29. Bibcode:2002PhLB..545...23C. arXiv:astro-ph/9908168. doi:10.1016/S0370-2693(02)02589-3.

[49] See dark fluid.

[50] Rafael J. F. Marcondes (5 October 2016). "Interacting dark energy models in Cosmology and large-scale structure observational tests".

[51] Exirifard, Q. (2010). "Phenomenological covariant approach to gravity". *General Relativity and Gravitation*. **43**: 93–106. Bibcode:2011GReGr..43...93E. arXiv:0808.1962. doi:10.1007/s10714-010-1073-6.

[52] Chevallier, M; Polarski, D (2001). "Accelerating Universes with Scaling Dark Matter". *International Journal of Modern Physics D*. **10**: 213–224. Bibcode:2001IJMPD..10..213C. arXiv:gr-qc/0009008. doi:10.1142/S0218271801000822.

[53] Linder, Eric V. (3 March 2003). "Exploring the Expansion History of the Universe". *Physical Review Letters*. **90** (9): 091301. Bibcode:2003PhRvL..90i1301L. PMID 12689209. arXiv:astro-ph/0208512v1. doi:10.1103/PhysRevLett.90.091301.

[54] Alcaniz, E.M.; Alcaniz, J.S. (2008). "A parametric model for dark energy". *Physics Letters B*. **666**: 415–419. Bibcode:2008PhLB..666..415B. arXiv:0805.1713. doi:10.1016/j.physletb.2008.08.012.

[55] Jassal, H.K; Bagla, J.S (2010). "Understanding the origin of CMB constraints on Dark Energy". *Monthly Notices of the Royal Astronomical Society*. **405**: 2639–2650. Bibcode:2010MNRAS.405.2639J. arXiv:astro-ph/0601389. doi:10.1111/j.1365-2966.2010.16647.x.

[56] Wetterich, C. (2004). "Phenomenological parameterization of quintessence". arXiv:astro-ph/0403289v1.

[57] Wiltshire, David L. (2007). "Exact Solution to the Averaging Problem in Cosmology". *Physical Review Letters*. **99** (25): 251101. Bibcode:2007PhRvL..99y1101W. PMID 18233512. arXiv:0709.0732. doi:10.1103/PhysRevLett.99.251101.

[58] Ishak, Mustapha; Richardson, James; Garred, David; Whittington, Delilah; Nwankwo, Anthony; Sussman, Roberto (2007). "Dark Energy or Apparent Acceleration Due to a Relativistic Cosmological Model More Complex than FLRW?". *Physical Review D*. **78** (12): 123531. Bibcode:2008PhRvD..78l3531I. arXiv:0708.2943. doi:10.1103/PhysRevD.78.123531.

[59] Mattsson, Teppo (2007). "Dark energy as a mirage". *Gen. Rel. Grav*. **42** (3): 567–599. Bibcode:2010GReGr..42..567M. arXiv:0711.4264. doi:10.1007/s10714-009-0873-z.

[60] Clifton, Timothy; Ferreira, Pedro (April 2009). "Does Dark Energy Really Exist?". *Scientific American*. **300** (4): 48–55. PMID 19363920. doi:10.1038/scientificamerican0409-48. Retrieved April 30, 2009.

[61] Wiltshire, D. (2008). "Cosmological equivalence principle and the weak-field limit". *Physical Review D*. **78** (8): 084032. Bibcode:2008PhRvD..78h4032W. arXiv:0809.1183. doi:10.1103/PhysRevD.78.084032.

[62] Gray, Stuart. "Dark questions remain over dark energy". ABC Science Australia. Retrieved 27 January 2013.

[63] Merali, Zeeya (March 2012). "Is Einstein's Greatest Work All Wrong—Because He Didn't Go Far Enough?". *Discover magazine*. Retrieved 27 January 2013.

[64] Wolchover, Natalie (27 September 2011) 'Accelerating universe' could be just an illusion, MSNBC

[65] Tsagas, Christos G. (2011). "Peculiar motions, accelerated expansion, and the cosmological axis". *Physical Review D*. **84** (6): 063503. Bibcode:2011PhRvD..84f3503T. arXiv:1107.4045. doi:10.1103/PhysRevD.84.063503.

[66] Krauss, Lawrence M.; Scherrer, Robert J. (March 2008). "The End of Cosmology?". *Scientific American*. **82**. Retrieved 2011-01-06.

[67] Is the universe expanding faster than the speed of light? (see the last two paragraphs)

[68] Lineweaver, Charles; Tamara M. Davis (2005). "Misconceptions about the Big Bang" (PDF). *Scientific American*. Retrieved 2008-11-06.

[69] Loeb, Abraham (2002). "The Long-Term Future of Extragalactic Astronomy". *Physical Review D*. **65** (4): 047301. Bibcode:2002PhRvD..65d7301L. arXiv:astro-ph/0107568. doi:10.1103/PhysRevD.65.047301.

[70] Krauss, Lawrence M.; Robert J. Scherrer (2007). "The Return of a Static Universe and the End of Cosmology". *General Relativity and Gravitation*. **39** (10): 1545–1550. Bibcode:2007GReGr..39.1545K. arXiv:0704.0221. doi:10.1007/s10714-007-0472-9.

[71] Using Tiny Particles To Answer Giant Questions. Science Friday, 3 Apr 2009. According to the transcript, Brian Greene makes the comment "And actually, in the far future, everything we now see, except for our local galaxy and a region of galaxies will have disappeared. The entire universe will disappear before our very eyes, and it's one of my arguments for actually funding cosmology. We've got to do it while we have a chance."

[72] *How the Universe Works 3*. End of the Universe. Discovery Channel. 2014.

[73] 'Cyclic universe' can explain cosmological constant, NewScientistSpace, 4 May 2006

[74] Steinhardt, P. J.; Turok, N. (2002-04-25). "A Cyclic Model of the Universe". *Science*. **296** (5572): 1436–1439. Bibcode:2002Sci...296.1436S. PMID 11976408. arXiv:hep-th/0111030v2. doi:10.1126/science.1070462. Retrieved 2012-04-29.

[75] Merritt, David "Cosmology and Convention", *Studies In History and Philosophy of Science Part B: Studies In History and Philosophy of Modern Physics*, **57**(1):41–52, February 2017.

9.9 External links

-

- Dark Energy on *In Our Time* at the BBC.
- Dark energy Eric Linder Scholarpedia 3(2):4900. doi:10.4249/scholarpedia.4900
- Dark energy: how the paradigm shifted Physicsworld.com
- Dennis Overbye (November 2006). "9 Billion-Year-Old 'Dark Energy' Reported". *The New York Times*.
- "Mysterious force's long presence" BBC News online (2006) More evidence for dark energy being the cosmological constant
- "Astronomy Picture of the Day" one of the images of the Cosmic Microwave Background which confirmed the presence of dark energy and dark matter
- SuperNova Legacy Survey home page The Canada-France-Hawaii Telescope Legacy Survey Supernova Program aims primarily at measuring the equation of state of Dark Energy. It is designed to precisely measure several hundred high-redshift supernovae.
- "Report of the Dark Energy Task Force"
- "HubbleSite.org – Dark Energy Website" Multimedia presentation explores the science of dark energy and Hubble's role in its discovery.
- "Surveying the dark side"
- "Dark energy and 3-manifold topology" Acta Physica Polonica 38 (2007), p. 3633–3639
- The Dark Energy Survey
- The Joint Dark Energy Mission
- Harvard: Dark Energy Found Stifling Growth in Universe, primary source
- April 2010 Smithsonian Magazine Article
- HETDEX Dark energy experiment
- Dark Energy FAQ
- "The Hunt for Dark Energy" George FR Ellis, Peter Cameron and David Tong discuss the presence of dark energy in the Universe
- Euclid ESA Satellite, a mission to map the geometry of the dark universe

Chapter 10

Dark matter

Not to be confused with antimatter, dark energy, dark fluid, or dark flow. For other uses, see Dark Matter (disambiguation).

Dark matter is a hypothetical type of matter distinct from baryonic matter (ordinary matter such as protons and neutrons), neutrinos and dark energy. The existence of dark matter would explain a number of otherwise puzzling astronomical observations.[1] The name refers to the fact that it does not emit or interact with electromagnetic radiation, such as light, and is thus invisible to the entire electromagnetic spectrum.[2] Although dark matter has not been directly observed, its existence and properties are inferred from its gravitational effects such as the motions of visible matter,[3] gravitational lensing, its influence on the universe's large-scale structure, on galaxies, and its effects in the cosmic microwave background.

The standard model of cosmology indicates that the total mass–energy of the universe contains 4.9% ordinary matter, 26.8% dark matter and 68.3% dark energy.[4][5][6][7] Thus, dark matter constitutes 84.5%[note 1] of total mass, while dark energy plus dark matter constitute 95.1% of total mass–energy content.[8][9][10][11] The great majority of ordinary matter in the universe is also unseen, since visible stars and gas inside galaxies and clusters account for less than 10% of the ordinary matter contribution to the mass-energy density of the universe.[12] The most widely accepted hypothesis on the form for dark matter is that it is composed of weakly interacting massive particles (WIMPs) that interact only through gravity and the weak force.[13] The dark matter hypothesis plays a central role in current modeling of cosmic structure formation, galaxy formation and evolution, and on explanations of the anisotropies observed in the cosmic microwave background (CMB). All these lines of evidence suggest that galaxies, galaxy clusters, and the universe as a whole contain far more matter than that which is observable via electromagnetic signals.[14] Many experiments to detect proposed dark matter particles through non-gravitational means are under way;[15] however, no dark matter particle has been conclusively identified.

Although the existence of dark matter is generally accepted by most of the astronomical community, a minority of astronomers,[16] motivated by the lack of conclusive identification of dark matter, argue for various modifications of the standard laws of general relativity, such as MOND, TeVeS, and conformal gravity[17] that attempt to account for the observations without invoking additional matter.[18]

Lord Kelvin estimated the number of dark bodies in the Milky Way galaxy from the observed velocity dispersion of the stars, the speed the stars were orbiting around the center of the galaxy, which he used to estimate the mass of the galaxy, which was different from the mass of stars which can be seen, and concluded that "many of our stars, perhaps a great majority of them, may be dark bodies."[19]

In 1906 Henri Poincaré in "The Milky Way and Theory of Gases" used "dark matter," or "matière obscure" in French in discussing Kelvin's work.[19]

The first to suggest the existence of dark matter (using stellar velocities) was Dutch astronomer Jacobus Kapteyn in 1922.[20][21] Fellow Dutchman and radio astronomy pioneer Jan Oort also hypothesized the existence of dark matter in 1932.[21][22][23] Oort was studying stellar motions in the local galactic neighborhood and found that the mass in the galactic plane must be greater than what was observed, but this measurement was later determined to be erroneous.[24]

In 1933, Swiss astrophysicist Fritz Zwicky, who studied galactic clusters while working at the California Institute of Technology, made a similar inference.[25][26][27] Zwicky applied the virial theorem to the Coma galaxy cluster and obtained evidence of unseen mass that he called *dunkle Materie* 'dark matter'. Zwicky estimated its mass based on the motions of galaxies near its edge and compared that to an estimate based on its brightness and number of galaxies. He estimated that the cluster had about 400 times more mass than was visually observable. The gravity effect of the visible galaxies was far too small for such fast orbits, thus mass must be hidden from view. Based on these conclusions, Zwicky inferred that some unseen matter provided the mass and associated gravitation attraction to hold the cluster together. This was the first formal inference about the existence of dark matter.[28] Zwicky's estimates were off by more than an order of magnitude, mainly due to an obsolete value of the Hubble constant;[29] the same calculation today shows a smaller fraction, using greater values for luminous mass. However, Zwicky did correctly infer that the bulk of the matter was dark.[28]

The first robust indications that the mass to light ratio was anything other than unity came from measurements of galaxy rotation curves. In 1939, Horace W. Babcock reported the rotation curve for the Andromeda nebula, which suggested that the mass-to-luminosity ratio increases radially.[30] He attributed it to either light absorption within the galaxy or modified dynamics in the outer portions of the spiral and not to missing matter.

Vera Rubin and Kent Ford in the 1960s–1970s provided further strong evidence, also using galaxy rotation curves.[31][32][33] Rubin worked with a new spectrograph to measure the velocity curve of edge-on spiral galaxies with greater accuracy.[33] This result was confirmed in 1978.[34] An influential paper presented Rubin's results in 1980.[35] Rubin found that most galaxies must contain about six times as much dark as visible mass;[36] thus, by around 1980 the apparent need for dark matter was widely recognized as a major unsolved problem in astronomy.[31]

At the same time that Rubin and Ford were exploring optical rotation curves, radio astronomers were making use of new radio telescopes to map the 21 cm line of atomic hydrogen in nearby galaxies. The radial distribution of interstellar atomic hydrogen often extends to much larger galactic radii than those accessible by optical studies, allowing the sampling of rotation curves - and thus of the total mass distribution - to a new dynamical regime. Early mapping of Andromeda with the 300-foot telescope at Green Bank [37] and the 250-foot dish at Jodrell Bank [38] already showed that the HI rotation curve did not trace the expected Keplerian decline. As more sensitive receivers became available, Morton Roberts and Robert Whitehurst [39] were able to trace the rotational velocity of Andromeda to 30 kpc, much beyond the optical measurements. Illustrating the advantage of tracing the gas disk at large radii, Figure 16 of that paper [39] combines the optical data [33] (the cluster of points at radii of less than 15 kpc with a single point further out) with the HI data between 20 and 30 kpc, exhibiting the flatness of the outer galaxy rotation curve; the solid curve peaking at the center is the optical surface density, while the other curve shows the cumulative mass, still rising linearly at the outermost measurement. In parallel, the use of interferometric arrays for extragalactic HI spectroscopy was being developed. In 1972, David Rogstad and Seth Shostak [40] published HI rotation curves of five spirals mapped with the Owens Valley interferometer; the rotation curves of all five were very flat, suggesting very large values of mass-to-light ratio in the outer parts of their extended HI disks.

A stream of observations in the 1980s indicated its presence, including gravitational lensing of background objects by galaxy clusters,[41] the temperature distribution of hot gas in galaxies and clusters, and the pattern of anisotropies in the cosmic microwave background. According to consensus among cosmologists, dark matter is composed primarily of a not yet characterized type of subatomic particle.[13][42] The search for this particle, by a variety of means, is one of the major efforts in particle physics.[15]

10.1.1 Cosmic microwave background radiation (CMB)

In cosmology, the CMB is explained as relic radiation which has travelled freely since the era of recombination, around 375,000 years after the Big Bang. The CMB's anisotropies are explained as the result of small primordial density fluctuations, and subsequent acoustic oscillations in the photon-baryon plasma whose restoring force is gravity.[43]

The Cosmic Background Explorer (COBE) satellite found the CMB spectrum to be a very precise blackbody spectrum with a temperature of 2.726 K. In 1992, COBE detected CMB fluctuations (anisotropies) at a level of about one part in 10^5.[44]

In the following decade, CMB anisotropies were investigated by ground-based and balloon experiments. Their primary goal was to measure the angular scale of the first acoustic peak of the anisotropies' power spectrum, for which COBE had insufficient resolution. During the 1990s, the first peak was measured with increasing sensitivity, and in 2000 the BOOMERanG experiment[45] reported that the highest power fluctuations occur at scales of approximately one degree, showing that the Universe is close to flat. These measurements were able to rule out cosmic strings as the leading theory of cosmic structure formation, and suggested cosmic inflation was the correct theory.

Ground-based interferometers provided fluctuation measurements with higher accuracy, including the Very Small Array, the Degree Angular Scale Interferometer (DASI) and the Cosmic Background Imager (CBI). DASI first detected the CMB polarization,[46][47] and CBI provided the first E-mode polarization spectrum with compelling evidence that it is out of phase with the T-mode spectrum.[48] COBE's successor, the Wilkinson Microwave Anisotropy Probe (WMAP) provided the most detailed measurements of (large-scale) anisotropies in the CMB in 2003–2010.[49] ESA's Planck spacecraft returned more detailed results in 2013–2015.

WMAP's measurements played the key role in establishing the Standard Model of Cosmology, namely the Lambda-CDM model, which posits a dark energy-dominated flat universe, supplemented by dark matter and atoms with density fluctuations seeded by a Gaussian, adiabatic, nearly scale invariant process. Its basic properties are determined by six adjustable parameters: dark matter density, baryon (atom) density, the universe's age (or equivalently, the Hubble constant), the initial fluctuation amplitude and their scale dependence.

10.2 Technical definition

See also: Friedmann equations

In standard cosmology, matter is anything whose energy density scales with the inverse cube of the scale factor, i.e. $\rho \propto a^{-3}$. This is in contrast to radiation, which scales to the inverse fourth power of the scale factor $\rho \propto a^{-4}$, and dark energy, which is unaffected $\rho \propto a^{0}$. This can be understood intuitively: for an ordinary particle in a square box, doubling the length of a side of the box decreases the density (and hence energy density) by a factor of eight (2^3). For radiation, the decrease in energy density is greater, because an increase in spatial distance also causes a redshift. Dark energy, as an intrinsic property of space, has a constant energy density regardless of the volume under consideration.[50]

Dark matter is that component of the universe which is not ordinary matter, but still obeys $\rho \propto a^{-3}$.

10.3 Observational evidence

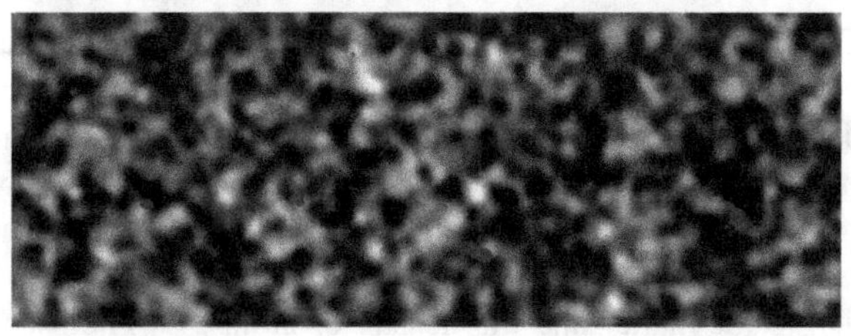

Dark matter map of KiDS survey region (region G12).[52]

10.3.1 Galaxy rotation curves

Main article: Galaxy rotation curve

The arms of spiral galaxies rotate around the galactic centre. The luminous mass density of a spiral galaxy decreases as one goes from the centre to the outskirts. If luminous mass were all the matter, then we can model the galaxy as a point mass in the centre and test masses orbiting around it (similar to the solar system). From Kepler's Second Law, we expect that the rotation velocities will decrease with distance from the centre, similar to our solar system. This is not observed.[53] Instead, the galaxy rotation curve remains flat as distant from the centre as the data is available.

If we assume the validity of Kepler's laws, then the obvious way to resolve this discrepancy is to conclude that the mass distribution in spiral galaxies is not similar to that of the solar system. In particular, there is a lot of non-luminous matter in the outskirts of the galaxy ("dark matter").

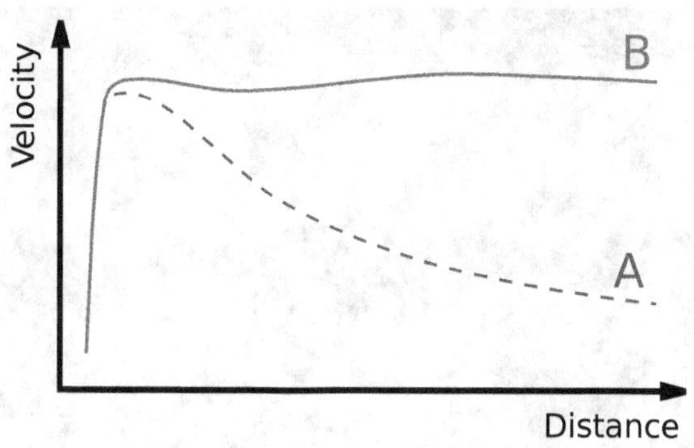

*Rotation curve of a typical spiral galaxy: predicted (**A**) and observed (**B**). Dark matter can explain the 'flat' appearance of the velocity curve out to a large radius.*

10.3.2 Velocity dispersions

Main article: Velocity dispersion

Stars in bound systems must obey the virial theorem. The theorem, together with the measured velocity distribution, can be used to measure the mass distribution in a bound system, such as elliptical galaxies or globular clusters. With some exceptions, velocity dispersion estimates of elliptical galaxies[54] do not match the predicted velocity dispersion from the observed mass distribution, even assuming complicated distributions of stellar orbits.[55]

As with galaxy rotation curves, the obvious way to resolve the discrepancy is to postulate the existence of non-luminous matter.

10.3.3 Galaxy clusters

Galaxy clusters are particularly important for dark matter studies since their masses can be estimated in three independent ways:

- From the scatter in radial velocities of the galaxies within clusters
- From X-rays emitted by hot gas in the clusters. From the X-ray energy spectrum and flux, the gas temperature and density can be estimated, hence giving the pressure; assuming pressure and gravity balance determines the cluster's mass profile.
- Gravitational lensing (usually of more distant galaxies) can measure cluster masses without relying on observations of dynamics (e.g., velocity).

Generally, these three methods are in reasonable agreement that dark matter outweighs visible matter by approximately

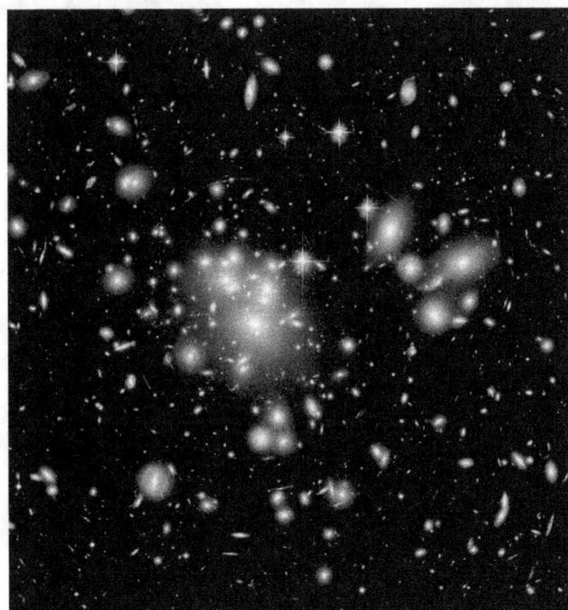

Strong gravitational lensing as observed by the Hubble Space Telescope in Abell 1689 indicates the presence of dark matter—enlarge the image to see the lensing arcs.

5 to 1.

10.3.4 Gravitational lensing

One of the consequences of general relativity is that massive objects should act as a lens to bend the light from a more distant source (such as a quasar) around a massive object (such as a cluster of galaxies) lying between the source and the observer. The more massive an object, the more lensing is observed.

Strong lensing is the observed distortion of background galaxies into arcs when their light passes through such a gravitational lens. It has been observed around many distant clusters including Abell 1689.[56] By measuring the distortion geometry, the mass of the intervening cluster can be obtained. In the dozens of cases where this has been done, the mass-to-light ratios obtained correspond to the dynamical dark matter measurements of clusters.[57] Lensing can lead to multiple copies of an image. By analyzing the distribution of multiple image copies, scientists have been able to deduce and map the distribution of dark matter around the MACS J0416.1-2403 galaxy cluster.[58][59]

Weak gravitational lensing investigates minute distortions of galaxies, using statistical analyses from vast galaxy surveys. By examining the apparent shear deformation of the adjacent background galaxies, the mean distribution of dark matter can be characterized. The mass-to-light ratios correspond to dark matter densities predicted by other large-scale structure measurements.[60]

10.3.5 Cosmic microwave background

Main article: Cosmic microwave background

Although both dark matter and ordinary matter are "matter", they do not behave in the same way. In particular, ordinary matter interacts with radiation, while dark matter does not. They therefore leave different imprints on the cosmic microwave background (CMB).

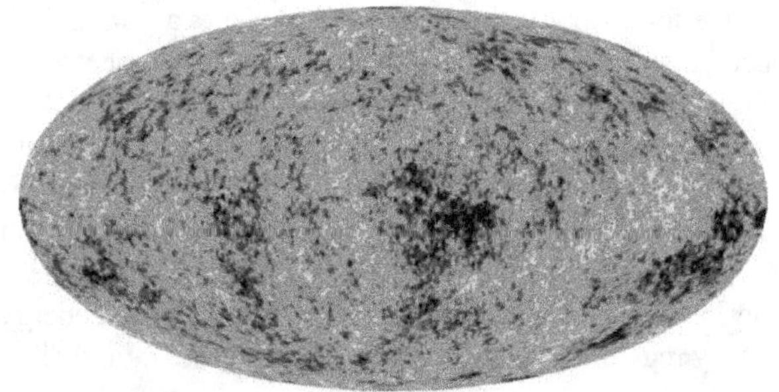

The cosmic microwave background by WMAP

The cosmic microwave background is very close to a perfect blackbody, but contains anisotropies. The anisotropies can be decomposed into a power spectrum, whose peaks constrain cosmological parameters.[61] The first peak mostly shows the density of baryonic matter, while the third peak relates mostly to the density of dark matter, measuring the density of matter and the density of atoms.[61] The spectrum was first observed by WMAP, and the results support the Lambda-CDM model.[49]

The CMB angular power spectrum provides powerful evidence in support of dark matter, as its precise structure is difficult to reproduce with any competing model such as MOND.[62][63]

10.3.6 Sky surveys and baryon acoustic oscillations

Main article: Baryon acoustic oscillations

Baryon acoustic oscillations (BAO) are regular, periodic fluctuations in the density of the visible baryonic matter (normal matter) of the universe. These are predicted to arise in the Lambda-CDM model due to the early universe's acoustic oscillations in the photon-baryon fluid and can be observed in the cosmic microwave background angular power spectrum. BAOs set up a preferred length scale for baryons. As the dark matter and baryons clumped together after recombination, the effect is much weaker in the galaxy distribution in the nearby universe, but is detectable as a subtle (~ 1 percent) preference for pairs of galaxies to be separated by 147 Mpc, compared to those separated by 130 or 160 Mpc. This feature was predicted theoretically in the 1990s and then discovered in 2005, in two large galaxy redshift surveys, the Sloan Digital Sky Survey and the 2dF Galaxy Redshift Survey.[64] Combining the CMB observations with BAO measurements from galaxy redshift surveys provides a precise estimate of the Hubble constant and the average matter density in the Universe.[43] The results support the Lambda-CDM model.

10.3.7 Redshift-space distortions

Large galaxy redshift surveys may be used to make a three-dimensional map of the galaxy distribution. These maps are slightly distorted because distances are estimated from observed redshifts; the redshift contains a contribution from the galaxy's so-called peculiar velocity in addition to the dominant Hubble expansion term. On average, superclusters are expanding but more slowly than the cosmic mean due to their gravity, while voids are expanding faster than average. In a redshift map, galaxies in front of a supercluster have excess radial velocities towards it and have redshifts slightly higher than their distance would imply, while galaxies behind the supercluster have redshifts slightly low for their distance. This effect causes superclusters to appear "squashed" in the radial direction, and likewise voids are "stretched"; angular positions are unaffected. The effect is not detectable for any one structure since the true shape is not known, but can be measured by averaging over many structures assuming we are not at a special location in the Universe.

The effect was predicted quantitatively by Nick Kaiser in 1987, and first decisively measured in 2001 by the 2dF Galaxy Redshift Survey.[65] Results are in agreement with the Lambda-CDM model.

10.3.8 Type Ia supernova distance measurements

Main articles: Type Ia supernova and Shape of the universe

Type Ia supernovae can be used as "standard candles" to measure extragalactic distances, which can in turn be used to measure how fast the universe has expanded in the past. The data indicates that the universe is expanding at an accelerating rate, the cause of which is usually ascribed to dark energy.[66] Since observations indicate the universe is almost flat,[67][68][69] we expect the total energy density of everything in the universe to sum to 1 ($\Omega_{tot} \sim 1$). The measured dark energy density is $\Omega\Lambda = \sim 0.690$; the observed ordinary matter energy density is $\Omega_m = \sim 0.0482$ and the energy density of radiation is negligible. This leaves a missing $\Omega_{dm} = \sim 0.258$ that nonetheless behaves like matter (see technical definition section above) – dark matter.[70]

10.3.9 Lyman-alpha forest

Main article: Lyman-alpha forest

In astronomical spectroscopy, the Lyman-alpha forest is the sum of the absorption lines arising from the Lyman-alpha transition of neutral hydrogen in the spectra of distant galaxies and quasars. Lyman-alpha forest observations can also constrain cosmological models.[71] These constraints agree with those obtained from WMAP data.

10.3.10 Structure formation

Main article: Structure formation

Structure formation refers to the period after the Big Bang when density perturbations collapsed to form stars, galaxies, and clusters. Prior to structure formation, the Friedmann solutions to general relativity describe a homogeneous universe. Later, small anisotropies gradually grew and condensed the homogeneous universe into stars, galaxies and larger structures. Ordinary matter is affected by radiation, which is the dominant element of the universe at very early times. As a result, its density perturbations are washed out and unable to condense into structure.[73] If there were only ordinary matter in the universe, there would not have been enough time for density perturbations to grow into the galaxies and clusters that we see today.

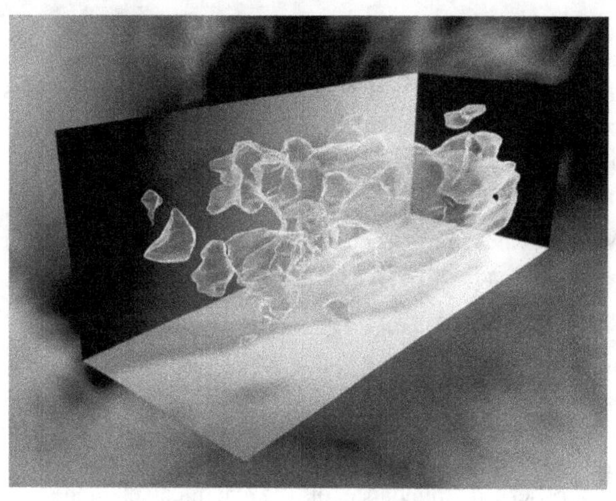

3D map of the large-scale distribution of dark matter, reconstructed from measurements of weak gravitational lensing with the Hubble Space Telescope.[72]

Dark matter provides a solution to this problem because it is unaffected by radiation. Therefore, its density perturbations can grow first. The resulting gravitational potential acts as an attractive potential well for ordinary matter collapsing later, speeding up the structure formation process.[73][74]

10.4 Composition of dark matter: Baryonic vs. nonbaryonic

Dark matter can refer to any substance which interacts predominantly via gravity with visible matter (e.g. stars and planets). Hence in principle it need not be composed of a new type of fundamental particle but could, at least in part, be made up of standard baryonic matter, such as protons or electrons. However, for the reasons outlined below, most scientists consider the dark matter to be dominated by a non-baryonic component, which is likely composed of a new fundamental particle (or similar exotic state).

10.4.1 Baryonic matter

Baryonic matter is made of baryons (protons and neutrons) that make up stars and planets. It also encompasses less common black holes, neutron stars, faint old white dwarfs and brown dwarfs, collectively known as massive compact halo objects (MACHOs).[75]

Multiple lines of evidence suggest the majority of dark matter is not made of baryons:

- Sufficient diffuse, baryonic gas or dust would be visible when backlit by stars.
- The theory of Big Bang nucleosynthesis predicts the observed abundance of the chemical elements. If there are more baryons, then there should also be more helium, lithium and heavier elements synthesized during the Big Bang.[76][77] Agreement with observed abundances requires that baryonic matter makes up between 4–5% of the universe's critical density. In contrast, large-scale structure and other observations indicate that the total matter density is about 30% of the critical density.[70]
- Astronomical searches for gravitational microlensing in the Milky Way found that at most a small fraction of the dark matter may be in dark, compact, conventional objects (MACHOs, etc.); the excluded range of object masses is from half the Earth's mass up to 30 solar masses, which covers nearly all the plausible candidates.[78][79][80][81][82][83]
- Detailed analysis of the small irregularities (anisotropies) in the cosmic microwave background.[84] Observations by WMAP and Planck indicate that around five-sixths of the total matter is in a form that interacts significantly with ordinary matter or photons only through gravitational effects.

10.4.2 Non-baryonic matter

Candidates for nonbaryonic dark matter are hypothetical particles such as axions, sterile neutrinos or WIMPs (e.g. supersymmetric particles). The three neutrino types already observed are indeed abundant, and "dark", and matter, but because their individual masses – however uncertain they may be – are almost certainly tiny, they can only supply a small fraction of dark matter, due to limits derived from large-scale structure and high-redshift galaxies.[85]

Unlike baryonic matter, nonbaryonic matter did not contribute to the formation of the elements in the early universe ("Big Bang nucleosynthesis")[13] and so its presence is revealed only via its gravitational effects. In addition, if the particles of which it is composed are supersymmetric, they can undergo annihilation interactions with themselves, possibly resulting in observable by-products such as gamma rays and neutrinos ("indirect detection").[85]

10.5 Classification of dark matter: cold, warm or hot

Dark matter can be divided into *cold*, *warm* and *hot* categories.[86] These categories refer to velocity rather than an actual temperature, indicating how far corresponding objects moved due to random motions in the early universe, before they slowed due to cosmic expansion – this is an important distance called the "free streaming length" (FSL). Primordial density fluctuations smaller than this length get washed out as particles spread from overdense to underdense regions, while larger fluctuations are unaffected; therefore this length sets a minimum scale for later structure formation. The categories are set with respect to the size of a protogalaxy (an object that later evolves into a dwarf galaxy): dark matter particles are classified as cold, warm, or hot according as their FSL; much smaller (cold), similar (warm), or much larger (hot) than a protogalaxy.[87][88]

Mixtures of the above are also possible: a theory of mixed dark matter was popular in the mid-1990s, but was rejected following the discovery of dark energy.

Cold dark matter leads to a "bottom-up" formation of structure while hot dark matter would result in a "top-down" formation scenario; the latter is excluded by high-redshift galaxy observations.[15]

10.5.1 Alternative definitions

These categories also correspond to fluctuation spectrum effects and the interval following the Big Bang at which each type became non-relativistic. Davis *et al.* wrote in 1985:

> Candidate particles can be grouped into three categories on the basis of their effect on the fluctuation spectrum (Bond *et al.* 1983). If the dark matter is composed of abundant light particles which remain relativistic until shortly before recombination, then it may be termed “hot”. The best candidate for hot dark matter is a neutrino ... A second possibility is for the dark matter particles to interact more weakly than neutrinos, to be less abundant, and to have a mass of order 1 keV. Such particles are termed “warm dark matter”, because they have lower thermal velocities than massive neutrinos ... there are at present few candidate particles which fit this description. Gravitinos and photinos have been suggested (Pagels and Primack 1982; Bond, Szalay and Turner 1982) ... Any particles which became nonrelativistic very early, and so were able to diffuse a negligible distance, are termed “cold” dark matter (CDM). There are many candidates for CDM including supersymmetric particles.[89]

Another approximate dividing line is that “warm” dark matter became non-relativistic when the universe was approximately 1 year old and 1 millionth of its present size and in the radiation-dominated era (photons and neutrinos), with a photon temperature 2.7 million K. Standard physical cosmology gives the particle horizon size as $2ct$ (speed of light multiplied by time) in the radiation-dominated era, thus 2 light-years. A region of this size would expand to 2 million light years today (absent structure formation). The actual FSL is roughly 5 times the above length, since it continues to grow slowly as particle velocities decrease inversely with the scale factor after they become non-relativistic. In this example the FSL would correspond to 10 million light-years or 3 Mpc today, around the size containing an average large galaxy.

The 2.7 million K photon temperature gives a typical photon energy of 250 electron-volts, thereby setting a typical mass scale for “warm” dark matter: particles much more massive than this, such as GeV – TeV mass WIMPs, would become non-relativistic much earlier than 1 year after the Big Bang and thus have FSLs much smaller than a protogalaxy, making them “cold”. Conversely, much lighter particles, such as neutrinos with masses of only a few eV, have FSLs much larger than a protogalaxy, thus qualifying them as “hot”.

10.5.2 Cold dark matter

Main article: Cold dark matter

“Cold” dark matter offers the simplest explanation for most cosmological observations. It is dark matter composed of constituents with an FSL much smaller than a protogalaxy. This is the focus for dark matter research, as hot dark matter does not seem to be capable of supporting galaxy or galaxy cluster formation, and most particle candidates slowed early.

The constituents of “cold” dark matter are unknown. Possibilities range from large objects like MACHOs (such as black holes[90]) or RAMBOs (such as clusters of brown dwarfs), to new particles such as WIMPs and axions.

Studies of Big Bang nucleosynthesis and gravitational lensing convinced most cosmologists[15][91][92][93][94][95] that MACHOs[91][93] cannot make up more than a small fraction of dark matter.[13][91] According to A. Peter: "... the only *really plausible* dark-matter candidates are new particles.”[92]

The 1997 DAMA/NaI experiment and its successor DAMA/LIBRA in 2013, claimed to directly detect dark matter particles passing through the Earth, but many researchers remain skeptical, as negative results from similar experiments seem incompatible with the DAMA results.

Many supersymmetric models offer dark matter candidates in the form of the WIMPy Lightest Supersymmetric Particle (LSP).[96] Separately, heavy sterile neutrinos exist in non-supersymmetric extensions to the standard model that explain the small neutrino mass through the seesaw mechanism.

10.5.3 Warm dark matter

Main article: Warm dark matter

“Warm” dark matter refers to particles with an FSL comparable to the size of a protogalaxy. Predictions based on warm dark matter are similar to those for cold dark matter on large scales, but with less small-scale density perturbations. This reduces the predicted abundance of dwarf galaxies and may lead to lower density of dark matter in the central parts of large galaxies; some researchers consider this to be a better fit to observations. A challenge for this model is the lack of particle candidates with the required mass ~ 300 eV to 3000 eV.

No known particles can be categorized as "warm" dark matter. A postulated candidate is the sterile neutrino: a heavier, slower form of neutrino that does not interact through the weak force, unlike other neutrinos. Some modified gravity theories, such as scalar-tensor-vector gravity, require "warm" dark matter to make their equations work.

10.5.4 Hot dark matter

Main article: Hot dark matter

"Hot" dark matter consists of particles whose FSL is much larger than the size of a protogalaxy. The neutrino qualifies as such particle. They were discovered independently, long before the hunt for dark matter: they were postulated in 1930, and detected in 1956. Neutrinos' mass is less than 10^{-6} that of an electron. Neutrinos interact with normal matter only via gravity and the weak force, making them difficult to detect (the weak force only works over a small distance, thus a neutrino triggers a weak force event only if it hits a nucleus head-on). This makes them 'weakly interacting light particles' (WILPs), as opposed to WIMPs.

The three known flavours of neutrinos are the *electron*, *muon*, and *tau*. Their masses are slightly different. Neutrinos oscillate among the flavours as they move. It is hard to determine an exact upper bound on the collective average mass of the three neutrinos (or for any of the three individually). For example, if the average neutrino mass were over 50 eV/c^2 (less than 10^{-5} of the mass of an electron), the universe would collapse. CMB data and other methods indicate that their average mass probably does not exceed 0.3 eV/c^2. Thus, observed neutrinos cannot explain dark matter.[97]

Because galaxy-size density fluctuations get washed out by free-streaming, "hot" dark matter implies that the first objects that can form are huge supercluster-size pancakes, which then fragment into galaxies. Deep-field observations show instead that galaxies formed first, followed by clusters and superclusters as galaxies clump together.

10.6 Detection of dark matter particles

If dark matter is made up of sub-atomic particles, then millions, possibly billions, of such particles must pass through every square centimeter of the Earth each second.[98][99] Many experiments aim to test this hypothesis. Although WIMPs are popular search candidates,[15] the Axion Dark Matter eXperiment (ADMX) searches for axions. Another candidate is heavy hidden sector particles that only interact with ordinary matter via gravity.

These experiments can be divided into two classes: direct detection experiments, which search for the scattering of dark matter particles off atomic nuclei within a detector; and indirect detection, which look for the products of dark matter particle annihilations or decays.[85]

10.6.1 Direct detection

For more details on this topic, see Weakly interacting massive particles § Direct detection.

Direct detection experiments aim to observe low-energy recoils (typically a few keVs) of nuclei induced by interactions with particles of dark matter, which (in theory) are passing through the Earth. After such a recoil the nucleus will emit energy as e.g. scintillation light or phonons, which is then detected by sensitive apparatus. In order to do this effectively it is crucial to maintain a low background, and so such experiments operate deep underground to reduce the interference from cosmic rays. Examples of underground laboratories which house direct detection experiments include the Stawell mine, the Soudan mine, the SNOLAB underground laboratory at Sudbury, the Gran Sasso National Laboratory, the Canfranc Underground Laboratory, the Boulby Underground Laboratory, the Deep Underground Science and Engineering Laboratory and the China Jinping Underground Laboratory.

These experiments mostly use either cryogenic or noble liquid detector technologies. Cryogenic detectors operating at temperatures below 100mK, detect the heat produced when a particle hits an atom in a crystal absorber such as germanium. Noble liquid detectors detect scintillation produced by a particle collision in liquid xenon or argon. Cryogenic detector experiments include: CDMS, CRESST, EDELWEISS, EURECA. Noble liquid experiments include ZEPLIN, XENON, DEAP, ArDM, WARP, DarkSide, PandaX, and LUX, the Large Underground Xenon experiment. Both of these techniques focus strongly on their ability to distinguish background particles (which predominantly scatter off electrons) from dark matter particles (that scatter off nuclei). Other experiments include SIMPLE and PICASSO.

Currently there has been no well-established claim of dark matter detection from a direct detection experiment, leading instead to strong upper limits on the mass and interaction cross section with nucleons of such dark matter particles.[100] The DAMA/NaI and more recent DAMA/LIBRA experimental collaborations claim to have detected an annual modulation in the rate of events in their detectors,[101][102] which they claim is due to dark matter.

This results from the expectation that as the Earth orbits the Sun, the velocity of the detector relative to the dark matter halo will vary by a small amount. This claim is so far unconfirmed and in contradiction with negative results from other experiments such as LUX and SuperCDMS.[103]

A special case of direct detection experiments covers those with directional sensitivity. This is a search strategy based on the motion of the Solar System around the Galactic Center.[104][105][106][107] A low pressure time projection chamber makes it possible to access information on recoiling tracks and constrain WIMP-nucleus kinematics. WIMPs coming from the direction in which the Sun is travelling (roughly towards Cygnus) may then be separated from background, which should be isotropic. Directional dark matter experiments include DMTPC, DRIFT, Newage and MIMAC.

10.6.2 Indirect detection

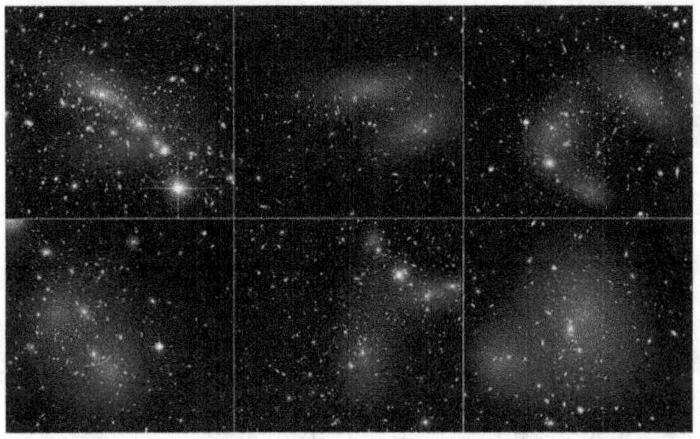

Collage of six cluster collisions with dark matter maps. The clusters were observed in a study of how dark matter in clusters of galaxies behaves when the clusters collide.[108]

Indirect detection experiments search for the products of the self-annihilation or decay of dark matter particles in outer space. For example, in regions of high dark matter density (e.g. the centre of our galaxy) two dark matter particles could annihilate to produce gamma rays or Standard Model particle-antiparticle pairs.[109] Alternatively if the dark matter particle is unstable, it could decay into standard model (or other) particles. These processes could be detected indirectly through an excess of gamma rays, antiprotons or positrons emanating from high density regions in our galaxy or others.[110] A major difficulty inherent in such searches is that there are various astrophysical sources which can mimic the signal expected from dark matter, and so multiple signals will likely be required for a conclusive discovery.[15][85]

A few of the dark matter particles passing through the Sun or Earth may scatter off atoms and lose energy. Thus dark matter may accumulate at the center of these bodies, increasing the chance of collision/annihilation. This could produce a distinctive signal in the form of high-energy neutrinos.[111] Such a signal would be strong indirect proof of WIMP dark matter.[15] High-energy neutrino telescopes such as AMANDA, IceCube and ANTARES are searching for this signal.[112] The detection by LIGO in September 2015 of gravitational waves, opens the possibility of observing dark matter in a new way, particularly if it is the form of primordial black holes.[113][114][115]

Many experimental searches have been undertaken to look for such emission from dark matter annihilation or decay, examples of which follow. The EGRET gamma ray telescope observed more gamma rays in 2008 than expected from the Milky Way, but scientists concluded that this was most likely due to incorrect estimation of the telescope's sensitivity.[116]

The Fermi Gamma-ray Space Telescope is searching for similar gamma rays.[117] In April 2012, an analysis of previously available data from its Large Area Telescope instrument produced statistical evidence of a 130 GeV signal in the gamma radiation coming from the center of the Milky Way.[118] WIMP annihilation was seen as the most probable explanation.[119]

At higher energies, ground-based gamma-ray telescopes have set limits on the annihilation of dark matter in dwarf spheroidal galaxies[120] and in clusters of galaxies.[121]

The PAMELA experiment (launched 2006) detected excess positrons. They could be from dark matter annihilation or from pulsars. No excess antiprotons were observed.[122]

In 2013 results from the Alpha Magnetic Spectrometer on the International Space Station indicated excess high-energy cosmic rays that could be due to dark matter annihilation.[123][124][125][126][127][128]

10.6.3 Collider searches for dark matter

An alternative approach to the detection of dark matter particles in nature is to produce them in a laboratory. Experiments with the Large Hadron Collider (LHC) may be able to detect dark matter particles produced in collisions of the LHC proton beams. Because a dark matter particle should have negligible interactions with normal visible matter, it may be detected indirectly as (large amounts of) missing energy and momentum that escape the detectors, provided other (non-negligible) collision products are detected.[129] Constraints on dark matter also exist from the LEP experiment using a similar principle, but probing the interaction of dark matter particles with electrons rather than quarks.[130] It is important to note that any discovery from collider searches will need to be corroborated by discoveries in the indirect or direct detection sectors, in order to prove that the particle discovered is in fact the dark matter of our Universe.

10.7 Alternative theories

For more details on this topic, see Alternatives to general relativity.

Because dark matter remains to be conclusively identified, many theories that aim to explain the observational evidence without invoking dark matter have emerged. The obvious way to do this is to modify general relativity. General relativity is well-tested on solar-system scales, but its validity on galactic or cosmological scales is less certain. A suitable modification to general relativity can conceivably eliminate the need for dark matter. The most well-known theories of this class are MOND and its relativistic generalization TeVeS,[131] and f(R) gravity.[132] Alternative theories abound.[133][134] Entropic gravity, invented by Erik Verlinde, provides a theoretical basis for MOND.

A problem with alternative theories is that the observational evidence for dark matter comes from so many independent angles (see the "observational evidence" section above). Any alternative theory not only has to explain all the evidence, it also has to explain individual cases such as the Bullet Cluster,[135] wherein two colliding clusters of galaxies gave rise to an 8σ statistical significance spatial offset of the center of the total mass from the center of the baryonic mass peaks.[136] Nonetheless, there has been some scattered successes for alternative theories, such as a 2016 test of gravitational lensing in entropic gravity.[137][138][139]

The prevailing opinion among most astrophysicists is that while modifications to general relativity can conceivably explain part of the observational evidence, there is probably enough data to conclude there must still be some dark matter.[140]

10.8 In philosophy of science

In philosophy of science, dark matter is an example of an "auxiliary hypothesis", an ad hoc postulate that is added to a theory in response to observations that falsify it. It has been argued that the dark matter hypothesis is a conventionalist hypothesis, that is, a hypothesis that adds no empirical content and hence is unfalsifiable in the sense defined by Karl Popper.[141]

10.9 In popular culture

Main article: Dark matter in fiction

Mention of dark matter is made in works of fiction. In such cases, it is usually attributed extraordinary physical or magical properties. Such descriptions are often inconsistent with the hypothesized properties of dark matter in physics and cosmology.

10.10 See also

Related theories

- Dark energy
- Conformal gravity
- Entropic gravity
- Dark electromagnetism
- Massive gravity
- Unparticle physics

Experiments

- DEAP, a search apparatus
- DAMPE, a space mission
- General Antiparticle Spectrometer
- Multidark, a research program
- Illustris project, astrophysical simulations

Dark matter candidates

- Light dark matter
- Mirror matter
- Exotic matter
- Neutralino
- Dark galaxy
- Scalar field dark matter
- Self-interacting dark matter
- WIMP hypothetical particles
- SIMP hypothetical particles
- Chameleon particle

10.11 Notes

[1] Since dark energy, by convention, does not count as "matter", this is 26.8/(4.9 + 26.8)=0.845

10.12 References

[1] Overbye, Dennis (20 February 2017). "Cosmos Controversy: The Universe Is Expanding, but How Fast?". *New York Times*. Retrieved 21 February 2017.

[2] "Dark Matter". *CERN Physics*. 20 January 2012.

[3] Trimble, V. (1987). "Existence and nature of dark matter in the universe". *Annual Review of Astronomy and Astrophysics*. **25**: 425–472. Bibcode:1987ARA&A..25..425T. doi:10.1146/annurev.aa.25.090187.002233.

[4] "Planck Mission Brings Universe Into Sharp Focus". *NASA Mission Pages*. 21 March 2013.

[5] "Dark Energy, Dark Matter". *NASA Science: Astrophysics*. 5 June 2015.

[6] Ade, P. A. R.; Aghanim, N.; Armitage-Caplan, C.; (Planck Collaboration); et al. (22 March 2013). "Planck 2013 results. I. Overview of products and scientific results – Table 9". *Astronomy and Astrophysics*. **1303**: 5062. Bibcode:2014A&A...571A...1P. arXiv:1303.5062. doi:10.1051/0004-6361/201321529.

[7] Francis, Matthew (22 March 2013). "First Planck results: the Universe is still weird and interesting". *Arstechnica*.

[8] "Planck captures portrait of the young Universe, revealing earliest light". University of Cambridge. 21 March 2013. Retrieved 21 March 2013.

[9] Sean Carroll, Ph.D., Caltech, 2007, The Teaching Company, *Dark Matter, Dark Energy: The Dark Side of the Universe*, Guidebook Part 2 page 46, Accessed 7 October 2013, "...dark matter: An invisible, essentially collisionless component of matter that makes up about 25 percent of the energy density of the universe... it's a different kind of particle... something not yet observed in the laboratory..."

[10] Ferris, Timothy. "Dark Matter". Retrieved 10 June 2015.

[11] Jarosik, N.; et al. (2011). "Seven-Year Wilson Microwave Anisotropy Probe (WMAP) Observations: Sky Maps, Systematic Errors, and Basic Results". *Astrophysical Journal Supplement*. **192** (2): 14. Bibcode:2011ApJS..192...14J. arXiv:1001.4744. doi:10.1088/0067-0049/192/2/14.

[12] Persic, Massimo; Salucci, Paolo (1 September 1992). "The baryon content of the Universe". *Monthly Notices of the Royal Astronomical Society*. **258** (1): 14P–18P. Bibcode:1992MNRAS.258P..14P. ISSN 0035-8711. arXiv:astro-ph/0502178. doi:10.1093/mnras/258.1.14P.

[13] Copi, C. J.; Schramm, D. N.; Turner, M. S. (1995). "Big-Bang Nucleosynthesis and the Baryon Density of the Universe". *Science*. **267** (5195): 192–199. Bibcode:1995Sci...267..192C. PMID 7809624. arXiv:astro-ph/9407006. doi:10.1126/science.7809624.

[14] Siegfried, T. (5 July 1999). "Hidden Space Dimensions May Permit Parallel Universes, Explain Cosmic Mysteries". *The Dallas Morning News*.

[15] Bertone, G.; Hooper, D.; Silk, J. (2005). "Particle dark matter: Evidence, candidates and constraints". *Physics Reports*. **405** (5–6): 279–390. Bibcode:2005PhR...405..279B. arXiv:hep-ph/0404175. doi:10.1016/j.physrep.2004.08.031.

[16] Kroupa, P.; et al. (2010). "Local-Group tests of dark-matter Concordance Cosmology: Towards a new paradigm for structure formation". *Astronomy and Astrophysics*. **523**: 32–54. Bibcode:2010A&A...523A..32K. arXiv:1006.1647. doi:10.1051/0004-6361/201014892.

[17] Conformal theory: New light on dark matter, dark energy, and dark galactic halos." (PDF) Robert K. Nesbet. IBM Almaden Research Center, 17 June 2014.

[18] Angus, G. (2013). "Cosmological simulations in MOND: the cluster scale halo mass function with light sterile neutrinos". *Monthly Notices of the Royal Astronomical Society*. **436**: 202–211. Bibcode:2013MNRAS.436..202A. arXiv:1309.6094. doi:10.1093/mnras/stt1564.

[19] "A history of dark matter". *Ars Technica*. Retrieved 2017-02-08.

[20] Kapteyn, Jacobus Cornelius (1922). "First attempt at a theory of the arrangement and motion of the sidereal system". *Astrophysical Journal*. **55**: 302–327. Bibcode:1922ApJ....55..302K. doi:10.1086/142670. It is incidentally suggested that when the theory is perfected it may be possible to determine *the amount of dark matter* from its gravitational effect. (emphasis in original)

[21] Rosenberg, Leslie J (30 June 2014). *Status of the Axion Dark-Matter Experiment (ADMX)* (PDF). 10th PATRAS Workshop on Axions, WIMPs and WISPs. p. 2.

[22] Oort, J.H. (1932) "The force exerted by the stellar system in the direction perpendicular to the galactic plane and some related problems," *Bulletin of the Astronomical Institutes of the Netherlands*, **6** : 249–287.

[23] "The Hidden Lives of Galaxies: Hidden Mass". *Imagine the Universe!*. NASA/GSFC.

[24] Kuijken, K.; Gilmore, G. (July 1989). "The Mass Distribution in the Galactic Disc – Part III – the Local Volume Mass Density" (PDF). *Monthly Notices of the Royal Astronomical Society*. **239** (2): 651–664. Bibcode:1989MNRAS.239..651K. doi:10.1093/mnras/239.2.651.

[25] Zwicky, F. (1933). "Die Rotverschiebung von extragalaktischen Nebeln". *Helvetica Physica Acta*. **6**: 110–127. Bibcode:1933AcHPh...6..110Z.

[26] Zwicky, F. (1937). "On the Masses of Nebulae and of Clusters of Nebulae". *The Astrophysical Journal*. **86**: 217. Bibcode:1937ApJ....86..217Z. doi:10.1086/143864.

[27] Zwicky, F. (1933), "Die Rotverschiebung von extragalaktischen Nebeln", *Helvetica Physica Acta*, **6**: 110–127, Bibcode:1933AcHPh...6..110Z See also Zwicky, F. (1937), "On the Masses of Nebulae and of Clusters of Nebulae", *Astrophysical Journal*, **86**: 217, Bibcode:1937ApJ....86..217Z, doi:10.1086/143864

[28] Some details of Zwicky's calculation and of more modern values are given in Richmond, M., *Using the virial theorem: the mass of a cluster of galaxies*, retrieved 10 July 2007

[29] Freese, Katherine (4 May 2014). *The Cosmic Cocktail: Three Parts Dark Matter*. Princeton University Press. ISBN 978-1-4008-5007-5.

[30] Babcock, H, 1939, "The rotation of the Andromeda Nebula", Lick Observatory bulletin ; no. 498

[31] Overbye, Dennis (December 27, 2016). "Vera Rubin, 88, Dies; Opened Doors in Astronomy, and for Women". *New York Times*. Retrieved December 27, 2016.

[32] First observational evidence of dark matter. Darkmatterphysics.com. Retrieved 6 August 2013.

[33] Rubin, Vera C.; Ford, W. Kent, Jr. (February 1970). "Rotation of the Andromeda Nebula from a Spectroscopic Survey of Emission Regions". *The Astrophysical Journal*. **159**: 379–403. Bibcode:1970ApJ...159..379R. doi:10.1086/150317.

[34] Bosma, A. (1978). "The distribution and kinematics of neutral hydrogen in spiral galaxies of various morphological types" (Ph.D. Thesis). Rijksuniversiteit Groningen.

[35] Rubin, V.; Thonnard, W. K. Jr.; Ford, N. (1980). "Rotational Properties of 21 Sc Galaxies with a Large Range of Luminosities and Radii from NGC 4605 (R = 4kpc) to UGC 2885 (R = 122kpc)". *The Astrophysical Journal*. **238**: 471. Bibcode:1980ApJ...238..471R. doi:10.1086/158003.

[36] Randall 2015, pp. 13–14.

[37] Roberts, Morton S. (May 1966). "A High-Resolution 21-cm Hydrogen-Line Survey of the Andromeda Nebula". *The Astrophysical Journal*. **159**: 639–656. Bibcode:1966ApJ...144..639R. doi:10.1086/148645.

[38] Gottesman, S. T.; Davies, R. D.; Reddish, V. C. (1966). "A neutral hydrogen survey of the southern regions of the Andromeda nebula". *Monthly Notices of the Royal Astronomical Society*. **133**: 359–387. Bibcode:1966MNRAS.133..359G. doi:10.1093/mnras/133.4.359.

[39] Roberts, Morton S.; Whitehurst, Robert N. (October 1975). "The rotation curve and geometry of M31 at large galactocentric distances". *The Astrophysical Journal*. **201**: 327–346. Bibcode:1975ApJ...201..327R. doi:10.1086/153889.

[40] Rogstad, D. H.; Shostak, G. Seth (September 1972). "Gross Properties of Five Scd Galaxies as Determined from 21-centimeter Observations". *The Astrophysical Journal*. **176**: 315–321. Bibcode:1972ApJ...176..315R. doi:10.1086/151636.

[41] Randall 2015, pp. 14–16.

[42] Bergstrom, L. (2000). "Non-baryonic dark matter: Observational evidence and detection methods". *Reports on Progress in Physics*. **63** (5): 793–841. Bibcode:2000RPPh...63..793B. arXiv:hep-ph/0002126 . doi:10.1088/0034-4885/63/5/2r3.

[43] Komatsu, E.; et al. (2009). "Five-Year Wilkinson Microwave Anisotropy Probe Observations: Cosmological Interpretation". *The Astrophysical Journal Supplement*. **180** (2): 330–376. Bibcode:2009ApJS..180..330K. arXiv:0803.0547 . doi:10.1088/0067-0049/180/2/330.

[44] Boggess, N. W.; et al. (1992). "The COBE Mission: Its Design and Performance Two Years after the launch". *The Astrophysical Journal*. **397**: 420. Bibcode:1992ApJ...397..420B. doi:10.1086/171797.

[45] Melchiorri, A.; et al. (2000). "A Measurement of Ω from the North American Test Flight of Boomerang". *The Astrophysical Journal Letters*. **536** (2): L63–L66. Bibcode:2000ApJ...536L..63M. arXiv:astro-ph/9911445 . doi:10.1086/312744.

[46] Leitch, E. M.; et al. (2002). "Measurement of polarization with the Degree Angular Scale Interferometer". *Nature*. **420** (6917): 763–771. Bibcode:2002Natur.420..763L. PMID 12490940. arXiv:astro-ph/0209476 . doi:10.1038/nature01271.

[47] Leitch, E. M.; et al. (2005). "Degree Angular Scale Interferometer 3 Year Cosmic Microwave Background Polarization Results". *The Astrophysical Journal*. **624** (1): 10–20. Bibcode:2005ApJ...624...10L. arXiv:astro-ph/0409357 . doi:10.1086/428825.

[48] Readhead, A. C. S.; et al. (2004). "Polarization Observations with the Cosmic Background Imager". *Science*. **306** (5697): 836–844. Bibcode:2004Sci...306..836R. PMID 15472038. arXiv:astro-ph/0409569 . doi:10.1126/science.1105598.

[49] Hinshaw, G.; et al. (2009). "Five-Year Wilkinson Microwave Anisotropy Probe Observations: Data Processing, Sky Maps, and Basic Results". *The Astrophysical Journal Supplement*. **180** (2): 225–245. Bibcode:2009ApJS..180..225H. arXiv:0803.0732 . doi:10.1088/0067-0049/180/2/225.

[50] Daniel Baumann. "Cosmology: Part III Mathematical Tripos, Cambridge University" (PDF). p. 21–22.

[51] "Serious Blow to Dark Matter Theories?" (Press release). European Southern Observatory. 18 April 2012.

[52] "Dark Matter May be Smoother than Expected – Careful study of large area of sky imaged by VST reveals intriguing result". *www.eso.org*. Retrieved 8 December 2016.

[53] Corbelli, E. & Salucci, P. (2000). "The extended rotation curve and the dark matter halo of M33". *Monthly Notices of the Royal Astronomical Society*. **311** (2): 441–447. Bibcode:2000MNRAS.311..441C. arXiv:astro-ph/9909252 . doi:10.1046/j.1365-8711.2000.03075.x.

[54] Faber, S. M.; Jackson, R. E. (1976). "Velocity dispersions and mass-to-light ratios for elliptical galaxies". *The Astrophysical Journal*. **204**: 668–683. Bibcode:1976ApJ...204..668F. doi:10.1086/154215.

[55] Binny, James; Merrifield, Michael (1998). *Galactic Astronomy*. Princeton University Press. p. 712-713.

[56] Taylor, A. N.; et al. (1998). "Gravitational Lens Magnification and the Mass of Abell 1689". *The Astrophysical Journal*. **501** (2): 539–553. Bibcode:1998ApJ...501..539T. arXiv:astro-ph/9801158 . doi:10.1086/305827.

[57] Wu, X.; Chiueh, T.; Fang, L.; Xue, Y. (1998). "A comparison of different cluster mass estimates: consistency or discrepancy?". *Monthly Notices of the Royal Astronomical Society*. **301** (3): 861–871. Bibcode:1998MNRAS.301..861W. arXiv:astro-ph/9808179 . doi:10.1046/j.1365-8711.1998.02055.x.

[58] Cho, Adrian (2017). "Scientists unveil the most detailed map of dark matter to date". *Science*. doi:10.1126/science.aal0847.

[59] Natarajan, Priyamvada; Chadayammuri, Urmila; Jauzac, Mathilde; Richard, Johan; Kneib, Jean-Paul; Ebeling, Harald; Jiang, Fangzhou; Bosch, Frank van den; Limousin, Marceau; Jullo, Eric; Atek, Hakim; Pillepich, Annalisa; Popa, Cristina; Marinacci, Federico; Hernquist, Lars; Meneghetti, Massimo; Vogelsberger, Mark (2017). "Mapping substructure in the HST Frontier Fields cluster lenses and in cosmological simulations". *Monthly Notices of the Royal Astronomical Society*. doi:10.1093/mnras/stw3385.

[60] Refregier, A. (2003). "Weak gravitational lensing by large-scale structure". *Annual Review of Astronomy and Astrophysics*. **41** (1): 645–668. Bibcode:2003ARA&A..41..645R. arXiv:astro-ph/0307212 . doi:10.1146/annurev.astro.41.111302.102207.

[61] The details are technical. See Wayne Hu (2001). "Intermediate Guide to the Acoustic Peaks and Polarization". for an intermediate-level introduction.

[62] Skordis, C.; et al. (2006). "Large Scale Structure in Bekenstein's theory of relativistic Modified Newtonian Dynamics" (PDF). *Phys.Rev.Lett.* **96**: 011301. Bibcode:2006PhRvL..96a1301S. doi:10.1103/PhysRevLett.96.011301.

[63] Ade, P.A.R.; et al. (2015). "Planck 2015 results. XIII. Cosmological parameters" (PDF). *Astron.Astrophys.* **594**: A13. Bibcode:2016A&A...594A..13P. doi:10.1051/0004-6361/201525830.

[64] Percival, W. J.; et al. (2007). "Measuring the Baryon Acoustic Oscillation scale using the Sloan Digital Sky Survey and 2dF Galaxy Redshift Survey". *Monthly Notices of the Royal Astronomical Society*. **381** (3): 1053–1066. Bibcode:2007MNRAS.381.1053P. arXiv:0705.3323 . doi:10.1111/j.1365-2966.2007.12268.x.

[65] Peacock, J.; et al. (2001). "A measurement of the cosmological mass density from clustering in the 2dF Galaxy Redshift Survey". *Nature*. **410**: 169–73. Bibcode:2001Natur.410..169P. PMID 11242069. arXiv:astro-ph/0103143 . doi:10.1038/35065528.

[66] Kowalski, M.; et al. (2008). "Improved Cosmological Constraints from New, Old, and Combined Supernova Data Sets". *The Astrophysical Journal*. **686** (2): 749–778. Bibcode:2008ApJ...686..749K. arXiv:0804.4142 . doi:10.1086/589937.

[67] "Will the Universe expand forever?". NASA. 24 January 2014. Retrieved 16 March 2015.

[68] "Our universe is Flat". FermiLab/SLAC. 7 April 2015.

[69] Marcus Y. Yoo (2011). "Unexpected connections". *Engineering & Science*. Caltech. LXXIV1: 30.

[70] "Planck Publications: Planck 2015 Results". European Space Agency. February 2015. Retrieved 9 February 2015.

[71] Viel, M.; Bolton, J. S.; Haehnelt, M. G. (2009). "Cosmological and astrophysical constraints from the Lyman α forest flux probability distribution function". *Monthly Notices of the Royal Astronomical Society*. **399** (1): L39–L43. Bibcode:2009MNRAS.399L..39V. arXiv:0907.2927 . doi:10.1111/j.1745-3933.2009.00720.x.

[72] "Hubble Maps the Cosmic Web of "Clumpy" Dark Matter in 3-D" (Press release). NASA. 7 January 2007.

[73] A. H. Jaffe. "Cosmology 2012: Lecture Notes" (PDF).

[74] L. F. Low (12 October 2016). ""Constraints on the composite photon theory". *Modern Physics Letters A*. doi:10.1142/S021773231675002X.

[75] Randall 2015, p. 286.

[76] Achim Weiss, "Big Bang Nucleosynthesis: Cooking up the first light elements" in: Einstein Online Vol. 2 (2006), 1017

[77] Raine, D.; Thomas, T. (2001). *An Introduction to the Science of Cosmology*. IOP Publishing. p. 30. ISBN 0-7503-0405-7.

[78] Tisserand, P.; Le Guillou, L.; Afonso, C.; Albert, J. N.; Andersen, J.; Ansari, R.; Aubourg, É.; Bareyre, P.; Beaulieu, J. P.; Charlot, X.; Coutures, C.; Ferlet, R.; Fouqué, P.; Glicenstein, J. F.; Goldman, B.; Gould, A.; Graff, D.; Gros, M.; Haissinski, J.; Hamadache, C.; De Kat, J.; Lasserre, T.; Lesquoy, É.; Loup, C.; Magneville, C.; Marquette, J. B.; Maurice, É.; Maury, A.; Milsztajn, A.; Moniez, M. (2007). "Limits on the Macho content of the Galactic Halo from the EROS-2 Survey of the Magellanic Clouds". *Astronomy and Astrophysics*. **469** (2): 387–404. Bibcode:2007A&A...469..387T. arXiv:astro-ph/0607207. doi:10.1051/0004-6361:20066017.

[79] Graff, D. S.; Freese, K. (1996). "Analysis of a *Hubble Space Telescope* Search for Red Dwarfs: Limits on Baryonic Matter in the Galactic Halo". *The Astrophysical Journal*. **456**. Bibcode:1996ApJ...456L..49G. arXiv:astro-ph/9507097. doi:10.1086/309850.

[80] Najita, J. R.; Tiede, G. P.; Carr, J. S. (2000). "From Stars to Superplanets: The Low-Mass Initial Mass Function in the Young Cluster IC 348". *The Astrophysical Journal*. **541** (2): 977–1003. Bibcode:2000ApJ...541..977N. arXiv:astro-ph/0005290. doi:10.1086/309477.

[81] Wyrzykowski, Lukasz et al. (2011) The OGLE view of microlensing towards the Magellanic Clouds – IV. OGLE-III SMC data and final conclusions on MACHOs, MNRAS, 416, 2949

[82] Freese, Katherine; Fields, Brian; Graff, David (2000). "Death of Stellar Baryonic Dark Matter Candidates". arXiv:astro-ph/0007444 [astro-ph].

[83] Freese, Katherine; Fields, Brian; Graff, David (2000). "Death of Stellar Baryonic Dark Matter". *The First Stars*. ESO Astrophysics Symposia. p. 18. Bibcode:2000fist.conf...18F. ISBN 3-540-67222-2. arXiv:astro-ph/0002058. doi:10.1007/10719504_3.

[84] Canetti, L.; Drewes, M.; Shaposhnikov, M. (2012). "Matter and Antimatter in the Universe". *New J.Phys*. **14**: 095012. Bibcode:2012NJPh...14i5012C. arXiv:1204.4186. doi:10.1088/1367-2630/14/9/095012.

[85] Bertone, G.; Merritt, D. (2005). "Dark Matter Dynamics and Indirect Detection". *Modern Physics Letters A*. **20** (14): 1021–1036. Bibcode:2005MPLA...20.1021B. arXiv:astro-ph/0504422. doi:10.1142/S0217732305017391.

[86] Silk, Joseph (6 December 2000). "IX". *The Big Bang: Third Edition*. Henry Holt and Company. ISBN 978-0-8050-7256-3.

[87] Vittorio, N.; J. Silk (1984). "Fine-scale anisotropy of the cosmic microwave background in a universe dominated by cold dark matter". *Astrophysical Journal Letters*. **285**: L39–L43. Bibcode:1984ApJ...285L..39V. doi:10.1086/184361.

[88] Umemura, Masayuki; Satoru Ikeuchi (1985). "Formation of Subgalactic Objects within Two-Component Dark Matter". *Astrophysical Journal*. **299**: 583–592. Bibcode:1985ApJ...299..583U. doi:10.1086/163726.

[89] Davis, M.; Efstathiou, G., Frenk, C. S., & White, S. D. M. (May 15, 1985). "The evolution of large-scale structure in a universe dominated by cold dark matter". *Astrophysical Journal*. **292**: 371–394. Bibcode:1985ApJ...292..371D. doi:10.1086/163168.

[90] Hawkins, M. R. S. (2011). "The case for primordial black holes as dark matter". *Monthly Notices of the Royal Astronomical Society*. **415** (3): 2744–2757. Bibcode:2011MNRAS.415.2744H. arXiv:1106.3875. doi:10.1111/j.1365-2966.2011.18890.x.

[91] Carr, B. J.; et al. (May 2010). "New cosmological constraints on primordial black holes" (PDF). *Physical Review D*. **81** (10): 104019. Bibcode:2010PhRvD..81j4019C. arXiv:0912.5297. doi:10.1103/PhysRevD.81.104019.

[92] Peter, A. H. G. (2012). "Dark Matter: A Brief Review". arXiv:1201.3942 [astro-ph.CO].

[93] Garrett, Katherine; Dūda, Gintaras (2011). "Dark Matter: A Primer". *Advances in Astronomy*. **2011**: 1–22. Bibcode:2011AdAst2011E...8G. arXiv:1006.2483. doi:10.1155/2011/968283. MACHOs can only account for a very small percentage of the nonluminous mass in our galaxy, revealing that most dark matter cannot be strongly concentrated or exist in the form of baryonic astrophysical objects. Although microlensing surveys rule out baryonic objects like brown dwarfs, black holes, and neutron stars in our galactic halo, can other forms of baryonic matter make up the bulk of dark matter? The answer, surprisingly, is no...

[94] Bertone, G. (2010). "The moment of truth for WIMP dark matter". *Nature*. **468** (7322): 389–393. Bibcode:2010Natur.468..389B. PMID 21085174. arXiv:1011.3532. doi:10.1038/nature09509.

[95] Olive, Keith A. (2003). "TASI Lectures on Dark Matter". p. 21

[96] Jungman, Gerard; Kamionkowski, Marc; Griest, Kim (1 March 1996). "Supersymmetric dark matter". *Physics Reports*. **267** (5–6): 195–373. Bibcode:1996PhR...267..195J. arXiv:hep-ph/9506380. doi:10.1016/0370-1573(95)00058-5.

[97] "Neutrinos as Dark Matter". Astro.ucla.edu. 21 September 1998. Retrieved 6 January 2011.

[98] Gaitskell, Richard J. (2004). "Direct Detection of Dark Matter". *Annual Review of Nuclear and Particle Science*. **54**: 315–359. Bibcode:2004ARNPS..54..315G. doi:10.1146/annurev.nucl.54.070103.181244.

[99] "NEUTRALINO DARK MATTER". Retrieved 26 December 2011. Griest, Kim. "WIMPs and MACHOs" (PDF). Retrieved 26 December 2011.

[100] Drees, M.; Gerbier, G. (2015). "Dark Matter" (PDF). *Chin. Phys. C.* **38**: 090001.

[101] Bernabei, R.; Belli, P.; Cappella, F.; Cerulli, R.; Dai, C. J.; d'Angelo, A.; He, H. L.; Incicchitti, A.; Kuang, H. H.; Ma, J. M.; Montecchia, F.; Nozzoli, F.; Prosperi, D.; Sheng, X. D.; Ye, Z. P. (2008). "First results from DAMA/LIBRA and the combined results with DAMA/NaI". *Eur. Phys. J. C.* **56** (3): 333–355. arXiv:0804.2741. doi:10.1140/epjc/s10052-008-0662-y.

[102] Drukier, A.; Freese, K.; Spergel, D. (1986). "Detecting Cold Dark Matter Candidates". *Physical Review D.* **33** (12): 3495–3508. Bibcode:1986PhRvD..33.3495D. doi:10.1103/PhysRevD.33.3495.

[103] Davis, Jonathan H. (2015). "The Past and Future of Light Dark Matter Direct Detection". *Int.J.Mod.Phys. A.* **30** (15): 1530038. doi:10.1142/S0217751X15300380.

[104] Stonebraker, Alan (3 January 2014). "Synopsis: Dark-Matter Wind Sways through the Seasons". *Physics – Synopses.* American Physical Society. Retrieved 6 January 2014.

[105] Lee, Samuel K.; Mariangela Lisanti, Annika H. G. Peter, and Benjamin R. Safdi (3 January 2014). "Effect of Gravitational Focusing on Annual Modulation in Dark-Matter Direct-Detection Experiments". *Phys. Rev. Lett.* American Physical Society. **112** (1): 011301 (2014) [5 pages]. Bibcode:2014PhRvL.112a1301L. PMID 24483881. arXiv:1308.1953. doi:10.1103/PhysRevLett.112.011301.

[106] The Dark Matter Group. "An Introduction to Dark Matter". *Dark Matter Research.* Sheffield, UK: University of Sheffield. Retrieved 7 January 2014.

[107] "Blowing in the Wind". *Kavli News.* Sheffield, UK: Kavli Foundation. Retrieved 7 January 2014. Scientists at Kavli MIT are working on...a tool to track the movement of dark matter.

[108] "Dark matter even darker than once thought". Retrieved 16 June 2015.

[109] Bertone, Gianfranco (2010). "Dark Matter at the Centers of Galaxies". *Particle Dark Matter: Observations, Models and Searches.* Cambridge University Press. pp. 83–104. ISBN 978-0-521-76368-4. arXiv:1001.3706.

[110] Ellis, J.; Flores, R. A.; Freese, K.; Ritz, S.; Seckel, D.; Silk, J. (1988). "Cosmic ray constraints on the annihilations of relic particles in the galactic halo". *Physics Letters B.* **214** (3): 403–412. Bibcode:1988PhLB..214..403E. doi:10.1016/0370-2693(88)91385-8.

[111] Freese, K. (1986). "Can Scalar Neutrinos or Massive Dirac Neutrinos be the Missing Mass?". *Physics Letters B.* **167** (3): 295–300. Bibcode:1986PhLB..167..295F. doi:10.1016/0370-2693(86)90349-7.

[112] Randall 2015, p. 298.

[113] Sokol, Joshua; et al. (20 February 2016). "Surfing Gravity's Waves". *New Scientist* (3061).

[114] "Did Gravitational Wave Detector Find Dark Matter?". Johns Hopkins University. 15 June 2016. Retrieved 20 June 2015. While their existence has not been established with certainty, primordial black holes have in the past been suggested as a possible solution to the dark matter mystery. Because there's so little evidence of them, though, the primordial black hole-dark matter hypothesis has not gained a large following among scientists. The LIGO findings, however, raise the prospect anew, especially as the objects detected in that experiment conform to the mass predicted for dark matter. Predictions made by scientists in the past held that conditions at the birth of the universe would produce lots of these primordial black holes distributed roughly evenly in the universe, clustering in halos around galaxies. All this would make them good candidates for dark matter.

[115] Bird, Simeon; Cholis, Illian (19 May 2016). "Did LIGO Detect Dark Matter?". *Physical Review Letters.* 116, 201301. Bibcode:2016PhRvL.116t1301B. doi:10.1103/PhysRevLett.116.201301. Retrieved 20 June 2016.

[116] Stecker, F.W.; Hunter, S; Kniffen, D (2008). "The likely cause of the EGRET GeV anomaly and its implications". *Astroparticle Physics.* **29** (1): 25–29. Bibcode:2008APh....29...25S. arXiv:0705.4311. doi:10.1016/j.astropartphys.2007.11.002.

[117] Atwood, W.B.; Abdo, A. A.; Ackermann, M.; Althouse, W.; Anderson, B.; Axelsson, M.; Baldini, L.; Ballet, J.; et al. (2009). "The large area telescope on the Fermi Gamma-ray Space Telescope Mission". *Astrophysical Journal.* **697** (2): 1071–1102. Bibcode:2009ApJ...697.1071A. arXiv:0902.1089. doi:10.1088/0004-637X/697/2/1071.

[118] Weniger, Christoph (2012). "A Tentative Gamma-Ray Line from Dark Matter Annihilation at the Fermi Large Area Telescope". *Journal of Cosmology and Astroparticle Physics.* **2012** (8): 7. Bibcode:2012JCAP...08..007W. arXiv:1204.2797v2. doi:10.1088/1475-7516/2012/08/007.

[119] Cartlidge, Edwin (24 April 2012). "Gamma rays hint at dark matter". Institute Of Physics. Retrieved 23 April 2013.

[120] Albert, J.; Aliu, E.; Anderhub, H.; Antoranz, P.; Backes, M.; Baixeras, C.; Barrio, J. A.; Bartko, H.; Bastieri, D.; Becker, J. K.; Bednarek, W.; Berger, K.; Bigongiari, C.; Biland, A.; Bock, R. K.; Bordas, P.; Bosch-Ramon, V.; Bretz, T.; Britvitch, I.; Camara, M.; Carmona, E.; Chilingarian, A.; Commichau, S.; Contreras, J. L.; Cortina, J.; Costado, M. T.; Curtef, V.; Danielyan, V.; Dazzi, F.; De Angelis, A. (2008). "Upper Limit for γ-Ray Emission above 140 GeV from the Dwarf Spheroidal Galaxy Draco". *The Astrophysical Journal.* **679**: 428–431. Bibcode:2008ApJ...679..428A. arXiv:0711.2574. doi:10.1086/529135.

[121] Aleksić, J.; Antonelli, L. A.; Antoranz, P.; Backes, M.; Baixeras, C.; Balestra, S.; Barrio, J. A.; Bastieri, D.; González, J. B.; Bednarek, W.; Berdyugin, A.; Berger, K.; Bernardini, E.; Biland, A.; Bock, R. K.; Bonnoli, G.; Bordas, P.; Tridon, D. B.; Bosch-Ramon, V.; Bose, D.; Braun, I.; Bretz, T.; Britzger, D.; Camara, M.; Carmona, E.; Carosi, A.; Colin, P.; Commichau, S.; Contreras, J. L.; Cortina, J. (2010). "Magic Gamma-Ray Telescope Observation of the Perseus Cluster of Galaxies: Implications for Cosmic Rays, Dark Matter, and Ngc 1275". *The Astrophysical Journal*. **710**: 634–647. Bibcode:2010ApJ...710..634A. arXiv:0909.3267. doi:10.1088/0004-637X/710/1/634.

[122] Adriani, O.; Barbarino, G. C.; Bazilevskaya, G. A.; Bellotti, R.; Boezio, M.; Bogomolov, E. A.; Bonechi, L.; Bongi, M.; Bonvicini, V.; Bottai, S.; Bruno, A.; Cafagna, F.; Campana, D.; Carlson, P.; Casolino, M.; Castellini, G.; De Pascale, M. P.; De Rosa, G.; De Simone, N.; Di Felice, V.; Galper, A. M.; Grishantseva, L.; Hofverberg, P.; Koldashov, S. V.; Krutkov, S. Y.; Kvashnin, A. N.; Leonov, A.; Malvezzi, V.; Marcelli, L.; Menn, W. (2009). "An anomalous positron abundance in cosmic rays with energies 1.5–100 GeV". *Nature*. **458** (7238): 607–609. Bibcode:2009Natur.458..607A. PMID 19340076. arXiv:0810.4995. doi:10.1038/nature07942.

[123] Aguilar, M. (AMS Collaboration); et al. (3 April 2013). "First Result from the Alpha Magnetic Spectrometer on the International Space Station: Precision Measurement of the Positron Fraction in Primary Cosmic Rays of 0.5–350 GeV". *Physical Review Letters*. **110**: 141102. Bibcode:2013PhRvL.110n1102A. PMID 25166975. doi:10.1103/PhysRevLett.110.141102. Retrieved 3 April 2013.

[124] "First Result from the Alpha Magnetic Spectrometer Experiment". *AMS Collaboration*. 3 April 2013. Retrieved 3 April 2013.

[125] Heilprin, John; Borenstein, Seth (3 April 2013). "Scientists find hint of dark matter from cosmos". Associated Press. Retrieved 3 April 2013.

[126] Amos, Jonathan (3 April 2013). "Alpha Magnetic Spectrometer zeroes in on dark matter". *BBC*. Retrieved 3 April 2013.

[127] Perrotto, Trent J.; Byerly, Josh (2 April 2013). "NASA TV Briefing Discusses Alpha Magnetic Spectrometer Results". *NASA*. Retrieved 3 April 2013.

[128] Overbye, Dennis (3 April 2013). "New Clues to the Mystery of Dark Matter". *New York Times*. Retrieved 3 April 2013.

[129] Kane, G.; Watson, S. (2008). "Dark Matter and LHC:. what is the Connection?". *Modern Physics Letters A*. **23** (26): 2103–2123. Bibcode:2008MPLA...23.2103K. arXiv:0807.2244. doi:10.1142/S0217732308028314.

[130] Fox, P.J.; Harnik, R.; Kopp, J.; Tsai, Y. (2011). "LEP Shines Light on Dark Matter" (PDF). *Phys. Rev. D*. **84**: 014028. doi:10.1103/PhysRevD.84.014028.

[131] For a review, see: Pavel Kroupa; et al. (December 2012). "The failures of the Standard Model of Cosmology require a new paradigm". *International Journal of Modern Physics D*. **21** (4). doi:10.1142/S0218271812300030.

[132] For a review, see: Salvatore Capozziello & Mariafelicia De Laurentis (October 2012). "The dark matter problem from f(R) gravity viewpoint". *Annalen Der Physik*. **524** (9–10). doi:10.1002/andp.201200109.

[133] Phillip D. Mannheim (April 2006). "Alternatives to Dark Matter and Dark Energy". *Progress in Particle and Nuclear Physics*. **56** (2). arXiv:astro-ph/0505266. doi:10.1016/j.ppnp.2005.08.001.

[134] Austin Joyce; et al. (March 2015). "Beyond the Cosmological Standard Model". *Physics Reports*. **568**. arXiv:1407.0059. doi:10.1016/j.physrep.2014.12.002.

[135] M. Markevitch; S. Randall; D. Clowe; A. Gonzalez; et al. (16–23 July 2006). "Dark Matter and the Bullet Cluster" (PDF). *36th COSPAR Scientific Assembly*. Beijing, China. Abstract only

[136] Clowe, Douglas; et al. (2006). "A Direct Empirical Proof of the Existence of Dark Matter". *The Astrophysical Journal Letters*. **648** (2): L109–L113. Bibcode:2006ApJ...648L.109C. arXiv:astro-ph/0608407. doi:10.1086/508162.

[137] "Verlinde's new theory of gravity passes first test". December 16, 2016.

[138] Brouwer, Margot M.; et al. (11 December 2016). "First test of Verlinde's theory of Emergent Gravity using Weak Gravitational Lensing measurements". *Monthly Notices of the Royal Astronomical Society*. **466** (to appear): 2547–2559. arXiv:1612.03034. doi:10.1093/mnras/stw3192.

[139] "First test of rival to Einstein's gravity kills off dark matter". 15 December 2016. Retrieved 20 February 2017.

[140] Sean Carroll (9 May 2012). "Dark Matter vs. Modified Gravity: a Trialogue". Retrieved 14 February 2017.

[141] Merritt, David "Cosmology and Convention", *Studies In History and Philosophy of Science Part B: Studies In History and Philosophy of Modern Physics*, **57**(1):41–52, February 2017.

10.13 External links

- Dark matter at DMOZ
- Dark matter (Astronomy) at *Encyclopædia Britannica*
- A history of dark matter (February 2017), *Ars Technica*
- What is dark matter?, *CosmosMagazine.com*

- The Dark Matter Crisis 18 August 2010 by Pavel Kroupa, posted in General
- The European astroparticle physics network
- Helmholtz Alliance for Astroparticle Physics
- "NASA Finds Direct Proof of Dark Matter" (Press release). NASA. 21 August 2006.
- Tuttle, Kelen (22 August 2006). "Dark Matter Observed". SLAC (Stanford Linear Accelerator Center) Today.
- "Astronomers claim first 'dark galaxy' find". New Scientist. 23 February 2005.
- Sample, Ian (17 December 2009). "Dark Matter Detected". London: Guardian. Retrieved 1 May 2010.
- Video lecture on dark matter by Scott Tremaine, IAS professor
- Science Daily story "Astronomers' Doubts About the Dark Side ..."
- Gray, Meghan; Merrifield, Mike; Copeland, Ed (2010). "Dark Matter". *Sixty Symbols*. Brady Haran for the University of Nottingham.
- The Physicist Who Denies that Dark Matter Exists By Oded Carmeli

Chapter 11

Dark fluid

Not to be confused with dark energy, dark flow, or dark matter.

In astronomy and cosmology, **dark fluid** is an alternative theory to both dark matter and dark energy and attempts to explain both phenomena in a single framework.[1][2]

Dark fluid proposes that dark matter and dark energy are not separate physical phenomena as previously thought, nor do they have separate origins, but that they are strongly linked together and can be considered as two facets of a single fluid. At galactic scales, the dark fluid behaves like dark matter, and at larger scales its behavior becomes similar to dark energy. Our observations within the scales of the Earth and the Solar System are currently insufficient to explain the gravitational effects observed at such larger scales.

11.1 Overview

Two major conundrums have arisen in astrophysics and cosmology in recent times, both dealing with the laws of gravity. The first was the realization that there aren't enough visible stars or gas inside galaxies to account for their high rate of rotation. The theory of dark matter was created to explain this phenomenon. It theorizes that the galaxies are spinning as fast as they are because there is more matter in those galaxies (including our own Milky Way) than can be seen by counting the mass of stars and gas alone, and that this unseen (dark) matter is invisible because it doesn't interact with the electromagnetic force from which all forms of light come.

The second conundrum came from the observations of a very specific kind of supernova, known as a Type Ia supernova used as a standard candle: when they were compared in distant vs. nearby galaxies, it was found that the distant supernova were fainter, and thus farther away than expected. This implied that the Universe was not only expanding, but accelerating its expansion. The theory of dark energy was created to explain this phenomenon.

In the traditional approach to modeling effects of gravity, general relativity is assumed to be valid at cosmological scales as well as in the Solar System, where its predictions have been more accurately tested. Not changing the rules of gravity, however, implies the presence of dark matter and dark energy in parts of the Universe where the curvature of the space-time manifold is far less than that in the Solar System. It is phenomenologically possible to alter the equations of gravity in regions of low space-time curvature such that the dynamics of the space-time causes what we assign to the presence of dark matter and dark energy.[3] Dark fluid theory hypothesizes that the dark fluid is a specific kind of fluid whose attractive and repulsive behaviors depend on the local energy density. In this theory, the dark fluid behaves like dark matter in the regions of space where the baryon density is high. The idea is that when the dark fluid is in the presence of matter, it slows down and coagulates around it; this then attracts more dark fluid to coagulate around it, thus amplifying the force of gravity near it. The effect is always present, but only becomes noticeable in the presence of a very large mass such as a galaxy. This description is similar to theories of dark matter, and a special case of the equations of dark fluid reproduces dark matter.

On the other hand, in places where there is relatively little matter, as in the voids between galactic superclusters, this theory predicts that the dark fluid relaxes and acquires a negative pressure. Thus dark fluid becomes a repulsive force, with an effect similar to that of dark energy.

Dark fluid goes beyond dark matter and dark energy in that it predicts a continuous range of attractive and repulsive qualities under various matter density cases. Indeed, special cases of various other gravitational theories are reproduced by dark fluid, e.g. inflation, quintessence, k-essence, f(R), Generalized Einstein-Aether f(K), MOND, TeVeS, BSTV, etc. Dark fluid theory also suggests new models, such as a certain f(K+R) model which suggests interesting corrections to MOND that depend on redshift and density.

11.2 Simplifying assumptions

Dark fluid theory is not treated like a standard fluid mechanics model, because many of the fluid mechanics equations are too difficult to solve completely. A formalized fluid mechanical approach, like the generalized Chaplygin gas model, would be an ideal method for modeling this theory, but it requires too many observational data points to be computationally feasible at present, and there are not enough such data points available to cosmologists yet. So a simplification step was undertaken by modeling the theory through scalar field models instead, as is done in other alternative approaches to dark energy and dark matter.[2][4]

11.3 References

[1] Alexandre Arbey (2005) “Is it possible to consider Dark Energy and Dark Matter as a same and unique Dark Fluid? http://arxiv.org/abs/astro-ph/0506732"

[2] Alexandre Arbey (2006) “Dark Fluid: a complex scalar field to unify dark energy and dark matter http://arxiv.org/abs/astro-ph/0601274"

[3] Exirifard, Q. (2010). “Phenomenological covariant approach to gravity”. *General Relativity and Gravitation*. **43**: 93–106. Bibcode:2011GReGr..43...93E. arXiv:0808.1962 . doi:10.1007/s10714-010-1073-6.

[4] Arbey, A.; Mahmoudi, F. (2007). “One-loop quantum corrections to cosmological scalar field potentials”. *Physical Review D*. **75**. doi:10.1103/PhysRevD.75.063513.

11.4 External links

- Arbey, A. (2008). “Cosmological constraints on unifying Dark Fluid models”. *The Open Astronomy Journal*. **1**: 27–38. doi:10.2174/1874381100801010027.
- Zong-Kuan Guo, Yuan-Zhong Zhang, Cosmology with a Variable Chaplygin Gas" (2005).
- Anaelle Halle, HongSheng Zhao, Baojiu Li Perturbations in a non-uniform dark energy fluid: equations reveal effects of modified gravity and dark matter" (2008)

Chapter 12

Cold dark matter

In cosmology and physics, **cold dark matter** (**CDM**) is a hypothetical form of dark matter whose particles moved slowly compared to the speed of light (the *cold* in CDM) since the universe was approximately one year old (a time when the cosmic particle horizon contained the mass of one typical galaxy); and interact very weakly with ordinary matter and electromagnetic radiation (the *dark* in CDM). It is believed that approximately 84.54% of matter in the Universe is dark matter, with only a small fraction being the ordinary baryonic matter that composes stars, planets and living organisms.

12.1 History

The theory was originally published in 1982 by three independent groups of cosmologists; James Peebles, [1] J. Richard Bond, Alex Szalay and Michael Turner;[2] and George Blumenthal, H. Pagels and Joel Primack.[3] An influential review article in 1984 by Blumenthal, Sandra Moore Faber, Primack and Martin Rees developed the details of the theory.[4]

12.2 Structure formation

In the cold dark matter theory, structure grows hierarchically, with small objects collapsing under their self-gravity first and merging in a continuous hierarchy to form larger and more massive objects. In the hot dark matter paradigm, popular in the early 1980s, structure does not form hierarchically (*bottom-up*), but rather forms by fragmentation (*top-down*), with the largest superclusters forming first in flat pancake-like sheets and subsequently fragmenting into smaller pieces like our galaxy the Milky Way. Predictions of the cold dark matter paradigm are in general agreement with observations of cosmological large scale structure.

12.2.1 Lambda CDM model

Main article: Lambda-CDM model

Since the late 1980s or 1990s, most cosmologists favor the cold dark matter theory (specifically the modern Lambda-CDM model) as a description of how the Universe went from a smooth initial state at early times (as shown by the cosmic microwave background radiation) to the lumpy distribution of galaxies and their clusters we see today — the large-scale structure of the Universe. The theory sees the role that dwarf galaxies played as crucial, as they are thought to be natural building blocks that form larger structures, created by small-scale density fluctuations in the early Universe.[5]

12.3 Composition

Dark matter is detected through its gravitational interactions with ordinary matter and radiation. As such, it is very difficult to determine what the constituents of cold dark matter are. The candidates fall roughly into three categories:

- Axions are very light particles with a specific type of self-interaction that makes them a suitable CDM candidate.[6][7] Axions have the theoretical advantage that their existence solves the Strong CP problem in QCD, but have not been detected.

- MACHOs or *Massive Compact Halo Objects* are large, condensed objects such as black holes, neutron stars, white dwarfs, very faint stars, or non-luminous objects like planets. The search for these consists of using gravitational lensing to see the effect of these objects on background galaxies. Most experts believe that the constraints from those searches rule out MACHOs as a viable dark matter candidate.[8][9][10][11][12][13]

- WIMPs: Dark matter is composed of *Weakly Interacting Massive Particles*. There is no currently known particle with the required properties, but many extensions of the standard model of particle physics predict such particles. The search for WIMPs involves attempts at direct detection by highly sensitive detectors, as well as attempts at production by particle accelerators. WIMPs are generally regarded as the most promising dark matter candidates.[9][11][13] The DAMA/NaI experiment and its successor DAMA/LIBRA have claimed to directly detect dark matter particles passing through the Earth, but many scientists remain skeptical, as no results from similar experiments seem compatible with the DAMA results.

12.4 Challenges

Several discrepancies between the predictions of the particle cold dark matter paradigm and observations of galaxies and their clustering have arisen:

- The cuspy halo problem: the density distributions of dark matter halos in cold dark matter simulations are much more peaked than what is observed in galaxies by investigating their rotation curves.[14]
- The missing satellites problem: cold dark matter simulations predict much larger numbers of small dwarf galaxies than are observed around galaxies like the Milky Way.[15]
- The disk of satellites problem: dwarf galaxies around the Milky Way and Andromeda galaxies are observed to be orbiting in thin, planar structures whereas the simulations predict that they should be distributed randomly about their parent galaxies.[16]
- Galaxy morphology problem: If galaxies grew hierarchically, then massive galaxies required many mergers. Major mergers indelibly create a classical bulge. On the contrary, about 80% of observed galaxies are bulgeless, and giant pure disc galaxies are commonplace.[17] The bulgeless fraction was nearly constant for 8 billion years.[18]

Some of these problems have proposed solutions, but it remains unclear whether they can be solved without abandoning the CDM paradigm.[19]

12.5 See also

- Fuzzy cold dark matter
- Meta-cold dark matter
- Dark matter
  - Hot dark matter (HDM)
  - Warm dark matter (WDM)
  - Self-interacting dark matter (SIDM)
- Lambda-CDM model
- Modified Newtonian dynamics

12.6 References

[1] Peebles, P. J. E. (December 1982). "Large-scale background temperature and mass fluctuations due to scale-invariant primeval perturbations". *The Astrophysical Journal*. **263**: L1. Bibcode:1982ApJ...263L...1P. doi:10.1086/183911.

[2] "Formation of galaxies in a gravitino-dominated universe". *Physical Review Letters*. **48**: 1636–1639. Bibcode:1982PhRvL..48.1636B. doi:10.1103/PhysRevLett.48.1636.

[3] Blumenthal, George R.; Pagels, Heinz; Primack, Joel R. (2 September 1982). "Galaxy formation by dissipationless particles heavier than neutrinos". *Nature*. **299** (5878): 37–38. Bibcode:1982Natur.299...37B. doi:10.1038/299037a0.

[4] Blumenthal, G. R.; Faber, S. M.; Primack, J. R.; Rees,, M. J. (1984). "Formation of galaxies and large-scale structure with cold dark matter". *Nature*. **311** (517): 517–525. Bibcode:1984Natur.311..517B. doi:10.1038/311517a0.

[5] Battinelli, P.; S. Demers (2005-10-06). "The C star population of DDO 190: 1. Introduction" (PDF). *Astronomy and Astrophysics*. Astronomy & Astrophysics. **447**: 1. Bibcode:2006A&A...447..473B. doi:10.1051/0004-6361:20052829. Archived from the original on 2005-10-06. Retrieved 2012-08-19. Dwarf galaxies play a crucial role in the CDM scenario for galaxy formation, having been suggested to be the natural building blocks from which larger structures are built up by merging processes. In this scenario dwarf galaxies are formed from small-scale density fluctuations in the primeval Universe.

[6] e.g. M. Turner (2010). "Axions 2010 Workshop". U. Florida, Gainesville, USA.

[7] e.g. Pierre Sikivie (2008). "Axion Cosmology". Lect. Notes Phys. 741, 19-50.

[8] Carr, B. J.; et al. (May 2010). "New cosmological constraints on primordial black holes". *Physical Review D*. **81** (10): 104019. Bibcode:2010PhRvD..81j4019C. arXiv:0912.5297. doi:10.1103/PhysRevD.81.104019.

[9] Peter, A. H. G. (2012). "Dark Matter: A Brief Review". arXiv:1201.3942.

[10] Bertone, Gianfranco; Hooper, Dan; Silk, Joseph (January 2005). "Particle dark matter: evidence, candidates and constraints". *Physics Reports*. **405** (5–6): 279–390. Bibcode:2005PhR...405..279B. arXiv:hep-ph/0404175 . doi:10.1016/j.physrep.2004.08.031.

[11] Garrett, Katherine; Dūda, Gintaras. "Dark Matter: A Primer". *Advances in Astronomy*. **2011**: 968283. Bibcode:2011AdAst2011E...8G. arXiv:1006.2483 . doi:10.1155/2011/968283.. p. 3: "MACHOs can only account for a very small percentage of the nonluminous mass in our galaxy, revealing that most dark matter cannot be strongly concentrated or exist in the form of baryonic astrophysical objects. Although microlensing surveys rule out baryonic objects like brown dwarfs, black holes, and neutron stars in our galactic halo, can other forms of baryonic matter make up the bulk of dark matter? The answer, surprisingly, is no..."

[12] Gianfranco Bertone, "The moment of truth for WIMP dark matter," Nature 468, 389–393 (18 November 2010)

[13] Olive, Keith A (2003). "TASI Lectures on Dark Matter". *Physics*. **54**: 21.

[14] Gentile, G.; P., Salucci (2004). "The cored distribution of dark matter in spiral galaxies". *Monthly Notices of the Royal Astronomical Society*. **351**: 903–922. Bibcode:2004MNRAS.351..903G. arXiv:astro-ph/0403154 . doi:10.1111/j.1365-2966.2004.07836.x.

[15] Klypin, Anatoly; Kravtsov, Andrey V.; Valenzuela, Octavio; Prada, Francisco (1999). "Where Are the Missing Galactic Satellites?". *ApJ*. **522**: 82–92. Bibcode:1999ApJ...522...82K. arXiv:astro-ph/9901240 . doi:10.1086/307643.

[16] Marcel Pawlowski et al., "Co-orbiting satellite galaxy structures are still in conflict with the distribution of primordial dwarf galaxies" MNRAS (2014) http://arxiv.org/abs/1406.1799

[17] Kormendy, J.; Drory, N.; Bender, R.; Cornell, M. E. (2010). "Bulgeless Giant Galaxies Challenge Our Picture of Galaxy Formation by Hierarchical Clustering". *The Astrophysical Journal*. **723**: 54–80. Bibcode:2010ApJ...723...54K. arXiv:1009.3015 . doi:10.1088/0004-637X/723/1/54.

[18] Sachdeva, S.; Saha, K. (2016). "Survival of Pure Disk Galaxies over the Last 8 Billion Years". *The Astrophysical Journal Letters*. **820**: L4. Bibcode:2016ApJ...820L...4S. arXiv:1602.08942 . doi:10.3847/2041-8205/820/1/L4.

[19] Kroupa, P.; Famaey, B.; de Boer, Klaas S.; Dabringhausen, Joerg; Pawlowski, Marcel; Boily, Christian; Jerjen, Helmut; Forbes, Duncan; Hensler, Gerhard (2010). "Local-Group tests of dark-matter Concordance Cosmology: Towards a new paradigm for structure formation". *Astronomy and Astrophysics*. **523**: 32–54. Bibcode:2010A&A...523A..32K. arXiv:1006.1647 . doi:10.1051/0004-6361/201014892.

12.7 Further reading

- Bertone, Gianfranco (2010). *Particle Dark Matter: Observations, Models and Searches*. Cambridge University Press. p. 762. ISBN 978-0-521-76368-4.

Chapter 13

Baryon

Not to be confused with Baryonyx.

A **baryon** is a composite subatomic particle made up of three quarks (a **triquark**, as distinct from mesons, which are composed of one quark and one antiquark). Baryons and mesons belong to the hadron family of particles, which are the quark-based particles. The name "baryon" comes from the Greek word for "heavy" (βαρύς, *barys*), because, at the time of their naming, most known elementary particles had lower masses than the baryons.

As quark-based particles, baryons participate in the strong interaction, whereas leptons, which are not quark-based, do not. The most familiar baryons are the protons and neutrons that make up most of the mass of the visible matter in the universe. Electrons (the other major component of the atom) are leptons.

Each baryon has a corresponding antiparticle (antibaryon) where quarks are replaced by their corresponding antiquarks. For example, a proton is made of two up quarks and one down quark; and its corresponding antiparticle, the antiproton, is made of two up antiquarks and one down antiquark.

13.1 Background

Baryons are strongly interacting fermions, that is, they are acted on by the strong nuclear force and are described by Fermi–Dirac statistics, which apply to all particles obeying the Pauli exclusion principle. This is in contrast to the bosons, which do not obey the exclusion principle.

Baryons, along with mesons, are hadrons, meaning they are particles composed of quarks. Quarks have baryon numbers of $B = 1/3$ and antiquarks have baryon number of $B = -1/3$. The term "baryon" usually refers to *triquarks*—baryons made of three quarks ($B = 1/3 + 1/3 + 1/3 = 1$).

Other exotic baryons have been proposed, such as pentaquarks—baryons made of four quarks and one antiquark ($B = 1/3 + 1/3 + 1/3 + 1/3 - 1/3 = 1$),[1][2] but their existence is not generally accepted. The particle physics community as a whole did not view their existence as likely in 2006,[3] and in 2008, considered evidence to be overwhelmingly against the existence of the reported pentaquarks.[4] However, in July 2015, the LHCb experiment observed two resonances consistent with pentaquark states in the Λ0
b → J/ψK−
p decay, with a combined statistical significance of 15σ.[5][6]

In theory, heptaquarks (5 quarks, 2 antiquarks), nonaquarks (6 quarks, 3 antiquarks), etc. could also exist.

13.2 Baryonic matter

Nearly all matter that may be encountered or experienced in everyday life is baryonic matter, which includes atoms of any sort, and provides those with the property of mass. Non-baryonic matter, as implied by the name, is any sort of matter that is not composed primarily of baryons. This might include neutrinos and free electrons, dark matter, such as supersymmetric particles, axions, and black holes.

The very existence of baryons is also a significant issue in cosmology, because it is assumed that the Big Bang produced a state with equal amounts of baryons and antibaryons. The process by which baryons came to outnumber their antiparticles is called baryogenesis.

13.3 Baryogenesis

Main article: Baryogenesis

Experiments are consistent with the number of quarks in the universe being a constant and, to be more specific, the number of baryons being a constant ; in technical language, the total baryon number appears to be *conserved*. Within the prevailing Standard Model of particle physics, the number of baryons may change in multiples of three due to the action of sphalerons, although this is rare and has not been observed under experiment. Some grand unified theories of particle physics also predict that a single proton can decay, changing the baryon number by one; however, this has not yet been observed under experiment. The excess of baryons over antibaryons in the present universe is thought to be due to non-conservation of baryon number in the very early universe, though this is not well understood.

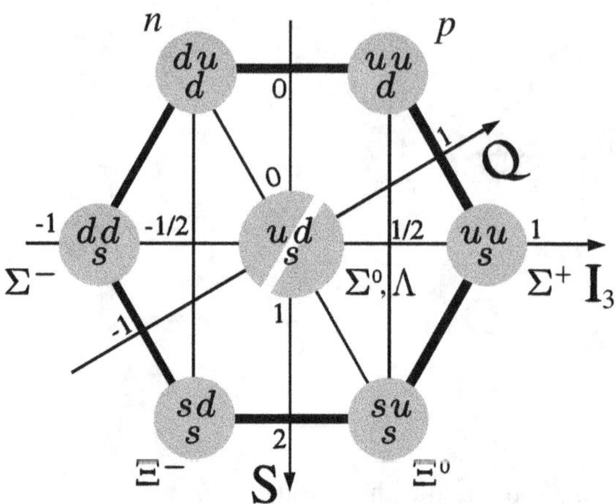

*Combinations of three **u**, **d** or s quarks forming baryons with a spin-1/2 form the* uds baryon octet

13.4 Properties

13.4.1 Isospin and charge

Main article: Isospin

The concept of isospin was first proposed by Werner

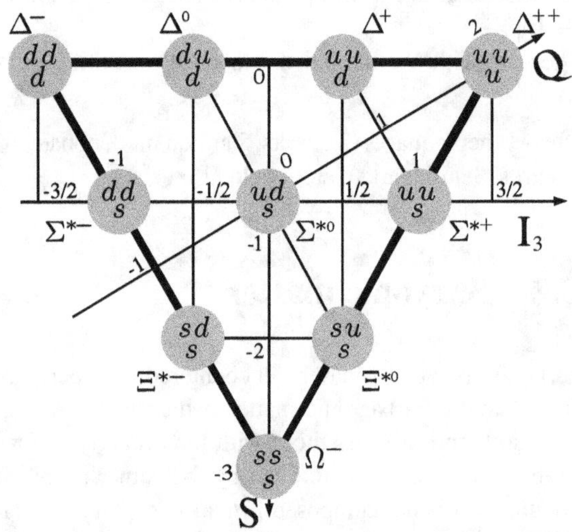

*Combinations of three **u**, **d** or s quarks forming baryons with a spin-3/2 form the* uds baryon decuplet

Heisenberg in 1932 to explain the similarities between protons and neutrons under the strong interaction.[7] Although they had different electric charges, their masses were so similar that physicists believed they were the same particle. The different electric charges were explained as being the result of some unknown excitation similar to spin. This unknown excitation was later dubbed *isospin* by Eugene Wigner in 1937.[8]

This belief lasted until Murray Gell-Mann proposed the quark model in 1964 (containing originally only the u, d, and s quarks).[9] The success of the isospin model is now understood to be the result of the similar masses of the u and d quarks. Since the u and d quarks have similar masses, particles made of the same number then also have similar masses. The exact specific u and d quark composition determines the charge, as u quarks carry charge +2/3 while d quarks carry charge −1/3. For example, the four Deltas all have different charges(Δ++ (uuu), Δ+ (uud), Δ0 (udd), Δ−
(ddd)), but have similar masses (~1,232 MeV/c^2) as they are each made of a combination of three u and d quarks. Under the isospin model, they were considered to be a single particle in different charged states.

The mathematics of isospin was modeled after that of spin. Isospin projections varied in increments of 1 just like those of spin, and to each projection was associated a "charged state". Since the "Delta particle" had four "charged states", it was said to be of isospin I = 3/2. Its "charged states" Δ
++ , Δ+ , Δ0 ,and Δ−
, corresponded to the isospin projections I_3 = +3/2, I_3 = +1/2, I_3 = −1/2, and I_3 = −3/2, respectively. Another example is the "nucleon particle". As there were two nucleon "charged states", it was said to be of isospin 1/2. The positive nucleon
N+
(proton) was identified with I_3 = +1/2 and the neutral

nucleon
N0
(neutron) with I_3 = −1/2.[10] It was later noted that the isospin projections were related to the up and down quark content of particles by the relation:

$$I_3 = \frac{1}{2}[(n_u - n_{\bar{u}}) - (n_d - n_{\bar{d}})],$$

where the n's are the number of up and down quarks and antiquarks.

In the "isospin picture", the four Deltas and the two nucleons were thought to be the different states of two particles. However, in the quark model, Deltas are different states of nucleons (the N^{++} or N^- are forbidden by Pauli's exclusion principle). Isospin, although conveying an inaccurate picture of things, is still used to classify baryons, leading to unnatural and often confusing nomenclature.

13.4.2 Flavour quantum numbers

Main article: Flavour (particle physics) § Flavour quantum numbers

The strangeness flavour quantum number S (not to be confused with spin) was noticed to go up and down along with particle mass. The higher the mass, the lower the strangeness (the more s quarks). Particles could be described with isospin projections (related to charge) and strangeness (mass) (see the uds octet and decuplet figures on the right). As other quarks were discovered, new quantum numbers were made to have similar description of udc and udb octets and decuplets. Since only the u and d mass are similar, this description of particle mass and charge in terms of isospin and flavour quantum numbers works well only for octet and decuplet made of one u, one d, and one other quark, and breaks down for the other octets and decuplets (for example, ucb octet and decuplet). If the quarks all had the same mass, their behaviour would be called *symmetric*, as they would all behave in exactly the same way with respect to the strong interaction. Since quarks do not have the same mass, they do not interact in the same way (exactly like an electron placed in an electric field will accelerate more than a proton placed in the same field because of its lighter mass), and the symmetry is said to be broken.

It was noted that charge (Q) was related to the isospin projection (I_3), the baryon number (B) and flavour quantum numbers (S, C, B', T) by the Gell-Mann–Nishijima formula:[10]

$$Q = I_3 + \frac{1}{2}(B + S + C + B' + T),$$

where S, C, B', and T represent the strangeness, charm, bottomness and topness flavour quantum numbers, respectively. They are related to the number of strange, charm, bottom, and top quarks and antiquark according to the relations:

$$S = -(n_s - n_{\bar{s}}),$$

$$C = +(n_c - n_{\bar{c}}),$$

$$B' = -(n_b - n_{\bar{b}}),$$

$$T = +(n_t - n_{\bar{t}}),$$

meaning that the Gell-Mann–Nishijima formula is equivalent to the expression of charge in terms of quark content:

$$Q = \frac{2}{3}[(n_u - n_{\bar{u}}) + (n_c - n_{\bar{c}}) + (n_t - n_{\bar{t}})] - \frac{1}{3}[(n_d - n_{\bar{d}}) + (n_s - n_{\bar{s}}) + (n_b - n_{\bar{b}})].$$

13.4.3 Spin, orbital angular momentum, and total angular momentum

Main articles: Spin (physics), Angular momentum operator, Quantum numbers, and Clebsch–Gordan coefficients

Spin (quantum number S) is a vector quantity that represents the "intrinsic" angular momentum of a particle. It comes in increments of 1/2 ħ (pronounced "h-bar"). The ħ is often dropped because it is the "fundamental" unit of spin, and it is implied that "spin 1" means "spin 1 ħ". In some systems of natural units, ħ is chosen to be 1, and therefore does not appear anywhere.

Quarks are fermionic particles of spin 1/2 (S = 1/2). Because spin projections vary in increments of 1 (that is 1 ħ), a single quark has a spin vector of length 1/2, and has two spin projections (S_z = +1/2 and S_z = −1/2). Two quarks can have their spins aligned, in which case the two spin vectors add to make a vector of length S = 1 and three spin projections (S_z = +1, S_z = 0, and S_z = −1). If two quarks have unaligned spins, the spin vectors add up to make a vector of length S = 0 and has only one spin projection (S_z = 0), etc. Since baryons are made of three quarks, their spin vectors can add to make a vector of length S = 3/2, which has four spin projections (S_z = +3/2, S_z = +1/2, S_z = −1/2, and S_z = −3/2), or a vector of length S = 1/2 with two spin projections (S_z = +1/2, and S_z = −1/2).[11]

There is another quantity of angular momentum, called the orbital angular momentum, (azimuthal quantum number *L*), that comes in increments of 1 ħ, which represent the angular moment due to quarks orbiting around each other. The total angular momentum (total angular momentum quantum number *J*) of a particle is therefore the combination of intrinsic angular momentum (spin) and orbital angular momentum. It can take any value from $J = |L - S|$ to $J = |L + S|$, in increments of 1.

Particle physicists are most interested in baryons with no orbital angular momentum ($L = 0$), as they correspond to ground states—states of minimal energy. Therefore, the two groups of baryons most studied are the $S = 1/2$; $L = 0$ and $S = 3/2$; $L = 0$, which corresponds to $J = 1/2^+$ and $J = 3/2^+$, respectively, although they are not the only ones. It is also possible to obtain $J = 3/2^+$ particles from $S = 1/2$ and $L = 2$, as well as $S = 3/2$ and $L = 2$. This phenomenon of having multiple particles in the same total angular momentum configuration is called *degeneracy*. How to distinguish between these degenerate baryons is an active area of research in baryon spectroscopy.[12][13]

13.4.4 Parity

Main article: Parity (physics)

If the universe were reflected in a mirror, most of the laws of physics would be identical—things would behave the same way regardless of what we call "left" and what we call "right". This concept of mirror reflection is called *intrinsic parity* or *parity* (*P*). Gravity, the electromagnetic force, and the strong interaction all behave in the same way regardless of whether or not the universe is reflected in a mirror, and thus are said to conserve parity (P-symmetry). However, the weak interaction *does* distinguish "left" from "right", a phenomenon called parity violation (P-violation).

Based on this, one might think that, if the wavefunction for each particle (in more precise terms, the quantum field for each particle type) were simultaneously mirror-reversed, then the new set of wavefunctions would perfectly satisfy the laws of physics (apart from the weak interaction). It turns out that this is not quite true: In order for the equations to be satisfied, the wavefunctions of certain types of particles have to be multiplied by −1, in addition to being mirror-reversed. Such particle types are said to have *negative* or *odd* parity ($P = -1$, or alternatively $P = -$), while the other particles are said to have *positive* or *even* parity ($P = +1$, or alternatively $P = +$).

For baryons, the parity is related to the orbital angular momentum by the relation:[14]

$$P = (-1)^L.$$

As a consequence, baryons with no orbital angular momentum ($L = 0$) all have even parity ($P = +$).

13.5 Nomenclature

Baryons are classified into groups according to their isospin (*I*) values and quark (*q*) content. There are six groups of baryons—nucleon(N),Delta(Δ),Lambda(Λ),Sigma(Σ),Xi(Ξ),andOmega(Ω).The rules for classifi cation are defined by theParticle Data Group.
These rules consider the up(u),down(d)and strange(s)quarks to be *light* and the charm(c),bottom(b), and top(t)quarks to be*heavy*.
The rules cover all the particles that can be made from three of each of the six quarks, even
though baryons made of top quarks are not expected to exist because of thetop quark's short lifetime. The rules do not cover pentaquarks.[15]

- Baryons with three
  u and/or d quarks are N
  's ($I = 1/2$) or
  Δ
  's ($I = 3/2$).

- Baryons with two
u
and/or
d
quarks are
Λ
's (I = 0) or
Σ
's (I = 1). If the third quark is heavy, its identity is given by a subscript.
- Baryons with one
u
or
d
quark are
Ξ
's (I = 1/2). One or two subscripts are used if one or both of the remaining quarks are heavy.
- Baryons with no
u
or
d
quarks are
Ω
's (I = 0), and subscripts indicate any heavy quark content.
- Baryons that decay strongly have their masses as part of their names. For example, Σ^0 does not decay strongly, but $\Delta^{++}(1232)$ does.

It is also a widespread (but not universal) practice to follow some additional rules when distinguishing between some states that would otherwise have the same symbol.[10]

- Baryons in total angular momentum J = 3/2 configuration that have the same symbols as their J = 1/2 counterparts are denoted by an asterisk (*).
- Two baryons can be made of three different quarks in J = 1/2 configuration. In this case, a prime (′) is used to distinguish between them.
    - *Exception*: When two of the three quarks are one up and one down quark, one baryon is dubbed Λ while the other is dubbed Σ.

Quarks carry charge, so knowing the charge of a particle indirectly gives the quark content. For example, the rules above say that a
Λ+
c contains a c quark and some combination of two u and/or d quarks. The c quark has a charge of (Q = +2/3), therefore the other two must be a u quark (Q = +2/3), and a d quark (Q = −1/3) to have the correct total charge (Q = +1).

13.6 See also

- Eightfold way
- List of baryons
- Meson
- Timeline of particle discoveries

13.7 Notes

[1] H. Muir (2003)

[2] K. Carter (2003)

[3] W.-M. Yao *et al.* (2006): Particle listings – Θ^+

[4] C. Amsler *et al.* (2008): Pentaquarks

[5] LHCb (14 July 2015). "Observation of particles composed of five quarks, pentaquark-charmonium states, seen in Λ0 b → J/ψpK⁻ decays.". CERN. Retrieved 2015-07-14.

[6] R. Aaij et al. (LHCb collaboration) (2015). "Observation of J/ψp resonances consistent with pentaquark states in Λ0 b→J/ψK⁻p decays". *Physical Review Letters.* **115** (7). Bibcode:2015PhRvL.115g2001A. arXiv:1507.03414. doi:10.1103/PhysRevLett.115.072001.

[7] W. Heisenberg (1932)

[8] E. Wigner (1937)

[9] M. Gell-Mann (1964)

[10] S.S.M. Wong (1998a)

[11] R. Shankar (1994)

[12] H. Garcilazo *et al.* (2007)

[13] D.M. Manley (2005)

[14] S.S.M. Wong (1998b)

[15] C. Amsler *et al.* (2008): Naming scheme for hadrons

13.8 References

- C. Amsler *et al.* (Particle Data Group) (2008). "Review of Particle Physics". *Physics Letters B.* **667** (1): 1–1340. Bibcode:2008PhLB..667....1A. doi:10.1016/j.physletb.2008.07.018.
- H. Garcilazo; J. Vijande & A. Valcarce (2007). "Faddeev study of heavy-baryon spectroscopy". *Journal of Physics G.* **34** (5): 961–976. doi:10.1088/0954-3899/34/5/014.

- K. Carter (2006). "The rise and fall of the pentaquark". Fermilab and SLAC. Retrieved 2008-05-27.
- W.-M. Yao *et al.*(Particle Data Group) (2006). "Review of Particle Physics". *Journal of Physics G*. **33**: 1–1232. Bibcode:2006JPhG...33....1Y. arXiv:astro-ph/0601168 ට. doi:10.1088/0954-3899/33/1/001.
- D.M. Manley (2005). "Status of baryon spectroscopy". *Journal of Physics: Conference Series*. **5**: 230–237. Bibcode:2005JPhCS...9..230M. doi:10.1088/1742-6596/9/1/043.
- H. Muir (2003). "Pentaquark discovery confounds sceptics". New Scientist. Retrieved 2008-05-27.
- S.S.M. Wong (1998a). "Chapter 2—Nucleon Structure". *Introductory Nuclear Physics* (2nd ed.). New York (NY): John Wiley & Sons. pp. 21–56. ISBN 0-471-23973-9.
- S.S.M. Wong (1998b). "Chapter 3—The Deuteron". *Introductory Nuclear Physics* (2nd ed.). New York (NY): John Wiley & Sons. pp. 57–104. ISBN 0-471-23973-9.
- R. Shankar (1994). *Principles of Quantum Mechanics* (2nd ed.). New York (NY): Plenum Press. ISBN 0-306-44790-8.
- E. Wigner (1937). "On the Consequences of the Symmetry of the Nuclear Hamiltonian on the Spectroscopy of Nuclei". *Physical Review*. **51** (2): 106–119. Bibcode:1937PhRv...51..106W. doi:10.1103/PhysRev.51.106.
- M. Gell-Mann (1964). "A Schematic of Baryons and Mesons". *Physics Letters*. **8** (3): 214–215. Bibcode:1964PhL.....8..214G. doi:10.1016/S0031-9163(64)92001-3.
- W. Heisenberg (1932). "Über den Bau der Atomkerne I". *Zeitschrift für Physik* (in German). **77**: 1–11. Bibcode:1932ZPhy...77....1H. doi:10.1007/BF01342433.
- W. Heisenberg (1932). "Über den Bau der Atomkerne II". *Zeitschrift für Physik* (in German). **78** (3–4): 156–164. Bibcode:1932ZPhy...78..156H. doi:10.1007/BF01337585.
- W. Heisenberg (1932). "Über den Bau der Atomkerne III". *Zeitschrift für Physik* (in German). **80** (9–10): 587–596. Bibcode:1933ZPhy...80..587H. doi:10.1007/BF01335696.

13.9 External links

- Particle Data Group—Review of Particle Physics (2008).
- Georgia State University—HyperPhysics
- Baryons made thinkable, an interactive visualisation allowing physical properties to be compared

Chapter 14

Equation of state (cosmology)

In cosmology, the **equation of state** of a perfect fluid is characterized by a dimensionless number w , equal to the ratio of its pressure p to its energy density ρ :

$$w = \frac{p}{\rho}$$

It is closely related to the thermodynamic equation of state and ideal gas law.

14.1 The equation

The perfect gas equation of state may be written as

$$p = \rho_m RT = \rho_m C^2$$

where ρ_m is the mass density, R is the particular gas constant, T is the temperature and $C = \sqrt{RT}$ is a characteristic thermal speed of the molecules. Thus

$$w = \frac{p}{\rho} = \frac{\rho_m C^2}{\rho_m c^2} = \frac{C^2}{c^2} \approx 0$$

where $\rho = \rho_m c^2$ and $C \ll c$ for a "cold" gas, c = speed of light.

14.1.1 FLRW equations and the equation of state

The equation of state may be used in Friedmann–Lemaître–Robertson–Walker equations to describe the evolution of an isotropic universe filled with a perfect fluid. If a is the scale factor then

$$\rho \propto a^{-3(1+w)}.$$

If the fluid is the dominant form of matter in a flat universe, then

$$a \propto t^{\frac{2}{3(1+w)}},$$

where t is the proper time.

In general the Friedmann acceleration equation is

$$3\frac{\ddot{a}}{a} = \Lambda - 4\pi G(\rho + 3p)$$

where Λ is the cosmological constant and G is Newton's constant, and $\ddot{a}$ is the second proper time derivative of the scale factor.

If we define (what might be called "effective") energy density and pressure as

$$\rho' \equiv \rho + \frac{\Lambda}{8\pi G}$$

$$p' \equiv p - \frac{\Lambda}{8\pi G}$$

and

$$p' = w'\rho'$$

the acceleration equation may be written as

$$\frac{\ddot{a}}{a} = -\frac{4}{3}\pi G\left(\rho' + 3p'\right) = -\frac{4}{3}\pi G(1 + 3w')\rho'$$

14.1.2 Non-relativistic matter

The equation of state of ordinary non-relativistic matter (e.g. cold dust) is $w = 0$, which means that it is diluted as $\rho \propto a^{-3} = V^{-1}$, where V is the volume. This means that the energy density red-shifts as the volume, which is natural for ordinary non-relativistic matter.

14.1.3 Ultra-relativistic matter

The equation of state of ultra-relativistic matter (e.g. radiation, but also matter in the very early universe) is $w = 1/3$ which means that it is diluted as $\rho \propto a^{-4}$. In an expanding universe, the energy density decreases more quickly than the volume expansion, because radiation has momentum and, by the de Broglie hypothesis a wavelength, which is red-shifted.

14.1.4 Acceleration of cosmic inflation

Cosmic inflation and the accelerated expansion of the universe can be characterized by the equation of state of dark energy. In the simplest case, the equation of state of the cosmological constant is $w = -1$. In this case, the above expression for the scale factor is not valid and $a \propto e^{Ht}$, where the constant H is the Hubble parameter. More generally, the expansion of the universe is accelerating for any equation of state $w < -1/3$. The accelerated expansion of the Universe was indeed observed.[1] According to observations, the value of equation of state of cosmological constant is near -1.

Hypothetical phantom energy would have an equation of state $w < -1$, and would cause a Big Rip. Using the existing data, it is still impossible to distinguish between phantom $w < -1$ and non-phantom $w \geq -1$.

14.1.5 Fluids

In an expanding universe, fluids with larger equations of state disappear more quickly than those with smaller equations of state. This is the origin of the flatness and monopole problems of the big bang: curvature has $w = -1/3$ and monopoles have $w = 0$, so if they were around at the time of the early big bang, they should still be visible today. These problems are solved by cosmic inflation which has $w \approx -1$. Measuring the equation of state of dark energy is one of the largest efforts of observational cosmology. By accurately measuring w , it is hoped that the cosmological constant could be distinguished from quintessence which has $w \neq -1$.

14.1.6 Scalar modeling

A scalar field ϕ can be viewed as a sort of perfect fluid with equation of state

$$w = \frac{\frac{1}{2}\dot{\phi}^2 - V(\phi)}{\frac{1}{2}\dot{\phi}^2 + V(\phi)},$$

where $\dot{\phi}$ is the time-derivative of ϕ and $V(\phi)$ is the potential energy. A free ($V = 0$) scalar field has $w = 1$, and one with vanishing kinetic energy is equivalent to a cosmological constant: $w = -1$. Any equation of state in between, but not crossing the $w = -1$ barrier known as the Phantom Divide Line (PDL),[2] is achievable, which makes scalar fields useful models for many phenomena in cosmology.

14.2 Notes

[1] Hogan, Jenny. "Welcome to the Dark Side." Nature 448.7151 (2007): 240-245. http://www.nature.com/nature/journal/v448/n7151/full/448240a.html

[2] Vikman, Alexander (2005). "Can dark energy evolve to the Phantom?". *Phys. Rev. D.* **71**: 023515. Bibcode:2005PhRvD..71b3515V. arXiv:astro-ph/0407107 . doi:10.1103/PhysRevD.71.023515.

Chapter 15

Lambda-CDM model

"Standard cosmological model" redirects here. For other uses, see Standard model (disambiguation).

The **ΛCDM** (**Lambda cold dark matter**) or **Lambda-CDM** model is a parametrization of the Big Bang cosmological model in which the universe contains a cosmological constant, denoted by Lambda (Greek **Λ**), associated with dark energy, and cold dark matter (abbreviated **CDM**). It is frequently referred to as the **standard model** of Big Bang cosmology because it is the simplest model that provides a reasonably good account of the following properties of the cosmos:

- the existence and structure of the cosmic microwave background
- the large-scale structure in the distribution of galaxies
- the abundances of hydrogen (including deuterium), helium, and lithium
- the accelerating expansion of the universe observed in the light from distant galaxies and supernovae

The model assumes that general relativity is the correct theory of gravity on cosmological scales. It emerged in the late 1990s as a **concordance cosmology**, after a period of time when disparate observed properties of the universe appeared mutually inconsistent, and there was no consensus on the makeup of the energy density of the universe.

The ΛCDM model can be extended by adding cosmological inflation, quintessence and other elements that are current areas of speculation and research in cosmology.

Some alternative models challenge the assumptions of the ΛCDM model. Examples of these are modified Newtonian dynamics, modified gravity and theories of large-scale variations in the matter density of the universe.[1]

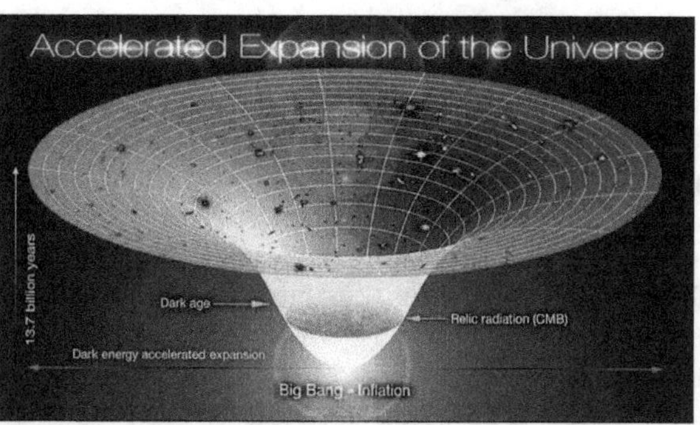

Lambda-CDM, accelerated expansion of the universe. The timeline in this schematic diagram extends from the Big Bang/inflation era 13.7 Gyr ago to the present cosmological time.

15.1 Overview

Most modern cosmological models are based on the cosmological principle, which states that our observational location in the universe is not unusual or special; on a large-enough scale, the universe looks the same in all directions (isotropy) and from every location (homogeneity).[2]

The model includes an expansion of metric space that is well documented both as the red shift of prominent spectral absorption or emission lines in the light from distant galaxies and as the time dilation in the light decay of supernova luminosity curves. Both effects are attributed to a Doppler shift in electromagnetic radiation as it travels across expanding space. Although this expansion increases the distance between objects that are not under shared gravitational influence, it does not increase the size of the objects (e.g. galaxies) in space. It also allows for distant galaxies to recede from each other at speeds greater than the speed of light; local expansion is less than the speed of light, but expansion summed across great distances can collectively exceed the speed of light.

The letter Λ (lambda) represents the cosmological constant, which is currently associated with a vacuum energy or dark energy in empty space that is used to explain the contemporary accelerating expansion of space against the attractive effects of gravity. A cosmological constant has negative pressure, $p = -\rho c^2$, which contributes to the stress-energy tensor that, according to the general theory of relativity, causes accelerating expansion. The fraction of the total energy density of our (flat or almost flat) universe that is dark energy, Ω_Λ , is currently [2015] estimated to be 0.692 ± 0.012, or even 0.6911 ± 0.0062 based on Planck satellite data.[3]

Cold dark matter is a form of matter introduced in order to account for gravitational effects observed in very large-scale structures (the "flat" rotation curves of galaxies; the gravitational lensing of light by galaxy clusters; and enhanced clustering of galaxies) that cannot be accounted for by the quantity of observed matter. Dark matter is described as being cold (i.e. its velocity is far less than the speed of light at the epoch of radiation-matter equality); non-baryonic (i.e. consisting of matter other than protons and neutrons); dissipationless (i.e. cannot cool by radiating photons); and collisionless (i.e. the dark matter particles interact with each other and other particles only through gravity and possibly the weak force). The dark matter component is currently [2013] estimated to constitute about 26.8% of the mass-energy density of the universe.

The remaining 4.9% [2013] comprises all ordinary matter observed as atoms, chemical elements, gas and plasma, the stuff of which visible planets, stars and galaxies are made. As a matter of fact, the great majority of ordinary matter in the universe is unseen, since visible stars and gas inside galaxies and clusters account for less than 10 per cent of the ordinary matter contribution to the mass-energy density of the universe.[4]

Also, the energy density includes a very small fraction (~ 0.01%) in cosmic microwave background radiation, and not more than 0.5% in relic neutrinos. Although very small today, these were much more important in the distant past, dominating the matter at redshift > 3200.

The model includes a single originating event, the "Big Bang", which was not an explosion but the abrupt appearance of expanding space-time containing radiation at temperatures of around 10^{15} K. This was immediately (within 10^{-29} seconds) followed by an exponential expansion of space by a scale multiplier of 10^{27} or more, known as cosmic inflation. The early universe remained hot (above 10,000 K) for several hundred thousand years, a state that is detectable as a residual cosmic microwave background, or CMB, a very low energy radiation emanating from all parts of the sky. The "Big Bang" scenario, with cosmic inflation and standard particle physics, is the only current cosmological model consistent with the observed continuing expansion of space, the observed distribution of lighter elements in the universe (hydrogen, helium, and lithium), and the spatial texture of minute irregularities (anisotropies) in the CMB radiation. Cosmic inflation also addresses the "horizon problem" in the CMB; indeed, it seems likely that the universe is larger than the observable particle horizon.

The model uses the FLRW metric, the Friedmann equations and the cosmological equations of state to describe the observable universe from right after the inflationary epoch to present and future.

15.2 Cosmic expansion history

The expansion of the universe is parametrized by a dimensionless scale factor $a = a(t)$ (with time t counted from the birth of the universe), defined relative to the present day, so $a_0 = a(t_0) = 1$; the usual convention in cosmology is that subscript 0 denotes present-day values, so t_0 is the current age of the universe. The scale factor is related to the observed redshift[5] z of the light emitted at time $t_{\rm em}$ by

$$\frac{1}{a(t_{\rm em})} = 1 + z.$$

The expansion rate is described by the time-dependent Hubble parameter, $H(t)$, defined as

$$H(t) \equiv \frac{\dot{a}}{a},$$

where $\dot{a}$ is the time-derivative of the scale factor. The first Friedmann equation gives the expansion rate in terms of the matter+radiation density ρ , the curvature k , the cosmological constant Λ , and the gravitational constant G .[5]

$$H^2 = \left(\frac{\dot{a}}{a}\right)^2 = \frac{8\pi G}{3}\rho - \frac{kc^2}{a^2} + \frac{\Lambda c^2}{3}.$$

A critical density $\rho_{\rm crit}$ is the present-day density, which gives zero curvature k , assuming the cosmological constant Λ is zero, regardless of its actual value. Substituting these conditions to the Friedmann equation gives

$$\rho_{\rm crit} = \frac{3H_0^2}{8\pi G} = 1.878\ 47(23) \times 10^{-26}\ h^2\ {\rm kg\ m^{-3}},$$ [6]

where $h \equiv H_0/(100\ {\rm km\ s^{-1}\ Mpc^{-1}})$ is the reduced Hubble constant. If the cosmological constant were actually zero, the critical density would also be the dividing line between eventual recollapse of the universe to a Big Crunch, or unlimited expansion. Because it is not, the universe is predicted to expand forever regardless of whether the total density is slightly above or below the critical density, though this may not apply if the cosmological constant is time-dependent.

It is standard to define the present-day **density parameter** Ω_x for various species as the dimensionless ratio

$$\Omega_x \equiv \frac{\rho_x(t=t_0)}{\rho_{\text{crit}}} = \frac{8\pi G \rho_x(t=t_0)}{3H_0^2}$$

where the subscript x is one of “b” for baryons, “c” for cold dark matter, “rad” for radiation (photons plus relativistic neutrinos), and “DE” or "Λ" for dark energy.

Since the densities of various species scale as different powers of a , e.g. a^{-3} for matter etc., the Friedmann equation can be conveniently rewritten in terms of the various density parameters as

$$H(a) \equiv \frac{\dot{a}}{a} = H_0 \quad \text{X}$$

$$\sqrt{(\Omega_c + \Omega_b)a^{-3} + \Omega_{rad}a^{-4} + \Omega_k a^{-2} + \Omega_{DE}a^{-3(1+w)}}$$

where w is the equation of state of dark energy, and assuming negligible neutrino mass (significant neutrino mass requires a more complex equation). The various Ω parameters add up to 1 by construction. In the general case this is integrated by computer to give the expansion history a(t) and also observable distance-redshift relations for any chosen values of the cosmological parameters, which can then be compared with observations such as supernovae and baryon acoustic oscillations.

In the minimal 6-parameter Lambda-CDM model, it is assumed that curvature Ω_k is zero and $w = -1$, so this simplifies to

$$H(a) = H_0\sqrt{\Omega_m a^{-3} + \Omega_{rad}a^{-4} + \Omega_\Lambda}$$

Observations show that the radiation density is very small today, $\Omega_{rad} \sim 10^{-4}$; if this term is neglected the above has an analytic solution[7]

$$a(t) = (\Omega_m/\Omega_\Lambda)^{1/3}\sinh^{2/3}(t/t_\Lambda)$$

where $t_\Lambda \equiv 2/(3H_0\sqrt{\Omega_\Lambda})$; this is fairly accurate for a > 0.01 or t > 10 Myr. Solving for $a(t) = 1$ gives the present age of the universe t_0 in terms of the other parameters.

It follows that the transition from decelerating to accelerating expansion (the second derivative $\ddot{a}$ crossing zero) occurred when

$$a = (\Omega_m/2\Omega_\Lambda)^{1/3}$$

which evaluates to a ~ 0.6 or z ~ 0.66 for the Planck best-fit parameters.

15.3 Historical development

The discovery of the Cosmic Microwave Background (CMB) in 1965 confirmed a key prediction of the Big Bang cosmology. From that point on, it was generally accepted that the universe started in a hot, dense state and has been expanding over time. The rate of expansion depends on the types of matter and energy present in the universe, and in particular, whether the total density is above or below the so-called critical density. During the 1970s, most attention focused on pure-baryonic models, but there were serious challenges explaining the formation of galaxies, given the small anisotropies in the CMB (upper limits at that time). In the early 1980s, it was realized that this could be resolved if cold dark matter dominated over the baryons, and the theory of cosmic inflation motivated models with critical density. During the 1980s, most research focused on cold dark matter with critical density in matter, around 95% CDM and 5% baryons: these showed success at forming galaxies and clusters of galaxies, but problems remained; notably, the model required a Hubble constant lower than preferred by observations, and observations around 1988-1990 showed more large-scale galaxy clustering than predicted. These difficulties sharpened with the discovery of CMB anisotropy by COBE in 1992, and several modified CDM models, including ΛCDM and mixed cold+hot dark matter, came under active consideration through the mid-1990s. The ΛCDM model then became the leading model following the observations of accelerating expansion in 1998, and was quickly supported by other observations: in 2000, the BOOMERanG microwave background experiment measured the total (matter+energy) density to be close to 100% of critical, whereas in 2001 the 2dFGRS galaxy redshift survey measured the matter density to be near 25%; the large difference between these values supports a positive Λ or dark energy. Much more precise spacecraft measurements of the microwave background from WMAP in 2003 – 2010 and Planck in 2013 - 2015 have continued to support the model and pin down the parameter values, most of which are now constrained below 1 percent uncertainty.

There is currently active research into many aspects of the ΛCDM model, both to refine the parameters and possibly detect deviations. In addition, ΛCDM has no explicit physical theory for the origin or physical nature of dark matter or dark energy; the nearly scale-invariant spectrum of the CMB perturbations, and their image across the celestial sphere, are believed to result from very small thermal and acoustic irregularities at the point of recombination. A large majority of astronomers and astrophysicists support the ΛCDM model or close relatives of it, but Milgrom, McGaugh, and Kroupa are leading critics, attacking the dark matter portions of the theory from the perspective of galaxy formation models and supporting the alternative MOND theory, which requires a modification of the Einstein field equations and the Friedmann equations as seen in proposals such as MOG theory or TeVeS theory. Other proposals by theoretical astrophysicists of cosmological alternatives to Einstein's general relativity that attempt to account for dark energy or dark matter include f(R) gravity, scalar–tensor theories such as galileon theories, brane cosmologies, the DGP model, and massive gravity and its extensions such as bimetric gravity.

15.4 Successes

In addition to explaining pre-2000 observations, the model has made a number of successful predictions: notably the existence of the baryon acoustic oscillation feature, discovered in 2005 in the predicted location; and the statistics of weak gravitational lensing, first observed in 2000 by several teams. The polarization of the CMB, discovered in 2002 by DASI [8] is now a dramatic success: in the 2015 Planck data release,[9] there are seven observed peaks in the temperature (TT) power spectrum, six peaks in the temperature-polarization (TE) cross spectrum, and five peaks in the polarization (EE) spectrum. The six free parameters can be well constrained by the TT spectrum alone, and then the TE and EE spectra can be predicted theoretically to few-percent precision with no further adjustments allowed: comparison of theory and observations shows an excellent match.

15.5 Challenges

Extensive searches for dark matter particles have so far shown no well-agreed detection; the dark energy may be almost impossible to detect in a laboratory, and its value is unnaturally small compared to naive theoretical predictions.

Comparison of the model with observations is very successful on large scales (larger than galaxies, up to the observable horizon), but may have some problems on sub-galaxy scales, possibly predicting too many dwarf galaxies and too much dark matter in the innermost regions of galaxies. These small scales are harder to resolve in computer simulations, so it is not yet clear whether the problem is the simulations, non-standard properties of dark matter, or a more radical error in the model.

It has been argued that the ΛCDM model is built upon a foundation of conventionalist stratagems, rendering it unfalsifiable in the sense defined by Karl Popper.[10]

15.6 Parameters

The simple ΛCDM model is based on six parameters: physical baryon density parameter; physical dark matter density parameter; the age of the universe; scalar spectral index; curvature fluctuation amplitude; and reionization optical depth.[17] In accordance with Occam's razor, six is the smallest number of parameters needed to give an acceptable fit to current observations; other possible parameters are fixed at "natural" values, e.g. total density parameter = 1.00, dark energy equation of state = −1. (See below for extended models that allow these to vary.)

The values of these six parameters are mostly not predicted by current theory (though, ideally, they may be related by a future "Theory of Everything"), except that most versions of cosmic inflation predict the scalar spectral index should be slightly smaller than 1, consistent with the estimated value 0.96. The parameter values, and uncertainties, are estimated using large computer searches to locate the region of parameter space providing an acceptable match to cosmological observations. From these six parameters, the other model values, such as the Hubble constant and the dark energy density, can be readily calculated.

Commonly, the set of observations fitted includes the cosmic microwave background anisotropy, the brightness/redshift relation for supernovae, and large-scale galaxy clustering including the baryon acoustic oscillation feature. Other observations, such as the Hubble constant, the abundance of galaxy clusters, weak gravitational lensing and globular cluster ages, are generally consistent with these, providing a check of the model, but are less precisely measured at present.

Parameter values listed below are from the Planck Collaboration Cosmological parameters 68% confidence limits for the base ΛCDM model from Planck CMB power spectra, in combination with lensing reconstruction and external data (BAO+JLA+H_0).[12] See also Planck (spacecraft).

[1] The "physical baryon density parameter" Ω_b h^2 is the "baryon density parameter" Ω_b multiplied by the square of the reduced Hubble constant h = H_0 / (100 km s^{-1} Mpc^{-1}).[13][14] Likewise for the difference between "physical dark matter density parameter" and "dark matter density parameter".

[2] A density $\rho_x = \Omega_x \rho_{crit}$ is expressed in terms of the criti-

cal density ρ_{crit}, which is the total density of matter/energy needed for the universe to be spatially flat. Measurements indicate that the actual total density ρ_{tot} is very close if not equal to this value, see below.

[3] This is the minimal value allowed by solar and terrestrial neutrino oscillation experiments.

[4] from the Standard Model of particle physics

[5] Calculated from $\Omega_b h^2$ and h = H_0 / (100 km s^{-1} Mpc^{-1}).

[6] Calculated from $\Omega_c h^2$ and h = H_0 / (100 km s^{-1} Mpc^{-1}).

[7] Calculated from h = H_0 / (100 km s^{-1} Mpc^{-1}) per ρ_{crit} = 1.87847×10^{-26} h^2 kg m^{-3}.[6]

15.7 Extended models

Extended models allow one or more of the "fixed" parameters above to vary, in addition to the basic six; so these models join smoothly to the basic six-parameter model in the limit that the additional parameter(s) approach the default values. For example, possible extensions of the simplest ΛCDM model allow for spatial curvature (Ω_{tot} may be different from 1); or quintessence rather than a cosmological constant where the equation of state of dark energy is allowed to differ from −1. Cosmic inflation predicts tensor fluctuations (gravitational waves). Their amplitude is parameterized by the tensor-to-scalar ratio (denoted r), which is determined by the unknown energy scale of inflation. Other modifications allow hot dark matter in the form of neutrinos more massive than the minimal value, or a running spectral index; the latter is generally not favoured by simple cosmic inflation models.

Allowing additional variable parameter(s) will generally *increase* the uncertainties in the standard six parameters quoted above, and may also shift the central values slightly. The Table below shows results for each of the possible "6+1" scenarios with one additional variable parameter; this indicates that, as of 2015, there is no convincing evidence that any additional parameter is different from its default value.

Some researchers have suggested that there is a running spectral index, but no statistically significant study has revealed one. Theoretical expectations suggest that the tensor-to-scalar ratio r should be between 0 and 0.3, and the latest results are now within those limits.

15.8 See also

- Bolshoi Cosmological Simulation
- Dark matter
- Galaxy formation and evolution
- Illustris project
- List of cosmological computation software
- Millennium Run
- Nature timeline
- WIMPs
- The ΛCDM model is also known as the standard model of cosmology, but is not related to the Standard Model of particle physics.

15.9 References

[1] P. Kroupa, B. Famaey, K.S. de Boer, J. Dabringhausen, M. Pawlowski, C.M. Boily, H. Jerjen, D. Forbes, G. Hensler, M. Metz, "Local-Group tests of dark-matter concordance cosmology. Towards a new paradigm for structure formation" A&A 523, 32 (2010).

[2] Andrew Liddle. *An Introduction to Modern Cosmology (2nd ed.)*. London: Wiley, 2003.

[3] Camille M. Carlisle, *Planck Upholds Standard Cosmology*, Sky & Telescope, February 10, 2015

[4] Persic, Massimo; Salucci, Paolo (1992-09-01). "The baryon content of the Universe". *Monthly Notices of the Royal Astronomical Society*. **258** (1): 14P–18P. Bibcode:1992MNRAS.258P..14P. ISSN 0035-8711. arXiv:astro-ph/0502178. doi:10.1093/mnras/258.1.14P.

[5] Dodelson, Scott (2008). *Modern cosmology* (4. [print.]. ed.). San Diego, CA [etc.]: Academic Press. ISBN 978-0122191411.

[6] K.A. Olive et al. (Particle Data Group) (2015). "The Review of Particle Physics. 2. Astrophysical constants and parameters" (PDF). *Particle Data Group*. Archived from the original on 3 December 2015. Retrieved 10 January 2016. External link in |website= (help)

[7] Frieman, Joshua A.; Turner, Michael S.; Huterer, Dragan (September 2008). "Dark Energy and the Accelerating Universe". *Annual Review of Astronomy and Astrophysics*. **46** (1): 385–432. Bibcode:2008ARA&A..46..385F. arXiv:0803.0982. doi:10.1146/annurev.astro.46.060407.145243.

[8] Kovac, J. M.; Leitch, E. M.; Pryke, C.; Carlstrom, J. E.; Halverson, N. W.; Holzapfel, W. L. (19 December 2002). "Detection of polarization in the cosmic microwave background using DASI". *Nature*. **420** (6917): 772–787. Bibcode:2002Natur.420..772K. PMID 12490941. arXiv:astro-ph/0209478. doi:10.1038/nature01269.

[9] Collaboration, Planck; Ade, P. A. R.; Aghanim, N.; Arnaud, M.; Ashdown, M.; Aumont, J.; Baccigalupi, C.; Banday, A. J.; Barreiro, R. B.; Bartlett, J. G.; Bartolo, N.; Battaner, E.; Battye, R.; Benabed, K.; Benoit, A.; Benoit-Levy, A.; Bernard, J. -P.; Bersanelli, M.; Bielewicz, P.; Bonaldi, A.; Bonavera, L.; Bond, J. R.; Borrill, J.; Bouchet, F. R.; Boulanger, F.; Bucher, M.; Burigana, C.; Butler, R. C.; Calabrese, E.; et al. (2015). "Planck 2015 Results. XIII. Cosmological Parameters". arXiv:1502.01589 [astro-ph.CO].

[10] Merritt, David "Cosmology and Convention", *Studies In History and Philosophy of Science Part B: Studies In History and Philosophy of Modern Physics*, **57**(1):41-52, February 2017.

[11] Planck 2015,[12] p. 32, table 4, last column.

[12] Planck Collaboration. "Planck 2015 results. XIII. Cosmological parameters". arXiv:1502.01589.

[13] Appendix A of the LSST Science Book Version 2.0

[14] p. 7 of Findings of the Joint Dark Energy Mission Figure of Merit Science Working Group

[15] Table 8 on p. 39 of Jarosik, N. et al. (WMAP Collaboration). "Seven-Year Wilkinson Microwave Anisotropy Probe (WMAP) Observations: Sky Maps, Systematic Errors, and Basic Results" (PDF). nasa.gov. Retrieved 2010-12-04. (from NASA's WMAP Documents page)

[16] Planck Collaboration; Adam, R.; Aghanim, N.; Ashdown, M.; Aumont, J.; Baccigalupi, C.; Ballardini, M.; Banday, A. J.; Barreiro, R. B. (2016-05-11). "Planck intermediate results. XLVII. Planck constraints on reionization history". arXiv:1605.03507 [astro-ph].

[17] Spergel, D. N. (2015). "The dark side of the cosmology: dark matter and dark energy". *Science*. **347** (6226): 1100–1102. Bibcode:2015Sci...347.1100S. PMID 25745164. doi:10.1126/science.aaa0980.

15.10 Further reading

- Rebolo, R.; et al. (2004). "Cosmological parameter estimation using Very Small Array data out to ℓ= 1500". *Monthly Notices of the Royal Astronomical Society*. **353** (3): 747–759. Bibcode:2004MNRAS.353..747R. arXiv:astro-ph/0402466. doi:10.1111/j.1365-2966.2004.08102.x.
- Ostriker, J. P.; Steinhardt, P. J. (1995). "Cosmic Concordance". arXiv:astro-ph/9505066.
- Ostriker, Jeremiah P.; Mitton, Simon (2013). *Heart of Darkness: Unraveling the mysteries of the invisible universe*. Princeton, NJ: Princeton University Press. ISBN 978-0-691-13430-7.

15.11 External links

- Bolshoi Simulation
- Cosmology tutorial/NedWright
- Millennium Simulation
- WMAP estimated cosmological parameters/Latest Summary

Chapter 16

Cosmological constant

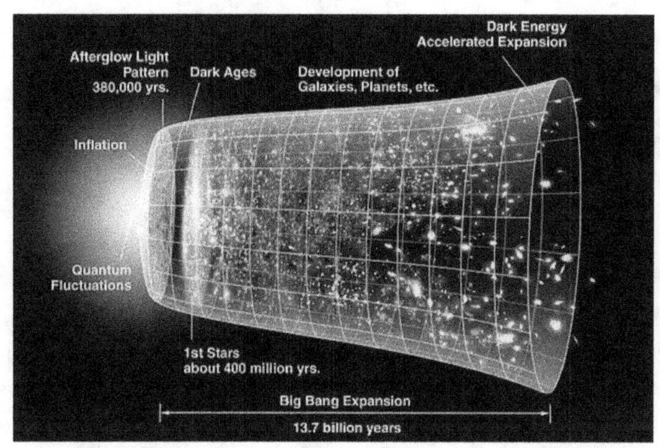

Sketch of the timeline of the Universe in the ΛCDM model. The accelerated expansion in the last third of the timeline represents the dark-energy dominated era.

In cosmology, the **cosmological constant** (usually denoted by the Greek capital letter lambda: Λ) is the value of the energy density of the vacuum of space. It was originally introduced by Albert Einstein in 1917[1] as an addition to his theory of general relativity to "hold back gravity" and achieve a static universe, which was the accepted view at the time. Einstein abandoned the concept after Hubble's 1929 discovery that all galaxies outside the Local Group (the group that contains the Milky Way Galaxy) are moving away from each other, implying an overall expanding universe. From 1929 until the early 1990s, most cosmology researchers assumed the cosmological constant to be zero.

Since the 1990s, several developments in observational cosmology, especially the discovery of the accelerating universe from distant supernovae in 1998 (in addition to independent evidence from the cosmic microwave background and large galaxy redshift surveys), have shown that around 68% of the mass–energy density of the universe can be attributed to dark energy.[2] While dark energy is poorly understood at a fundamental level, the main required properties of dark energy are that it functions as a type of antigravity, it dilutes much more slowly than matter as the universe expands, and it clusters much more weakly than matter, or perhaps not at all. The cosmological constant is the simplest possible form of dark energy since it is constant in both space and time, and this leads to the current standard model of cosmology known as the Lambda-CDM model, which provides a good fit to many cosmological observations.

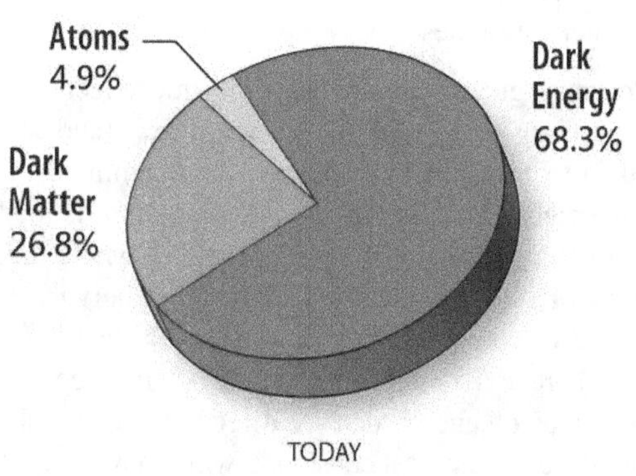

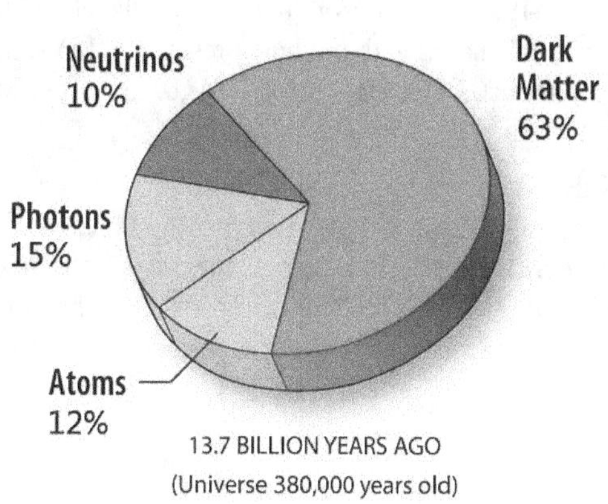

Estimated ratios of dark matter and dark energy (which may be the cosmological constant) in the universe. According to current theories of physics, dark energy now dominates as the largest source of energy of the universe, in contrast to earlier epochs when it was insignificant.

16.1 Equation

The cosmological constant Λ appears in Einstein's field equation in the form of

$$R_{\mu\nu} - \frac{1}{2}R\,g_{\mu\nu} + \Lambda\,g_{\mu\nu} = \frac{8\pi G}{c^4}T_{\mu\nu},$$

where R and g describe the structure of spacetime, T pertains to matter and energy affecting that structure, and G and c are conversion factors that arise from using traditional units of measurement. When Λ is zero, this reduces to the original field equation of general relativity. When T is zero, the field equation describes empty space (the vacuum).

The cosmological constant has the same effect as an intrinsic energy density of the vacuum, ϱ_{vac} (and an associated pressure). In this context, it is commonly moved onto the right-hand side of the equation, and defined with a proportionality factor of 8π: $\Lambda = 8\pi\varrho_{\text{vac}}$, where unit conventions of general relativity are used (otherwise factors of G and c would also appear, i.e. $\Lambda = 8\pi(G/c^2)\varrho_{\text{vac}} = \kappa\varrho_{\text{vac}}$, where κ is Einstein's constant). It is common to quote values of energy density directly, though still using the name "cosmological constant", with convention $8\pi G = 1$. (The true dimension of Λ is a length^{-2} and it has the value of 1.19×10^{-52} m^{-2} or in reduced Planck units : ~ 3×10^{-122}, calculated with the best present (2015) values of $\Omega\Lambda = 0.6911\pm0.0062$ and $H_o = 67.74\pm0.46$ (km/s)/Mpc = $(2.195\pm0.015)\times10^{-18}$ s^{-1}).

A positive vacuum energy density resulting from a cosmological constant implies a negative pressure, and vice versa. If the energy density is positive, the associated negative pressure will drive an accelerated expansion of the universe, as observed. (See dark energy and cosmic inflation for details.)

16.1.1 ΩΛ (Omega Lambda)

Instead of the cosmological constant itself, cosmologists often refer to the ratio between the energy density due to the cosmological constant and the critical density of the universe, the tipping point for a sufficient density to stop the universe from expanding forever. This ratio is usually denoted $\Omega\Lambda$, and is estimated to be 0.6911 ± 0.0062, according to results published by the Planck Collaboration in 2015.[3]

In a flat universe $\Omega\Lambda$ is the fraction of the energy of the universe due to the cosmological constant, i.e., what we would intuitively call the fraction of the universe that is made up of dark energy. Note that this value changes over time: the critical density changes with cosmological time, but the energy density due to the cosmological constant remains unchanged throughout the history of the universe: the amount of dark energy increases as the universe grows, while the amount of matter does not.

16.1.2 Equation of state

Another ratio that is used by scientists is the equation of state, usually denoted w, which is the ratio of pressure that dark energy puts on the universe to the energy per unit volume.[4] This ratio is $w = -1$ for a true cosmological constant, and is generally different for alternative time-varying forms of vacuum energy such as quintessence.

16.2 History

Einstein included the cosmological constant as a term in his field equations for general relativity because he was dissatisfied that otherwise his equations did not allow, apparently, for a static universe: gravity would cause a universe that was initially at dynamic equilibrium to contract. To counteract this possibility, Einstein added the cosmological constant.[5] However, soon after Einstein developed his static theory, observations by Edwin Hubble indicated that the universe appears to be expanding; this was consistent with a cosmological solution to the *original* general relativity equations that had been found by the mathematician Friedmann, working on the Einstein equations of general relativity. Einstein later reputedly referred to his failure to accept the validation of his equations—when they had predicted the expansion of the universe in theory, before it was demonstrated in observation of the cosmological red shift—as the "biggest blunder" of his life.[6][7]

In fact, adding the cosmological constant to Einstein's equations does not lead to a static universe at equilibrium because the equilibrium is unstable: if the universe expands slightly, then the expansion releases vacuum energy, which causes yet more expansion. Likewise, a universe that contracts slightly will continue contracting.[8]:59

However, the cosmological constant remained a subject of theoretical and empirical interest. Empirically, the onslaught of cosmological data in the past decades strongly suggests that our universe has a positive cosmological constant.[5] The explanation of this small but positive value is an outstanding theoretical challenge (*see the section below*).

Finally, it should be noted that some early generalizations of Einstein's gravitational theory, known as classical unified field theories, either introduced a cosmological constant on theoretical grounds or found that it arose naturally from the mathematics. For example, Sir Arthur Stanley Eddington claimed that the cosmological constant version of the vacuum field equation expressed the "epistemological" property that the universe is “self-gauging", and Erwin Schrödinger's pure-affine theory using a simple variational principle produced the field equation with a cosmological term.

16.3 Positive value

Observations announced in 1998 of distance–redshift relation for Type Ia supernovae[9][10] indicated that the expansion of the universe is accelerating. When combined with measurements of the cosmic microwave background radiation these implied a value of $\Omega\Lambda \approx 0.7$,[11] a result which has been supported and refined by more recent measurements. There are other possible causes of an accelerating universe, such as quintessence, but the cosmological constant is in most respects the simplest solution. Thus, the current standard model of cosmology, the Lambda-CDM model, includes the cosmological constant, which is measured to be on the order of 10^{-52} m^{-2}, in metric units. Multiplied by other constants that appear in the equations, it is often expressed as 10^{-52} m^{-2}, 10^{-35} s^{-2}, 10^{-47} GeV^4, 10^{-29} g/cm^3.[12] More recent measurements give the value as $(1.501 \pm 0.043) \times 10^{-25}$kg/m ^{3} .[13] In terms of Planck units, and as a natural dimensionless value, the cosmological constant, Λ, is on the order of 10^{-122}.[14] Modern calculations considering the vacuum energy of all known scalar and vector fields leads to 10^{-54} orders of magnitude smaller than the prediction.[15]

As was only recently seen, by works of 't Hooft, Susskind[16] and others, a positive cosmological constant has surprising consequences, such as a finite maximum entropy of the observable universe (see the holographic principle).

16.4 Predictions

16.4.1 Quantum field theory

See also: Cosmological constant problem

A major outstanding problem is that most quantum field theories predict a huge value for the quantum vacuum. A common assumption is that the quantum vacuum is equivalent to the cosmological constant. Although no theory exists that supports this assumption, arguments can be made in its favor.[17]

Such arguments are usually based on dimensional analysis and effective field theory. If the universe is described by an effective local quantum field theory down to the Planck scale, then we would expect a cosmological constant of the order of $M_{\rm pl}^4$. As noted above, the measured cosmological constant is smaller than this by a factor of 10^{-120}. This discrepancy has been called “the worst theoretical prediction in the history of physics!".[18]

Some supersymmetric theories require a cosmological constant that is exactly zero, which further complicates things. This is the *cosmological constant problem*, the worst problem of fine-tuning in physics: there is no known natural way to derive the tiny cosmological constant used in cosmology from particle physics.

16.4.2 Anthropic principle

One possible explanation for the small but non-zero value was noted by Steven Weinberg in 1987 following the anthropic principle.[19] Weinberg explains that if the vacuum energy took different values in different domains of the universe, then observers would necessarily measure values similar to that which is observed: the formation of life-supporting structures would be suppressed in domains where the vacuum energy is much larger. Specifically, if the vacuum energy is negative and its absolute value is substantially larger than it appears to be in the observed universe (say, a factor of 10 larger), holding all other variables (e.g. matter density) constant, that would mean that the universe is closed; furthermore, its lifetime would be shorter than the age of our universe, possibly too short for intelligent life to form. On the other hand, a universe with a large positive cosmological constant would expand too fast, preventing galaxy formation. According to Weinberg, domains where the vacuum energy is compatible with life would be comparatively rare. Using this argument, Weinberg predicted that the cosmological constant would have a value of less than a hundred times the currently accepted value.[20] In 1992, Weinberg refined this prediction of the cosmological constant to 5 to 10 times the matter density.[21]

This argument depends on a lack of a variation of the distribution (spatial or otherwise) in the vacuum energy density, as would be expected if dark energy were the cosmological constant. There is no evidence that the vacuum energy does vary, but it may be the case if, for example, the vacuum energy is (even in part) the potential of a scalar field such as the residual inflaton (also see quintessence). Another theoretical approach that deals with the issue is that of multiverse theories, which predict a large number of “parallel” universes with different laws of physics and/or values of fundamental constants. Again, the anthropic principle states that we can only live in one of the universes that is compatible with some form of intelligent life. Critics claim that these theories, when used as an explanation for fine-tuning, commit the inverse gambler’s fallacy.

In 1995, Weinberg’s argument was refined by Alexander

Vilenkin to predict a value for the cosmological constant that was only ten times the matter density,[22] i.e. about three times the current value since determined.

16.5 See also

- Higgs mechanism
- Lambdavacuum solution
- Naturalness (physics)
- Quantum electrodynamics
- de Sitter relativity
- Unruh effect

16.6 References

[1] Einstein, A (1917). "Kosmologische Betrachtungen zur allgemeinen Relativitaetstheorie". *Sitzungsberichte der Königlich Preussischen Akademie der Wissenschaften Berlin*. part 1: 142–152.

[2] What is Dark Energy?, Space.com, 1 May 2013

[3] Collaboration, Planck, PAR Ade, N Aghanim, C Armitage-Caplan, M Arnaud, et al., Planck 2015 results. XIII. Cosmological parameters. arXiv preprint 1502.1589v2 , 6 Feb 2015.

[4] Hogan, Jenny (2007). "Welcome to the Dark Side". *Nature*. **448** (7151): 240–245. Bibcode:2007Natur.448..240H. PMID 17637630. doi:10.1038/448240a.

[5] Urry, Meg (2008). *The Mysteries of Dark Energy*. Yale Science. Yale University.

[6] Gamov, George (1970). *My World Line*. Viking Press. p. 44. ISBN 978-0670503766

[7] Rosen, Rebecca J. "Einstein Likely Never Said One of His Most Oft-Quoted Phrases". *The Atlantic*. The Atlantic Media Company. Retrieved 10 August 2013.

[8] Barbara Sue Ryden (2003). *Introduction to cosmology*. Addison-Wesley. ISBN 978-0-8053-8912-8.

[9] Riess, A.; et al. (September 1998). "Observational Evidence from Supernovae for an Accelerating Universe and a Cosmological Constant". *The Astronomical Journal*. **116** (3): 1009–1038. Bibcode:1998AJ....116.1009R. arXiv:astro-ph/9805201. doi:10.1086/300499.

[10] Perlmutter, S.; et al. (June 1999). "Measurements of Omega and Lambda from 42 High-Redshift Supernovae". *The Astrophysical Journal*. **517** (2): 565–586. Bibcode:1999ApJ...517..565P. arXiv:astro-ph/9812133. doi:10.1086/307221.

[11] See e.g. Baker, Joanne C.; et al. (1999). "Detection of cosmic microwave background structure in a second field with the Cosmic Anisotropy Telescope". *Monthly Notices of the Royal Astronomical Society*. **308** (4): 1173–1178. Bibcode:1999MNRAS.308.1173B. arXiv:astro-ph/9904415. doi:10.1046/j.1365-8711.1999.02829.x.

[12] Tegmark, Max; et al. (2004). "Cosmological parameters from SDSS and WMAP". *Physical Review D*. **69** (103501): 103501. Bibcode:2004PhRvD..69j3501T. arXiv:astro-ph/0310723. doi:10.1103/PhysRevD.69.103501.

[13] http://physics.stackexchange.com/a/314883/92058

[14] John D. Barrow The Value of the Cosmological Constant

[15] Martin, Jerome. "Everything You Always Wanted To Know About The Cosmological Constant Problem (But Were Afraid To Ask)". *Comptes Rendus Physique*. **13** (6-7): 566–665. ISSN 1631-0705. doi:10.1016/j.crhy.2012.04.008.

[16] Lisa Dyson, Matthew Kleban, Leonard Susskind: "Disturbing Implications of a Cosmological Constant"

[17] Rugh, S; Zinkernagel, H. (2001). "The Quantum Vacuum and the Cosmological Constant Problem". *Studies in History and Philosophy of Modern Physics*. **33** (4): 663–705. doi:10.1016/S1355-2198(02)00033-3.

[18] MP Hobson; GP Efstathiou; AN Lasenby (2006). *General Relativity: An introduction for physicists* (Reprinted with corrections 2007 ed.). Cambridge University Press. p. 187. ISBN 978-0-521-82951-9.

[19] Weinberg, S (1987). "Anthropic Bound on the Cosmological Constant". *Phys. Rev. Lett.* **59** (22): 2607–2610. Bibcode:1987PhRvL..59.2607W. PMID 10035596. doi:10.1103/PhysRevLett.59.2607.

[20] Alexander Vilenkin, *Many Worlds in One: The Search for Other Universes*, ISBN 978-0-8090-9523-0, pp. 138–9

[21] Weinberg, Steven (1993). *Dreams of a Final Theory: the search for the fundamental laws of nature*. Vintage Press. p. 182. ISBN 0-09-922391-0.

[22] Alexander Vilenkin, *Many Worlds in One: The Search for Other universes*, ISBN 978-0-8090-9523-0, p. 146, which references Vilenkin' *Predictions from quantum cosmology*, Physical Review Letters, vol 74, p. 846 (1995)

- Michael, E., University of Colorado, Department of Astrophysical and Planetary Sciences, "The Cosmological Constant"
- Ferguson, Kitty (1991). *Stephen Hawking: Quest For A Theory of Everything*, Franklin Watts. ISBN 0-553-29895-X.
- John D. Barrow; John K. Webb (June 2005). "Inconstant Constants". *Scientific American*.
- *Beyond the Cosmological Standard Model*[1] (2014)

16.7 External links

- Cosmological constant (astronomy) at *Encyclopædia Britannica*
- Carroll, Sean M., *"The Cosmological Constant"* (short), *"The Cosmological Constant"* (extended).
- News story: More evidence for dark energy being the cosmological constant
- Cosmological constant article from Scholarpedia
- Copeland, Ed; Merrifield, Mike. "Λ – Cosmological Constant". *Sixty Symbols*. Brady Haran for the University of Nottingham.

[1] Austin Joyce; Bhuvnesh Jain; Justin Khoury; Mark Trodden (2014). "Beyond the Cosmological Standard Model". arXiv:1407.0059 .

Chapter 17

Friedmann–Lemaître–Robertson–Walker metric

The **Friedmann–Lemaître–Robertson–Walker (FLRW) metric** is an exact solution of Einstein's field equations of general relativity; it describes a homogeneous, isotropic expanding or contracting universe that is path connected, but not necessarily simply connected.[1][2][3] The general form of the metric follows from the geometric properties of homogeneity and isotropy; Einstein's field equations are only needed to derive the scale factor of the universe as a function of time. Depending on geographical or historical preferences, the set of the four scientists — Alexander Friedmann, Georges Lemaître, Howard P. Robertson and Arthur Geoffrey Walker are customarily grouped as **Friedmann–Robertson–Walker (FRW)** or **Robertson–Walker (RW)** or **Friedmann–Lemaître (FL)**). This model is sometimes called the *Standard Model* of modern cosmology,[4] although such a description is also associated with the further developed Lambda-CDM model. The FLRW model was developed independently by the named authors in the 1920s and 1930s.

17.1 General metric

The FLRW metric starts with the assumption of homogeneity and isotropy of space. It also assumes that the spatial component of the metric can be time-dependent. The generic metric which meets these conditions is

$$-c^2 d\tau^2 = -c^2 dt^2 + a(t)^2 d⍰^2$$

where ⍰ ranges over a 3-dimensional space of uniform curvature, that is, elliptical space, Euclidean space, or hyperbolic space. It is normally written as a function of three spatial coordinates, but there are several conventions for doing so, detailed below. d⍰ does not depend on *t* — all of the time dependence is in the function *a*(*t*), known as the "scale factor".

17.1.1 Reduced-circumference polar coordinates

In reduced-circumference polar coordinates the spatial metric has the form

$$d⍰^2 = \frac{dr^2}{1-kr^2} + r^2 d⍰^2, \quad \text{where} d⍰^2 = d\theta^2 + \sin^2\theta \, d\phi^2.$$

k is a constant representing the curvature of the space. There are two common unit conventions:

- *k* may be taken to have units of length^{-2}, in which case *r* has units of length and *a*(*t*) is unitless. *k* is then the Gaussian curvature of the space at the time when *a*(*t*) = 1. *r* is sometimes called the reduced circumference because it is equal to the measured circumference of a circle (at that value of *r*), centered at the origin, divided by 2π (like the *r* of Schwarzschild coordinates). Where appropriate, *a*(*t*) is often chosen to equal 1 in the present cosmological era, so that d⍰ measures comoving distance.

- Alternatively, *k* may be taken to belong to the set {−1,0,+1} (for negative, zero, and positive curvature respectively). Then *r* is unitless and *a*(*t*) has units of length. When *k* = ±1, *a*(*t*) is the radius of curvature of the space, and may also be written *R*(*t*).

A disadvantage of reduced circumference coordinates is that they cover only half of the 3-sphere in the case of positive curvature—circumferences beyond that point begin to decrease, leading to degeneracy. (This is not a problem if space is elliptical, i.e. a 3-sphere with opposite points identified.)

17.1.2 Hyperspherical coordinates

In *hyperspherical* or *curvature-normalized* coordinates the coordinate r is proportional to radial distance; this gives

$$\mathrm{d}\boxed{?}^2 = \mathrm{d}r^2 + S_k(r)^2 \, \mathrm{d}\boxed{?}^2$$

where $\mathrm{d}\boxed{?}$ is as before and

$$S_k(r) = \begin{cases} \sqrt{k}^{\,-1} \sin(r\sqrt{k}), & k > 0 \\ r, & k = 0 \\ \sqrt{|k|}^{\,-1} \sinh(r\sqrt{|k|}), & k < 0. \end{cases}$$

As before, there are two common unit conventions:

- k may be taken to have units of length^{-2}, in which case r has units of length and $a(t)$ is unitless. k is then the Gaussian curvature of the space at the time when $a(t) = 1$. Where appropriate, $a(t)$ is often chosen to equal 1 in the present cosmological era, so that $\mathrm{d}\boxed{?}$ measures comoving distance.
- Alternatively, as before, k may be taken to belong to the set $\{-1,0,+1\}$ (for negative, zero, and positive curvature respectively). Then r is unitless and $a(t)$ has units of length. When $k = \pm 1$, $a(t)$ is the radius of curvature of the space, and may also be written $R(t)$. Note that, when $k = +1$, r is essentially a third angle along with θ and φ. The letter χ may be used instead of r.

Though it is usually defined piecewise as above, S is an analytic function of both k and r. It can also be written as a power series

$$S_k(r) = \sum_{n=0}^{\infty} \frac{(-1)^n k^n r^{2n+1}}{(2n+1)!} = r - \frac{kr^3}{6} + \frac{k^2 r^5}{120} - \cdots$$

or as

$$S_k(r) = r \, \mathrm{sinc}\,(r\sqrt{k})$$

where sinc is the unnormalized sinc function and $\sqrt{k}$ is one of the imaginary, zero or real square roots of k. These definitions are valid for all k.

17.1.3 Cartesian coordinates

When $k = 0$ one may write simply

$$\mathrm{d}\boxed{?}^2 = \mathrm{d}x^2 + \mathrm{d}y^2 + \mathrm{d}z^2.$$

This can be extended to $k \neq 0$ by defining

$$x = r\cos\theta$$

$$y = r\sin\theta\cos\phi$$

$$z = r\sin\theta\sin\phi$$

where r is one of the radial coordinates defined above, but this is rare.

17.1.4 Curvature

Cartesian coordinates

In flat (k=0) FRW space using Cartesian coordinates, the surviving components of the Ricci tensor are[5]

$$R_{tt} = -3\frac{\ddot{a}}{a}, \quad R_{xx} = R_{yy} = R_{zz} = c^{-2}(a\ddot{a} + 2\dot{a}^2)$$

and the Ricci scalar is

$$R = 6c^{-2}\left(\frac{\ddot{a}(t)}{a(t)} + \frac{\dot{a}^2(t)}{a^2(t)}\right).$$

Spherical coordinates

In more general FRW space using spherical coordinates (called "reduced-circumference polar coordinates" above), the surviving components of the Ricci tensor are[6]

$$R_{tt} = -3\frac{\ddot{a}}{a},$$

$$R_{rr} = \frac{c^{-2}(a(t)\ddot{a}(t) + 2\dot{a}^2(t)) + 2k}{1 - kr^2}$$

$$R_{\theta\theta} = r^2(c^{-2}(a(t)\ddot{a}(t) + 2\dot{a}^2(t)) + 2k)$$

$$R_{\phi\phi} = r^2(c^{-2}(a(t)\ddot{a}(t) + 2\dot{a}^2(t)) + 2k)\sin^2(\theta)$$

and the Ricci scalar is

$$R = 6\left(\frac{\ddot{a}(t)}{c^2 a(t)} + \frac{\dot{a}^2(t)}{c^2 a^2(t)} + \frac{k}{a^2(t)}\right).$$

17.2 Solutions

Main article: Friedmann equations

Einstein's field equations are not used in deriving the general form for the metric: it follows from the geometric properties of homogeneity and isotropy. However, determining the time evolution of $a(t)$ does require Einstein's field equations together with a way of calculating the density, $\rho(t)$, such as a cosmological equation of state.

This metric has an analytic solution to Einstein's field equations $G_{\mu\nu} + \Lambda g_{\mu\nu} = \frac{8\pi G}{c^4}T_{\mu\nu}$ giving the Friedmann equations when the energy-momentum tensor is similarly assumed to be isotropic and homogeneous. The resulting equations are:[7]

$$\left(\frac{\dot{a}}{a}\right)^2 + \frac{kc^2}{a^2} - \frac{\Lambda c^2}{3} = \frac{8\pi G}{3}\rho$$

$$2\frac{\ddot{a}}{a} + \left(\frac{\dot{a}}{a}\right)^2 + \frac{kc^2}{a^2} - \Lambda c^2 = -\frac{8\pi G}{c^2}p.$$

These equations are the basis of the standard big bang cosmological model including the current ΛCDM model. Because the FLRW model assumes homogeneity, some popular accounts mistakenly assert that the big bang model cannot account for the observed lumpiness of the universe. In a strictly FLRW model, there are no clusters of galaxies, stars or people, since these are objects much denser than a typical part of the universe. Nonetheless, the FLRW model is used as a first approximation for the evolution of the real, lumpy universe because it is simple to calculate, and models which calculate the lumpiness in the universe are added onto the FLRW models as extensions. Most cosmologists agree that the observable universe is well approximated by an *almost FLRW model*, i.e., a model which follows the FLRW metric apart from primordial density fluctuations. As of 2003, the theoretical implications of the various extensions to the FLRW model appear to be well understood, and the goal is to make these consistent with observations from COBE and WMAP.

If the spacetime is multiply connected, then each event will be represented by more than one tuple of coordinates.

17.2.1 Interpretation

The pair of equations given above is equivalent to the following pair of equations

$$\dot{\rho} = -3\frac{\dot{a}}{a}\left(\rho + \frac{p}{c^2}\right)$$

$$\frac{\ddot{a}}{a} = -\frac{4\pi G}{3}\left(\rho + \frac{3p}{c^2}\right) + \frac{\Lambda c^2}{3}$$

with k , the spatial curvature index, serving as a constant of integration for the first equation.

The first equation can be derived also from thermodynamical considerations and is equivalent to the first law of thermodynamics, assuming the expansion of the universe is an adiabatic process (which is implicitly assumed in the derivation of the Friedmann–Lemaître–Robertson–Walker metric).

The second equation states that both the energy density and the pressure cause the expansion rate of the universe $\dot{a}$ to decrease, i.e., both cause a deceleration in the expansion of the universe. This is a consequence of gravitation, with pressure playing a similar role to that of energy (or mass) density, according to the principles of general relativity. The cosmological constant, on the other hand, causes an acceleration in the expansion of the universe.

17.2.2 Cosmological constant

The cosmological constant term can be omitted if we make the following replacements

$$\rho \to \rho + \frac{\Lambda c^2}{8\pi G}$$

$$p \to p - \frac{\Lambda c^4}{8\pi G}.$$

Therefore, the cosmological constant can be interpreted as arising from a form of energy which has negative pressure, equal in magnitude to its (positive) energy density:

$$p = -\rho c^2.$$

Such form of energy—a generalization of the notion of a cosmological constant—is known as dark energy.

In fact, in order to get a term which causes an acceleration of the universe expansion, it is enough to have a scalar field which satisfies

$$p < -\frac{\rho c^2}{3}.$$

Such a field is sometimes called quintessence.

17.2.3 Newtonian interpretation

This is due to McCrea and Milne [8] although sometimes incorrectly ascribed to Friedmann. The Friedmann equations are equivalent to this pair of equations:

$$-a^3\dot{\rho} = 3a^2\dot{a}\rho + \frac{3a^2p\dot{a}}{c^2}$$

$$\frac{\dot{a}^2}{2} - \frac{G\frac{4\pi a^3}{3}\rho}{a} = -\frac{kc^2}{2}.$$

The first equation says that the decrease in the mass contained in a fixed cube (whose side is momentarily *a*) is the amount which leaves through the sides due to the expansion of the universe plus the mass equivalent of the work done by pressure against the material being expelled. This is the conservation of mass-energy (first law of thermodynamics) contained within a part of the universe.

The second equation says that the kinetic energy (seen from the origin) of a particle of unit mass moving with the expansion plus its (negative) gravitational potential energy (relative to the mass contained in the sphere of matter closer to the origin) is equal to a constant related to the curvature of the universe. In other words, the energy (relative to the origin) of a co-moving particle in free-fall is conserved. General relativity merely adds a connection between the spatial curvature of the universe and the energy of such a particle: positive total energy implies negative curvature and negative total energy implies positive curvature.

The cosmological constant term is assumed to be treated as dark energy and thus merged into the density and pressure terms.

During the Planck epoch, one cannot neglect quantum effects. So they may cause a deviation from the Friedmann equations.

17.3 Name and history

The main results of the FLRW model were first derived by the Soviet mathematician Alexander Friedmann in 1922 and 1924. Although his work was published in the prestigious physics journal Zeitschrift für Physik, it remained relatively unnoticed by his contemporaries. Friedmann was in direct communication with Albert Einstein, who, on behalf of Zeitschrift für Physik, acted as the scientific referee of Friedmann's work. Eventually Einstein acknowledged the correctness of Friedmann's calculations, but failed to appreciate the physical significance of Friedmann's predictions.

Friedmann died in 1925. In 1927, Georges Lemaître, a Belgian priest, astronomer and periodic professor of physics at the Catholic University of Leuven, arrived independently at similar results as Friedmann had and published them in Annals of the Scientific Society of Brussels. In the face of the observational evidence for the expansion of the universe obtained by Edwin Hubble in the late 1920s, Lemaître's results were noticed in particular by Arthur Eddington, and in 1930–31 his paper was translated into English and published in the Monthly Notices of the Royal Astronomical Society.

Howard P. Robertson from the US and Arthur Geoffrey Walker from the UK explored the problem further during the 1930s. In 1935 Robertson and Walker rigorously proved that the FLRW metric is the only one on a spacetime that is spatially homogeneous and isotropic (as noted above, this is a geometric result and is not tied specifically to the equations of general relativity, which were always assumed by Friedmann and Lemaître).

Because the dynamics of the FLRW model were derived by Friedmann and Lemaître, the latter two names are often omitted by scientists outside the US. Conversely, US physicists often refer to it as simply "Robertson–Walker". The full four-name title is the most democratic and it is frequently used. Often the "Robertson–Walker" *metric*, so-called since they proved its generic properties, is distinguished from the dynamical "Friedmann-Lemaître" *models*, specific solutions for $a(t)$ which assume that the only contributions to stress-energy are cold matter ("dust"), radiation, and a cosmological constant.

17.4 Einstein's radius of the universe

Einstein's radius of the universe is the radius of curvature of space of Einstein's universe, a long-abandoned static model that was supposed to represent our universe in idealized form. Putting

$$\dot{a} = \ddot{a} = 0$$

in the Friedmann equation, the radius of curvature of space of this universe (Einstein's radius) is

$$R_E = c/\sqrt{4\pi G\rho}\,,$$

where c is the speed of light, G is the Newtonian gravitational constant, and ρ is the density of space of this universe. The numerical value of Einstein's radius is of the order of 10^{10} light years.

17.5 Evidence

By combining the observation data from some experiments such as WMAP and Planck with theoretical results of Ehlers–Geren–Sachs theorem and its generalization,[9] astrophysicists now agree that the universe is almost homogeneous and isotropic (when averaged over a very large scale) and thus nearly a FLRW spacetime.

17.6 Notes

[1] For an early reference, see Robertson (1935); Robertson *assumes* multiple connectedness in the positive curvature case and says that "we are still free to restore" simple connectedness.

[2] M. Lachieze-Rey; J.-P. Luminet (1995), "Cosmic Topology", *Physics Reports*, **254** (3): 135–214, Bibcode:1995PhR...254..135L, arXiv:gr-qc/9605010, doi:10.1016/0370-1573(94)00085-H

[3] G. F. R. Ellis; H. van Elst (1999). "Cosmological models (Cargèse lectures 1998)". In Marc Lachièze-Rey. *Theoretical and Observational Cosmology*. NATO Science Series C. **541**. pp. 1–116. Bibcode:1999toc..conf....1E. ISBN 978-0792359463. arXiv:gr-qc/9812046.

[4] L. Bergström, A. Goobar (2006), *Cosmology and Particle Astrophysics* (2nd ed.), Sprint, p. 61, ISBN 3-540-32924-2

[5] Wald, Robert. *General Relativity*. p. 97.

[6] "Cosmology" (PDF). p. 23.

[7] P. Ojeda and H. Rosu (2006), "Supersymmetry of FRW barotropic cosmologies", *International Journal of Theoretical Physics*, **45** (6): 1191–1196, Bibcode:2006IJTP...45.1152R, arXiv:gr-qc/0510004, doi:10.1007/s10773-006-9123-2

[8] McCrea, W. H.; Milne, E. A. (1934). "Newtonian universes and the curvature of space". Quarterly Journal of Mathematics. 5: 73–80.

[9] See pp. 351ff. in Hawking, Stephen W.; Ellis, George F. R. (1973), *The large scale structure of space-time*, Cambridge University Press, ISBN 0-521-09906-4. The original work is Ehlers, J., Geren, P., Sachs, R.K.: Isotropic solutions of Einstein-Liouville equations. J. Math. Phys. 9, 1344 (1968). For the generalization, see Stoeger, W. R.; Maartens, R; Ellis, George (2007), "Proving Almost-Homogeneity of the Universe: An Almost Ehlers-Geren-Sachs Theorem", *Ap. J.*, **39**: 1–5, Bibcode:1995ApJ...443....1S, doi:10.1086/175496.

17.7 References

- Friedmann, Alexander (1922), "Über die Krümmung des Raumes", *Zeitschrift für Physik A*, **10** (1): 377–386, Bibcode:1922ZPhy...10..377F, doi:10.1007/BF01332580
- Friedmann, Alexander (1924), "Über die Möglichkeit einer Welt mit konstanter negativer Krümmung des Raumes", *Zeitschrift für Physik A*, **21** (1): 326–332, Bibcode:1924ZPhy...21..326F, doi:10.1007/BF01328280 English trans. in 'General Relativity and Gravitation' 1999 vol.31, 31–
- Lemaître, Georges (1931), "Expansion of the universe, A homogeneous universe of constant mass and increasing radius accounting for the radial velocity of extra-galactic nebulæ", *Monthly Notices of the Royal Astronomical Society*, **91**: 483–490, Bibcode:1931MNRAS..91..483L, doi:10.1093/mnras/91.5.483 *translated from* Lemaître, Georges (1927), "Un univers homogène de masse constante et de rayon croissant rendant compte de la vitesse radiale des nébuleuses extra-galactiques", *Annales de la Société Scientifique de Bruxelles*, **A47**: 49–56, Bibcode:1927ASSB...47...49L
- Lemaître, Georges (1933), "l'Univers en expansion", *Annales de la Société Scientifique de Bruxelles*, **A53**: 51–85, Bibcode:1933ASSB...53...51L
- Robertson, H. P. (1935), "Kinematics and world structure", *Astrophysical Journal*, **82**: 284–301, Bibcode:1935ApJ....82..284R, doi:10.1086/143681
- Robertson, H. P. (1936), "Kinematics and world structure II", *Astrophysical Journal*, **83**: 187–201, Bibcode:1936ApJ....83..187R, doi:10.1086/143716
- Robertson, H. P. (1936), "Kinematics and world structure III", *Astrophysical Journal*, **83**: 257–271, Bibcode:1936ApJ....83..257R, doi:10.1086/143726
- Walker, A. G. (1937), "On Milne's theory of world-structure", *Proceedings of the London Mathematical Society* 2, **42** (1): 90–127, doi:10.1112/plms/s2-42.1.90
- North J D:(1965)*The Measure of the Universe - a history of modern cosmology*, Oxford Univ. Press, Dover reprint 1990, ISBN 0-486-66517-8
- Harrison, E. R. (1967), "Classification of uniform cosmological models", *Monthly Notices of the Royal Astronomical Society*, **137**: 69–79, Bibcode:1967MNRAS.137...69H, doi:10.1093/mnras/137.1.69
- d'Inverno, Ray (1992), *Introducing Einstein's Relativity*, Oxford: Oxford University Press, ISBN 0-19-859686-3. *(See Chapter 23 for a particularly clear and concise introduction to the FLRW models.)*

Chapter 18

Friedmann equations

Alexander Friedmann

The **Friedmann equations** are a set of equations in physical cosmology that govern the expansion of space in homogeneous and isotropic models of the universe within the context of general relativity. They were first derived by Alexander Friedmann in 1922[1] from Einstein's field equations of gravitation for the Friedmann–Lemaître–Robertson–Walker metric and a perfect fluid with a given mass density ρ and pressure p. The equations for negative spatial curvature were given by Friedmann in 1924.[2]

18.1 Assumptions

Main article: Friedmann–Lemaître–Robertson–Walker metric

The Friedmann equations start with the simplifying assumption that the universe is spatially homogeneous and isotropic, i.e. the cosmological principle; empirically, this is justified on scales larger than ~100 Mpc. The cosmological principle implies that the metric of the universe must be of the form

$$ds^2 = a(t)^2 \, ds_3^2 - c^2 \, dt^2$$

where ds_3^2 is a three-dimensional metric that must be one of **(a)** flat space, **(b)** a sphere of constant positive curvature or **(c)** a hyperbolic space with constant negative curvature. The parameter k discussed below takes the value 0, 1, −1, or the Gaussian curvature, in these three cases respectively. It is this fact that allows us to sensibly speak of a "scale factor", $a(t)$.

Einstein's equations now relate the evolution of this scale factor to the pressure and energy of the matter in the universe. From FLRW metric we compute Christoffel symbols, then the Ricci tensor. With the stress–energy tensor for a perfect fluid, we substitute them into Einstein's field equations and the resulting equations are described below.

18.2 Equations

There are two independent Friedmann equations for modeling a homogeneous, isotropic universe. The first is:

$$\frac{\dot{a}^2 + kc^2}{a^2} = \frac{8\pi G\rho + \Lambda c^2}{3}$$

which is derived from the 00 component of Einstein's field equations. The second is:

$$\frac{\ddot{a}}{a} = -\frac{4\pi G}{3}\left(\rho + \frac{3p}{c^2}\right) + \frac{\Lambda c^2}{3}$$

which is derived from the first together with the trace of Einstein's field equations. a is the scale factor, $H \equiv \frac{\dot{a}}{a}$ is the Hubble parameter. G, Λ, and c are universal constants (G is Newton's gravitational constant, Λ is the cosmological constant, and c is the speed of light in vacuum). k is constant throughout a particular solution, but may vary from one solution to another. a, H, ρ, and p are functions of time. ρ, and p are the density and pressure, respectively. $\frac{k}{a^2}$ is the spatial curvature in any time slice of the universe; it is equal to one-sixth of the spatial Ricci curvature scalar R since $R = \frac{6}{c^2a^2}(\ddot{a}a + \dot{a}^2 + kc^2)$ in the Friedmann model. We see that in the Friedmann equations, a(t) depends only on ρ, p, Λ, and intrinsic curvature k. It does not depend on which coordinate system we chose for spatial slices. There are two commonly used choices for a and k which describe the same physics:

- k = +1, 0 or −1 depending on whether the shape of the universe is a closed 3-sphere, flat (i.e. Euclidean space) or an open 3-hyperboloid, respectively.[3] If k = +1, then a is the radius of curvature of the universe. If k = 0, then a may be fixed to any arbitrary positive number at one particular time. If k = −1, then (loosely speaking) one can say that $i \cdot a$ is the radius of curvature of the universe.
- a is the scale factor which is taken to be 1 at the present time. k is the spatial curvature when $a = 1$ (i.e. today). If the shape of the universe is hyperspherical and R_t is the radius of curvature (R_0 in the present-day), then $a = R_t/R_0$. If k is positive, then the universe is hyperspherical. If k is zero, then the universe is flat. If k is negative, then the universe is hyperbolic.

Using the first equation, the second equation can be re-expressed as

$$\dot{\rho} = -3H\left(\rho + \frac{p}{c^2}\right),$$

which eliminates Λ and expresses the conservation of mass-energy $T^{\alpha\beta}{}_{;\beta} = 0$.

These equations are sometimes simplified by replacing

$$\rho \to \rho - \frac{\Lambda c^2}{8\pi G}$$

$$p \to p + \frac{\Lambda c^4}{8\pi G}$$

to give:

$$H^2 = \left(\frac{\dot{a}}{a}\right)^2 = \frac{8\pi G}{3}\rho - \frac{kc^2}{a^2}$$

$$\dot{H} + H^2 = \frac{\ddot{a}}{a} = -\frac{4\pi G}{3}\left(\rho + \frac{3p}{c^2}\right).$$

The simplified form of the second equation is invariant under this transformation.

The Hubble parameter can change over time if other parts of the equation are time dependent (in particular the mass density, the vacuum energy, or the spatial curvature). Evaluating the Hubble parameter at the present time yields Hubble's constant which is the proportionality constant of Hubble's law. Applied to a fluid with a given equation of state, the Friedmann equations yield the time evolution and geometry of the universe as a function of the fluid density.

Some cosmologists call the second of these two equations the **Friedmann acceleration equation** and reserve the term *Friedmann equation* for only the first equation.

18.3 Density parameter

The **density parameter,** Ω , is defined as the ratio of the actual (or observed) density ρ to the critical density ρ_c of the Friedmann universe. The relation between the actual density and the critical density determines the overall geometry of the universe; when they are equal, the geometry of the universe is flat (Euclidean). In earlier models, which did not include a cosmological constant term, critical density was initially defined as the watershed point between an expanding and a contracting Universe.

To date, the critical density is estimated to be approximately five atoms (of monatomic hydrogen) per cubic metre, whereas the average density of ordinary matter in the Universe is believed to be 0.2–0.25 atoms per cubic metre.[4][5]

A much greater density comes from the unidentified dark matter; both ordinary and dark matter contribute in favour of contraction of the universe. However, the largest part comes from so-called dark energy, which accounts for the cosmological constant term. Although the total density is equal to the critical density (exactly, up to measurement error), the dark energy does not lead to contraction of the universe but rather may accelerate its expansion. Therefore, the universe may expand forever.

An expression for the critical density is found by assuming Λ to be zero (as it is for all basic Friedmann universes) and setting the normalised spatial curvature, k, equal to zero.

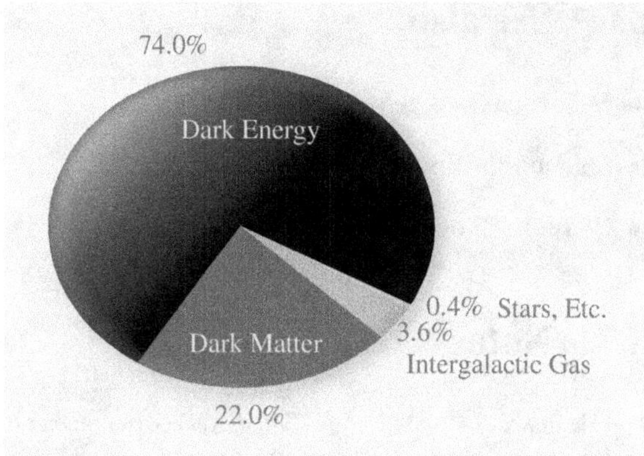

Estimated relative distribution for components of the energy density of the Universe. Dark energy dominates the total energy (74%) while dark matter (22%) constitutes most of the mass. Of the remaining baryonic matter (4%), only one tenth is compact. In February 2015, the European-led research team behind the Planck cosmology probe released new data refining these values to 4.9% ordinary matter, 25.9% dark matter and 69.1% dark energy.

When the substitutions are applied to the first of the Friedmann equations we find:

$$\rho_c = \frac{3H^2}{8\pi G}.$$

The density parameter (useful for comparing different cosmological models) is then defined as:

$$\Omega \equiv \frac{\rho}{\rho_c} = \frac{8\pi G\rho}{3H^2}.$$

This term originally was used as a means to determine the spatial geometry of the universe, where ρ_c is the critical density for which the spatial geometry is flat (or Euclidean). Assuming a zero vacuum energy density, if Ω is larger than unity, the space sections of the universe are closed; the universe will eventually stop expanding, then collapse. If Ω is less than unity, they are open; and the universe expands forever. However, one can also subsume the spatial curvature and vacuum energy terms into a more general expression for Ω in which case this density parameter equals exactly unity. Then it is a matter of measuring the different components, usually designated by subscripts. According to the ΛCDM model, there are important components of Ω due to baryons, cold dark matter and dark energy. The spatial geometry of the universe has been measured by the WMAP spacecraft to be nearly flat. This means that the universe can be well approximated by a model where the spatial curvature parameter k is zero; however, this does not necessarily imply that the universe is infinite: it might merely be that the universe is much larger than the part we see. (Similarly, the fact that Earth is approximately flat at the scale of the Netherlands does not imply that the Earth is flat: it only implies that it is much larger than the Netherlands.)

The first Friedmann equation is often seen in terms of the present values of the density parameters, that is[6]

$$\frac{H^2}{H_0^2} = \Omega_{0,R}a^{-4} + \Omega_{0,M}a^{-3} + \Omega_{0,k}a^{-2} + \Omega_{0,\Lambda}.$$

Here $\Omega_{0,R}$ is the radiation density today (i.e. when $a = 1$), $\Omega_{0,M}$ is the matter (dark plus baryonic) density today, $\Omega_{0,k} = 1 - \Omega_0$ is the "spatial curvature density" today, and $\Omega_{0,\Lambda}$ is the cosmological constant or vacuum density today.

18.4 Useful solutions

The Friedmann equations can be solved exactly in presence of a perfect fluid with equation of state

$$p = w\rho c^2,$$

where p is the pressure, ρ is the mass density of the fluid in the comoving frame and w is some constant.

In spatially flat case ($k = 0$), the solution for the scale factor is

$$a(t) = a_0\, t^{\frac{2}{3(w+1)}}$$

where a_0 is some integration constant to be fixed by the choice of initial conditions. This family of solutions labelled by w is extremely important for cosmology. E.g. $w = 0$ describes a matter-dominated universe, where the pressure is negligible with respect to the mass density. From the generic solution one easily sees that in a matter-dominated universe the scale factor goes as

$$a(t) \propto t^{2/3}$$

Another important example is the case of a radiation-dominated universe, i.e., when $w = 1/3$. This leads to

$$a(t) \propto t^{1/2}$$

Note that this solution is not valid for domination of the cosmological constant, which corresponds to an $w = -1$. In this case the energy density is constant and the scale factor grows exponentially.

Solutions for other values of k can be found at Tersic, Balsa. "Lecture Notes on Astrophysics" (PDF). Retrieved 20 July 2011..

18.4.1 Mixtures

If the matter is a mixture of two or more non-interacting fluids each with such an equation of state, then

$$\dot{\rho}_f = -3H\left(\rho_f + \frac{p_f}{c^2}\right)$$

holds separately for each such fluid f. In each case,

$$\dot{\rho}_f = -3H\left(\rho_f + w_f\rho_f\right)$$

from which we get

$$\rho_f \propto a^{-3(1+w_f)}.$$

For example, one can form a linear combination of such terms

$$\rho = Aa^{-3} + Ba^{-4} + Ca^{0}$$

where: A is the density of "dust" (ordinary matter, $w = 0$) when $a = 1$; B is the density of radiation ($w = 1/3$) when $a = 1$; and C is the density of "dark energy" ($w=-1$). One then substitutes this into

$$\left(\frac{\dot{a}}{a}\right)^2 = \frac{8\pi G}{3}\rho - \frac{kc^2}{a^2}$$

and solves for a as a function of time.

18.5 Rescaled Friedmann equation

Set $\tilde{a} = \frac{a}{a_0}$, $\rho_c = \frac{3H_0^2}{8\pi G}$, $\Omega = \frac{\rho}{\rho_c}$, $t = \frac{\tilde{t}}{H_0}$, $\Omega_c = -\frac{kc^2}{H_0^2 a_0^2}$, where a_0 and H_0 are separately the scale factor and the Hubble parameter today. Then we can have

$$\frac{1}{2}\left(\frac{d\tilde{a}}{d\tilde{t}}\right)^2 + U_{\text{eff}}(\tilde{a}) = \frac{1}{2}\Omega_c$$

where $U_{\text{eff}}(\tilde{a}) = \frac{-\Omega\tilde{a}^2}{2}$. For any form of the effective potential $U_{\text{eff}}(\tilde{a})$, there is an equation of state $p = p(\rho)$ that will produce it.

18.6 See also

- Mathematics of general relativity
- Solutions of Einstein's field equations
- Warm inflation

18.7 Notes

[1] Friedman, A (1922). "Über die Krümmung des Raumes". *Z. Phys.* (in German). **10** (1): 377–386. Bibcode:1922ZPhy...10..377F. doi:10.1007/BF01332580. (English translation: Friedman, A (1999). "On the Curvature of Space". *General Relativity and Gravitation.* **31** (12): 1991–2000. Bibcode:1999GReGr..31.1991F. doi:10.1023/A:1026751225741.). The original Russian manuscript of this paper is preserved in the Ehrenfest archive.

[2] Friedmann, A (1924). "Über die Möglichkeit einer Welt mit konstanter negativer Krümmung des Raumes". *Z. Phys.* (in German). **21** (1): 326–332. Bibcode:1924ZPhy...21..326F. doi:10.1007/BF01328280. (English translation: Friedmann, A (1999). "On the Possibility of a World with Constant Negative Curvature of Space". *General Relativity and Gravitation.* **31** (12): 2001–2008. Bibcode:1999GReGr..31.2001F. doi:10.1023/A:1026755309811.)

[3] Ray A d'Inverno, *Introducing Einstein's Relativity*, ISBN 0-19-859686-3.

[4] Rees, M., Just Six Numbers, (2000) Orion Books, London, p. 81, p. 82

[5] "Universe 101". NASA. Retrieved September 9, 2015. The actual density of atoms is equivalent to roughly 1 proton per 4 cubic meters.

[6] http://adsabs.harvard.edu/cgi-bin/bib_query?arXiv: astro-ph/0703739

Chapter 19

Hubble's law

Hubble's law is the name for the observation in physical cosmology that:

1. Objects observed in deep space (extragalactic space, 10 megaparsecs (Mpc) or more) are found to have a Doppler shift interpretable as relative velocity away from Earth;
2. This Doppler-shift-measured velocity, of various galaxies receding from the Earth, is approximately proportional to their distance from the Earth for galaxies up to a few hundred megaparsecs away.[1][2]

Hubble's law is considered the first observational basis for the expansion of the universe and today serves as one of the pieces of evidence most often cited in support of the Big Bang model.[3][4] The motion of astronomical objects due solely to this expansion is known as the **Hubble flow**.[5]

Although widely attributed to Edwin Hubble, the law was first derived from the general relativity equations by Georges Lemaître in a 1927 article where he proposed the expansion of the universe and suggested an estimated value of the rate of expansion, now called the **Hubble constant**.[3][6][7][8] Two years later Edwin Hubble confirmed the existence of that law and determined a more accurate value for the constant that now bears his name.[9] See also: Alexander Friedmann § Relativity. Hubble inferred the recession velocity of the objects from their redshifts, many of which were earlier measured and related to velocity by Vesto Slipher in 1917.[10][11][12][13]

The law is often expressed by the equation $v = H_0D$, with H_0 the constant of proportionality (Hubble constant) between the "proper distance" D to a galaxy (which can change over time, unlike the comoving distance) and its velocity v (i.e. the derivative of proper distance with respect to cosmological time coordinate; see *Uses of the proper distance* for some discussion of the subtleties of this definition of 'velocity'). The SI unit of H_0 is s^{-1} but it is most frequently quoted in (km/s)/Mpc, thus giving the speed in km/s of a galaxy 1 megaparsec (3.09×10^{19} km) away. The reciprocal of H_0 is the Hubble time.

19.1 Observed values

19.2 Discovery

A decade before Hubble made his observations, a number of physicists and mathematicians had established a consistent theory of the relationship between space and time by using Einstein's field equations of general relativity. Applying the most general principles to the nature of the universe yielded a dynamic solution that conflicted with the then-prevailing notion of a static universe.

19.2.1 FLRW equations

In 1922, Alexander Friedmann derived his Friedmann equations from Einstein's field equations, showing that the Universe might expand at a rate calculable by the equations.[36] The parameter used by Friedmann is known today as the scale factor which can be considered as a scale invariant form of the proportionality constant of Hubble's law. Georges Lemaître independently found a similar solution in 1927. The Friedmann equations are derived by inserting the metric for a homogeneous and isotropic universe into Einstein's field equations for a fluid with a given density and pressure. This idea of an expanding spacetime would eventually lead to the Big Bang and Steady State theories of cosmology.

19.2.2 Lemaitre's Equation

In 1927, two years before Hubble published his own article, the Belgian priest and astronomer Georges Lemaître was the first to publish research deriving what is now known as Hubble's Law. According to the Canadian astronomer Sidney van den Bergh, "The 1927 discovery of the expansion of the Universe by Lemaitre was published in French in a low-impact journal. In the 1931 high-impact English translation of this article a critical equation was changed by omitting reference to what is now known as the Hubble constant.".[37] It is now known that the alterations in the translated paper were carried out by Lemaitre himself.[7][38]

19.2.3 Shape of the universe

Before the advent of modern cosmology, there was considerable talk about the size and shape of the universe. In 1920, the famous Shapley-Curtis debate took place between Harlow Shapley and Heber D. Curtis over this issue. Shapley argued for a small universe the size of the Milky Way galaxy and Curtis argued that the Universe was much larger. The issue was resolved in the coming decade with Hubble's improved observations.

19.2.4 Cepheid variable stars outside of the Milky Way

Edwin Hubble did most of his professional astronomical observing work at Mount Wilson Observatory, home to the world's most powerful telescope at the time. His observations of Cepheid variable stars in spiral nebulae enabled him to calculate the distances to these objects. Surprisingly, these objects were discovered to be at distances which placed them well outside the Milky Way. They continued to be called "nebulae" and it was only gradually that the term "galaxies" took over.

19.2.5 Combining redshifts with distance measurements

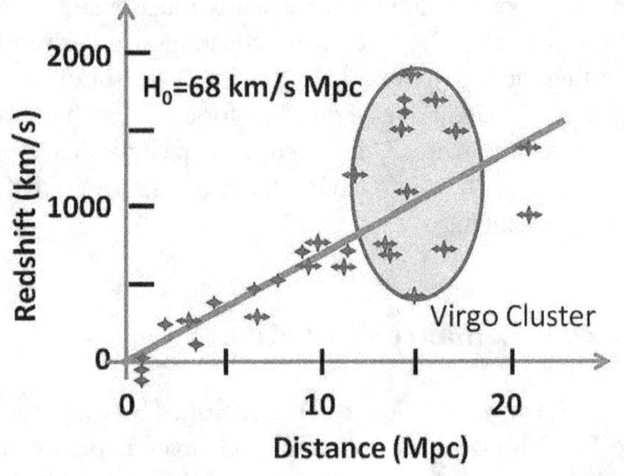

Fit of redshift velocities to Hubble's law.[39] Various estimates for the Hubble constant exist. The HST Key H_0 *Group fitted type Ia supernovae for redshifts between 0.01 and 0.1 to find that* $H_0 = 71 \pm 2$ *(statistical)* ± 6 *(systematic) km* s^{-1} Mpc^{-1}*,[30] while Sandage* et al. *find* $H_0 = 62.3 \pm 1.3$ *(statistical)* ± 5 *(systematic) km* s^{-1} Mpc^{-1}*.[40]*

The parameters that appear in Hubble's law: velocities and distances, are not directly measured. In reality we determine, say, a supernova brightness, which provides information about its distance, and the redshift $z = \Delta\lambda/\lambda$ of its spectrum of radiation. Hubble correlated brightness and parameter z.

Combining his measurements of galaxy distances with Vesto Slipher and Milton Humason's measurements of the redshifts associated with the galaxies, Hubble discovered a rough proportionality between redshift of an object and its distance. Though there was considerable scatter (now known to be caused by peculiar velocities – the 'Hubble flow' is used to refer to the region of space far enough out that the recession velocity is larger than local peculiar velocities), Hubble was able to plot a trend line from the 46 galaxies he studied and obtain a value for the Hubble constant of 500 km/s/Mpc (much higher than the currently accepted value due to errors in his distance calibrations). (See cosmic distance ladder for details.)

At the time of discovery and development of Hubble's law, it was acceptable to explain redshift phenomenon as a Doppler shift in the context of special relativity, and use the Doppler formula to associate redshift z with velocity. Today, the velocity-distance relationship of Hubble's law is viewed as a theoretical result with velocity to be connected with observed redshift not by the Doppler effect, but by a cosmological model relating recessional velocity to the expansion of the Universe. Even for small z the velocity entering the Hubble law is no longer interpreted as a Doppler effect, although at small z the velocity-redshift relation for both interpretations is the same.

Hubble Diagram

Hubble's law can be easily depicted in a "Hubble Diagram" in which the velocity (assumed approximately proportional to the redshift) of an object is plotted with respect to its distance from the observer.[41] A straight line of positive slope on this diagram is the visual depiction of Hubble's law.

19.2.6 Cosmological constant abandoned

Main article: Cosmological constant

After Hubble's discovery was published, Albert Einstein abandoned his work on the cosmological constant, which he had designed to modify his equations of general relativity to allow them to produce a static solution, which he thought was the correct state of the universe. The Einstein equations in their simplest form model generally either an expanding or contracting universe, so Einstein's cosmological constant was artificially created to counter the expansion or contraction to get a perfect static and flat universe.[42] After Hubble's discovery that the Universe was, in fact, expanding, Einstein called his faulty assumption that the Universe is static his "biggest mistake".[42] On its own, general relativity could predict the expansion of the Universe, which (through observations such as the bending of light by large masses, or the precession of the orbit of Mercury) could be experimentally observed and compared to his theoretical calculations using particular solutions of the equations he had originally formulated.

In 1931, Einstein made a trip to Mount Wilson to thank Hubble for providing the observational basis for modern cosmology.[43]

The cosmological constant has regained attention in recent decades as a hypothesis for dark energy.[44]

19.3 Interpretation

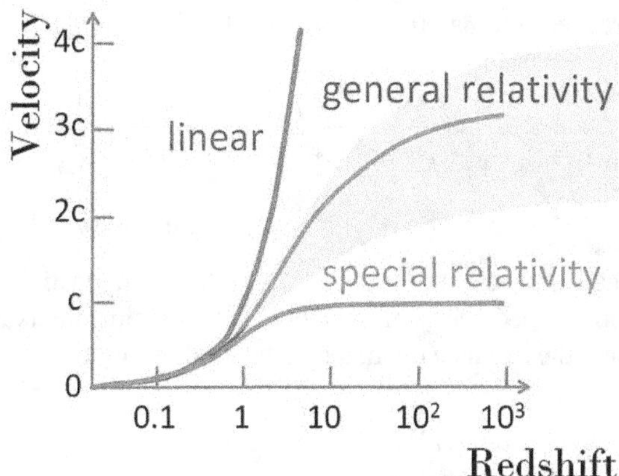

A variety of possible recessional velocity vs. redshift functions including the simple linear relation v = cz*; a variety of possible shapes from theories related to general relativity; and a curve that does not permit speeds faster than light in accordance with special relativity. All curves are linear at low redshifts. See Davis and Lineweaver.*[45]

The discovery of the linear relationship between redshift and distance, coupled with a supposed linear relation between recessional velocity and redshift, yields a straightforward mathematical expression for Hubble's Law as follows:

$$v = H_0 \, D$$

where

- v is the recessional velocity, typically expressed in km/s.
- H_0 is Hubble's constant and corresponds to the value of H (often termed the **Hubble parameter** which is a value that is time dependent and which can be expressed in terms of the scale factor) in the Friedmann equations taken at the time of observation denoted by the subscript *0*. This value is the same throughout the Universe for a given comoving time.
- D is the proper distance (which can change over time, unlike the comoving distance, which is constant) from the galaxy to the observer, measured in mega parsecs (Mpc), in the 3-space defined by given cosmological time. (Recession velocity is just $v = dD/dt$).

Hubble's law is considered a fundamental relation between recessional velocity and distance. However, the relation between recessional velocity and redshift depends on the cosmological model adopted, and is not established except for small redshifts.

For distances D larger than the radius of the Hubble sphere *r*HS , objects recede at a rate faster than the speed of light (*See* Uses of the proper distance for a discussion of the significance of this):

$$r_{HS} = \frac{c}{H_0} \ .$$

Since the Hubble "constant" is a constant only in space, not in time, the radius of the Hubble sphere may increase or decrease over various time intervals. The subscript '0' indicates the value of the Hubble constant today.[39] Current evidence suggests that the expansion of the Universe is accelerating (*see* Accelerating universe), meaning that, for any given galaxy, the recession velocity dD/dt is increasing over time as the galaxy moves to greater and greater distances; however, the Hubble parameter is actually thought to be decreasing with time, meaning that if we were to look at some *fixed* distance D and watch a series of different galaxies pass that distance, later galaxies would pass that distance at a smaller velocity than earlier ones.[46]

19.3.1 Redshift velocity and recessional velocity

Redshift can be measured by determining the wavelength of a known transition, such as hydrogen α-lines for distant quasars, and finding the fractional shift compared to a stationary reference. Thus redshift is a quantity unambiguous for experimental observation. The relation of redshift to recessional velocity is another matter. For an extensive discussion, see Harrison.[47]

Redshift velocity

The redshift z is often described as a *redshift velocity*, which is the recessional velocity that would produce the same redshift *if* it were caused by a linear Doppler effect (which, however, is not the case, as the shift is caused in part by a cosmological expansion of space, and because the velocities involved are too large to use a non-relativistic formula for Doppler shift). This redshift velocity can easily exceed the speed of light.[48] In other words, to determine the redshift velocity v_{rs}, the relation:

$$v_{rs} \equiv cz ,$$

is used.[49][50] That is, there is *no fundamental difference* between redshift velocity and redshift: they are rigidly proportional, and not related by any theoretical reasoning. The motivation behind the "redshift velocity" terminology is that the redshift velocity agrees with the velocity from a low-velocity simplification of the so-called Fizeau-Doppler formula[51]

$$z = \frac{\lambda_o}{\lambda_e} - 1 = \sqrt{\frac{1+v/c}{1-v/c}} - 1 \approx \frac{v}{c} .$$

Here, λ_o, λ_e are the observed and emitted wavelengths respectively. The "redshift velocity" v_{rs} is not so simply related to real velocity at larger velocities, however, and this terminology leads to confusion if interpreted as a real velocity. Next, the connection between redshift or redshift velocity and recessional velocity is discussed. This discussion is based on Sartori.[52]

Recessional velocity

Suppose *R(t)* is called the *scale factor* of the Universe, and increases as the Universe expands in a manner that depends upon the cosmological model selected. Its meaning is that all measured proper distances *D(t)* between co-moving points increase proportionally to *R*. (The co-moving points are not moving relative to each other except as a result of the expansion of space.) In other words:

$$\frac{D(t)}{D(t_0)} = \frac{R(t)}{R(t_0)} ,$$ [53]

where t_0 is some reference time. If light is emitted from a galaxy at time *te* and received by us at t_0, it is red shifted due to the expansion of space, and this redshift z is simply:

$$z = \frac{R(t_0)}{R(t_e)} - 1 .$$

Suppose a galaxy is at distance *D*, and this distance changes with time at a rate *dtD* . We call this rate of recession the "recession velocity" *vr*:

$$v_r = d_tD = \frac{d_tR}{R}D .$$

We now define the Hubble constant as

$$H \equiv \frac{d_tR}{R} ,$$

and discover the Hubble law:

$$v_r = HD .$$

From this perspective, Hubble's law is a fundamental relation between (i) the recessional velocity contributed by the expansion of space and (ii) the distance to an object; the connection between redshift and distance is a crutch used to connect Hubble's law with observations. This law can be related to redshift z approximately by making a Taylor series expansion:

$$z = \frac{R(t_0)}{R(t_e)} - 1 \approx \frac{R(t_0)}{R(t_0)\left(1 + (t_e - t_0)H(t_0)\right)} - 1 \approx (t_0 - t_e)H(t_0) ,$$

If the distance is not too large, all other complications of the model become small corrections and the time interval is simply the distance divided by the speed of light:

$z \approx (t_0 - t_e)H(t_0) \approx \frac{D}{c}H(t_0)$, or $cz \approx DH(t_0) = v_r$.

According to this approach, the relation $cz = v_r$ is an approximation valid at low redshifts, to be replaced by a relation at large redshifts that is model-dependent. See velocity-redshift figure.

19.3.2 Observability of parameters

Strictly speaking, neither v nor D in the formula are directly observable, because they are properties *now* of a galaxy, whereas our observations refer to the galaxy in the past, at the time that the light we currently see left it.

For relatively nearby galaxies (redshift z much less than unity), v and D will not have changed much, and v can be estimated using the formula $v = zc$ where c is the speed of light. This gives the empirical relation found by Hubble.

For distant galaxies, v (or D) cannot be calculated from z without specifying a detailed model for how H changes with time. The redshift is not even directly related to the recession velocity at the time the light set out, but it does have a simple interpretation: $(1+z)$ is the factor by which the Universe has expanded while the photon was travelling towards the observer.

19.3.3 Expansion velocity vs relative velocity

In using Hubble's law to determine distances, only the velocity due to the expansion of the Universe can be used. Since gravitationally interacting galaxies move relative to each other independent of the expansion of the Universe, these relative velocities, called peculiar velocities, need to be accounted for in the application of Hubble's law.

The Finger of God effect is one result of this phenomenon. In systems that are gravitationally bound, such as galaxies or our planetary system, the expansion of space is a much weaker effect than the attractive force of gravity.

19.3.4 Idealized Hubble's Law

The mathematical derivation of an idealized Hubble's Law for a uniformly expanding universe is a fairly elementary theorem of geometry in 3-dimensional Cartesian/Newtonian coordinate space, which, considered as a metric space, is entirely homogeneous and isotropic (properties do not vary with location or direction). Simply stated the theorem is this:

> *Any two points which are moving away from the origin, each along straight lines and with speed proportional to distance from the origin, will be moving away from each other with a speed proportional to their distance apart.*

In fact this applies to non-Cartesian spaces as long as they are locally homogeneous and isotropic; specifically to the negatively and positively curved spaces frequently considered as cosmological models (see shape of the universe).

An observation stemming from this theorem is that seeing objects recede from us on Earth is not an indication that Earth is near to a center from which the expansion is occurring, but rather that *every* observer in an expanding universe will see objects receding from them.

19.3.5 Ultimate fate and age of the universe

The value of the Hubble parameter changes over time, either increasing or decreasing depending on the value of the so-called deceleration parameter q , which is defined by

$$q = -\left(1 + \frac{\dot{H}}{H^2}\right).$$

In a universe with a deceleration parameter equal to zero, it follows that $H = 1/t$, where t is the time since the Big Bang. A non-zero, time-dependent value of q simply requires integration of the Friedmann equations backwards from the present time to the time when the comoving horizon size was zero.

It was long thought that q was positive, indicating that the expansion is slowing down due to gravitational attraction. This would imply an age of the Universe less than $1/H$ (which is about 14 billion years). For instance, a value for q of 1/2 (once favoured by most theorists) would give the age of the Universe as $2/(3H)$. The discovery in 1998 that q is apparently negative means that the Universe could actually be older than $1/H$. However, estimates of the age of the universe are very close to $1/H$.

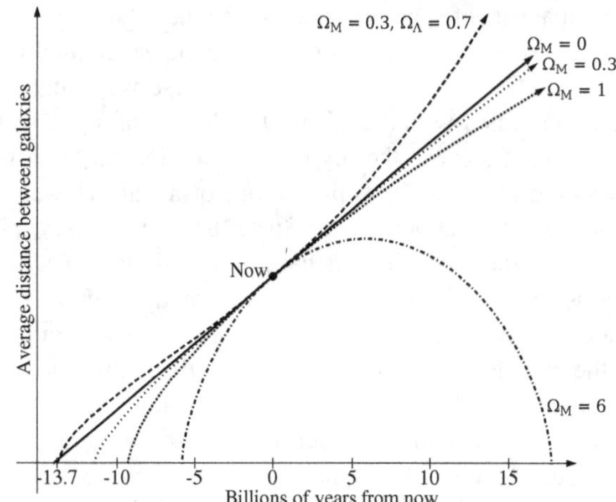

The age and ultimate fate of the universe can be determined by measuring the Hubble constant today and extrapolating with the observed value of the deceleration parameter, uniquely characterized by values of density parameters (ΩM for matter and ΩΛ for dark energy). A "closed universe" with ΩM > 1 and ΩΛ = 0 comes to an end in a Big Crunch and is considerably younger than its Hubble age. An "open universe" with ΩM ≤ 1 and ΩΛ = 0 expands forever and has an age that is closer to its Hubble age. For the accelerating universe with nonzero ΩΛ that we inhabit, the age of the universe is coincidentally very close to the Hubble age.

19.3.6 Olbers' paradox

Main article: Olbers' paradox

The expansion of space summarized by the Big Bang interpretation of Hubble's Law is relevant to the old conundrum known as Olbers' paradox: if the Universe were infinite, static, and filled with a uniform distribution of stars, then every line of sight in the sky would end on a star, and the sky would be as bright as the surface of a star. However, the night sky is largely dark. Since the 17th century, astronomers and other thinkers have proposed many possible ways to resolve this paradox, but the currently accepted resolution depends in part on the Big Bang theory and in part on the Hubble expansion. In a universe that exists for a finite amount of time, only the light of a finite number of stars has had a chance to reach us yet, and the paradox is resolved. Additionally, in an expanding universe, distant objects recede from us, which causes the light emanating from them to be redshifted and diminished in brightness.[54]

19.3.7 Dimensionless Hubble parameter

Instead of working with Hubble's constant, a common practice is to introduce the **dimensionless Hubble parameter**, usually denoted by h, and to write the Hubble's parameter H_0 as $h \times 100$ km s^{-1} Mpc^{-1}, all the uncertainty relative of the value of H_0 being then relegated to h.[55] If a subscript is presented after h, it refers to the value of h used in that text's preceding calculation, and is equal to H_0 / 100. Currently $h = 0.678$, which can be represented as $h_{0.678}$. This should not be confused with the dimensionless value of Hubble's constant, usually expressed in terms of Planck units, with current value of $H_0 \times t$P = 1.18×10^{-61}.

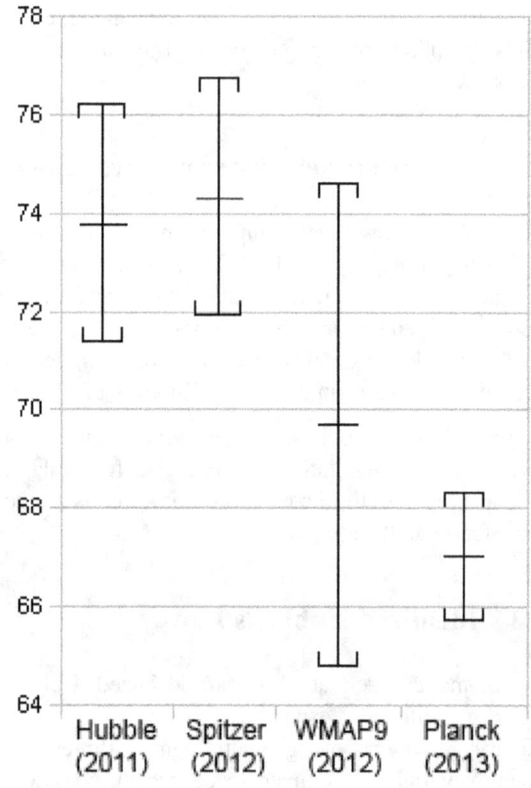

Value of the Hubble Constant including measurement uncertainty for recent surveys.[19]

19.4 Determining the Hubble constant

The value of the Hubble constant is estimated by measuring the redshift of distant galaxies and then determining the distances to the same galaxies (by some other method than Hubble's law). Uncertainties in the physical assumptions used to determine these distances have caused varying estimates of the Hubble constant.[3]

19.4.1 Earlier measurement and discussion approaches

For most of the second half of the 20th century the value of H_0 was estimated to be between 50 and 90 (km/s)/Mpc.

The value of the Hubble constant was the topic of a long and rather bitter controversy between Gérard de Vaucouleurs, who claimed the value was around 100, and Allan Sandage, who claimed the value was near 50.[31] In 1996, a debate moderated by John Bahcall between Sidney van den Bergh and Gustav Tammann was held in similar fashion to the earlier Shapley-Curtis debate over these two competing values.

This previously wide variance in estimates was partially resolved with the introduction of the ΛCDM model of the Universe in the late 1990s. With the ΛCDM model observations of high-redshift clusters at X-ray and microwave wavelengths using the Sunyaev-Zel'dovich effect, measurements of anisotropies in the cosmic microwave background radiation, and optical surveys all gave a value of around 70 for the constant.

More recent measurements from the Planck mission indicate a lower value of around 67.[19]

See table of measurements above for many recent and older measurements.

19.4.2 Acceleration of the expansion

Main article: Accelerating universe

A value for q measured from standard candle observations of Type Ia supernovae, which was determined in 1998 to be negative, surprised many astronomers with the implication that the expansion of the Universe is currently "accelerating"[56] (although the Hubble factor is still decreasing with time, as mentioned above in the Interpretation section; see the articles on dark energy and the ΛCDM model).

19.5 Derivation of the Hubble parameter

Start with the Friedmann equation:

$$H^2 \equiv \left(\frac{\dot{a}}{a}\right)^2 = \frac{8\pi G}{3}\rho - \frac{kc^2}{a^2} + \frac{\Lambda c^2}{3},$$

where H is the Hubble parameter, a is the scale factor, G is the gravitational constant, k is the normalised spatial curvature of the Universe and equal to −1, 0, or +1, and Λ is the cosmological constant.

19.5.1 Matter-dominated universe (with a cosmological constant)

If the Universe is matter-dominated, then the mass density of the Universe ρ can just be taken to include matter so

$$\rho = \rho_m(a) = \frac{\rho_{m_0}}{a^3},$$

where ρ_{m_0} is the density of matter today. We know for nonrelativistic particles that their mass density decreases proportional to the inverse volume of the Universe, so the equation above must be true. We can also define (see density parameter for Ω_m)

$$\rho_c = \frac{3H^2}{8\pi G};$$

$$\Omega_m \equiv \frac{\rho_{m_0}}{\rho_c} = \frac{8\pi G}{3H_0^2}\rho_{m_0};$$

so $\rho = \rho_c\Omega_m/a^3$. Also, by definition,

$$\Omega_k \equiv \frac{-kc^2}{(a_0 H_0)^2}$$

and

$$\Omega_\Lambda \equiv \frac{\Lambda c^2}{3H_0^2},$$

where the subscript nought refers to the values today, and $a_0 = 1$. Substituting all of this into the Friedmann equation at the start of this section and replacing a with $a = 1/(1+z)$ gives

$$H^2(z) = H_0^2\left(\Omega_M(1+z)^3 + \Omega_k(1+z)^2 + \Omega_\Lambda\right).$$

19.5.2 Matter- and dark energy-dominated universe

If the Universe is both matter-dominated and dark energy-dominated, then the above equation for the Hubble parameter will also be a function of the equation of state of dark energy. So now:

$$\rho = \rho_m(a) + \rho_{de}(a),$$

where ρ_{de} is the mass density of the dark energy. By definition, an equation of state in cosmology is $P = w\rho c^2$, and if this is substituted into the fluid equation, which describes how the mass density of the Universe evolves with time, then

$$\dot{\rho} + 3\frac{\dot{a}}{a}\left(\rho + \frac{P}{c^2}\right) = 0;$$

$$\frac{d\rho}{\rho} = -3\frac{da}{a}(1+w).$$

If w is constant, then

$$\ln\rho = -3\,(1+w)\ln a;$$

$$\rho = a^{-3(1+w)}.$$

Therefore, for dark energy with a constant equation of state w, $\rho_{de}(a) = \rho_{de0}a^{-3(1+w)}$. If this is substituted into the Friedman equation in a similar way as before, but this time set $k = 0$, which assumes a spatially flat universe, then (see Shape of the universe)

$$H^2(z) = H_0^2 \left(\Omega_M(1+z)^3 + \Omega_{de}(1+z)^{3(1+w)}\right).$$

If the dark energy derives from a cosmological constant such as that introduced by Einstein, it can be shown that $w = -1$. The equation then reduces to the last equation in the matter-dominated universe section, with Ω_k set to zero. In that case the initial dark energy density ρ_{de0} is given by[57]

$$\rho_{de0} = \frac{\Lambda c^2}{8\pi G} \text{ and } \Omega_{de} = \Omega_\Lambda.$$

If dark energy does not have a constant equation-of-state w, then

$$\rho_{de}(a) = \rho_{de0} e^{-3\int \frac{da}{a}(1+w(a))},$$

and to solve this, $w(a)$ must be parametrized, for example if $w(a) = w_0 + w_a(1-a)$, giving

$$H^2(z) = H_0^2 \left(\Omega_M a^{-3} + \Omega_{de} a^{-3(1+w_0+w_a)} e^{-3w_a(1-a)}\right).$$

Other ingredients have been formulated recently.[58][59][60]

19.6 Units derived from the Hubble constant

19.6.1 Hubble time

The Hubble constant H_0 has units of inverse time; the **Hubble time** tH is simply defined as the inverse of the Hubble constant,[61] i.e. $t_H \equiv \frac{1}{H_0} = \frac{1}{67.8\text{(km/s)/Mpc}} = 4.55 \cdot 10^{17}\text{s}$ = 14.4 billion years. This is slightly different from the age of the universe $t_0 \approx$ 13.8 billion years. The Hubble time is the age it would have had if the expansion had been linear, and it is different from the real age of the universe because the expansion isn't linear; they are related by a dimensionless factor which depends on the mass-energy content of the universe, which is around 0.96 in the standard Lambda-CDM model.

We currently appear to be approaching a period where the expansion is exponential due to the increasing dominance of vacuum energy. In this regime, the Hubble parameter is constant, and the universe grows by a factor e each Hubble time:

$$H \equiv \frac{\dot{a}}{a} = \text{const.} \Rightarrow a \propto \mathrm{e}^{Ht} = \mathrm{e}^{t/t_H}$$

Over long periods of time, the dynamics are complicated by general relativity, dark energy, inflation, etc., as explained above.

19.6.2 Hubble length

The Hubble length or Hubble distance is a unit of distance in cosmology, defined as cH_0^{-1} — the speed of light multiplied by the Hubble time. It is equivalent to 4,550 million parsecs or 14.4 billion light years. (The numerical value of the Hubble length in light years is, by definition, equal to that of the Hubble time in years.) The Hubble distance would be the distance between the Earth and the galaxies which are *currently* receding from us at the speed of light, as can be seen by substituting $D = c/H_0$ into the equation for Hubble's law, $v = H_0D$.

19.6.3 Hubble volume

Main article: Hubble volume

The Hubble volume is sometimes defined as a volume of the Universe with a comoving size of c/H_0. The exact definition varies: it is sometimes defined as the volume of a sphere with radius c/H_0, or alternatively, a cube of side c/H_0. Some cosmologists even use the term Hubble volume to refer to the volume of the observable universe, although this has a radius approximately three times larger.

19.7 See also

- Cosmology
- Dark energy
- Dark matter
- Tests of general relativity

19.8 Notes

[1] Riess, A.; et al. (September 1998). "Observational Evidence from Supernovae for an Accelerating Universe and a Cosmological Constant". *The Astronomical Journal.* **116** (3): 1009–1038. Bibcode:1998AJ....116.1009R. arXiv:astro-ph/9805201. doi:10.1086/300499.

[2] Perlmutter, S.; et al. (June 1999). "Measurements of Omega and Lambda from 42 High-Redshift Supernovae". *The Astrophysical Journal.* **517** (2): 565–586. Bibcode:1999ApJ...517..565P. arXiv:astro-ph/9812133. doi:10.1086/307221.

[3] Overbye, Dennis (20 February 2017). "Cosmos Controversy: The Universe Is Expanding, but How Fast?". *New York Times*. Retrieved 21 February 2017.

[4] Coles, P., ed. (2001). *Routledge Critical Dictionary of the New Cosmology*. Routledge. p. 202. ISBN 0-203-16457-1.

[5] "Hubble Flow". *The Swinburne Astronomy Online Encyclopedia of Astronomy*. Swinburne University of Technology. Retrieved 2013-05-14.

[6] Lemaître, G. (1927). "Un univers homogène de masse constante et de rayon croissant rendant compte de la vitesse radiale des nébuleuses extra-galactiques". *Annales de la Société Scientifique de Bruxelles A* (47): 49–59. Bibcode:1927ASSB...47...49L. Partially translated in Lemaître, G. (1931). "Expansion of the universe, A homogeneous universe of constant mass and increasing radius accounting for the radial velocity of extra-galactic nebulae". *Monthly Notices of the Royal Astronomical Society*. **91**: 483–490. Bibcode:1931MNRAS..91..483L. doi:10.1093/mnras/91.5.483.

[7] Livio, M. (2011). "Lost in translation: Mystery of the missing text solved". *Nature*. **479** (7372): 171. Bibcode:2011Natur.479..171L. PMID 22071745. doi:10.1038/479171a.

[8] Livio, M.; Riess, A. (2013). "Measuring the Hubble constant". *Physics Today*. **66** (10): 41. Bibcode:2013PhT....66j..41L. doi:10.1063/PT.3.2148.

[9] Hubble, E. (1929). "A relation between distance and radial velocity among extra-galactic nebulae". *Proceedings of the National Academy of Sciences*. **15** (3): 168–73. Bibcode:1929PNAS...15..168H. PMC 522427. PMID 16577160. doi:10.1073/pnas.15.3.168.

[10] Slipher, V.M. (1917). "Radial velocity observations of spiral nebulae". *The Observatory*. **40**: 304–306.

[11] Longair, M. S. (2006). *The Cosmic Century*. Cambridge University Press. p. 109. ISBN 0-521-47436-1.

[12] Nussbaumer, Harry (2013). *'Slipher's redshifts as support for de Sitter's model and the discovery of the dynamic universe' In Origins of the Expanding Universe: 1912-1932*. Astronomical Society of the Pacific. pp. 25–38.Physics ArXiv preprint

[13] O'Raifeartaigh, Cormac (2013). *The Contribution of V.M. Slipher to the discovery of the expanding universe in 'Origins of the Expanding Universe'*. Astronomical Society of the Pacific. pp. 49–62.Physics ArXiv preprint

[14] Bonvin, Vivien; Courbin, Frédéric; Suyu, Sherry H.; et al. (2016-11-22). "H0LiCOW – V. New COSMOGRAIL time delays of HE 0435−1223: H_0 to 3.8 per cent precision from strong lensing in a flat ΛCDM model". *MNRAS*. **465** (4): 4914–4930. arXiv:1607.01790. doi:10.1093/mnras/stw3006.

[15] Grieb, Jan N.; Sánchez, Ariel G.; Salazar-Albornoz, Salvador (2016-07-13). "The clustering of galaxies in the completed SDSS-III Baryon Oscillation Spectroscopic Survey: Cosmological implications of the Fourier space wedges of the final sample". arXiv:1607.03143.

[16] Riess, Adam G.; Macri, Lucas M.; Hoffmann, Samantha L.; Scolnic, Dan; Casertano, Stefano; Filippenko, Alexei V.; Tucker, Brad E.; Reid, Mark J.; Jones, David O. (2016-04-05). "A 2.4% Determination of the Local Value of the Hubble Constant". arXiv:1604.01424.

[17] "Planck Publications: Planck 2015 Results". European Space Agency. February 2015. Retrieved 9 February 2015.

[18] Cowen, Ron; Castelvecchi, Davide (2 December 2014). "European probe shoots down dark-matter claims". *Nature*. doi:10.1038/nature.2014.16462. Retrieved 6 December 2014.

[19] Bucher, P. A. R.; *et al.* (Planck Collaboration) (2013). "Planck 2013 results. I. Overview of products and scientific Results". arXiv:1303.5062 [astro-ph.CO].

[20] "Planck reveals an almost perfect universe". ESA. 21 March 2013. Retrieved 2013-03-21.

[21] "Planck Mission Brings Universe Into Sharp Focus". JPL. 21 March 2013. Retrieved 2013-03-21.

[22] Overbye, D. (21 March 2013). "An infant universe, born before we knew". *New York Times*. Retrieved 2013-03-21.

[23] Boyle, A. (21 March 2013). "Planck probe's cosmic 'baby picture' revises universe's vital statistics". *NBC News*. Retrieved 2013-03-21.

[24] Bennett, C. L.; et al. (2013). "Nine-year Wilkinson Microwave Anisotropy Probe (WMAP) observations: Final maps and results". *The Astrophysical Journal Supplement Series*. **208** (2): 20. Bibcode:2013ApJS..208...20B. arXiv:1212.5225. doi:10.1088/0067-0049/208/2/20.

[25] Jarosik, N.; et al. (2011). "Seven-year Wilkinson Microwave Anisotropy Probe (WMAP) observations: Sky maps, systematic errors, and basic results". *The Astrophysical Journal Supplement Series*. **192** (2): 14. Bibcode:2011ApJS..192...14J. arXiv:1001.4744. doi:10.1088/0067-0049/192/2/14.

[26] Results for H_0 and other cosmological parameters obtained by fitting a variety of models to several combinations of WMAP and other data are available at the NASA's LAMBDA website.

[27] Hinshaw, G.; et al. (WMAP Collaboration) (2009). "Five-year Wilkinson Microwave Anisotropy Probe observations: Data processing, sky maps, and basic results". *The Astrophysical Journal Supplement*. **180** (2): 225–245. Bibcode:2009ApJS..180..225H. arXiv:0803.0732. doi:10.1088/0067-0049/180/2/225.

[28] Spergel, D. N.; et al. (WMAP Collaboration) (2007). "Three-year Wilkinson Microwave Anisotropy Probe (WMAP) Observations: Implications for cosmology". *The Astrophysical Journal Supplement Series*. **170** (2): 377–408. Bibcode:2007ApJS..170..377S. arXiv:astro-ph/0603449. doi:10.1086/513700.

[29] Bonamente, M.; Joy, M. K.; Laroque, S. J.; Carlstrom, J. E.; Reese, E. D.; Dawson, K. S. (2006). "Determination of the cosmic distance scale from Sunyaev–Zel'dovich effect and Chandra X-ray measurements of high-redshift galaxy clusters". *The Astrophysical Journal*. **647**: 25. Bibcode:2006ApJ...647...25B. arXiv:astro-ph/0512349. doi:10.1086/505291.

[30] Freedman, W. L.; et al. (2001). "Final results from the Hubble Space Telescope Key Project to measure the Hubble constant". *The Astrophysical Journal*. **553** (1): 47–72. Bibcode:2001ApJ...553...47F. arXiv:astro-ph/0012376. doi:10.1086/320638.

[31] Overbye, D. (1999). "Prologue". *Lonely Hearts of the Cosmos* (2nd ed.). HarperCollins. p. 1*ff*. ISBN 978-0-316-64896-7.

[32] John P. Huchra (2008). "The Hubble Constant". *Harvard Center for Astrophysics*.

[33] Sandage, A. R. (1958). "Current problems in the extragalactic distance scale". *The Astrophysical Journal*. **127** (3): 513–526. Bibcode:1958ApJ...127..513S. doi:10.1086/146483.

[34] Edwin Hubble, *A Relation between Distance and Radial Velocity among Extra-Galactic Nebulae*, Proceedings of the National Academy of Sciences, vol. 15, no. 3, pp. 168-173, March 1929

[35] "Hubble's Constant". *Skywise Unlimited - Western Washington University*.

[36] Friedman, A. (1922). "Über die Krümmung des Raumes". *Zeitschrift für Physik*. **10** (1): 377–386. Bibcode:1922ZPhy...10..377F. doi:10.1007/BF01332580. Translated in Friedmann, A. (1999). "On the Curvature of Space". *General Relativity and Gravitation*. **31** (12): 1991–2000. Bibcode:1999GReGr..31.1991F. doi:10.1023/A:1026751225741.

[37] van den Bergh, Sydney. "The Curious Case of Lemaitre's Equation No. 24". arXiv:1106.1195.

[38] Block, David (2012). *'Georges Lemaitre and Stigler's law of eponymy' in Georges Lemaitre: Life, Science and Legacy* (Holder and Mitton ed.). Springer. pp. 89–96.

[39] Keel, W. C. (2007). *The Road to Galaxy Formation* (2nd ed.). Springer. pp. 7–8. ISBN 3-540-72534-2.

[40] Weinberg, S. (2008). *Cosmology*. Oxford University Press. p. 28. ISBN 0-19-852682-2.

[41] Kirshner, R. P. (2003). "Hubble's diagram and cosmic expansion". *Proceedings of the National Academy of Sciences*. **101** (1): 8–13. Bibcode:2003PNAS..101....8K. PMC 314128. PMID 14695886. doi:10.1073/pnas.2536799100.

[42] "What is a Cosmological Constant?". Goddard Space Flight Center. Retrieved 2013-10-17.

[43] Isaacson, W. (2007). *Einstein: His Life and Universe*. Simon & Schuster. p. 354. ISBN 0-7432-6473-8.

[44] "Einstein's Biggest Blunder? Dark Energy May Be Consistent With Cosmological Constant". Science Daily. 28 November 2007. Retrieved 2013-06-02.

[45] Davis, T. M.; Lineweaver, C. H. (2001). "Superluminal Recessional Velocities". *AIP Conference Proceedings*. **555**: 348–351. Bibcode:2001AIPC..555..348D. arXiv:astro-ph/0011070. doi:10.1063/1.1363540.

[46] "Is the universe expanding faster than the speed of light?". *Ask an Astronomer at Cornell University*. Archived from the original on 23 November 2003. Retrieved 5 June 2015.

[47] Harrison, E. (1992). "The redshift-distance and velocity-distance laws". *The Astrophysical Journal*. **403**: 28–31. Bibcode:1993ApJ...403...28H. doi:10.1086/172179.

[48] Madsen, M. S. (1995). *The Dynamic Cosmos*. CRC Press. p. 35. ISBN 0-412-62300-5.

[49] Dekel, A.; Ostriker, J. P. (1999). *Formation of Structure in the Universe*. Cambridge University Press. p. 164. ISBN 0-521-58632-1.

[50] Padmanabhan, T. (1993). *Structure formation in the universe*. Cambridge University Press. p. 58. ISBN 0-521-42486-0.

[51] Sartori, L. (1996). *Understanding Relativity*. University of California Press. p. 163, Appendix 5B. ISBN 0-520-20029-2.

[52] Sartori, L. (1996). *Understanding Relativity*. University of California Press. pp. 304–305. ISBN 0-520-20029-2.

[53] "Introduction to Cosmology", Matts Roos

[54] Chase, S. I.; Baez, J. C. (2004). "Olbers' Paradox". *The Original Usenet Physics FAQ*. Retrieved 2013-10-17. See also Asimov, I. (1974). "The Black of Night". *Asimov on Astronomy*. Doubleday. ISBN 0-385-04111-X.

[55] Peebles, P. J. E. (1993). *Principles of Physical Cosmology*. Princeton University Press.

[56] Perlmutter, S. (2003). "Supernovae, Dark Energy, and the Accelerating Universe" (PDF). *Physics Today*. **56** (4): 53–60. Bibcode:2003PhT....56d..53P. doi:10.1063/1.1580050.

[57] Carroll, Sean (2004). *Spacetime and Geometry: An Introduction to General Relativity* (illustrated ed.). San Fraancisco: Addison-Wesley. p. 328. ISBN 978-0-8053-8732-2.

[58] Tawfik, A.; Harko, T. (2012). "Quark-hadron phase transitions in the viscous early universe". *Physical Review D*. **85** (8): 084032. Bibcode:2012PhRvD..85h4032T. arXiv:1108.5697. doi:10.1103/PhysRevD.85.084032.

[59] Tawfik, A. (2011). "The Hubble parameter in the early universe with viscous QCD matter and finite cosmological constant". *Annalen der Physik*. **523** (5): 423. Bibcode:2011AnP...523..423T. arXiv:1102.2626. doi:10.1002/andp.201100038.

[60] Tawfik, A.; Wahba, M.; Mansour, H.; Harko, T. (2011). "Viscous quark-gluon plasma in the early universe". *Annalen der Physik*. **523** (3): 194. Bibcode:2011AnP...523..194T. arXiv:1001.2814. doi:10.1002/andp.201000052.

[61] Hawley, John F.; Holcomb, Katherine A. (2005). *Foundations of modern cosmology* (2nd ed.). Oxford [u.a.]: Oxford Univ. Press. p. 304. ISBN 0-19-853096-X.

19.9 References

- Hubble, E. P. (1937). *The Observational Approach to Cosmology*. Clarendon Press. LCCN 38011865.
- Kutner, M. (2003). *Astronomy: A Physical Perspective*. Cambridge University Press. ISBN 0-521-52927-1.
- Liddle, A. R. (2003). *An Introduction to Modern Cosmology* (2nd ed.). John Wiley & Sons. ISBN 0-470-84835-9.

19.10 Further reading

- Freedman, W. L.; Madore, B. F. (2010). "The Hubble Constant". *Annual Review of Astronomy and Astrophysics*. **48**: 673. Bibcode:2010ARA&A..48..673F. arXiv:1004.1856. doi:10.1146/annurev-astro-082708-101829.

19.11 External links

- NASA's WMAP - Big Bang Expansion: the Hubble Constant
- The Hubble Key Project
- The Hubble Diagram Project
- Merrifield, Michael (2009). "Hubble Constant". *Sixty Symbols*. Brady Haran for the University of Nottingham.

Chapter 20

Matter

This article is about the concept in the physical sciences. For other uses, see Matter (disambiguation).

In the classical physics observed in everyday life, **matter** is any substance that has mass and takes up space; this includes atoms and anything made up of these, but not other energy phenomena or waves such as light or sound.[1][2] More generally, however, in (modern) physics, matter is not a fundamental concept because a universal definition of it is elusive; for example, the elementary constituents of atoms may be point particles, each having no volume individually.

All the everyday objects that we can bump into, touch or squeeze are ultimately composed of atoms. This ordinary atomic matter is in turn made up of interacting subatomic particles—usually a nucleus of protons and neutrons, and a cloud of orbiting electrons.[3][4] Typically, science considers these composite particles matter because they have both rest mass and volume. By contrast, massless particles, such as photons, are not considered matter, because they have neither rest mass nor volume. However, not all particles with rest mass have a classical volume, since fundamental particles such as quarks and leptons (sometimes equated with matter) are considered "point particles" with no effective size or volume. Nevertheless, quarks and leptons together make up "ordinary matter", and their interactions contribute to the effective volume of the composite particles that make up ordinary matter.

Matter exists in *states* (or *phases*): the classical solid, liquid, and gas; as well as the more exotic plasma, Bose–Einstein condensates, fermionic condensates, and quark–gluon plasma.[5]

For much of the history of the natural sciences people have contemplated the exact nature of matter. The idea that matter was built of discrete building blocks, the so-called *particulate theory of matter*, was first put forward by the Greek philosophers Leucippus (~490 BC) and Democritus (~470–380 BC).[6]

20.1 Comparison with mass

Matter should not be confused with mass, as the two are not the same in modern physics.[7] Matter is itself a physical substance of which systems may be composed, while mass is not a substance but rather a quantitative **property** of matter and other substances or systems. While there are different views on what should be considered matter, the mass of a substance or system is the same irrespective of any such definition of matter. Another difference is that matter has an "opposite" called antimatter, but mass has no opposite—there is no such thing as "anti-mass" or negative mass. Antimatter has the same (i.e. positive) mass property as its normal matter counterpart.

Different fields of science use the term matter in different, and sometimes incompatible, ways. Some of these ways are based on loose historical meanings, from a time when there was no reason to distinguish mass from simply a quantity of matter. As such, there is no single universally agreed scientific meaning of the word "matter". Scientifically, the term "mass" is well-defined, but "matter" can be defined in several ways. Sometimes in the field of physics "matter" is simply equated with particles that exhibit rest mass (i.e., that cannot travel at the speed of light), such as quarks and leptons. However, in both physics and chemistry, matter exhibits both wave-like and particle-like properties, the so-called wave–particle duality.[8][9][10]

20.2 Definition

20.2.1 Based on atoms

A definition of "matter" based on its physical and chemical structure is: *matter is made up of atoms.*[11] Such **atomic matter** is also sometimes termed **ordinary matter**. As an example, deoxyribonucleic acid molecules (DNA) are matter under this definition because they are made of atoms. This definition can extend to include charged atoms and molecules, so as to include plasmas (gases of ions) and electrolytes (ionic solutions), which are not obviously included in the atoms definition. Alternatively, one can adopt the *protons, neutrons, and electrons* definition.

20.2.2 Based on protons, neutrons and electrons

A definition of "matter" more fine-scale than the atoms and molecules definition is: *matter is made up of what atoms and molecules are made of*, meaning anything made of positively charged protons, neutral neutrons, and negatively charged electrons.[12] This definition goes beyond atoms and molecules, however, to include substances made from these building blocks that are *not* simply atoms or molecules, for example electron beams in an old cathode ray tube television, or white dwarf matter—typically, carbon and oxygen nuclei in a sea of degenerate electrons. At a microscopic level, the constituent "particles" of matter such as protons, neutrons, and electrons obey the laws of quantum mechanics and exhibit wave–particle duality. At an even deeper level, protons and neutrons are made up of quarks and the force fields (gluons) that bind them together, leading to the next definition.

20.2.3 Based on quarks and leptons

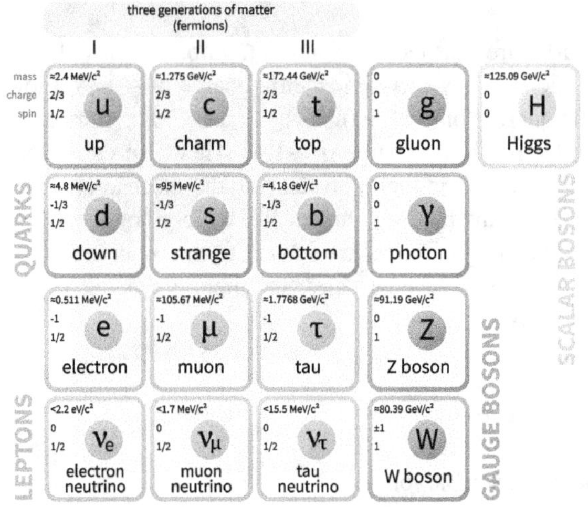

Under the "quarks and leptons" definition, the elementary and composite particles made of the quarks (in purple) and leptons (in green) would be matter—while the gauge bosons (in red) would not be matter. However, interaction energy inherent to composite particles (for example, gluons involved in neutrons and protons) contribute to the mass of ordinary matter.

As seen in the above discussion, many early definitions of what can be called *ordinary matter* were based upon its structure or *building blocks*. On the scale of elementary particles, a definition that follows this tradition can be stated as: *ordinary matter is everything that is composed of quarks and leptons*, or *ordinary matter is everything that is composed of any elementary fermions except antiquarks and antileptons*.[13][14][15] The connection between these formulations follows.

Leptons (the most famous being the electron), and quarks (of which baryons, such as protons and neutrons, are made) combine to form atoms, which in turn form molecules. Because atoms and molecules are said to be matter, it is natural to phrase the definition as: *ordinary matter is anything that is made of the same things that atoms and molecules are made of*. (However, notice that one also can make from these building blocks matter that is *not* atoms or molecules.) Then, because electrons are leptons, and protons, and neutrons are made of quarks, this definition in turn leads to the definition of matter as being *quarks and leptons*, which are two of the four types of elementary fermions (the other two being antiquarks and antileptons, which can be considered antimatter as described later). Carithers and Grannis state: *Ordinary matter is composed entirely of first-generation particles, namely the [up] and [down] quarks, plus the electron and its neutrino*.[14] (Higher generations particles quickly decay into first-generation particles, and thus are not commonly encountered.[16])

This definition of ordinary matter is more subtle than it first appears. All the particles that make up ordinary matter (leptons and quarks) are elementary fermions, while all the force carriers are elementary bosons.[17] The W and Z bosons that mediate the weak force are not made of quarks or leptons, and so are not ordinary matter, even if they have mass.[18] In other words, mass is not something that is exclusive to ordinary matter.

The quark–lepton definition of ordinary matter, however, identifies not only the elementary building blocks of matter, but also includes composites made from the constituents (atoms and molecules, for example). Such composites contain an interaction energy that holds the constituents together, and may constitute the bulk of the mass of the composite. As an example, to a great extent, the mass of an atom is simply the sum of the masses of its constituent protons, neutrons and electrons. However, digging deeper, the protons and neutrons are made up of quarks bound together by gluon fields (see dynamics of quantum chromodynamics) and these gluons fields contribute significantly to the mass of hadrons.[19] In other words, most of what composes the "mass" of ordinary matter is due to the binding energy of quarks within protons and neutrons.[20] For example, the sum of the mass of the three quarks in a nucleon is approximately 12.5 MeV/c^2, which is low compared to the mass of a nucleon (approximately 938 MeV/c^2).[21][22] The bottom line is that most of the mass of everyday objects comes from the interaction energy of its elementary components.

The Standard Model groups matter particles into three generations, where each generation consists of two quarks and two leptons. The first generation is the *up* and *down* quarks, the *electron* and the *electron neutrino*; the second includes the *charm* and *strange* quarks, the *muon* and the *muon neutrino*; the third generation consists of the *top* and *bottom* quarks and the *tau* and *tau neutrino*.[23] The most natural explanation for this would be that quarks and leptons of higher generations are excited states of the first generations. If this turns out to be the case, it would imply that quarks and leptons are composite particles, rather than elementary particles.[24]

This quark-lepton definition of matter also leads to what can be described as "conservation of (net) matter" laws—discussed later below. Alternatively, one could return to the mass-volume-space concept of matter, leading to the next definition, in which antimatter becomes included as a subclass of matter.

20.2.4 Based on elementary fermions (mass, volume, and space)

A common or traditional definition of matter is *anything that has mass and volume (occupies space)*.[25][26] For example, a car would be said to be made of matter, as it has mass and volume (occupies space).

The observation that matter occupies space goes back to antiquity. However, an explanation for why matter occupies space is recent, and is argued to be a result of the phenomenon described in the Pauli exclusion principle,[27][28] which applies to fermions. Two particular examples where the exclusion principle clearly relates matter to the occupation of space are white dwarf stars and neutron stars, discussed further below.

Thus, matter can be defined as everything composed of elementary fermions. Although we don't encounter them in everyday life, antiquarks (such as the antiproton) and antileptons (such as the positron) are the antiparticles of the quark and the lepton, are elementary fermions as well, and have essentially the same properties as quarks and leptons, including the applicability of the Pauli exclusion principle which can be said to prevent two particles from being in the same place at the same time (in the same state), i.e. makes each particle "take up space". This particular definition leads to matter being defined to include anything made of these antimatter particles as well as the ordinary quark and lepton, and thus also anything made of mesons, which are unstable particles made up of a quark and an antiquark.

20.2.5 In general relativity and cosmology

In the context of relativity, mass is not an additive quantity, in the sense that one can add the rest masses of particles in a system to get the total rest mass of the system.[1] Thus, in relativity usually a more general view is that it is not the sum of rest masses, but the energy–momentum tensor that quantifies the amount of matter. This tensor gives the rest mass for the entire system. "Matter" therefore is sometimes considered as anything that contributes to the energy–momentum of a system, that is, anything that is not purely gravity.[29][30] This view is commonly held in fields that deal with general relativity such as cosmology. In this view, light and other massless particles and fields are all part of "matter".

20.3 Structure

In particle physics, fermions are particles that obey Fermi–Dirac statistics. Fermions can be elementary, like the electron—or composite, like the proton and neutron. In the Standard Model, there are two types of elementary fermions: quarks and leptons, which are discussed next.

20.3.1 Quarks

Main article: Quark

Quarks are particles of spin- $\frac{1}{2}$, implying that they are fermions. They carry an electric charge of $-\frac{1}{3}$ e (down-type quarks) or $+\frac{2}{3}$ e (up-type quarks). For comparison, an electron has a charge of −1 e. They also carry colour charge, which is the equivalent of the electric charge for the strong interaction. Quarks also undergo radioactive decay, meaning that they are subject to the weak interaction. Quarks are massive particles, and therefore are also subject to gravity.

Baryonic matter

Main article: Baryon

Baryons are strongly interacting fermions, and so are subject to Fermi–Dirac statistics. Amongst the baryons are the protons and neutrons, which occur in atomic nuclei, but many other unstable baryons exist as well. The term baryon usually refers to triquarks—particles made of three quarks. "Exotic" baryons made of four quarks and one antiquark are known as the pentaquarks, but their existence is not generally accepted.

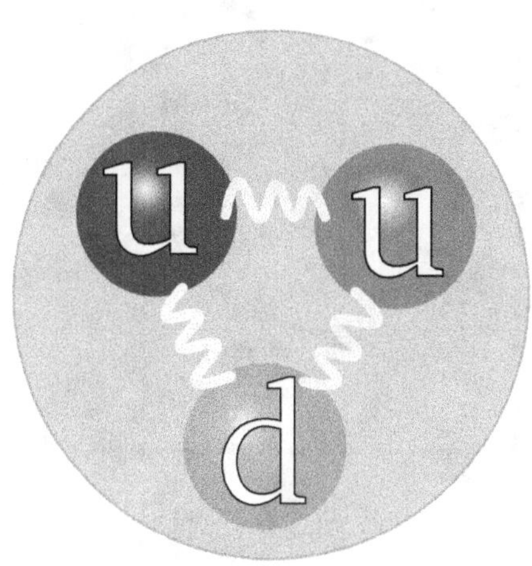

Quark structure of a proton: 2 up quarks and 1 down quark.

Baryonic matter is the part of the universe that is made of baryons (including all atoms). This part of the universe does not include dark energy, dark matter, black holes or various forms of degenerate matter, such as compose white dwarf stars and neutron stars. Microwave light seen by Wilkinson Microwave Anisotropy Probe (WMAP), suggests that only about 4.6% of that part of the universe within range of the best telescopes (that is, matter that may be visible because light could reach us from it), is made of baryonic matter. About 23% is dark matter, and about 72% is dark energy.[32]

As a matter of fact, the great majority of ordinary matter in the universe is unseen, since visible stars and gas inside galaxies and clusters account for less than 10 per cent of the ordinary matter contribution to the mass-energy density of the universe.[33]

Degenerate matter

Main article: Degenerate matter

In physics, **degenerate matter** refers to the ground state of a gas of fermions at a temperature near absolute zero.[34] The Pauli exclusion principle requires that only two fermions can occupy a quantum state, one spin-up and the other spin-down. Hence, at zero temperature, the fermions fill up sufficient levels to accommodate all the available fermions—and in the case of many fermions, the maximum kinetic energy (called the *Fermi energy*) and the pressure of the gas becomes very large, and depends on the number of fermions rather than the temperature, unlike normal states of matter.

A comparison between the white dwarf IK Pegasi B (center), its A-class companion IK Pegasi A (left) and the Sun (right). This white dwarf has a surface temperature of 35,500 K.

Degenerate matter is thought to occur during the evolution of heavy stars.[35] The demonstration by Subrahmanyan Chandrasekhar that white dwarf stars have a maximum allowed mass because of the exclusion principle caused a revolution in the theory of star evolution.[36]

Degenerate matter includes the part of the universe that is made up of neutron stars and white dwarfs.

Strange matter

Main article: Strange matter

Strange matter is a particular form of quark matter, usually thought of as a *liquid* of up, down, and strange quarks. It is contrasted with nuclear matter, which is a liquid of neutrons and protons (which themselves are built out of up and down quarks), and with non-strange quark matter, which is a quark liquid that contains only up and down quarks. At high enough density, strange matter is expected to be color superconducting. Strange matter is hypothesized to occur in the core of neutron stars, or, more speculatively, as isolated droplets that may vary in size from femtometers (strangelets) to kilometers (quark stars).

Two meanings of the term "strange matter" In particle physics and astrophysics, the term is used in two ways, one broader and the other more specific.

1. The broader meaning is just quark matter that contains three flavors of quarks: up, down, and strange. In this definition, there is a critical pressure and an associated critical density, and when nuclear matter (made of protons and neutrons) is compressed beyond this den-

sity, the protons and neutrons dissociate into quarks, yielding quark matter (probably strange matter).

2. The narrower meaning is quark matter that is *more stable than nuclear matter*. The idea that this could happen is the "strange matter hypothesis" of Bodmer[37] and Witten.[38] In this definition, the critical pressure is zero: the true ground state of matter is *always* quark matter. The nuclei that we see in the matter around us, which are droplets of nuclear matter, are actually metastable, and given enough time (or the right external stimulus) would decay into droplets of strange matter, i.e. strangelets.

20.3.2 Leptons

Main article: Lepton

Leptons are particles of spin- $1/2$, meaning that they are fermions. They carry an electric charge of −1 e (charged leptons) or 0 e (neutrinos). Unlike quarks, leptons do not carry colour charge, meaning that they do not experience the strong interaction. Leptons also undergo radioactive decay, meaning that they are subject to the weak interaction. Leptons are massive particles, therefore are subject to gravity.

20.4 Phases

Main article: Phase (matter)
See also: Phase diagram and State of matter
In bulk, matter can exist in several different forms, or states of aggregation, known as *phases*,[42] depending on ambient pressure, temperature and volume.[43] A phase is a form of matter that has a relatively uniform chemical composition and physical properties (such as density, specific heat, refractive index, and so forth). These phases include the three familiar ones (solids, liquids, and gases), as well as more exotic states of matter (such as plasmas, superfluids, supersolids, Bose–Einstein condensates, ...). A *fluid* may be a liquid, gas or plasma. There are also paramagnetic and ferromagnetic phases of magnetic materials. As conditions change, matter may change from one phase into another. These phenomena are called phase transitions, and are studied in the field of thermodynamics. In nanomaterials, the vastly increased ratio of surface area to volume results in matter that can exhibit properties entirely different from those of bulk material, and not well described by any bulk phase (see nanomaterials for more details).

Phases are sometimes called *states of matter*, but this term can lead to confusion with thermodynamic states. For example, two gases maintained at different pressures are in different *thermodynamic states* (different pressures), but in the same *phase* (both are gases).

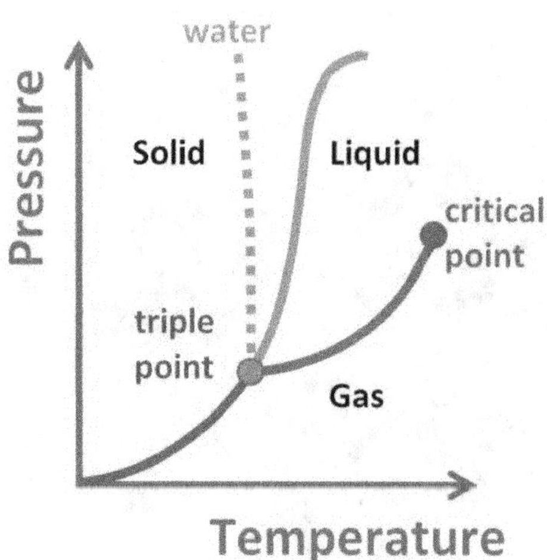

Phase diagram for a typical substance at a fixed volume. Vertical axis is Pressure, *horizontal axis is* Temperature. *The green line marks the freezing point (above the green line is* solid, *below it is* liquid*) and the blue line the boiling point (above it is* liquid *and below it is* gas*). So, for example, at higher* T, *a higher* P *is necessary to maintain the substance in liquid phase. At the triple point the three phases; liquid, gas and solid; can coexist. Above the critical point there is no detectable difference between the phases. The dotted line shows the anomalous behavior of water: ice melts at constant temperature with increasing pressure.[41]*

20.5 Antimatter

Main article: Antimatter

In particle physics and quantum chemistry, **antimatter** is matter that is composed of the antiparticles of those that constitute ordinary matter. If a particle and its antiparticle come into contact with each other, the two annihilate; that is, they may both be converted into other particles with equal energy in accordance with Einstein's equation $E = mc^2$. These new particles may be high-energy photons (gamma rays) or other particle–antiparticle pairs. The resulting particles are endowed with an amount of kinetic energy equal to the difference between the rest mass of the products of the annihilation and the rest mass of the original particle–antiparticle pair, which is often quite large. Depending on which definition of "matter" is adopted, antimatter can be said to be a particular subclass of matter, or the opposite of matter.

Antimatter is not found naturally on Earth, except very

briefly and in vanishingly small quantities (as the result of radioactive decay, lightning or cosmic rays). This is because antimatter that came to exist on Earth outside the confines of a suitable physics laboratory would almost instantly meet the ordinary matter that Earth is made of, and be annihilated. Antiparticles and some stable antimatter (such as antihydrogen) can be made in tiny amounts, but not in enough quantity to do more than test a few of its theoretical properties.

There is considerable speculation both in science and science fiction as to why the observable universe is apparently almost entirely matter (in the sense of quarks and leptons but not antiquarks or antileptons), and whether other places are almost entirely antimatter (antiquarks and antileptons) instead. In the early universe, it is thought that matter and antimatter were equally represented, and the disappearance of antimatter requires an asymmetry in physical laws called CP (charge-parity) symmetry violation, which can be obtained from the Standard Model,[44] but at this time the apparent asymmetry of matter and antimatter in the visible universe is one of the great unsolved problems in physics. Possible processes by which it came about are explored in more detail under baryogenesis.

Formally, antimatter particles can be defined by their negative baryon number or lepton number, while "normal" (non-antimatter) matter particles have positive baryon or lepton number.[45] These two classes of particles are the antiparticle partners of one another.

20.6 Conservation of matter

According to CP Symmetry, the two quantities that can define an amount of matter in the quark-lepton sense (and antimatter in an antiquark-antilepton sense), baryon number and lepton number, are conserved—or at least nearly so, considering CP violation. A baryon such as the proton or neutron has a baryon number of one, and a quark, because there are three in a baryon, is given a baryon number of 1/3. So the net amount of matter, as measured by the number of quarks (minus the number of antiquarks, which each have a baryon number of $-1/3$), which is proportional to baryon number, and number of leptons (minus antileptons), which is called the lepton number, is practically impossible to change in any process. Even in a nuclear bomb, none of the baryons (protons and neutrons of which the atomic nuclei are composed) are destroyed—there are as many baryons after as before the reaction, so none of these matter particles are actually destroyed and none are even converted to non-matter particles (like photons of light or radiation). Instead, nuclear (and perhaps chromodynamic) binding energy is released, as these baryons become bound into mid-size nuclei having less energy (and, equivalently, less mass) per nucleon compared to the original small (hydrogen) and large (plutonium etc.) nuclei. Even in electron–positron annihilation, there is actually no net matter being destroyed, because there was zero net matter (zero total lepton number and baryon number) to begin with before the annihilation—one lepton minus one antilepton equals zero net lepton number—and this net amount matter does not change as it simply remains zero after the annihilation.[46] So the only way to really "destroy" or "convert" ordinary matter is to pair it with the same amount of antimatter so that their "matterness" cancels out—but in practice there is almost no antimatter generally available in the universe (see baryon asymmetry and leptogenesis) with which to do so.

20.7 Other types

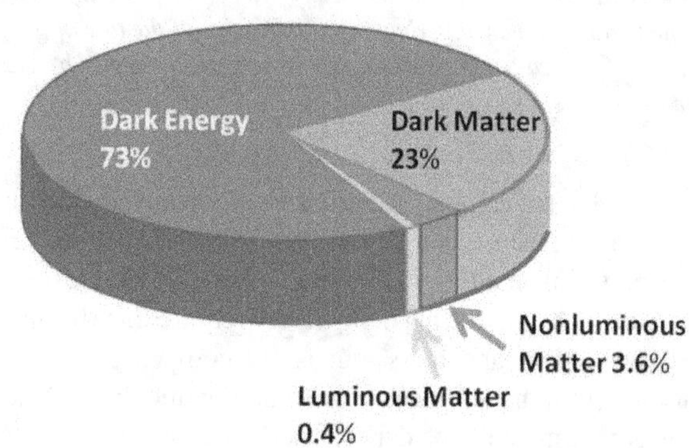

Pie chart showing the fractions of energy in the universe contributed by different sources. Ordinary matter *is divided into* luminous matter *(the stars and luminous gases and 0.005% radiation) and* nonluminous matter *(intergalactic gas and about 0.1% neutrinos and 0.04% supermassive black holes). Ordinary matter is uncommon. Modeled after Ostriker and Steinhardt.*[47] *For more information, see NASA.*

Ordinary matter, in the quarks and leptons definition, constitutes about 4% of the energy of the observable universe. The remaining energy is theorized to be due to exotic forms, of which 23% is dark matter[48][49] and 73% is dark energy.[50][51]

20.7.1 Dark matter

Main articles: Dark matter, Lambda-CDM model, and WIMPs
See also: Galaxy formation and evolution and Dark matter halo

In astrophysics and cosmology, **dark matter** is matter of unknown composition that does not emit or reflect

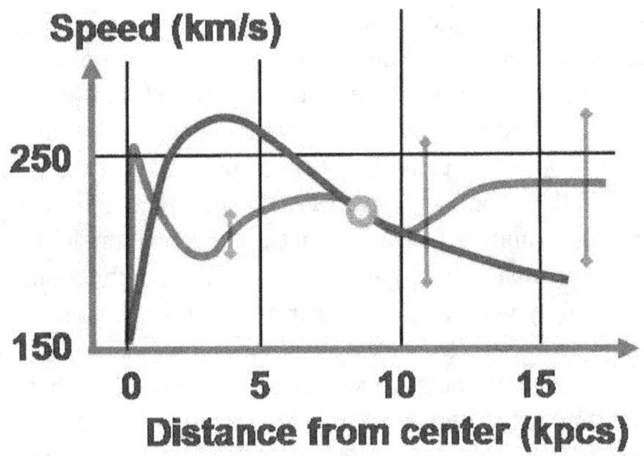

Galaxy rotation curve for the Milky Way. Vertical axis is speed of rotation about the galactic center. Horizontal axis is distance from the galactic center. The sun is marked with a yellow ball. The observed curve of speed of rotation is blue. The predicted curve based upon stellar mass and gas in the Milky Way is red. The difference is due to dark matter or perhaps a modification of the law of gravity.[52][53][54] *Scatter in observations is indicated roughly by gray bars.*

enough electromagnetic radiation to be observed directly, but whose presence can be inferred from gravitational effects on visible matter.[55][56] Observational evidence of the early universe and the big bang theory require that this matter have energy and mass, but is not composed ordinary baryons (protons and neutrons). The commonly accepted view is that most of the dark matter is non-baryonic in nature.[55] As such, it is composed of particles as yet unobserved in the laboratory. Perhaps they are supersymmetric particles,[57] which are not Standard Model particles, but relics formed at very high energies in the early phase of the universe and still floating about.[55]

20.7.2 Dark energy

Main article: Dark energy
See also: Big bang § Dark energy

In cosmology, **dark energy** is the name given to source of the repelling influence that is accelerating the rate of expansion of the universe. Its precise nature is currently a mystery, although its effects can reasonably be modeled by assigning matter-like properties such as energy density and pressure to the vacuum itself.[58][59]

> Fully 70% of the matter density in the universe appears to be in the form of dark energy. Twenty-six percent is dark matter. Only 4% is ordinary matter. So less than 1 part in 20 is made out of matter we have observed experimentally or described in the standard model of particle physics. Of the other 96%, apart from the properties just mentioned, we know absolutely nothing.
> — Lee Smolin: *The Trouble with Physics*, p. 16

20.7.3 Exotic matter

Main article: Exotic matter

Exotic matter is a concept of particle physics, which may include dark matter and dark energy but goes further to include any hypothetical material that violates one or more of the properties of known forms of matter. Some such materials might possess hypothetical properties like negative mass.

20.8 Historical development

20.8.1 Antiquity (c. 610 BC–c. 322 BC)

The pre-Socratics were among the first recorded speculators about the underlying nature of the visible world. Thales (c. 624 BC–c. 546 BC) regarded water as the fundamental material of the world. Anaximander (c. 610 BC–c. 546 BC) posited that the basic material was wholly characterless or limitless: the Infinite (*apeiron*). Anaximenes (flourished 585 BC, d. 528 BC) posited that the basic stuff was *pneuma* or air. Heraclitus (c. 535–c. 475 BC) seems to say the basic element is fire, though perhaps he means that all is change. Empedocles (c. 490–430 BC) spoke of four elements of which everything was made: earth, water, air, and fire.[60] Meanwhile, Parmenides argued that change does not exist, and Democritus argued that everything is composed of minuscule, inert bodies of all shapes called atoms, a philosophy called atomism. All of these notions had deep philosophical problems.[61]

Aristotle (384 BC – 322 BC) was the first to put the conception on a sound philosophical basis, which he did in his natural philosophy, especially in *Physics* book I.[62] He adopted as reasonable suppositions the four Empedoclean elements, but added a fifth, aether. Nevertheless, these elements are not basic in Aristotle's mind. Rather they, like everything else in the visible world, are composed of the basic *principles* matter and form.

> For my definition of matter is just this—the primary substratum of each thing, from which it comes to be without qualification, and which persists in the result.

— Aristotle, Physics I:9:192a32

The word Aristotle uses for matter, ὕλη (*hyle* or *hule*), can be literally translated as wood or timber, that is, "raw material" for building.[63] Indeed, Aristotle's conception of matter is intrinsically linked to something being made or composed. In other words, in contrast to the early modern conception of matter as simply occupying space, matter for Aristotle is definitionally linked to process or change: matter is what underlies a change of substance. For example, a horse eats grass: the horse changes the grass into itself; the grass as such does not persist in the horse, but some aspect of it—its matter—does. The matter is not specifically described (e.g., as atoms), but consists of whatever persists in the change of substance from grass to horse. Matter in this understanding does not exist independently (i.e., as a substance), but exists interdependently (i.e., as a "principle") with form and only insofar as it underlies change. It can be helpful to conceive of the relationship of matter and form as very similar to that between parts and whole. For Aristotle, matter as such can only *receive* actuality from form; it has no activity or actuality in itself, similar to the way that parts as such only have their existence *in* a whole (otherwise they would be independent wholes).

20.8.2 Seventeenth and eighteenth centuries

René Descartes (1596–1650) originated the modern conception of matter. He was primarily a geometer. Instead of, like Aristotle, deducing the existence of matter from the physical reality of change, Descartes arbitrarily postulated matter to be an abstract, mathematical substance that occupies space:

> So, extension in length, breadth, and depth, constitutes the nature of bodily substance; and thought constitutes the nature of thinking substance. And everything else attributable to body presupposes extension, and is only a mode of extended
> — René Descartes, Principles of Philosophy[64]

For Descartes, matter has only the property of extension, so its only activity aside from locomotion is to exclude other bodies:[65] this is the mechanical philosophy. Descartes makes an absolute distinction between mind, which he defines as unextended, thinking substance, and matter, which he defines as unthinking, extended substance.[66] They are independent things. In contrast, Aristotle defines matter and the formal/forming principle as complementary *principles* that together compose one independent thing (substance). In short, Aristotle defines matter (roughly speaking) as what things are actually made of (with a *potential* independent existence), but Descartes elevates matter to an actual independent thing in itself.

The continuity and difference between Descartes' and Aristotle's conceptions is noteworthy. In both conceptions, matter is passive or inert. In the respective conceptions matter has different relationships to intelligence. For Aristotle, matter and intelligence (form) exist together in an interdependent relationship, whereas for Descartes, matter and intelligence (mind) are definitionally opposed, independent substances.[67]

Descartes' justification for restricting the inherent qualities of matter to extension is its permanence, but his real criterion is not permanence (which equally applied to color and resistance), but his desire to use geometry to explain all material properties.[68] Like Descartes, Hobbes, Boyle, and Locke argued that the inherent properties of bodies were limited to extension, and that so-called secondary qualities, like color, were only products of human perception.[69]

Isaac Newton (1643–1727) inherited Descartes' mechanical conception of matter. In the third of his "Rules of Reasoning in Philosophy", Newton lists the universal qualities of matter as "extension, hardness, impenetrability, mobility, and inertia".[70] Similarly in *Optics* he conjectures that God created matter as "solid, massy, hard, impenetrable, movable particles", which were "...even so very hard as never to wear or break in pieces".[71] The "primary" properties of matter were amenable to mathematical description, unlike "secondary" qualities such as color or taste. Like Descartes, Newton rejected the essential nature of secondary qualities.[72]

Newton developed Descartes' notion of matter by restoring to matter intrinsic properties in addition to extension (at least on a limited basis), such as mass. Newton's use of gravitational force, which worked "at a distance", effectively repudiated Descartes' mechanics, in which interactions happened exclusively by contact.[73]

Though Newton's gravity would seem to be a *power* of bodies, Newton himself did not admit it to be an *essential* property of matter. Carrying the logic forward more consistently, Joseph Priestley (1733-1804) argued that corporeal properties transcend contact mechanics: chemical properties require the *capacity* for attraction.[73] He argued matter has other inherent powers besides the so-called primary qualities of Descartes, et al.[74]

20.8.3 Nineteenth and twentieth centuries

Since Priestley's time, there has been a massive expansion in knowledge of the constituents of the material world (viz., molecules, atoms, subatomic particles), but there has been no further development in the *definition* of matter. Rather the question has been set aside. Noam Chomsky (born 1928) summarizes the situation that has prevailed since that time:

> What is the concept of body that finally emerged?[...] The answer is that there is no clear and definite conception of body.[...] Rather, the material world is whatever we discover it to be, with whatever properties it must be assumed to have for the purposes of explanatory theory. Any intelligible theory that offers genuine explanations and that can be assimilated to the core notions of physics becomes part of the theory of the material world, part of our account of body. If we have such a theory in some domain, we seek to assimilate it to the core notions of physics, perhaps modifying these notions as we carry out this enterprise.
> — Noam Chomsky, *Language and problems of knowledge: the Managua lectures*, p. 144[73]

So matter is whatever physics studies and the object of study of physics is matter: there is no independent general definition of matter, apart from its fitting into the methodology of measurement and controlled experimentation. In sum, the boundaries between what constitutes matter and everything else remains as vague as the demarcation problem of delimiting science from everything else.[75]

In the 19th century, following the development of the periodic table, and of atomic theory, atoms were seen as being the fundamental constituents of matter; atoms formed molecules and compounds.[76]

The common definition in terms of occupying space and having mass is in contrast with most physical and chemical definitions of matter, which rely instead upon its structure and upon attributes not necessarily related to volume and mass. At the turn of the nineteenth century, the knowledge of matter began a rapid evolution.

Aspects of the Newtonian view still held sway. James Clerk Maxwell discussed matter in his work *Matter and Motion*.[77] He carefully separates "matter" from space and time, and defines it in terms of the object referred to in Newton's first law of motion.

However, the Newtonian picture was not the whole story. In the 19th century, the term "matter" was actively discussed by a host of scientists and philosophers, and a brief outline can be found in Levere.[78] A textbook discussion from 1870 suggests matter is what is made up of atoms:[79]

> Three divisions of matter are recognized in science: masses, molecules and atoms.
> A Mass of matter is any portion of matter appreciable by the senses.
> A Molecule is the smallest particle of matter into which a body can be divided without losing its identity.
> An Atom is a still smaller particle produced by division of a molecule.

Rather than simply having the attributes of mass and occupying space, matter was held to have chemical and electrical properties. In 1909 the famous physicist J. J. Thomson (1856-1940) wrote about the "constitution of matter" and was concerned with the possible connection between matter and electrical charge.[80]

There is an entire literature concerning the "structure of matter", ranging from the "electrical structure" in the early 20th century,[81] to the more recent "quark structure of matter", introduced today with the remark: *Understanding the quark structure of matter has been one of the most important advances in contemporary physics.*[82] In this connection, physicists speak of *matter fields*, and speak of particles as "quantum excitations of a mode of the matter field".[8][9] And here is a quote from de Sabbata and Gasperini: "With the word "matter" we denote, in this context, the sources of the interactions, that is spinor fields (like quarks and leptons), which are believed to be the fundamental components of matter, or scalar fields, like the Higgs particles, which are used to introduced mass in a gauge theory (and that, however, could be composed of more fundamental fermion fields)."[83]

In the late 19th century with the discovery of the electron, and in the early 20th century, with the discovery of the atomic nucleus, and the birth of particle physics, matter was seen as made up of electrons, protons and neutrons interacting to form atoms. Today, we know that even protons and neutrons are not indivisible, they can be divided into quarks, while electrons are part of a particle family called leptons. Both quarks and leptons are elementary particles, and are currently seen as being the fundamental constituents of matter.[84]

These quarks and leptons interact through four fundamental forces: gravity, electromagnetism, weak interactions, and strong interactions. The Standard Model of particle physics is currently the best explanation for all of physics, but despite decades of efforts, gravity cannot yet be accounted for at the quantum level; it is only described by classical physics (see quantum gravity and graviton).[85] Interactions between quarks and leptons are the result of an exchange of force-carrying particles (such as photons) between quarks and leptons.[86] The force-carrying particles are not themselves building blocks. As one consequence, mass and energy (which cannot be created or destroyed) cannot always be related to matter (which can be created out of non-matter particles such as photons, or even out of pure energy, such as kinetic energy). Force carriers are usually not considered matter: the carriers of the electric force (photons) possess energy (see Planck relation) and the carriers of the weak force (W and Z bosons) are massive, but neither are considered matter either.[87] However, while these particles are not considered matter, they do contribute to the total mass of atoms, subatomic particles, and all systems that contain them.[88][89]

20.9 Summary

The modern conception of matter has been refined many times in history, in light of the improvement in knowledge of just *what* the basic building blocks are, and in how they interact. The term "matter" is used throughout physics in a bewildering variety of contexts: for example, one refers to "condensed matter physics",[90] "elementary matter",[91] "partonic" matter, "dark" matter, "anti"-matter, "strange" matter, and "nuclear" matter. In discussions of matter and antimatter, normal matter has been referred to by Alfvén as *koinomatter* (Gk. *common matter*).[92] It is fair to say that in physics, there is no broad consensus as to a general definition of matter, and the term "matter" usually is used in conjunction with a specifying modifier.

The history of the concept of matter is a history of the fundamental *length scales* used to define matter. Different building blocks apply depending upon whether one defines matter on an atomic or elementary particle level. One may use a definition that matter is atoms, or that matter is hadrons, or that matter is leptons and quarks depending upon the scale at which one wishes to define matter.[93]

These quarks and leptons interact through four fundamental forces: gravity, electromagnetism, weak interactions, and strong interactions. The Standard Model of particle physics is currently the best explanation for all of physics, but despite decades of efforts, gravity cannot yet be accounted for at the quantum level; it is only described by classical physics (see quantum gravity and graviton).[85]

20.10 See also

20.11 References

[1] R. Penrose (1991). "The mass of the classical vacuum". In S. Saunders, H.R. Brown. *The Philosophy of Vacuum*. Oxford University Press. p. 21. ISBN 0-19-824449-5.

[2] "Matter (physics)". *McGraw-Hill's Access Science: Encyclopedia of Science and Technology Online*. Retrieved 2009-05-24.

[3] P. Davies (1992). *The New Physics: A Synthesis*. Cambridge University Press. p. 1. ISBN 0-521-43831-4.

[4] G. 't Hooft (1997). *In search of the ultimate building blocks*. Cambridge University Press. p. 6. ISBN 0-521-57883-3.

[5] "RHIC Scientists Serve Up "Perfect" Liquid" (Press release). Brookhaven National Laboratory. 18 April 2005. Retrieved 2009-09-15.

[6] J. Olmsted; G.M. Williams (1996). *Chemistry: The Molecular Science* (2nd ed.). Jones & Bartlett. p. 40. ISBN 0-8151-8450-6.

[7] J. Mongillo (2007). *Nanotechnology 101*. Greenwood Publishing. p. 30. ISBN 0-313-33880-9.

[8] P.C.W. Davies (1979). *The Forces of Nature*. Cambridge University Press. p. 116. ISBN 0-521-22523-X.

[9] S. Weinberg (1998). *The Quantum Theory of Fields*. Cambridge University Press. p. 2. ISBN 0-521-55002-5.

[10] M. Masujima (2008). *Path Integral Quantization and Stochastic Quantization*. Springer. p. 103. ISBN 3-540-87850-5.

[11] G. F. Barker (1870). "Divisions of matter". *A text-book of elementary chemistry: theoretical and inorganic*. John F Morton & Co. p. 2. ISBN 978-1-4460-2206-1.

[12] M. de Podesta (2002). *Understanding the Properties of Matter* (2nd ed.). CRC Press. p. 8. ISBN 0-415-25788-3.

[13] B. Povh; K. Rith; C. Scholz; F. Zetsche; M. Lavelle (2004). "Part I: Analysis: The building blocks of matter". *Particles and Nuclei: An Introduction to the Physical Concepts* (4th ed.). Springer. ISBN 3-540-20168-8. Ordinary matter is composed entirely of first-generation particles, namely the u and d quarks, plus the electron andits neutrino.

[14] B. Carithers; P. Grannis (1995). "Discovery of the Top Quark" (PDF). *Beam Line*. SLAC National Accelerator Laboratory. **25** (3): 4–16.

[15] Tsan, Ung Chan (2006). "WHAT IS A MATTER PARTICLE?". *International Journal of Modern Physics E*. **15**. doi:10.1142/S0218301306003916. *(From Abstract:)* Positive baryon numbers (A>0) and positive lepton numbers (L>0) characterize matter particles while negative baryon numbers and negative lepton numbers characterize antimatter particles. Matter particles and antimatter particles belong to two distinct classes of particles. Matter neutral particles are particles characterized by both zero baryon number and zero lepton number. This third class of particles includes mesons formed by a quark and an antiquark pair (a pair of matter particle and antimatter particle) and bosons which are messengers of known interactions (photons for electromagnetism, W and Z bosons for the weak interaction, gluons for

the strong interaction). The antiparticle of a matter particle belongs to the class of antimatter particles, the antiparticle of an antimatter particle belongs to the class of matter particles.

[16] D. Green (2005). *High PT physics at hadron colliders*. Cambridge University Press. p. 23. ISBN 0-521-83509-7.

[17] L. Smolin (2007). *The Trouble with Physics: The Rise of String Theory, the Fall of a Science, and What Comes Next*. Mariner Books. p. 67. ISBN 0-618-91868-X.

[18] The W boson mass is 80.398 GeV; see Figure 1 in C. Amsler *et al.* (Particle Data Group) (2008). "Review of Particle Physics: The Mass and Width of the W Boson" (PDF). *Physics Letters B*. **667**: 1. Bibcode:2008PhLB..667....1A. doi:10.1016/j.physletb.2008.07.018.

[19] I.J.R. Aitchison; A.J.G. Hey (2004). *Gauge Theories in Particle Physics*. CRC Press. p. 48. ISBN 0-7503-0864-8.

[20] B. Povh; K. Rith; C. Scholz; F. Zetsche; M. Lavelle (2004). *Particles and Nuclei: An Introduction to the Physical Concepts*. Springer. p. 103. ISBN 3-540-20168-8.

[21] A.M. Green (2004). *Hadronic Physics from Lattice QCD*. World Scientific. p. 120. ISBN 981-256-022-X.

[22] T. Hatsuda (2008). "Quark–gluon plasma and QCD". In H. Akai. *Condensed matter theories*. **21**. Nova Publishers. p. 296. ISBN 1-60021-501-7.

[23] K.W Staley (2004). "Origins of the Third Generation of Matter". *The Evidence for the Top Quark*. Cambridge University Press. p. 8. ISBN 0-521-82710-8.

[24] Y. Ne'eman; Y. Kirsh (1996). *The Particle Hunters* (2nd ed.). Cambridge University Press. p. 276. ISBN 0-521-47686-0. [T]he most natural explanation to the existence of higher generations of quarks and leptons is that they correspond to excited states of the first generation, and experience suggests that excited systems must be composite

[25] S.M. Walker; A. King (2005). *What is Matter?*. Lerner Publications. p. 7. ISBN 0-8225-5131-4.

[26] J.Kenkel; P.B. Kelter; D.S. Hage (2000). *Chemistry: An Industry-based Introduction with CD-ROM*. CRC Press. p. 2. ISBN 1-56670-303-4. All basic science textbooks define *matter* as simply the collective aggregate of all material substances that occupy space and have mass or weight.

[27] K.A. Peacock (2008). *The Quantum Revolution: A Historical Perspective*. Greenwood Publishing Group. p. 47. ISBN 0-313-33448-X.

[28] M.H. Krieger (1998). *Constitutions of Matter: Mathematically Modeling the Most Everyday of Physical Phenomena*. University of Chicago Press. p. 22. ISBN 0-226-45305-7.

[29] S.M. Caroll (2004). *Spacetime and Geometry*. Addison Wesley. pp. 163–164. ISBN 0-8053-8732-3.

[30] P. Davies (1992). *The New Physics: A Synthesis*. Cambridge University Press. p. 499. ISBN 0-521-43831-4. **Matter fields**: the fields whose quanta describe the elementary particles that make up the material content of the Universe (as opposed to the gravitons and their supersymmetric partners).

[31] C. Amsler *et al.* (Particle Data Group) (2008). "Reviews of Particle Physics: Quarks" (PDF). *Physics Letters B*. **667**: 1. Bibcode:2008PhLB..667....1A. doi:10.1016/j.physletb.2008.07.018.

[32] "Five Year Results on the Oldest Light in the Universe". NASA. 2008. Retrieved 2008-05-02.

[33] Persic, Massimo; Salucci, Paolo (1992-09-01). "The baryon content of the Universe". *Monthly Notices of the Royal Astronomical Society*. **258** (1): 14P–18P. Bibcode:1992MNRAS.258P..14P. ISSN 0035-8711. arXiv:astro-ph/0502178. doi:10.1093/mnras/258.1.14P.

[34] H.S. Goldberg; M.D. Scadron (1987). *Physics of Stellar Evolution and Cosmology*. Taylor & Francis. p. 202. ISBN 0-677-05540-4.

[35] H.S. Goldberg; M.D. Scadron (1987). *Physics of Stellar Evolution and Cosmology*. Taylor & Francis. p. 233. ISBN 0-677-05540-4.

[36] J.-P. Luminet; A. Bullough; A. King (1992). *Black Holes*. Cambridge University Press. p. 75. ISBN 0-521-40906-3.

[37] A. Bodmer (1971). "Collapsed Nuclei". *Physical Review D*. **4** (6): 1601. Bibcode:1971PhRvD...4.1601B. doi:10.1103/PhysRevD.4.1601.

[38] E. Witten (1984). "Cosmic Separation of Phases". *Physical Review D*. **30** (2): 272. Bibcode:1984PhRvD..30..272W. doi:10.1103/PhysRevD.30.272.

[39] C. Amsler *et al.* (Particle Data Group) (2008). "Review of Particle Physics: Leptons" (PDF). *Physics Letters B*. **667**: 1. Bibcode:2008PhLB..667....1A. doi:10.1016/j.physletb.2008.07.018.

[40] C. Amsler *et al.* (Particle Data Group) (2008). "Review of Particle Physics: Neutrinos Properties" (PDF). *Physics Letters B*. **667**: 1. Bibcode:2008PhLB..667....1A. doi:10.1016/j.physletb.2008.07.018.

[41] S. R. Logan (1998). *Physical Chemistry for the Biomedical Sciences*. CRC Press. pp. 110–111. ISBN 0-7484-0710-3.

[42] P.J. Collings (2002). "Chapter 1: States of Matter". *Liquid Crystals: Nature's Delicate Phase of Matter*. Princeton University Press. ISBN 0-691-08672-9.

[43] D.H. Trevena (1975). "Chapter 1.2: Changes of phase". *The Liquid Phase*. Taylor & Francis. ISBN 978-0-85109-031-3.

[44] National Research Council (US) (2006). *Revealing the hidden nature of space and time*. National Academies Press. p. 46. ISBN 0-309-10194-8.

[45] TSAN, U. C. (2012). "Negative Numbers And Antimatter Particles". *International Journal of Modern Physics E.* **21** (01): 1250005. doi:10.1142/S021830131250005X. *(From Abstract:)* Antimatter particles are characterized by negative baryonic number A or/and negative leptonic number L. Materialization and annihilation obey conservation of A and L (associated to all known interactions)

[46] Tsan, Ung Chan (2013). "MASS, MATTER, MATERIALIZATION, MATTERGENESIS AND CONSERVATION OF CHARGE". *International Journal of Modern Physics E.* **22** (05): 1350027. doi:10.1142/S0218301313500274. *(From Abstract:)* Matter conservation melans conservation of baryonic number A and leptonic number L, A and L being algebraic numbers. Positive A and L are associated to matter particles, negative A and L are associated to antimatter particles. All known interactions do conserve matter

[47] J.P. Ostriker; P.J. Steinhardt (2003). "New Light on Dark Matter". *Science.* **300** (5627): 1909–13. Bibcode:2003Sci...300.1909O. PMID 12817140. arXiv:astro-ph/0306402. doi:10.1126/science.1085976.

[48] K. Pretzl (2004). "Dark Matter, Massive Neutrinos and Susy Particles". *Structure and Dynamics of Elementary Matter.* Walter Greiner. p. 289. ISBN 1-4020-2446-0.

[49] K. Freeman; G. McNamara (2006). "What can the matter be?". *In Search of Dark Matter.* Birkhäuser Verlag. p. 105. ISBN 0-387-27616-5.

[50] J.C. Wheeler (2007). *Cosmic Catastrophes: Exploding Stars, Black Holes, and Mapping the Universe.* Cambridge University Press. p. 282. ISBN 0-521-85714-7.

[51] J. Gribbin (2007). *The Origins of the Future: Ten Questions for the Next Ten Years.* Yale University Press. p. 151. ISBN 0-300-12596-8.

[52] P. Schneider (2006). *Extragalactic Astronomy and Cosmology.* Springer. p. 4, Fig. 1.4. ISBN 3-540-33174-3.

[53] T. Koupelis; K.F. Kuhn (2007). *In Quest of the Universe.* Jones & Bartlett Publishers. p. 492; Fig. 16.13. ISBN 0-7637-4387-9.

[54] M. H. Jones; R. J. Lambourne; D. J. Adams (2004). *An Introduction to Galaxies and Cosmology.* Cambridge University Press. p. 21; Fig. 1.13. ISBN 0-521-54623-0.

[55] D. Majumdar (2007). "Dark matter — possible candidates and direct detection". arXiv:hep-ph/0703310 [hep-ph].

[56] K.A. Olive (2003). "Theoretical Advanced Study Institute lectures on dark matter". arXiv:astro-ph/0301505 [astro-ph].

[57] K.A. Olive (2009). "Colliders and Cosmology". *European Physical Journal C.* **59** (2): 269–295. Bibcode:2009EPJC...59..269O. arXiv:0806.1208. doi:10.1140/epjc/s10052-008-0738-8.

[58] J.C. Wheeler (2007). *Cosmic Catastrophes.* Cambridge University Press. p. 282. ISBN 0-521-85714-7.

[59] L. Smolin (2007). *The Trouble with Physics.* Mariner Books. p. 16. ISBN 0-618-91868-X.

[60] S. Toulmin; J. Goodfield (1962). *The Architecture of Matter.* University of Chicago Press. pp. 48–54.

[61] Discussed by Aristotle in *Physics*, esp. book I, but also later; as well as *Metaphysics* I–II.

[62] For a good explanation and elaboration, see R.J. Connell (1966). *Matter and Becoming.* Priory Press.

[63] H. G. Liddell; R. Scott; J. M. Whiton (1891). *A lexicon abridged from Liddell & Scott's Greek–English lexicon.* Harper and Brothers. p. 72.

[64] R. Descartes (1644). "The Principles of Human Knowledge". *Principles of Philosophy I.* p. 53.

[65] though even this property seems to be non-essential (René Descartes, *Principles of Philosophy* II [1644], "On the Principles of Material Things", no. 4.)

[66] R. Descartes (1644). "The Principles of Human Knowledge". *Principles of Philosophy I.* pp. 8, 54, 63.

[67] D.L. Schindler (1986). "The Problem of Mechanism". In D.L. Schindler. *Beyond Mechanism.* University Press of America.

[68] E.A. Burtt, *Metaphysical Foundations of Modern Science* (Garden City, New York: Doubleday and Company, 1954), 117–118.

[69] J.E. McGuire and P.M. Heimann, "The Rejection of Newton's Concept of Matter in the Eighteenth Century", *The Concept of Matter in Modern Philosophy* ed. Ernan McMullin (Notre Dame: University of Notre Dame Press, 1978), 104–118 (105).

[70] Isaac Newton, *Mathematical Principles of* Natural Philosophy, *trans. A. Motte, revised by F. Cajori (Berkeley: University of California Press, 1934), pp. 398–400. Further analyzed by Maurice A. Finocchiaro, "Newton's Third Rule of Philosophizing: A Role for Logic in Historiography",* Isis *65:1 (Mar. 1974), pp. 66–73.*

[71] Isaac Newton, *Optics*, Book III, pt. 1, query 31.

[72] McGuire and Heimann, 104.

[73] N. Chomsky (1988). *Language and problems of knowledge: the Managua lectures* (2nd ed.). MIT Press. p. 144. ISBN 0-262-53070-8.

[74] McGuire and Heimann, 113.

[75] Nevertheless, it remains true that the mathematization regarded as requisite for a modern physical theory carries its own implicit notion of matter, which is very like Descartes', despite the demonstrated vacuity of the latter's notions.

[76] M. Wenham (2005). *Understanding Primary Science: Ideas, Concepts and Explanations* (2nd ed.). Paul Chapman Educational Publishing. p. 115. ISBN 1-4129-0163-4.

[77] J.C. Maxwell (1876). *Matter and Motion*. Society for Promoting Christian Knowledge. p. 18. ISBN 0-486-66895-9.

[78] T.H. Levere (1993). "Introduction". *Affinity and Matter: Elements of Chemical Philosophy, 1800–1865*. Taylor & Francis. ISBN 2-88124-583-8.

[79] G.F. Barker (1870). "Introduction". *A Text Book of Elementary Chemistry: Theoretical and Inorganic*. John P. Morton and Company. p. 2.

[80] J. J. Thomson (1909). "Preface". *Electricity and Matter*. A. Constable.

[81] O.W. Richardson (1914). "Chapter 1". *The Electron Theory of Matter*. The University Press.

[82] M. Jacob (1992). *The Quark Structure of Matter*. World Scientific. ISBN 981-02-3687-5.

[83] V. de Sabbata; M. Gasperini (1985). *Introduction to Gravitation*. World Scientific. p. 293. ISBN 9971-5-0049-3.

[84] The history of the concept of matter is a history of the fundamental *length scales* used to define matter. Different building blocks apply depending upon whether one defines matter on an atomic or elementary particle level. One may use a definition that matter is atoms, or that matter is hadrons, or that matter is leptons and quarks depending upon the scale at which one wishes to define matter. B. Povh; K. Rith; C. Scholz; F. Zetsche; M. Lavelle (2004). "Fundamental constituents of matter". *Particles and Nuclei: An Introduction to the Physical Concepts* (4th ed.). Springer. ISBN 3-540-20168-8.

[85] J. Allday (2001). *Quarks, Leptons and the Big Bang*. CRC Press. p. 12. ISBN 0-7503-0806-0.

[86] B.A. Schumm (2004). *Deep Down Things: The Breathtaking Beauty of Particle Physics*. Johns Hopkins University Press. p. 57. ISBN 0-8018-7971-X.

[87] See for example, M. Jibu; K. Yasue (1995). *Quantum Brain Dynamics and Consciousness*. John Benjamins Publishing Company. p. 62. ISBN 1-55619-183-9., B. Martin (2009). *Nuclear and Particle Physics* (2nd ed.). John Wiley & Sons. p. 125. ISBN 0-470-74275-5. and K. W. Plaxco; M. Gross (2006). *Astrobiology: A Brief Introduction*. Johns Hopkins University Press. p. 23. ISBN 0-8018-8367-9.

[88] P. A. Tipler; R. A. Llewellyn (2002). *Modern Physics*. Macmillan. pp. 89–91, 94–95. ISBN 0-7167-4345-0.

[89] P. Schmüser; H. Spitzer (2002). "Particles". In L. Bergmann; et al. *Constituents of Matter: Atoms, Molecules, Nuclei*. CRC Press. pp. 773 *ff*. ISBN 0-8493-1202-7.

[90] P. M. Chaikin; T. C. Lubensky (2000). *Principles of Condensed Matter Physics*. Cambridge University Press. p. xvii. ISBN 0-521-79450-1.

[91] W. Greiner (2003). W. Greiner; M.G. Itkis; G. Reinhardt; M.C. Güçlü, eds. *Structure and Dynamics of Elementary Matter*. Springer. p. xii. ISBN 1-4020-2445-2.

[92] P. Sukys (1999). *Lifting the Scientific Veil: Science Appreciation for the Nonscientist*. Rowman & Littlefield. p. 87. ISBN 0-8476-9600-6.

[93] B. Povh; K. Rith; C. Scholz; F. Zetsche; M. Lavelle (2004). "Fundamental constituents of matter". *Particles and Nuclei: An Introduction to the Physical Concepts* (4th ed.). Springer. ISBN 3-540-20168-8.

20.12 Further reading

- Lillian Hoddeson; Michael Riordan, eds. (1997). *The Rise of the Standard Model*. Cambridge University Press. ISBN 0-521-57816-7.
- Timothy Paul Smith (2004). "The search for quarks in ordinary matter". *Hidden Worlds*. Princeton University Press. p. 1. ISBN 0-691-05773-7.
- Harald Fritzsch (2005). *Elementary Particles: Building blocks of matter*. World Scientific. p. 1. ISBN 981-256-141-2.
- Bertrand Russell (1992). "The philosophy of matter". *A Critical Exposition of the Philosophy of Leibniz* (Reprint of 1937 2nd ed.). Routledge. p. 88. ISBN 0-415-08296-X.
- Stephen Toulmin and June Goodfield, *The Architecture of Matter* (Chicago: University of Chicago Press, 1962).
- Richard J. Connell, *Matter and Becoming* (Chicago: The Priory Press, 1966).
- Ernan McMullin, *The Concept of Matter in Greek and Medieval Philosophy* (Notre Dame, Indiana: Univ. of Notre Dame Press, 1965).
- Ernan McMullin, *The Concept of Matter in Modern Philosophy* (Notre Dame, Indiana: University of Notre Dame Press, 1978).

20.13 External links

- Visionlearning Module on Matter
- Matter in the universe How much Matter is in the Universe?
- NASA on superfluid core of neutron star

- Matter and Energy: A False Dichotomy – Conversations About Science with Theoretical Physicist Matt Strassler

Chapter 21

Radiation

For other uses, see Radiation (disambiguation).
Not to be confused with Ionizing radiation.

In physics, **radiation** is the emission or transmission of energy in the form of waves or particles through space or through a material medium.[1][2] This includes:

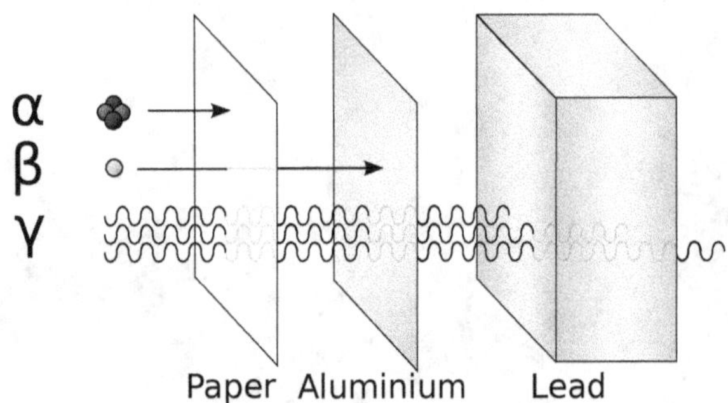

Illustration of the relative abilities of three different types of ionizing radiation to penetrate solid matter. Typical alpha particles (α) are stopped by a sheet of paper, while beta particles (β) are stopped by an aluminum plate. Gamma radiation (γ) is damped when it penetrates lead. Note caveats in the text about this simplified diagram.

The international symbol for types and levels of radiation that are unsafe for unshielded humans. Radiation in general exists throughout nature, such as in light and sound.

- electromagnetic radiation, such as radio waves, microwaves, visible light, x-rays, and gamma radiation (γ)
- particle radiation, such as alpha radiation (α), beta radiation (β), and neutron radiation (particles of non-zero rest energy)
- acoustic radiation, such as ultrasound, sound, and seismic waves (dependent on a physical transmission medium)
- gravitational radiation, radiation that takes the form of gravitational waves, or ripples in the curvature of spacetime.

Radiation is often categorized as either ionizing or non-ionizing depending on the energy of the radiated particles. Ionizing radiation carries more than 10 eV, which is enough to ionize atoms and molecules, and break chemical bonds. This is an important distinction due to the large difference in harmfulness to living organisms. A common source of ionizing radiation is radioactive materials that emit α, β, or γ radiation, consisting of helium nuclei, electrons or positrons, and photons, respectively. Other sources include X-rays from medical radiography examinations and muons, mesons, positrons, neutrons and other particles that constitute the secondary cosmic rays that are produced after primary cosmic rays interact with Earth's atmosphere.

Gamma rays, X-rays and the higher energy range of ultraviolet light constitute the ionizing part of the electromagnetic spectrum. The lower-energy, longer-wavelength part of the spectrum including visible light, infrared light, microwaves, and radio waves is non-ionizing; its main effect when interacting with tissue is heating. This type of radiation only damages cells if the intensity is high enough to cause excessive heating. Ultraviolet radiation has some features of both ionizing and non-ionizing radiation. While the part of the ultraviolet spectrum that penetrates the Earth's atmosphere is non-ionizing, this radiation does far more damage to many molecules in biological systems than can be accounted for by heating effects, sunburn being a well-known example. These properties derive from ultraviolet's power to alter chemical bonds, even without having quite enough energy to ionize atoms.

The word radiation arises from the phenomenon of waves *radiating* (i.e., traveling outward in all directions) from a source. This aspect leads to a system of measurements and physical units that are applicable to all types of radiation. Because such radiation expands as it passes through space, and as its energy is conserved (in vacuum). The intensity of all types of radiation from a point source follows an inverse-square law in relation to the distance from its source. Like any ideal law, the inverse-square law approximates a measured radiation intensity to the extent that the source approximates a geometric point.

21.1 Ionizing radiation

Radiation with sufficiently high energy can ionize atoms; that is to say it can knock electrons off atoms and create ions. Ionization occurs when an electron is stripped (or "knocked out") from an electron shell of the atom, which leaves the atom with a net positive charge. Because living cells and, more importantly, the DNA in those cells can be damaged by this ionization, exposure to ionizing radiation is considered to increase the risk of cancer. Thus "ionizing radiation" is somewhat artificially separated from particle radiation and electromagnetic radiation, simply due to its great potential for biological damage. While an individual cell is made of trillions of atoms, only a small fraction of those will be ionized at low to moderate radiation powers. The probability of ionizing radiation causing cancer is dependent upon the absorbed dose of the radiation, and is a function of the damaging tendency of the type of radiation (equivalent dose) and the sensitivity of the irradiated organism or tissue (effective dose).

If the source of the ionizing radiation is a radioactive material or a nuclear process such as fission or fusion, there is particle radiation to consider. Particle radiation is subatomic particles accelerated to relativistic speeds by nuclear reactions. Because of their momenta they are quite capable of knocking out electrons and ionizing materials, but since most have an electrical charge, they don't have the penetrating power of ionizing radiation. The exception is neutron particles; see below. There are several different kinds of these particles, but the majority are alpha particles, beta particles, neutrons, and protons. Roughly speaking, photons and particles with energies above about 10 electron volts (eV) are ionizing (some authorities use 33 eV, the ionization energy for water). Particle radiation from radioactive material or cosmic rays almost invariably carries enough energy to be ionizing.

Much ionizing radiation originates from radioactive materials and space (cosmic rays), and as such is naturally present in the environment, since most rock and soil has small concentrations of radioactive materials. The radiation is invisible and not directly detectable by human senses; as a result, instruments such as Geiger counters are usually required to detect its presence. In some cases, it may lead to secondary emission of visible light upon its interaction with matter, as in the case of Cherenkov radiation and radio-luminescence.

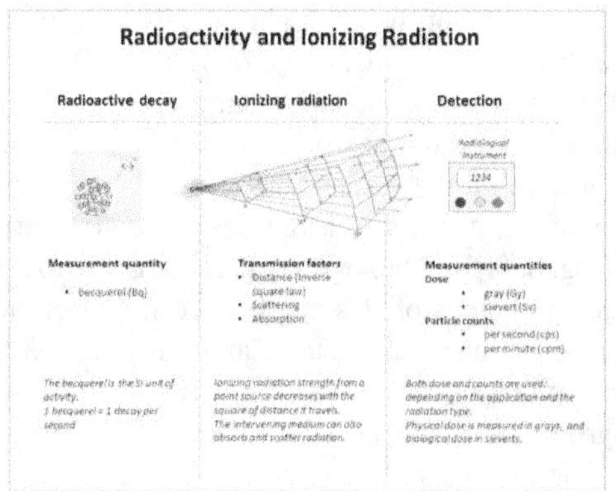

Graphic showing relationships between radioactivity and detected ionizing radiation

Ionizing radiation has many practical uses in medicine, research and construction, but presents a health hazard if used improperly. Exposure to radiation causes damage to living tissue; high doses result in Acute radiation syndrome (ARS), with skin burns, hair loss, internal organ failure and death, while any dose may result in an increased chance of cancer and genetic damage; a particular form of cancer, thyroid cancer, often occurs when nuclear weapons and reactors are the radiation source because of the biological proclivities of the radioactive iodine fission product, iodine-131.[3] However, calculating the exact risk and chance of cancer forming in cells caused by ionizing radiation is still not well understood and currently estimates are loosely determined by population based on data from the atomic bombing in Japan and from reactor accident follow-up, such as with the Chernobyl disaster. The International Commission on Radiological Protection states that "The Commission is aware of uncertainties and lack of precision of the models and parameter values", "Collective effective dose is not intended as a tool for epidemiological risk assessment, and it is inappropriate to use it in risk projections" and "in particular, the calculation of the number of cancer deaths based on collective effective doses from trivial individual doses should be avoided."[4]

21.1.1 Ultraviolet radiation

Main article: Ultraviolet

Ultraviolet, of wavelengths from 10 nm to 125 nm, ionizes air molecules, causing it to be strongly absorbed by air and by ozone (O_3) in particular. Ionizing UV therefore does not penetrate Earth's atmosphere to a significant degree, and is sometimes referred to as vacuum ultraviolet. Although present in space, this part of the UV spectrum is not of biological importance, because it does not reach living organisms on Earth.

There is a zone of the atmosphere in which ozone absorbs some 98% of non-ionizing but dangerous UV-C and UV-B. This so-called ozone layer, starts at about 20 miles (32 km) and extends upward. Some of the ultraviolet spectrum that does reach the ground (the part that begins above energies of 3.1 eV, a wavelength less than 400 nm) is non-ionizing, but is still biologically hazardous due to the ability of single photons of this energy to cause electronic excitation in biological molecules, and thus damage them by means of unwanted reactions. An example is the formation of pyrimidine dimers in DNA, which begins at wavelengths below 365 nm (3.4 eV), which is well below ionization energy. This property gives the ultraviolet spectrum some of the dangers of ionizing radiation in biological systems without actual ionization occurring. In contrast, visible light and longer-wavelength electromagnetic radiation, such as infrared, microwaves, and radio waves, consists of photons with too little energy to cause damaging molecular excitation, and thus this radiation is far less hazardous per unit of energy.

21.1.2 X-ray

Main article: X-ray

X-rays are electromagnetic waves with a wavelength less than about 10^{-9} m (greater than $3x10^{17}$ Hz and 1,240 eV). A smaller wavelength corresponds to a higher energy according to the equation $E=hc/\lambda$. ("E" is Energy; "h" is Planck's constant; "c" is the speed of light; "λ" is wavelength.) When an X-ray photon collides with an atom, the atom may absorb the energy of the photon and boost an electron to a higher orbital level or if the photon is very energetic, it may knock an electron from the atom altogether, causing the atom to ionize. Generally, larger atoms are more likely to absorb an X-ray photon since they have greater energy differences between orbital electrons. Soft tissue in the human body is composed of smaller atoms than the calcium atoms that make up bone, hence there is a contrast in the absorption of X-rays. X-ray machines are specifically designed to take advantage of the absorption difference between bone and soft tissue, allowing physicians to examine structure in the human body.

X-rays are also totally absorbed by the thickness of the earth's atmosphere, resulting in the prevention of the X-ray output of the sun, smaller in quantity than that of UV but nonetheless powerful, from reaching the surface.

21.1.3 Gamma radiation

Main article: Gamma ray

Gamma (γ) radiation consists of photons with a wavelength less than $3x10^{-11}$ meters (greater than 10^{19} Hz and 41.4 keV).[3] Gamma radiation emission is a nuclear process that occurs to rid an unstable nucleus of excess energy after most nuclear reactions. Both alpha and beta particles have an electric charge and mass, and thus are quite likely to interact with other atoms in their path. Gamma radiation, however, is composed of photons, which have neither mass nor electric charge and, as a result, penetrates much further through matter than either alpha or beta radiation.

Gamma rays can be stopped by a sufficiently thick or dense layer of material, where the stopping power of the material per given area depends mostly (but not entirely) on the total mass along the path of the radiation, regardless of whether the material is of high or low density. However, as is the case with X-rays, materials with high atomic number such as lead or depleted uranium add a modest (typically 20% to 30%) amount of stopping power over an equal mass of less dense and lower atomic weight materials (such as water or concrete). The atmosphere absorbs all gamma rays approaching Earth from space. Even air is capable of absorbing gamma rays, halving the energy of such waves by passing through, on the average, 500 ft (150 m).

21.1.4 Alpha radiation

Main article: Alpha decay

Alpha particles are helium-4 nuclei (two protons and two neutrons). They interact with matter strongly due to their charges and combined mass, and at their usual velocities only penetrate a few centimeters of air, or a few millimeters of low density material (such as the thin mica material which is specially placed in some Geiger counter tubes to allow alpha particles in). This means that alpha particles from ordinary alpha decay do not penetrate the outer layers of dead skin cells and cause no damage to the live tissues below. Some very high energy alpha particles compose about 10% of cosmic rays, and these are capable of penetrating the body and even thin metal plates. However, they are of danger only to astronauts, since they are deflected by the Earth's magnetic field and then stopped by its atmosphere.

Alpha radiation is dangerous when alpha-emitting radioisotopes are ingested or inhaled (breathed or swallowed). This brings the radioisotope close enough to sensitive live tissue for the alpha radiation to damage cells. Per unit of energy, alpha particles are at least 20 times more effective at cell-damage as gamma rays and X-rays. See relative biological effectiveness for a discussion of this. Examples of highly poisonous alpha-emitters are all isotopes of radium, radon, and polonium, due to the amount of decay that occur in these short half-life materials.

21.1.5 Beta radiation

Main article: Beta decay

Beta-minus (β^-) radiation consists of an energetic electron. It is more penetrating than alpha radiation, but less than gamma. Beta radiation from radioactive decay can be stopped with a few centimeters of plastic or a few millimeters of metal. It occurs when a neutron decays into a proton in a nucleus, releasing the beta particle and an antineutrino. Beta radiation from linac accelerators is far more energetic and penetrating than natural beta radiation. It is sometimes used therapeutically in radiotherapy to treat superficial tumors.

Beta-plus (β^+) radiation is the emission of positrons, which are the antimatter form of electrons. When a positron slows to speeds similar to those of electrons in the material, the positron will annihilate an electron, releasing two gamma photons of 511 keV in the process. Those two gamma photons will be traveling in (approximately) opposite direction. The gamma radiation from positron annihilation consists of high energy photons, and is also ionizing.

21.1.6 Neutron radiation

Main articles: Neutron radiation and Neutron temperature

Neutrons are categorized according to their speed/energy. Neutron radiation consists of free neutrons. These neutrons may be emitted during either spontaneous or induced nuclear fission. Neutrons are rare radiation particles; they are produced in large numbers only where chain reaction fission or fusion reactions are active; this happens for about 10 microseconds in a thermonuclear explosion, or continuously inside an operating nuclear reactor; production of the neutrons stops almost immediately in the reactor when it goes non-critical.

Neutrons are the only type of ionizing radiation that can make other objects, or material, radioactive. This process, called neutron activation, is the primary method used to produce radioactive sources for use in medical, academic, and industrial applications. Even comparatively low speed thermal neutrons cause neutron activation (in fact, they cause it more efficiently). Neutrons do not ionize atoms in the same way that charged particles such as protons and electrons do (by the excitation of an electron), because neutrons have no charge. It is through their absorption by nuclei which then become unstable that they cause ionization. Hence, neutrons are said to be "indirectly ionizing." Even neutrons without significant kinetic energy are indirectly ionizing, and are thus a significant radiation hazard. Not all materials are capable of neutron activation; in water, for example, the most common isotopes of both types atoms present (hydrogen and oxygen) capture neutrons and become heavier but remain stable forms of those atoms. Only the absorption of more than one neutron, a statistically rare occurrence, can activate a hydrogen atom, while oxygen requires two additional absorptions. Thus water is only very weakly capable of activation. The sodium in salt (as in sea water), on the other hand, need only absorb a single neutron to become Na-24, a very intense source of beta decay, with half-life of 15 hours.

In addition, high-energy (high-speed) neutrons have the ability to directly ionize atoms. One mechanism by which high energy neutrons ionize atoms is to strike the nucleus of an atom and knock the atom out of a molecule, leaving one or more electrons behind as the chemical bond is broken. This leads to production of chemical free radicals. In addition, very high energy neutrons can cause ionizing radiation by "neutron spallation" or knockout, wherein neutrons cause emission of high-energy protons from atomic nuclei (especially hydrogen nuclei) on impact. The last process imparts most of the neutron's energy to the proton, much like one billiard ball striking another. The charged protons, and other products from such reactions are directly ionizing.

High-energy neutrons are very penetrating and can travel great distances in air (hundreds or even thousands of meters) and moderate distances (several meters) in common solids. They typically require hydrogen rich shielding, such as concrete or water, to block them within distances of less than a meter. A common source of neutron radiation occurs inside a nuclear reactor, where a meters-thick water layer is used as effective shielding.

21.2 Cosmic radiation

Main article: Cosmic rays

There are two sources of high energy particles entering the Earth's atmosphere from outer space: the sun and deep space. The sun continuously emits particles, primarily free protons, in the solar wind, and occasionally augments the flow hugely with coronal mass ejections (CME).

The particles from deep space (inter- and extra-galactic) are much less frequent, but of much higher energies. These particles are also mostly protons, with much of the remainder consisting of helions (alpha particles). A few completely ionized nuclei of heavier elements are present. The origin of these galactic cosmic rays is not yet well understood, but they seem to be remnants of supernovae and especially gamma-ray bursts (GRB), which feature magnetic fields capable of the huge accelerations measured from these particles. They may also be generated by quasars, which are galaxy-wide jet phenomena similar to GRBs but known for their much larger size, and which seem to be a violent part of the universe's early history.

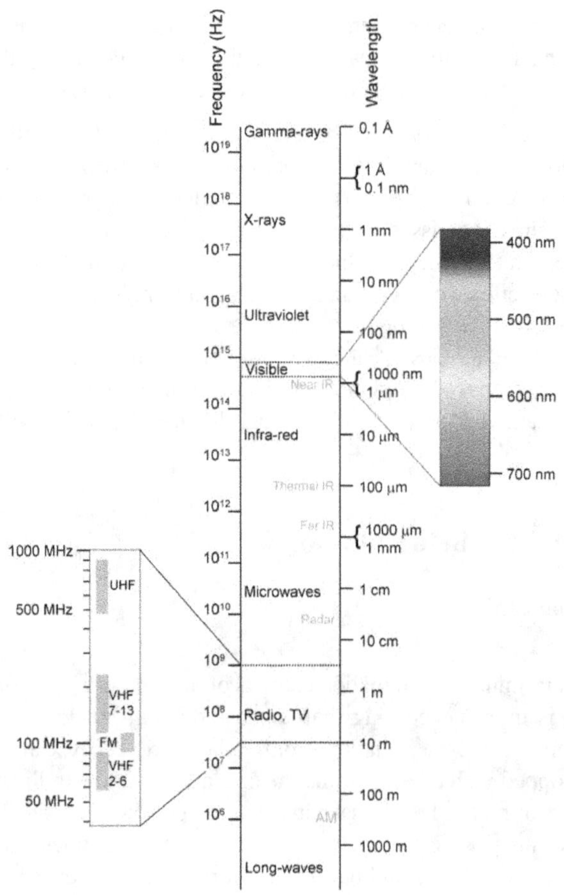

The electromagnetic spectrum

21.3 Non-ionizing radiation

Main articles: Non-ionizing radiation and Electromagnetic radiation

The kinetic energy of particles of non-ionizing radiation is too small to produce charged ions when passing through matter. For non-ionizing electromagnetic radiation (see types below), the associated particles (photons) have only sufficient energy to change the rotational, vibrational or electronic valence configurations of molecules and atoms. The effect of non-ionizing forms of radiation on living tissue has only recently been studied. Nevertheless, different biological effects are observed for different types of non-ionizing radiation.[3][5]

Even "non-ionizing" radiation is capable of causing thermal-ionization if it deposits enough heat to raise temperatures to ionization energies. These reactions occur at far higher energies than with ionization radiation, which requires only single particles to cause ionization. A familiar example of thermal ionization is the flame-ionization of a common fire, and the browning reactions in common food items induced by infrared radiation, during broiling-type cooking.

The electromagnetic spectrum is the range of all possible electromagnetic radiation frequencies.[3] The electromagnetic spectrum (usually just spectrum) of an object is the characteristic distribution of electromagnetic radiation emitted by, or absorbed by, that particular object.

The non-ionizing portion of electromagnetic radiation consists of electromagnetic waves that (as individual quanta or particles, see photon) are not energetic enough to detach electrons from atoms or molecules and hence cause their ionization. These include radio waves, microwaves, infrared, and (sometimes) visible light. The lower frequencies of ultraviolet light may cause chemical changes and molecular damage similar to ionization, but is technically not ionizing. The highest frequencies of ultraviolet light, as well as all X-rays and gamma-rays are ionizing.

The occurrence of ionization depends on the energy of the individual particles or waves, and not on their number. An intense flood of particles or waves will not cause ionization if these particles or waves do not carry enough energy to be ionizing, unless they raise the temperature of a body to a point high enough to ionize small fractions of atoms or molecules by the process of thermal-ionization (this, however, requires relatively extreme radiation intensities).

21.3.1 Ultraviolet light

Main article: Ultraviolet

As noted above, the lower part of the spectrum of ultraviolet, called soft UV, from 3 eV to about 10 eV, is non-ionizing. However, the effects of non-ionizing ultraviolet on chemistry and the damage to biological systems exposed to it (including oxidation, mutation, and cancer) are such that even this part of ultraviolet is often compared with ionizing radiation.

21.3.2 Visible light

Main article: Light

Light, or visible light, is a very narrow range of electromagnetic radiation of a wavelength that is visible to the human eye, or 380–750 nm which equates to a frequency range of 790 to 400 THz respectively.[3] More broadly, physicists use the term "light" to mean electromagnetic radiation of all wavelengths, whether visible or not.

21.3.3 Infrared

Main article: Infrared

Infrared (IR) light is electromagnetic radiation with a wavelength between 0.7 and 300 micrometers, which corresponds to a frequency range between 430 and 1 THz respectively. IR wavelengths are longer than that of visible light, but shorter than that of microwaves. Infrared may be detected at a distance from the radiating objects by "feel." Infrared sensing snakes can detect and focus infrared by use of a pinhole lens in their heads, called "pits". Bright sunlight provides an irradiance of just over 1 kilowatt per square meter at sea level. Of this energy, 53% is infrared radiation, 44% is visible light, and 3% is ultraviolet radiation.[3]

21.3.4 Microwave

Main article: Microwave

Microwaves are electromagnetic waves with wavelengths ranging from as short as one millimeter to as long as one meter, which equates to a frequency range of 300 MHz to 300 GHz. This broad definition includes both UHF and EHF (millimeter waves), but various sources use different other limits.[3] In all cases, microwaves include the entire super high frequency band (3 to 30 GHz, or 10 to 1 cm) at minimum, with RF engineering often putting the lower boundary at 1 GHz (30 cm), and the upper around 100 GHz (3mm).

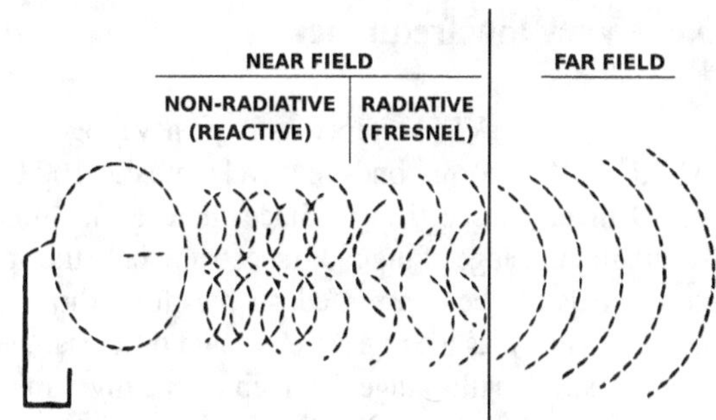

In electromagnetic radiation (such as microwaves from an antenna, shown here) the term "radiation" applies only to the parts of the electromagnetic field that radiate into infinite space and decrease in intensity by an inverse-square law of power so that the total radiation energy that crosses through an imaginary spherical surface is the same, no matter how far away from the antenna the spherical surface is drawn. Electromagnetic radiation includes the far field part of the electromagnetic field around a transmitter. A part of the "near-field" close to the transmitter, is part of the changing electromagnetic field, but does not count as electromagnetic radiation.

21.3.5 Radio waves

Main article: Radio waves

Radio waves are a type of electromagnetic radiation with wavelengths in the electromagnetic spectrum longer than infrared light. Like all other electromagnetic waves, they travel at the speed of light. Naturally occurring radio waves are made by lightning, or by certain astronomical objects. Artificially generated radio waves are used for fixed and mobile radio communication, broadcasting, radar and other navigation systems, satellite communication, computer networks and innumerable other applications. In addition, almost any wire carrying alternating current will radiate some of the energy away as radio waves; these are mostly termed interference. Different frequencies of radio waves have different propagation characteristics in the Earth's atmosphere; long waves may bend at the rate of the curvature of the Earth and may cover a part of the Earth very consistently, shorter waves travel around the world by multiple reflections off the ionosphere and the Earth. Much shorter wavelengths bend or reflect very little and travel along the line of sight.

21.3.6 Very low frequency

Very low frequency (VLF) refers to a frequency range of 30 Hz to 3 kHz which corresponds to wavelengths of 100,000 to 10,000 meters respectively. Since there is not much bandwidth in this range of the radio spectrum, only the very simplest signals can be transmitted, such as for radio navigation. Also known as the myriameter band or myriameter wave as the wavelengths range from ten to one myriameter (an obsolete metric unit equal to 10 kilometers).

21.3.7 Extremely low frequency

Main article: Extremely low frequency

Extremely low frequency (ELF) is radiation frequencies from 3 to 30 Hz (10^8 to 10^7 meters respectively). In atmosphere science, an alternative definition is usually given, from 3 Hz to 3 kHz.[3] In the related magnetosphere science, the lower frequency electromagnetic oscillations (pulsations occurring below ~3 Hz) are considered to lie in the ULF range, which is thus also defined differently from the ITU Radio Bands. A massive military ELF antenna in Michigan radiates very slow messages to otherwise unreachable receivers, such as submerged submarines.

21.3.8 Thermal radiation (heat)

Main article: Thermal radiation

Thermal radiation is a common synonym for infrared radiation emitted by objects at temperatures often encountered on Earth. Thermal radiation refers not only to the radiation itself, but also the process by which the surface of an object radiates its thermal energy in the form black body radiation. Infrared or red radiation from a common household radiator or electric heater is an example of thermal radiation, as is the heat emitted by an operating incandescent light bulb. Thermal radiation is generated when energy from the movement of charged particles within atoms is converted to electromagnetic radiation.

As noted above, even low-frequency thermal radiation may cause temperature-ionization whenever it deposits sufficient thermal energy to raises temperatures to a high enough level. Common examples of this are the ionization (plasma) seen in common flames, and the molecular changes caused by the "browning" during food-cooking, which is a chemical process that begins with a large component of ionization.

21.3.9 Black-body radiation

Main article: Black-body radiation

Black-body radiation is an idealized spectrum of radiation emitted by a body that is at a uniform temperature. The shape of the spectrum and the total amount of energy emitted by the body is a function of the absolute temperature of that body. The radiation emitted covers the entire electromagnetic spectrum and the intensity of the radiation (power/unit-area) at a given frequency is described by Planck's law of radiation. For a given temperature of a black-body there is a particular frequency at which the radiation emitted is at its maximum intensity. That maximum radiation frequency moves toward higher frequencies as the temperature of the body increases. The frequency at which the black-body radiation is at maximum is given by Wien's displacement law and is a function of the body's absolute temperature. A black-body is one that emits at any temperature the maximum possible amount of radiation at any given wavelength. A black-body will also absorb the maximum possible incident radiation at any given wavelength. A black-body with a temperature at or below room temperature would thus appear absolutely black, as it would not reflect any incident light nor would it emit enough radiation at visible wavelengths for our eyes to detect. Theoretically, a black-body emits electromagnetic radiation over the entire spectrum from very low frequency radio waves to x-rays, creating a continuum of radiation.

The color of a radiating black-body tells the temperature of its radiating surface. It is responsible for the color of stars, which vary from infrared through red (2,500K), to yellow (5,800K), to white and to blue-white (15,000K) as the peak radiance passes through those points in the visible spectrum. When the peak is below the visible spectrum the body is black, while when it is above the body is blue-white, since all the visible colors are represented from blue decreasing to red.

21.4 Discovery

Electromagnetic radiation of wavelengths other than visible light were discovered in the early 19th century. The discovery of infrared radiation is ascribed to William Herschel, the astronomer. Herschel published his results in 1800 before the Royal Society of London. Herschel, like Ritter, used a prism to refract light from the Sun and detected the infrared (beyond the red part of the spectrum), through an increase in the temperature recorded by a thermometer.

In 1801, the German physicist Johann Wilhelm Ritter made the discovery of ultraviolet by noting that the rays from a prism darkened silver chloride preparations more quickly than violet light. Ritter's experiments were an early precursor to what would become photography. Ritter noted that the UV rays were capable of causing chemical reactions.

The first radio waves detected were not from a natural source, but were produced deliberately and artificially by the German scientist Heinrich Hertz in 1887, using electrical circuits calculated to produce oscillations in the radio frequency range, following formulas suggested by the equations of James Clerk Maxwell.

Wilhelm Röntgen discovered and named X-rays. While experimenting with high voltages applied to an evacuated tube on 8 November 1895, he noticed a fluorescence on a nearby plate of coated glass. Within a month, he discovered the main properties of X-rays that we understand to this day.

In 1896, Henri Becquerel found that rays emanating from certain minerals penetrated black paper and caused fogging of an unexposed photographic plate. His doctoral student Marie Curie discovered that only certain chemical elements gave off these rays of energy. She named this behavior radioactivity.

Alpha rays (alpha particles) and beta rays (beta particles) were differentiated by Ernest Rutherford through simple experimentation in 1899. Rutherford used a generic pitchblende radioactive source and determined that the rays produced by the source had differing penetrations in materials. One type had short penetration (it was stopped by paper) and a positive charge, which Rutherford named *alpha rays.* The other was more penetrating (able to expose film through paper but not metal) and had a negative charge, and this type Rutherford named *beta.* This was the radiation that had been first detected by Becquerel from uranium salts. In 1900, the French scientist Paul Villard discovered a third neutrally charged and especially penetrating type of radiation from radium, and after he described it, Rutherford realized it must be yet a third type of radiation, which in 1903 Rutherford named gamma rays.

Henri Becquerel himself proved that beta rays are fast electrons, while Rutherford and Thomas Royds proved in 1909 that alpha particles are ionized helium. Rutherford and Edward Andrade proved in 1914 that gamma rays are like X-rays, but with shorter wavelengths.

Cosmic ray radiations striking the Earth from outer space were finally definitively recognized and proven to exist in 1912, as the scientist Victor Hess carried an electrometer to various altitudes in a free balloon flight. The nature of these radiations was only gradually understood in later years.

Neutron radiation was discovered with the neutron by Chadwick, in 1932. A number of other high energy particulate radiations such as positrons, muons, and pions were discovered by cloud chamber examination of cosmic ray reactions shortly thereafter, and others types of particle radiation were produced artificially in particle accelerators, through the last half of the twentieth century.

21.5 Uses

21.5.1 Medicine

Main articles: Medical radiography and Medical radiation scientist

Radiation and radioactive substances are used for diagnosis, treatment, and research. X-rays, for example, pass through muscles and other soft tissue but are stopped by dense materials. This property of X-rays enables doctors to find broken bones and to locate cancers that might be growing in the body.[6] Doctors also find certain diseases by injecting a radioactive substance and monitoring the radiation given off as the substance moves through the body.[7] Radiation used for cancer treatment is called ionizing radiation because it forms ions in the cells of the tissues it passes through as it dislodges electrons from atoms. This can kill cells or change genes so the cells cannot grow. Other forms of radiation such as radio waves, microwaves, and light waves are called non-ionizing. They don't have as much energy and are not able to ionize cells.

21.5.2 Communication

All modern communication systems use forms of electromagnetic radiation. Variations in the intensity of the radiation represent changes in the sound, pictures, or other information being transmitted. For example, a human voice can be sent as a radio wave or microwave by making the wave vary to correspond variations in the voice. Musicians have also experimented with gamma sonification, or using nuclear radiation, to produce sound and music.[8]

21.5.3 Science

Researchers use radioactive atoms to determine the age of materials that were once part of a living organism. The age of such materials can be estimated by measuring the amount of radioactive carbon they contain in a process called radiocarbon dating. Similarly, using other radioactive elements, the age of rocks and other geological features (even some man-made objects) can be determined; this is called Radiometric dating. Environmental scientists use radioactive atoms, known as tracer atoms, to identify the pathways taken by pollutants through the environment.

Radiation is used to determine the composition of materials in a process called neutron activation analysis. In this process, scientists bombard a sample of a substance with particles called neutrons. Some of the atoms in the sample absorb neutrons and become radioactive. The scientists can identify the elements in the sample by studying the emitted radiation.

21.6 See also

- Background radiation, which actually refers to the background ionizing radiation
- Čerenkov radiation
- Cosmic microwave background radiation, 3 K blackbody radiation that fills the Universe
- Electromagnetic spectrum
- Hawking radiation
- Ionizing radiation
- Banana equivalent dose
- Non-ionizing radiation
- Radiant energy, radiation by a source into the surrounding environment.
- Radiation damage – adverse effects on materials and devices
- Radiation hardening – making devices resistant to failure in high radiation environments
- Radiation hormesis – dosage threshold damage theory
- Radiation poisoning – adverse effects on life forms
- Radiation properties
- Radioactive contamination
- Radioactive decay
- Radiation Protection Convention, 1960 – by International Labour Organization

21.7 Notes and references

[1] Weisstein, Eric W. “Radiation”. *Eric Weisstein's World of Physics*. Wolfram Research. Retrieved 2014-01-11.

[2] “Radiation”. *The free dictionary by Farlex*. Farlex, Inc. Retrieved 2014-01-11.

[3] Kwan-Hoong Ng (20–22 October 2003). “Non-Ionizing Radiations – Sources, Biological Effects, Emissions and Exposures” (PDF). *Proceedings of the International Conference on Non-Ionizing Radiation at UNITEN ICNIR2003 Electromagnetic Fields and Our Health.*

[4] “ICRP Publication 103 The 2007 Recommendations of the International Commission on Protection” (PDF). ICRP. Retrieved 12 December 2013.

[5] Moulder, John E. “Static Electric and Magnetic Fields and Human Health”. Archived from the original on 14 July 2007.

[6] Radiography

[7] Nuclear medicine

[8] Dunn, Peter (2014). “Making Nuclear Music”. Slice of MIT. Retrieved 25 Aug 2014.

21.8 External links

-
- Radiation on *In Our Time* at the BBC.
- Health Physics Society Public Education Website
- Ionizing Radiation and Radon from World Health Organization
- Q&A: Health effects of radiation exposure, *BBC News*, 21 July 2011.

Chapter 22

Standard ruler

A **standard ruler** is an astronomical object for which the actual physical size is known. By measuring its angular size in the sky, one can use simple trigonometry to determine its distance from Earth. In simple terms, this is because objects of a fixed size appear smaller the further away they are.

Measuring distances is of great importance in cosmology, as the relationship between the distance and redshift of an object can be used to measure the expansion rate and geometry of the Universe. Distances can also be measured using standard candles; many different types of standard candles and rulers are needed to construct the cosmic distance ladder.

22.1 Relationship between angular size and distance

The relation between the angular diameter, θ, actual (physical) diameter, r, and distance, D, of an object from the observer is given by:

$\theta \approx \frac{r}{D}$

where θ is measured in radians.

Because space is expanding, there is no one, unique way of measuring the distance between source and observer. The distance measured by a standard ruler is what is known as the angular diameter distance. Standard candles measure another type of distance called the luminosity distance.

22.2 See also

- Standard candle
- Baryon acoustic oscillations
- Angular diameter distance
- Parallax
- Cosmic distance ladder

Chapter 23

Chandrasekhar limit

The **Chandrasekhar limit** (/tʃʌndrəˈʃeɪkər/) is the maximum mass of a stable white dwarf star. The limit was first indicated in papers published by Wilhelm Anderson and E. C. Stoner, and was named after Subrahmanyan Chandrasekhar, the Indian astrophysicist who independently discovered and improved upon the accuracy of the calculation in 1930, at the age of 19, in India. This limit was initially ignored by the community of scientists because such a limit would logically require the existence of black holes, which were considered a scientific impossibility at the time. White dwarfs resist gravitational collapse primarily through electron degeneracy pressure. (By comparison, main sequence stars resist collapse through thermal pressure.) The Chandrasekhar limit is the mass above which electron degeneracy pressure in the star's core is insufficient to balance the star's own gravitational self-attraction. Consequently, white dwarfs with masses greater than the limit would be subject to further gravitational collapse, evolving into a different type of stellar remnant, such as a neutron star or black hole. (However, white dwarfs generally avoid this fate by exploding before they undergo collapse.) Those with masses under the limit remain stable as white dwarfs.[1]

The currently accepted value of the limit is about 1.4 $M_{\odot}$ (2.765 × 10^{30} kg).[2][3][4]

23.1 Physics

Electron degeneracy pressure is a quantum-mechanical effect arising from the Pauli exclusion principle. Since electrons are fermions, no two electrons can be in the same state, so not all electrons can be in the minimum-energy level. Rather, electrons must occupy a band of energy levels. Compression of the electron gas increases the number of electrons in a given volume and raises the maximum energy level in the occupied band. Therefore, the energy of the electrons increases on compression, so pressure must be exerted on the electron gas to compress it, producing electron degeneracy pressure. With sufficient compression, electrons are forced into nuclei in the process of electron capture, relieving the pressure.

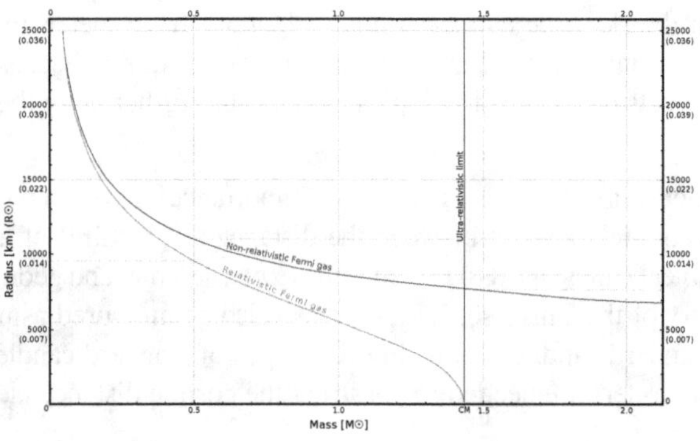

Radius–mass relations for a model white dwarf. The green curve uses the general pressure law for an ideal Fermi gas, while the blue curve is for a non-relativistic ideal Fermi gas. The black line marks the ultrarelativistic limit.

In the nonrelativistic case, electron degeneracy pressure gives rise to an equation of state of the form $P = K_1\rho^{\frac{5}{3}}$, where P is the pressure, ρ is the mass density, and K_1 is a constant. Solving the hydrostatic equation leads to a model white dwarf that is a polytrope of index 3/2—and therefore has radius inversely proportional to the cube root of its mass, and volume inversely proportional to its mass.[5]

As the mass of a model white dwarf increases, the typical energies to which degeneracy pressure forces the electrons are no longer negligible relative to their rest masses. The velocities of the electrons approach the speed of light, and special relativity must be taken into account. In the strongly relativistic limit, the equation of state takes the form $P = K_2\rho^{\frac{4}{3}}$. This yields a polytrope of index 3, which has a total mass, M_{limit} say, depending only on K_2.[6]

For a fully relativistic treatment, the equation of state used interpolates between the equations $P = K_1\rho^{\frac{5}{3}}$ for small ϱ and $P = K_2\rho^{\frac{4}{3}}$ for large ϱ. When this is done, the model radius still decreases with mass, but becomes zero at M_{limit}. This is the Chandrasekhar limit.[7] The curves of radius against mass for the non-relativistic and relativistic models are shown in the graph. They are colored blue and green, respectively. μ_e has been set equal to 2. Radius is measured in standard solar radii[8] or kilometers, and mass in standard solar masses.

Calculated values for the limit vary depending on the nuclear composition of the mass.[9] Chandrasekhar[10], eq. (36),[7], eq. (58),[11], eq. (43) gives the following expression, based on the equation of state for an ideal Fermi gas:

$$M_{\text{limit}} = \frac{\omega_3^0 \sqrt{3\pi}}{2} \left(\frac{\hbar c}{G} \right)^{3/2} \frac{1}{(\mu_e m_H)^2},$$

where:

- $\hbar$ is the reduced Planck constant
- c is the speed of light
- G is the gravitational constant
- μ_e is the average molecular weight per electron, which depends upon the chemical composition of the star.
- mH is the mass of the hydrogen atom.
- $\omega_3^0 \approx 2.018236$ is a constant connected with the solution to the Lane-Emden equation.

As $\sqrt{\hbar c/G}$ is the Planck mass, the limit is of the order of

$$\frac{M_{\text{Pl}}^3}{m_{\text{H}}^2}.$$

A more accurate value of the limit than that given by this simple model requires adjusting for various factors, including electrostatic interactions between the electrons and nuclei and effects caused by nonzero temperature.[9] Lieb and Yau[12] have given a rigorous derivation of the limit from a relativistic many-particle Schrödinger equation.

23.2 History

In 1926, the British physicist Ralph H. Fowler observed that the relationship between the density, energy, and temperature of white dwarfs could be explained by viewing them as a gas of nonrelativistic, non-interacting electrons and nuclei that obey Fermi–Dirac statistics.[13] This Fermi gas model was then used by the British physicist Edmund Clifton Stoner in 1929 to calculate the relationship among the mass, radius, and density of white dwarfs, assuming they were homogeneous spheres.[14] Wilhelm Anderson applied a relativistic correction to this model, giving rise to a maximum possible mass of approximately 1.37×10^{30} kg.[15] In 1930, Stoner derived the internal energy–density equation of state for a Fermi gas, and was then able to treat the mass–radius relationship in a fully relativistic manner, giving a limiting mass of approximately (for μ_e=2.5) 2.19 · 10^{30} kg.[16] Stoner went on to derive the pressure–density equation of state, which he published in 1932.[17] These equations of state were also previously published by the Soviet physicist Yakov Frenkel in 1928, together with some other remarks on the physics of degenerate matter.[18] Frenkel's work, however, was ignored by the astronomical and astrophysical community.[19]

A series of papers published between 1931 and 1935 had its beginning on a trip from India to England in 1930, where the Indian physicist Subrahmanyan Chandrasekhar worked on the calculation of the statistics of a degenerate Fermi gas.[20] In these papers, Chandrasekhar solved the hydrostatic equation together with the nonrelativistic Fermi gas equation of state,[5] and also treated the case of a relativistic Fermi gas, giving rise to the value of the limit shown above.[6][7][10][21] Chandrasekhar reviews this work in his Nobel Prize lecture.[11] This value was also computed in 1932 by the Soviet physicist Lev Davidovich Landau,[22] who, however, did not apply it to white dwarfs.

Chandrasekhar's work on the limit aroused controversy, owing to the opposition of the British astrophysicist Arthur Eddington. Eddington was aware that the existence of black holes was theoretically possible, and also realized that the existence of the limit made their formation possible. However, he was unwilling to accept that this could happen. After a talk by Chandrasekhar on the limit in 1935, he replied:

> The star has to go on radiating and radiating and contracting and contracting until, I suppose, it gets down to a few km radius, when gravity becomes strong enough to hold in the radiation, and the star can at last find peace. ... I think there should be a law of Nature to prevent a star from behaving in this absurd way![23]

Eddington's proposed solution to the perceived problem was to modify relativistic mechanics so as to make the law $P = K_1\rho^{5/3}$ universally applicable, even for large ρ.[24] Although Niels Bohr, Fowler, Wolfgang Pauli, and other physicists agreed with Chandrasekhar's analysis, at the time, owing to Eddington's status, they were unwilling to publicly support Chandrasekhar.[25], pp. 110–111 Through the rest of his life, Eddington held to his position in his writings,[26][27][28][29][30] including his work on his fundamental theory.[31] The drama associated with this disagreement is one of the main themes of *Empire of the Stars*, Arthur I. Miller's biography of Chandrasekhar.[25] In

Miller's view:

> Chandra's discovery might well have transformed and accelerated developments in both physics and astrophysics in the 1930s. Instead, Eddington's heavy-handed intervention lent weighty support to the conservative community astrophysicists, who steadfastly refused even to consider the idea that stars might collapse to nothing. As a result, Chandra's work was almost forgotten.[25]:150

23.3 Applications

The core of a star is kept from collapsing by the heat generated by the fusion of nuclei of lighter elements into heavier ones. At various stages of stellar evolution, the nuclei required for this process are exhausted, and the core collapses, causing it to become denser and hotter. A critical situation arises when iron accumulates in the core, since iron nuclei are incapable of generating further energy through fusion. If the core becomes sufficiently dense, electron degeneracy pressure will play a significant part in stabilizing it against gravitational collapse.[32]

If a main-sequence star is not too massive (less than approximately 8 solar masses), it eventually sheds enough mass to form a white dwarf having mass below the Chandrasekhar limit, which consists of the former core of the star. For more-massive stars, electron degeneracy pressure does not keep the iron core from collapsing to very great density, leading to formation of a neutron star, black hole, or, speculatively, a quark star. (For very massive, low-metallicity stars, it is also possible that instabilities destroy the star completely.)[33][34][35][36] During the collapse, neutrons are formed by the capture of electrons by protons in the process of electron capture, leading to the emission of neutrinos.[32], pp. 1046–1047. The decrease in gravitational potential energy of the collapsing core releases a large amount of energy on the order of 10^{46} joules (100 foes). Most of this energy is carried away by the emitted neutrinos.[37] This process is believed responsible for supernovae of types Ib, Ic, and II.[32]

Type Ia supernovae derive their energy from runaway fusion of the nuclei in the interior of a white dwarf. This fate may befall carbon–oxygen white dwarfs that accrete matter from a companion giant star, leading to a steadily increasing mass. As the white dwarf's mass approaches the Chandrasekhar limit, its central density increases, and, as a result of compressional heating, its temperature also increases. This eventually ignites nuclear fusion reactions, leading to an immediate carbon detonation, which disrupts the star and causes the supernova.[38], §5.1.2

A strong indication of the reliability of Chandrasekhar's formula is that the absolute magnitudes of supernovae of Type Ia are all approximately the same; at maximum luminosity, MV is approximately −19.3, with a standard deviation of no more than 0.3.[38], (1) A 1-sigma interval therefore represents a factor of less than 2 in luminosity. This seems to indicate that all type Ia supernovae convert approximately the same amount of mass to energy.

23.4 Super-Chandrasekhar mass supernovae

Main article: Champagne Supernova

In April 2003, the Supernova Legacy Survey observed a type Ia supernova, designated SNLS-03D3bb, in a galaxy approximately 4 billion light years away. According to a group of astronomers at the University of Toronto and elsewhere, the observations of this supernova are best explained by assuming that it arose from a white dwarf that grew to twice the mass of the Sun before exploding. They believe that the star, dubbed the "Champagne Supernova" by University of Oklahoma astronomer David R. Branch, may have been spinning so fast that a centrifugal tendency allowed it to exceed the limit. Alternatively, the supernova may have resulted from the merger of two white dwarfs, so that the limit was only violated momentarily. Nevertheless, they point out that this observation poses a challenge to the use of type Ia supernovae as standard candles.[39][40][41]

Since the observation of the Champagne Supernova in 2003, more very bright type Ia supernovae have been observed that are thought to have originated from white dwarfs whose masses exceeded the Chandrasekhar limit. These include SN 2006gz, SN 2007if and SN 2009dc.[42] The super-Chandrasekhar mass white dwarfs that gave rise to these supernovae are believed to have had masses up to 2.4–2.8 solar masses.[42] One way to potentially explain the problem of the Champagne Supernova was considering it the result of an aspherical explosion of a white dwarf. However, spectropolarimetric observations of SN 2009dc showed it had a polarization smaller than 0.3, making the large asphericity theory unlikely.[42]

23.5 Tolman–Oppenheimer–Volkoff limit

After a supernova explosion, a neutron star may be left behind. Like white dwarfs these objects are extremely compact and are supported by degeneracy pressure, but a neutron star is so massive and compressed that electrons and protons have combined to form neutrons, and the star is thus supported by neutron degeneracy pressure instead of electron degeneracy pressure. The limit of neutron degeneracy pressure, analogous to the Chandrasekhar limit, is known as the Tolman–Oppenheimer–Volkoff limit.

23.6 References

[1] Sean Carroll, Ph.D., Caltech, 2007, The Teaching Company, *Dark Matter, Dark Energy: The Dark Side of the Universe*, Guidebook Part 2 page 44, Accessed Oct. 7, 2013, "...Chandrasekhar limit: The maximum mass of a white dwarf star, about 1.4 times the mass of the Sun. Above this mass, the gravitational pull becomes too great, and the star must collapse to a neutron star or black hole..."

[2] Israel, edited by S.W. Hawking, W. (1989). *Three hundred years of gravitation* (1st pbk. ed., with corrections. ed.). Cambridge [Cambridgeshire]: Cambridge University Press. ISBN 0-521-37976-8.

[3] p. 55, How A Supernova Explodes, Hans A. Bethe and Gerald Brown, pp. 51–62 in *Formation And Evolution of Black Holes in the Galaxy: Selected Papers with Commentary*, Hans Albrecht Bethe, Gerald Edward Brown, and Chang-Hwan Lee, River Edge, New Jersey: World Scientific: 2003. ISBN 981-238-250-X.

[4] Mazzali, P. A.; Röpke, F. K.; Benetti, S.; Hillebrandt, W. (2007). "A Common Explosion Mechanism for Type Ia Supernovae". *Science* (PDF). **315** (5813): 825–828. Bibcode:2007Sci...315..825M. PMID 17289993. arXiv:astro-ph/0702351v1 . doi:10.1126/science.1136259.

[5] The Density of White Dwarf Stars, S. Chandrasekhar, *Philosophical Magazine* (7th series) **11** (1931), pp. 592–596.

[6] The Maximum Mass of Ideal White Dwarfs, S. Chandrasekhar, *Astrophysical Journal* **74** (1931), pp. 81–82.

[7] The Highly Collapsed Configurations of a Stellar Mass (second paper), S. Chandrasekhar, *Monthly Notices of the Royal Astronomical Society*, **95** (1935), pp. 207-–225.

[8] *Standards for Astronomical Catalogues, Version 2.0*, section 3.2.2, web page, accessed 12-I-2007.

[9] The Neutron Star and Black Hole Initial Mass Function, F. X. Timmes, S. E. Woosley, and Thomas A. Weaver, *Astrophysical Journal* **457** (February 1, 1996), pp. 834–843.

[10] The Highly Collapsed Configurations of a Stellar Mass, S. Chandrasekhar, *Monthly Notices of the Royal Astronomical Society* **91** (1931), 456–466.

[11] *On Stars, Their Evolution and Their Stability*, Nobel Prize lecture, Subrahmanyan Chandrasekhar, December 8, 1983.

[12] A rigorous examination of the Chandrasekhar theory of stellar collapse, Elliott H. Lieb and Horng-Tzer Yau, *Astrophysical Journal* **323** (1987), pp. 140–144.

[13] On Dense Matter, R. H. Fowler, *Monthly Notices of the Royal Astronomical Society* **87** (1926), pp. 114–122.

[14] The Limiting Density of White Dwarf Stars, Edmund C. Stoner, *Philosophical Magazine* (7th series) **7** (1929), pp. 63–70.

[15] Über die Grenzdichte der Materie und der Energie, Wilhelm Anderson, *Zeitschrift für Physik* **56**, #11–12 (November 1929), pp. 851–856. DOI 10.1007/BF01340146.

[16] The Equilibrium of Dense Stars, Edmund C. Stoner, *Philosophical Magazine* (7th series) **9** (1930), pp. 944–963.

[17] The minimum pressure of a degenerate electron gas, E. C. Stoner, *Monthly Notices of the Royal Astronomical Society* **92** (May 1932), pp. 651–661.

[18] Anwendung der Pauli-Fermischen Elektronengastheorie auf das Problem der Kohäsionskräfte, J. Frenkel, *Zeitschrift für Physik* **50**, #3–4 (March 1928), pp. 234–248. DOI 10.1007/BF01328867.

[19] The article by Ya I Frenkel' on `binding forces' and the theory of white dwarfs, D. G. Yakovlev, *Physics Uspekhi* **37**, #6 (1994), pp. 609–612.

[20] Chandrasekhar's biographical memoir at the National Academy of Sciences, web page, accessed 12-I-2007.

[21] Stellar Configurations with degenerate Cores, S. Chandrasekhar, *The Observatory* **57** (1934), pp. 373–377.

[22] On the Theory of Stars, in *Collected Papers of L. D. Landau*, ed. and with an introduction by D. ter Haar, New York: Gordon and Breach, 1965; originally published in *Phys. Z. Sowjet.* **1** (1932), 285.

[23] Meeting of the Royal Astronomical Society, Friday, 1935 January 11, *The Observatory* **58** (February 1935), pp. 33–41.

[24] On "Relativistic Degeneracy", Sir A. S. Eddington, *Monthly Notices of the Royal Astronomical Society* **95** (1935), 194–206.

[25] *Empire of the Stars: Obsession, Friendship, and Betrayal in the Quest for Black Holes*, Arthur I. Miller, Boston, New York: Houghton Mifflin, 2005, ISBN 0-618-34151-X; reviewed at *The Guardian*: The battle of black holes.

[26] The International Astronomical Union meeting in Paris, 1935, *The Observatory* **58** (September 1935), pp. 257–265, at p. 259.

[27] Note on "Relativistic Degeneracy", Sir A. S. Eddington, *Monthly Notices of the Royal Astronomical Society* **96** (November 1935), 20–21.

[28] The Pressure of a Degenerate Electron Gas and Related Problems, Arthur Eddington, *Proceedings of the Royal Society of London. Series A, Mathematical and Physical Sciences* **152** (November 1, 1935), pp. 253–272.

[29] *Relativity Theory of Protons and Electrons*, Sir Arthur Eddington, Cambridge: Cambridge University Press, 1936, chapter 13.

[30] The physics of white dwarf matter, Sir A. S. Eddington, *Monthly Notices of the Royal Astronomical Society* **100** (June 1940), pp. 582–594.

[31] *Fundamental Theory*, Sir A. S. Eddington, Cambridge: Cambridge University Press, 1946, §43–45.

[32] The evolution and explosion of massive stars, S. E. Woosley, A. Heger, and T. A. Weaver, *Reviews of Modern Physics* **74**, #4 (October 2002), pp. 1015–1071.

[33] White dwarfs in open clusters. VIII. NGC 2516: a test for the mass-radius and initial-final mass relations, D. Koester and D. Reimers, *Astronomy and Astrophysics* **313** (1996), pp. 810–814.

[34] An Empirical Initial-Final Mass Relation from Hot, Massive White Dwarfs in NGC 2168 (M35), Kurtis A. Williams, M. Bolte, and Detlev Koester, *Astrophysical Journal* **615**, #1 (2004), pp. L49–L52; also arXiv astro-ph/0409447.

[35] How Massive Single Stars End Their Life, A. Heger, C. L. Fryer, S. E. Woosley, N. Langer, and D. H. Hartmann, *Astrophysical Journal* **591**, #1 (2003), pp. 288–300.

[36] Strange quark matter in stars: a general overview, Jürgen Schaffner-Bielich, *Journal of Physics G: Nuclear and Particle Physics* **31**, #6 (2005), pp. S651–S657; also arXiv astro-ph/0412215.

[37] The Physics of Neutron Stars, by J. M. Lattimer and M. Prakash, *Science* **304**, #5670 (2004), pp. 536–542; also arXiv astro-ph/0405262.

[38] Type IA Supernova Explosion Models, Wolfgang Hillebrandt and Jens C. Niemeyer, *Annual Review of Astronomy and Astrophysics* **38** (2000), pp. 191–230.

[39] The weirdest Type Ia supernova yet, LBL press release, web page accessed 13-I-2007.

[40] Champagne Supernova Challenges Ideas about How Supernovae Work, web page, spacedaily.com, accessed 13-I-2007.

[41] The type Ia supernova SNLS-03D3bb from a super-Chandrasekhar-mass white dwarf star, D. Andrew Howell et al., *Nature* **443** (September 21, 2006), pp. 308–311; also, arXiv:astro-ph/0609616.

[42] Hachisu, Izumi; Kato, M.; et al. (2012). "A single degenerate progenitor model for type Ia supernovae highly exceeding the Chandrasekhar mass limit". *The Astrophysical Journal.* **744** (1): 76–79 (Article ID 69). Bibcode:2012ApJ...744...69H. arXiv:1106.3510 . doi:10.1088/0004-637X/744/1/69.

23.7 Further reading

- *On Stars, Their Evolution and Their Stability*, Nobel Prize lecture, Subrahmanyan Chandrasekhar, December 8, 1983.
- *White dwarf stars and the Chandrasekhar limit*, Masters' thesis, Dave Gentile, DePaul University, 1995.
- Estimating Stellar Parameters from Energy Equipartition, sciencebits.com. Discusses how to find mass-radius relations and mass limits for white dwarfs using simple energy arguments.

Chapter 24

Luminosity

For other uses, see Luminosity (disambiguation).

Image of galaxy NGC 4945 showing the huge luminosity of the central few star clusters, suggesting there is an AGN located in the center of the galaxy.

In astronomy, **luminosity** is the total amount of energy emitted by a star, galaxy, or other astronomical object per unit time.[1] It is related to the brightness, which is the luminosity of an object in a given spectral region.[1]

In SI units luminosity is measured in joules per second or watts. Values for luminosity are often given in the terms of the luminosity of the Sun, L⊙, which has a total power output of 3.846×10^{26} W.[2] Luminosity can also be given in terms of magnitude. The absolute bolometric magnitude (M_{bol}) of an object is a logarithmic measure of its total energy emission.

24.1 Measuring luminosity

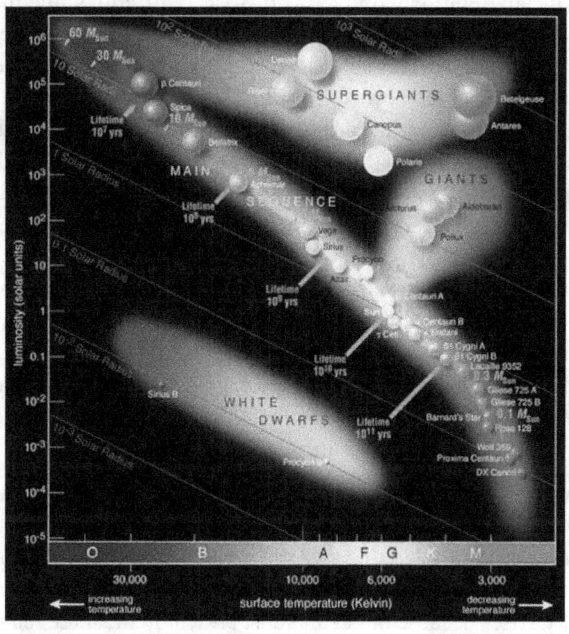

Hertzsprung–Russell diagram identifying stellar luminosity as a function of temperature for many stars in our solar neighborhood.

In astronomy, *luminosity* is the amount of electromagnetic energy a body radiates per unit of time.[3] It is most frequently measured in two forms: *visual* (visible light only) and *bolometric* (total radiant energy),[4] although luminosities at other wavelengths are increasingly being used as instruments become available to measure them. A bolometer is the instrument used to measure radiant energy over a wide band by absorption and measurement of heating. When not qualified, the term "luminosity" means bolometric luminosity, which is measured either in the SI units, watts, or in terms of solar luminosities. A star also radiates neutrinos, which carry off some energy (about 2% in the case of our Sun), contributing to the star's total luminosity.[5] While bolometers do exist, they cannot be used to measure even the apparent brightness of a star because they are insufficiently sensitive across the electromagnetic spectrum and because most wavelengths do not reach the surface of the Earth. In practice bolometric magnitudes are measured by taking measurements at certain wavelengths and constructing a model of the total spectrum that is most likely to match those measurements. In some cases, the process of estimation is extreme, with luminosities being calculated when less than 1% of the energy output is observed, for example with a hot Wolf-Rayet star observed only in the infra-red.

24.2 Stellar luminosity

A star's luminosity can be determined from two stellar characteristics: size and effective temperature.[3] The former is typically represented in terms of solar radii, R⊙, while the latter is represented in kelvins, but in most cases neither can be measured directly. To determine a star's radius, two other metrics are needed: the star's angular diameter and its distance from Earth, often calculated using parallax. Both can be measured with great accuracy in certain cases, with cool supergiants often having large angular diameters, and some cool evolved stars having masers in their atmospheres that can be used to measure the parallax using VLBI. However, for most stars the angular diameter or parallax, or both, are far below our ability to measure with any certainty. Since the effective temperature is merely a number that represents the temperature of a black body that would reproduce the luminosity, it obviously cannot be measured directly, but it can be estimated from the spectrum.

An alternative way to measure stellar luminosity is to measure the star's apparent brightness and distance. A third component needed to derive the luminosity is the degree of interstellar extinction that is present, a condition that usually arises because of gas and dust present in the interstellar medium (ISM), the Earth's atmosphere, and circumstellar matter. Consequently, one of astronomy's central challenges in determining a star's luminosity is to derive accurate measurements for each of these components, without which an accurate luminosity figure remains elusive.[6] Extinction can only be measured directly if the actual and observed luminosities are both known, but it can be estimated from the observed colour of a star, using models of the expected level of reddening from the interstellar medium.

In the current system of stellar classification, stars are grouped according to temperature, with the massive, very young and energetic Class O stars boasting temperatures in excess of 30,000 K while the less massive, typically older Class M stars exhibit temperatures less than 3,500 K. Because luminosity is proportional to temperature to the fourth power, the large variation in stellar temperatures produces an even vaster variation in stellar luminosity.[7] Because the luminosity depends on a high power of the stellar mass, high mass luminous stars have much shorter lifetimes. The most luminous stars are always young stars, no more than a few million years for the most extreme. In the Hertzsprung–Russell diagram, the x-axis represents temperature or spectral type while the y-axis represents luminosity or magnitude. The vast majority of stars are found along the main sequence with blue Class 0 stars found at the top left of the chart while red Class M stars fall to the bottom right. Certain stars like Deneb and Betelgeuse are found above and to the right of the main sequence, more luminous or cooler than their equivalents on the main sequence. Increased luminosity at the same temperature, or alternatively cooler temperature at the same luminosity, indicates that these stars are larger than those on the main sequence and they are called giants or supergiants.

Blue and white supergiants are high luminosity stars somewhat cooler than the most luminous main sequence stars. A star like Deneb, for example, has a luminosity around 200,000 L⊙, a spectral type of A2, and an effective temperature around 8,500 K, meaning it has a radius around 203 R⊙. For comparison, the red supergiant Betelgeuse has a luminosity around 100,000 L⊙, a spectral type of M2, and a temperature around 3,500 K, meaning its radius is about 1,000 R⊙. Red supergiants are the largest type of star, but the most luminous are much smaller and hotter, with temperatures up to 50,000 K and more and luminosities of several million L⊙, meaning their radii are just a few tens of R⊙. An example is R136a1, over 50,000 K and shining at over 8,000,000 L⊙ (mostly in the UV), it is only 35 R⊙.

24.3 Radio luminosity

The luminosity of a radio source is measured in W Hz^{-1}, to avoid having to specify a bandwidth over which it is measured. The observed strength, or flux density, of a radio source is measured in Jansky where 1 Jy = 10^{-26} W m^{-2} Hz^{-1}.

For example, consider a 10W transmitter at a distance of 1 million metres, radiating over a bandwidth of 1 MHz. By the time that power has reached the observer, the power is spread over the surface of a sphere with area $4\pi r^2$ or about 1.26×10^{13} m^2, so its flux density is 10 / 10^6 / 1.26×10^{13} W m^{-2} Hz^{-1} = 10^8 Jy.

More generally, for sources at cosmological distances, a k-correction must be made for the spectral index α of the source, and a relativistic correction must be made for the fact that the frequency scale in the emitted rest frame is different from that in the observer's rest frame. So the full expression for radio luminosity, assuming isotropic emission, is

$$L_\nu = \frac{S_{\rm obs} 4\pi {D_L}^2}{(1+z)^{1+\alpha}}$$

where $L\nu$ is the luminosity in W Hz^{-1}, S_{obs} is the observed flux density in W m^{-2} Hz^{-1}, DL is the luminosity distance in metres, z is the redshift, α is the spectral index (in the sense $I \propto \nu^\alpha$, and is typically -0.7).

For example, consider a 1 Jy signal from a radio source at a redshift of 1, at a frequency of 1.4 GHz. Ned Wright's cosmology calculator calculates a luminosity distance for a redshift of 1 to be 6701 Mpc = $2{\times}10^{26}$ m giving a radio luminosity of $10^{-26} \times 4\pi(2{\times}10^{26})^2 / (1+1)^{(1-0.7)} = 4{\times}10^{27}$ W Hz^{-1}.

To calculate the total radio power, this luminosity must be integrated over the bandwidth of the emission. A common assumption is to set the bandwidth to the observing frequency, which effectively assumes the power radiated has uniform intensity from zero frequency up to the observing frequency. In the case above, the total power is $4{\times}10^{27} \times 1.4{\times}10^9 = 5.7{\times}10^{36}$ W. This is sometimes expressed in terms of the total (i.e. integrated over all wavelengths) luminosity of the Sun which is $3.86{\times}10^{26}$ W, giving a radio power of $1.5{\times}10^{10}$ $L_\odot$.

24.4 Magnitude

Main article: Magnitude (astronomy)

Luminosity is an intrinsic measurable property of a star independent of distance. The concept of magnitude, on the other hand, incorporates distance. First conceived by the Greek astronomer Hipparchus in the second century BC, the original concept of magnitude grouped stars into six discrete categories depending on how bright they appeared. The brightest first magnitude stars were twice as bright as the next brightest stars, which were second magnitude; second was twice as bright as third, third twice as bright as fourth and so on down to the faintest stars, which Hipparchus categorized as sixth magnitude.[8] The system was but a simple delineation of stellar brightness into six distinct groups and made no allowance for the variations in brightness within a group. With the invention of the telescope at the beginning of the seventeenth century, researchers soon realized that there were subtle variations among stars and millions fainter than the sixth magnitude—hence the need for a more sophisticated system to describe a continuous range of values beyond what the naked eye could see.[8][9]

In 1856 Norman Pogson, noticing that photometric measurements had established first magnitude stars as being about 100 times brighter than sixth magnitude stars, formalized the Hipparchus system by creating a logarithmic scale, with every interval of one magnitude equating to a variation in brightness of $100^{1/5}$ or roughly 2.512 times. Consequently, a first magnitude star is about 2.5 times brighter than a second magnitude star, 2.5^2 brighter than a third magnitude star, 2.5^3 brighter than a fourth magnitude star, et cetera. Based on this continuous scale, any star with a magnitude between 5.5 and 6.5 is now considered to be sixth magnitude, a star with a magnitude between 4.5 and 5.5 is fifth magnitude and so on. With this new mathematical rigor, a first magnitude star should then have a magnitude in the range 0.5 to 1.5, thus excluding the nine brightest stars with magnitudes lower than 0.5, as well as the four brightest with negative values. It is customary therefore to extend the definition of a first magnitude star to any star with a magnitude less than 0.5, as can be seen in accompanying table.[8]

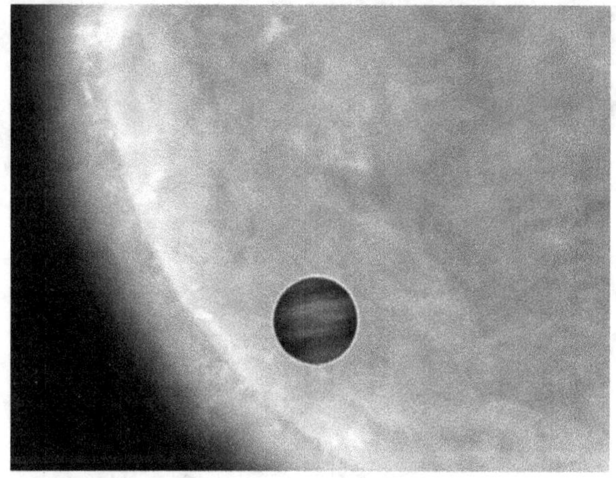

Artist impression of a transiting planet temporarily diminishing the star's brightness, leading to its discovery.[10]

The Pogson logarithmic scale is used to measure both apparent and absolute magnitudes, the latter corresponding to the brightness of a star or other celestial body as seen if it would be located at an interstellar distance of 10 parsecs. The apparent magnitude is a measure of the diminishing flux of light as a result of distance according to the inverse-square law.[11] In addition to this brightness decrease from increased distance, there is an extra decrease of brightness due to extinction from intervening interstellar dust.[9]

By measuring the width of certain absorption lines in the stellar spectrum, it is often possible to assign a certain luminosity class to a star without knowing its distance. Thus a fair measure of its absolute magnitude can be determined without knowing its distance nor the interstellar extinction, allowing astronomers to estimate a star's distance and extinction without parallax calculations. Since the stellar parallax is usually too small to be measured for many distant stars, this is a common method of determining such distances.

To conceptualize the range of magnitudes in our own galaxy, the smallest star to be identified has about 8% of the Sun's mass and glows feebly at absolute magnitude +19. Compared to the Sun, which has an absolute of +4.8, this faint star is 14 magnitudes or 400,000 times dimmer than our Sun. Our galaxy's most massive stars begin their lives with masses of roughly 100 times solar, radiating at upwards of absolute magnitude –8, over 160,000 times the solar luminosity. The total range of stellar luminosities, then, occupies a range of 27 magnitudes, or a factor of 60 billion.[7]

In measuring star brightnesses, absolute magnitude, apparent magnitude, and distance are interrelated parameters—if two are known, the third can be determined. Since the Sun's luminosity is the standard, comparing these parameters with the Sun's apparent magnitude and distance is the easiest way to remember how to convert between them.

24.5 Luminosity formula

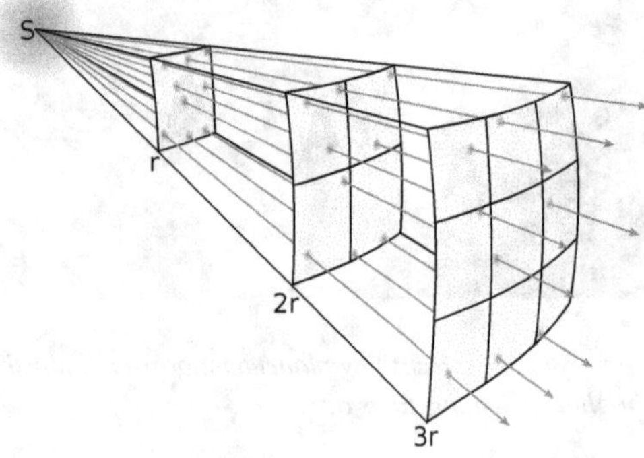

Point source S *is radiating light equally in all directions. The amount passing through an area* A *varies with the distance of the surface from the light.*

The Stefan–Boltzmann equation applied to a black body gives the value for luminosity for a black body, an idealized object which is perfectly opaque and non-reflecting:[3]

$$L = \sigma A T^4$$

where A is the area, T is the temperature (in Kelvins) and σ is the Stefan–Boltzmann constant, with a value of $5.670367(13)\times10^{-8}$ $\mathrm{W{\cdot}m^{-2}{\cdot}K^{-4}}$.[12]

Imagine a point source of light of luminosity L that radiates equally in all directions. A hollow sphere centered on the point would have its entire interior surface illuminated. As the radius increases, the surface area will also increase, and the constant luminosity has more surface area to illuminate, leading to a decrease in observed brightness.

$$F = \frac{L}{A}$$

where

A is the area of the illuminated surface.

F is the flux density of the illuminated surface.

The surface area of a sphere with radius r is $A = 4\pi r^2$, so for stars and other point sources of light:

$$F = \frac{L}{4\pi r^2}$$

where r is the distance from the observer to the light source.

It has been shown that the luminosity of a star L (assuming the star is a black body, which is a good approximation) is also related to temperature T and radius R of the star by the equation:[3]

$$L = 4\pi R^2 \sigma T^4$$

where

σ is the Stefan–Boltzmann constant 5.67×10^{-8} $\mathrm{W{\cdot}m^{-2}{\cdot}K^{-4}}$.

Dividing by the luminosity of the Sun $L_\odot$ and cancelling constants, we obtain the relationship:[3]

$$\frac{L}{L_\odot} = \left(\frac{R}{R_\odot}\right)^2 \left(\frac{T}{T_\odot}\right)^4$$

where $R_\odot$ and $T_\odot$ are the radius and temperature of the Sun, respectively.

For stars on the main sequence, luminosity is also related to mass:

$$\frac{L}{L_\odot} \approx \left(\frac{M}{M_\odot}\right)^{3.5}$$

24.6 Magnitude formulae

24.6.1 Apparent

The magnitude of a star is a logarithmic scale of observed visible brightness. The apparent magnitude is the observed visible brightness from Earth, and the absolute magnitude is the apparent magnitude at a distance of 10 parsecs. Given a visible luminosity (not total luminosity), one can calculate the apparent magnitude of a star from a given distance (ignoring extinction):

$$m_{\text{star}} = m_{\odot} - 2.5\log_{10}\left[\frac{L_{\text{star}}}{L_{\odot}}\left(\frac{d_{\odot}}{d_{\text{star}}}\right)^2\right]$$

where

m_{star} *is the apparent magnitude of the star (a pure number)*

$m_{\odot}$ *is the apparent magnitude of the Sun (also a pure number)*

L_{star} *is the visible luminosity of the star*

$L_{\odot}$ *is the solar visible luminosity*

d_{star} *is the distance to the star*

$d_{\odot}$ *is the distance to the Sun*

Or simplified, given $m_{\odot} = -26.73$, $d_{\odot} = 1.58\times10^{-5}$ lyr :

$m_{\text{star}} = -2.72 - 2.5\ \log(L_{\text{star}}/d_{\text{star}}^2)$, where L_{star} is measured in $L_{\odot}$.

24.6.2 Bolometric

The difference in bolometric magnitude is related to the luminosity ratio according to:

$$M_{\text{bol,star}} - M_{\text{bol},\odot} = -2.5\log_{10}\frac{L_{\text{star}}}{L_{\odot}}$$

which makes by inversion:

$$\frac{L_{\text{star}}}{L_{\odot}} = 10^{(M_{\text{bol},\odot} - M_{\text{bol,star}})/2.5}$$

where

$L_{\odot}$ *is the Sun's (sol) luminosity (bolometric luminosity)*

L_{star} is the star's luminosity (bolometric luminosity)

$M_{\text{bol},\odot}$ *is the bolometric magnitude of the Sun*

$M_{\text{bol,star}}$ *is the bolometric magnitude of the star.*

24.7 See also

- Orders of magnitude (power)
- List of most luminous stars
- List of brightest stars

24.8 References

[1] Hopkins, Jeanne (1980). *Glossary of Astronomy and Astrophysics* (2nd ed.). The University of Chicago Press. ISBN 0-226-35171-8.

[2] Williams, David R. (1 July 2013). "Sun Fact Sheet — Sun/Earth Comparison". National Aeronautics and Space Administration. Retrieved 13 April 2014.

[3] "Luminosity of Stars". Australia Telescope National Facility. 12 July 2004. Archived from the original on 9 August 2014.

[4] "Luminosity". Swinburne University of Technology. Retrieved 2 July 2012.

[5] Bahcall, John. "Solar Neutrino Viewgraphs". Institute for Advanced Study School of Natural Science. Retrieved 2012-07-03.

[6] Karttunen, Hannu (2003). *Fundamental Astronomy*. Springer-Verlag. p. 289. ISBN 978-3-540-00179-9.

[7] Ledrew, Glenn (February 2001). "The Real Starry Sky" (PDF). *Journal of the Royal Astronomical Society of Canada*. **95**: 32–33. Bibcode:2001JRASC..95...32L. Retrieved 2 July 2012.

[8] Nigel Foster. "THE FIRST MAGNITUDE STARS". Knowle Astronomical Society. Retrieved 25 September 2012.

[9] "Magnitude System". Astronomy Notes. 2 November 2010. Retrieved 2 July 2012.

[10] "COROT discovers its first exoplanet and catches scientists by surprise". European Space Agency. 3 May 2007. Retrieved 2 July 2012.

[11] Joshua E. Barnes (February 18, 2003). "The Inverse-Square Law". Institute for Astronomy - University of Hawaii. Retrieved 26 September 2012.

[12] "CODATA Value: Stefan-Boltzmann constant". *The NIST Reference on Constants, Units, and Uncertainty*. US National Institute of Standards and Technology. June 2015. Retrieved 2015-09-25. 2014 CODATA recommended values

24.9 Further reading

- Böhm-Vitense, Erika (1989). "Chapter 6. The luminosities of the stars". *Introduction to Stellar Astrophysics: Volume 1, Basic Stellar Observations and Data*. Cambridge University Press. pp. 41–48. ISBN 978-0-521-34869-0.

24.10 External links

- Ned Wright's cosmology calculator
- University of Southampton radio luminosity calculator at the Wayback Machine (archived 8 May 2015)

Chapter 25

Distance measures (cosmology)

Distance measures are used in physical cosmology to give a natural notion of the distance between two objects or events in the universe. They are often used to tie some *observable* quantity (such as the luminosity of a distant quasar, the redshift of a distant galaxy, or the angular size of the acoustic peaks in the CMB power spectrum) to another quantity that is not *directly* observable, but is more convenient for calculations (such as the comoving coordinates of the quasar, galaxy, etc.). The distance measures discussed here all reduce to the common notion of Euclidean distance at low redshift.

In accord with our present understanding of cosmology, these measures are calculated within the context of general relativity, where the Friedmann–Lemaître–Robertson–Walker solution is used to describe the universe.

25.1 Overview

There are a few different definitions of "distance" in cosmology which all coincide for sufficiently small redshifts. The expressions for these distances are most practical when written as functions of redshift z , since redshift is always the observable. They can easily be written as functions of scale factor $a = 1/(1+z)$, cosmic t or conformal time η as well by performing a simple transformation of variables. By defining the dimensionless Hubble parameter and the Hubble distance $d_H = c/H_0$, the relation between the different distances becomes apparent.

$$E(z) = \sqrt{\Omega_r(1+z)^4 + \Omega_m(1+z)^3 + \Omega_k(1+z)^2 + \Omega_\Lambda}$$

Here, Ω_m is the total matter density, Ω_Λ is the dark energy density, $\Omega_k = 1 - \Omega_m - \Omega_\Lambda$ represents the curvature, H_0 is the Hubble parameter today and c is the speed of light. The Hubble parameter at a given redshift is then $H(z) = H_0 E(z)$.

To compute the distance to an object from its redshift, we must integrate the above equation. Although for some limited choices of parameters (e.g. matter-only: $\Omega_m = \Omega_{\text{total}} = 1$) the comoving distance integral defined below has a closed analytic form, in general—and specifically for the parameters of our Universe—we can only find a solution numerically. Cosmologists commonly use the following measures for distances from the observer to an object at redshift z along the line of sight:[1]

Comoving distance:

$$d_C(z) = d_H \int_0^z \frac{dz'}{E(z')}$$

Transverse comoving distance:

$$d_M(z) = \begin{cases} \frac{d_H}{\sqrt{\Omega_k}} \sinh\left(\sqrt{\Omega_k} d_C(z)/d_H\right) & \text{for} \Omega_k > 0 \\ d_C(z) & \text{for} \Omega_k = 0 \\ \frac{d_H}{\sqrt{|\Omega_k|}} \sin\left(\sqrt{|\Omega_k|} d_C(z)/d_H\right) & \text{for} \Omega_k < 0 \end{cases}$$

Angular diameter distance:

$$d_A(z) = \frac{d_M(z)}{1+z}$$

Luminosity distance:

$$d_L(z) = (1+z) d_M(z)$$

Light-travel distance:

$$d_T(z) = d_H \int_0^z \frac{dz'}{(1+z')E(z')}$$

Note that the comoving distance is recovered from the transverse comoving distance by taking the limit $\Omega_k \to 0$, such that the two distance measures are equivalent in a flat universe.

Age of the universe is $\lim_{z\to\infty} d_T(z)/c$, and the time elapsed since redshift z until now is

$$t(z) = d_T(z)/c$$

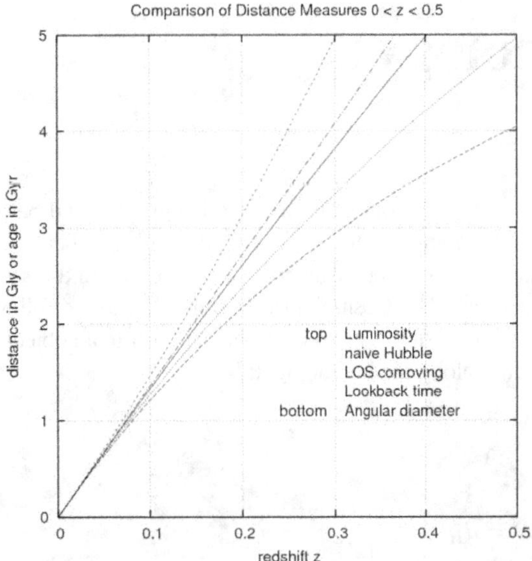

A comparison of cosmological distance measures, from redshift zero to redshift of 0.5. The background cosmology is Hubble parameter 72 km/s/Mpc, $\Omega_\Lambda = 0.732$ *,* $\Omega_{\text{matter}} = 0.266$ *,* $\Omega_{\text{radiation}} = 0.266/3454$ *, and* Ω_k *chosen so that the sum of Omega parameters is 1.*

25.2 Alternative terminology

Peebles (1993) calls the transverse comoving distance the "angular size distance", which is not to be mistaken for the angular diameter distance.[2] Even though it is not a matter of nomenclature, the comoving distance is equivalent to the proper motion distance, which is defined as the ratio of the transverse velocity and its proper motion in radians per time. Occasionally, the symbols χ or r are used to denote both the comoving and the angular diameter distance. Sometimes, the light-travel distance is also called the "lookback distance".

25.3 Details

25.3.1 Comoving distance

Main article: Comoving distance

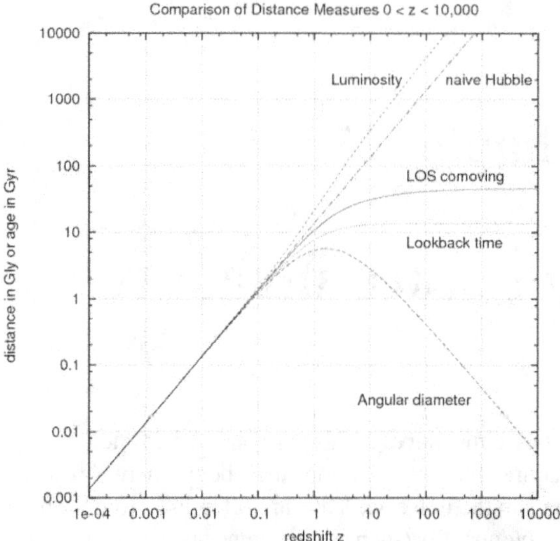

A comparison of cosmological distance measures, from redshift zero to redshift of 10,000, corresponding to the epoch of matter/radiation equality. The background cosmology is Hubble parameter 72 km/s/Mpc, $\Omega_\Lambda = 0.732$ *,* $\Omega_{\text{matter}} = 0.266$ *,* $\Omega_{\text{radiation}} = 0.266/3454$ *, and* Ω_k *chosen so that the sum of Omega parameters is one.*

The comoving distance between fundamental observers, i.e. observers that are both moving with the Hubble flow, does not change with time, as comoving distance accounts for the expansion of the universe. Comoving distance is obtained by integrating the proper distances of nearby fundamental observers along the line of sight (**LOS**), where the proper distance is what a measurement at constant cosmic time would yield.

In standard cosmology, **comoving distance** and **proper distance** are two closely related distance measures used by cosmologists to measure distances between objects; the comoving distance is the proper distance at the present time.

25.3.2 Proper distance

Proper distance roughly corresponds to where a distant object would be at a specific moment of cosmological time, which can change over time due to the expansion of the universe. *Comoving distance* factors out the expansion of the universe, which gives a distance that does not change in time due to the expansion of space (though this may change due to other, local factors, such as the motion of a galaxy within a cluster); the comoving distance is the proper distance at the present time.

25.3.3 Transverse comoving distance

Two comoving objects at constant redshift z that are separated by an angle $\delta\theta$ on the sky are said to have the distance $\delta\theta d_M(z)$, where the transverse comoving distance d_M is defined appropriately.

25.3.4 Angular diameter distance

Main article: Angular diameter distance

An object of size x at redshift z that appears to have angular size $\delta\theta$ has the angular diameter distance of $d_A(z) = x/\delta\theta$. This is commonly used to observe so called standard rulers, for example in the context of baryon acoustic oscillations.

25.3.5 Luminosity distance

Main article: Luminosity distance

If the intrinsic luminosity L of a distant object is known, we can calculate its luminosity distance by measuring the flux S and determine $d_L(z) = \sqrt{L/4\pi S}$, which turns out to be equivalent to the expression above for $d_L(z)$. This quantity is important for measurements of standard candles like type Ia supernovae, which were first used to discover the acceleration of the expansion of the universe.

25.3.6 Light-travel distance

This distance is the time (in years) that it took light to reach the observer from the object multiplied by the speed of light. For instance, the radius of the observable universe in this distance measure becomes the age of the universe multiplied by the speed of light (1 light year/year) i.e. 13.8 billion light years. Also see misconceptions about the size of the visible universe.

25.3.7 Etherington's distance duality

Main article: Etherington's reciprocity theorem

The Etherington's distance-duality equation [3] is the relationship between the luminosity distance of standard candles and the angular-diameter distance. It is expressed as follows: $d_L = (1+z)^2 d_A$

25.4 See also

- Big Bang
- Comoving distance
- Friedmann equations
- Parsec
- Physical cosmology
- Cosmic distance ladder
- Friedmann-Lemaître-Robertson-Walker metric
- Subatomic scale

25.5 References

[1] David W. Hogg (2000). "Distance measures in cosmology". arXiv:astro-ph/9905116v4.

[2] Peebles, P. J. E. (1993). *Principles of Physical Cosmology*. Princeton University Press. pp. 310–320. Bibcode:1993ppc..book.....P. ISBN 978-0-691-01933-8.

[3] I.M.H. Etherington, "LX. On the Definition of Distance in General Relativity", Philosophical Magazine, Vol. 15, S. 7 (1933), pp. 761-773.

- Scott Dodelson, *Modern Cosmology.* Academic Press (2003).

25.6 External links

- 'The Distance Scale of the Universe' compares different cosmological distance measures.
- 'Distance measures in cosmology' explains in detail how to calculate the different distance measures as a function of world model and redshift.
- iCosmos: Cosmology Calculator (With Graph Generation) calculates the different distance measures as a function of cosmological model and redshift, and generates plots for the model from redshift 0 to 20.

Chapter 26

Baryon acoustic oscillations

In cosmology, **baryon acoustic oscillations** (**BAO**) are regular, periodic fluctuations in the density of the visible baryonic matter (normal matter) of the universe. In the same way that supernovae provide a "standard candle" for astronomical observations,[1] BAO matter clustering provides a "standard ruler" for length scale in cosmology.[2] The length of this standard ruler (~490 million light years in today's universe[3]) can be measured by looking at the large scale structure of matter using astronomical surveys.[3] BAO measurements help cosmologists understand more about the nature of dark energy (which causes the apparent slight acceleration of the expansion of the universe) by constraining cosmological parameters.[2]

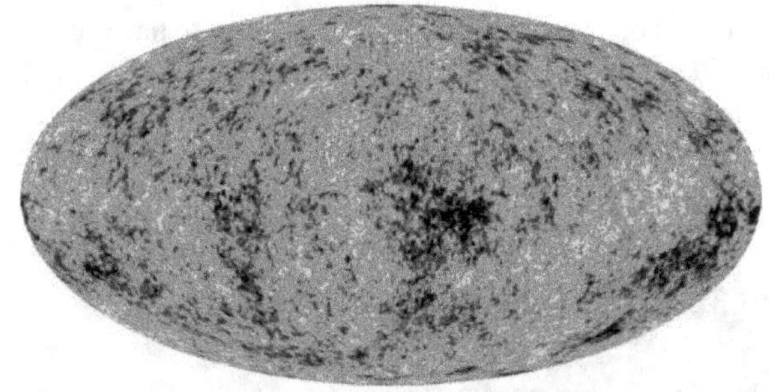

Figure 1: Temperature anisotropies of the CMB based on the nine year WMAP data (2012).[5][6][7]

26.1 The early universe

The early universe consisted of a hot, dense plasma of electrons and baryons (protons and neutrons). Photons (light particles) traveling in this universe were essentially trapped, unable to travel for any considerable distance before interacting with the plasma via Thomson scattering.[4] As the universe expanded, the plasma cooled to below 3000 K—a low enough energy such that the electrons and protons in the plasma could combine to form neutral hydrogen atoms. This recombination happened when the universe was around 379,000 years old, or at a redshift of $z = 1089$.[4] Photons interact to a much lesser degree with neutral matter, and therefore at recombination the universe became transparent to photons, allowing them to decouple from the matter and free-stream through the universe.[4] Technically speaking, the mean free path of the photons became on the order of the size of the universe. The cosmic microwave background (CMB) radiation is light that was emitted after recombination that is only now reaching our telescopes. Therefore, when we look at Wilkinson Microwave Anisotropy Probe (WMAP) data, we are looking back in time to see an image of the universe when it was only 379,000 years old.[4]

WMAP indicates (Figure 1) a smooth, homogeneous universe with density anisotropies of 10 parts per million.[4] However, when we observe the universe today we find large structure and density fluctuations. Galaxies, for instance, are a million times more dense than the universe's mean density.[2] The current belief is that the universe was built in a bottom-up fashion, meaning that the small anisotropies of the early universe acted as gravitational seeds for the structure we see today. Overdense regions attract more matter, whereas underdense regions attract less, and thus these small anisotropies we see in the CMB became the large scale structures we observe in the universe today.

26.2 Cosmic sound

Imagine an overdense region of the primordial plasma. While this region of overdensity gravitationally attracts matter towards it, the heat of photon-matter interactions creates a large amount of outward pressure. These counteracting forces of gravity and pressure created oscillations, analogous to sound waves created in air by pressure differences.[3]

Consider a single wave originating from this overdense region from the center of the plasma. This region contains dark matter, baryons and photons. The pressure results in a spherical sound wave of both baryons and photons moving with a speed slightly over half the speed of light[8][9] outwards from the overdensity. The dark matter interacts only gravitationally, and so it stays at the center of the sound wave, the origin of the overdensity. Before decoupling, the photons and baryons moved outwards together. After decoupling the photons were no longer interacting with the baryonic matter and they diffused away. That relieved the pressure on the system, leaving behind a shell of baryonic matter at a fixed radius. This radius is often referred to as the sound horizon.[3] Without the photo-baryon pressure driving the system outwards, the only remaining force on the baryons was gravitational. Therefore, the baryons and dark matter (left behind at the center of the perturbation) formed a configuration which included overdensities of matter both at the original site of the anisotropy and in the shell at the sound horizon for that anisotropy.[3]

Many such anisotropies created the ripples in the density of space that attracted matter and eventually galaxies formed in a similar pattern. Therefore, one would expect to see a greater number of galaxies separated by the sound horizon than at other length scales.[3] This particular configuration of matter occurred at each anisotropy in the early universe, and therefore the universe is not composed of one sound ripple,[10] but many overlapping ripples.[11] As an analogy, imagine dropping many pebbles into a pond and watching the resulting wave patterns in the water.[2] It is not possible to observe this preferred separation of galaxies on the sound horizon scale by eye, but one can measure this artifact statistically by looking at the separations of large numbers of galaxies.

26.3 Standard ruler

The physics of the propagation of the baryon waves in the early universe is fairly simple; as a result cosmologists can predict the size of the sound horizon at the time of recombination. In addition the CMB provides a measurement of this scale to high accuracy.[3] However, in the time between recombination and present day, the universe has been expanding. This expansion is well supported by observations and is one of the foundations of the Big Bang Model. In the late 1990s, observations of supernovae[1] determined that not only is the universe expanding, it is expanding at an increasing rate. A better understanding of the acceleration of the universe, or dark energy, has become one of the most important questions in cosmology today. In order to understand the nature of the dark energy, it is important to have a variety of ways of measuring the acceleration. BAO can add to the body of knowledge about this acceleration by comparing observations of the sound horizon today (using clustering of galaxies) to that of the sound horizon at the time of recombination (using the CMB).[3] Thus BAO provides a measuring stick with which to better understand the nature of the acceleration, completely independent from the supernova technique.

26.4 BAO signal in the Sloan Digital Sky Survey

The Sloan Digital Sky Survey (SDSS) is a 2.5-metre wide-angle optical telescope at Apache Point Observatory in New Mexico. The goal of this five-year survey was to take images and spectra of millions of celestial objects. The result of compiling the SDSS data is a three-dimensional map of objects in the nearby universe: the SDSS catalog. The SDSS catalog provides a picture of the distribution of matter in a large enough portion of the universe that one can search for a BAO signal by noting whether there is a statistically significant overabundance of galaxies separated by the predicted sound horizon distance.

The SDSS team looked at a sample of 46,748 luminous red galaxies (LRGs), over 3,816 square-degrees of sky (approximately five billion light years in diameter) and out to a redshift of $z = 0.47$.[3] They analyzed the clustering of these galaxies by calculating a two-point correlation function on the data.[12] The correlation function (ξ) is a function of comoving galaxy separation distance (s) and describes the probability that one galaxy will be found within a given distance of another.[13] One would expect a high correlation of galaxies at small separation distances (due to the clumpy nature of galaxy formation) and a low correlation at large separation distances. The BAO signal would show up as a bump in the correlation function at a comoving separation equal to the sound horizon. This signal was detected by the SDSS team in 2005.[3][14] SDSS confirmed the WMAP results that the sound horizon is ~150 Mpc in today's universe.[2][3]

26.5 Detection in other galaxy surveys

The 2dFGRS collaboration and the SDSS collaboration reported a detection of the BAO signal in the power spectrum at around the same time.[15] Both teams are credited and recognized for the discovery by the community as evidenced by the Shaw Prize in Astronomy[16] which was awarded to both groups. Since then, further detections have been reported in the 6dF Galaxy Survey (6dFGS),[17] WiggleZ[18] and BOSS.[19]

26.6 BAO and dark energy formalism

26.6.1 BAO constraints dark energy parameters

The BAO in the radial and tangential directions provide measurements of the Hubble parameter and angular diameter distance, respectively. The angular diameter distance and Hubble parameter can include different functions that explain dark energy behavior.[20][21] These functions have two parameters w_0 and w_1 and one can constrain them with chi-square technique.[22]

26.6.2 General relativity and dark energy

In general relativity, the expansion of the universe is parametrized by a scale factor $a(t)$ which is related to redshift:[4]

$$a(t) \equiv (1 + z(t))^{-1}$$

The Hubble parameter, $H(z)$, in terms of the scale factor is:

$$H(t) \equiv \frac{\dot{a}}{a}$$

where $\dot{a}$ is the time-derivative of the scale factor. The Friedmann equations express the expansion of the universe in terms of Newton's gravitational constant, G_N , the mean gauge pressure, P , the Universe's density ρ, the curvature, k , and the cosmological constant, Λ:[4]

$$H^2 = \left(\frac{\dot{a}}{a}\right)^2 = \frac{8\pi G}{3}\rho - \frac{kc^2}{a^2} + \frac{\Lambda c^2}{3}$$

$$\dot{H} + H^2 = \frac{\ddot{a}}{a} = -\frac{4\pi G}{3}\left(\rho + \frac{3p}{c^2}\right) + \frac{\Lambda c^2}{3}$$

Observational evidence of the acceleration of the universe implies that (at present time) $\ddot{a} > 0$. Therefore, the following are possible explanations:[23]

- The universe is dominated by some field or particle that has negative pressure such that the equation of state:

$$w = \frac{P}{\rho} < -1/3$$

- There is a non-zero cosmological constant, Λ.
- The Friedmann equations are incorrect since they contain over simplifications in order to make the general relativistic field equations easier to compute.

In order to differentiate between these scenarios, precise measurements of the Hubble parameter as a function of redshift are needed.

26.6.3 Measured observables of dark energy

The density parameter, Ω, of various components, x , of the universe can be expressed as ratios of the density of x to the critical density, ρ_c:[23]

$$\rho_c = \frac{3H^2}{8\pi G}$$

$$\Omega_x \equiv \frac{\rho_x}{\rho_c} = \frac{8\pi G\rho_x}{3H^2}$$

The Friedman equation can be rewritten in terms of the density parameter. For the current prevailing model of the universe, ΛCDM, this equation is as follows:[23]

$$H^2(a) =$$

$$\left(\frac{\dot{a}}{a}\right)^2 = H_0^2\left[\Omega_m a^{-3} + \Omega_r a^{-4} + \Omega_k a^{-2} + \Omega_\Lambda a^{-3(1+w)}\right]$$

where m is matter, r is radiation, k is curvature, Λ is dark energy, and w is the equation of state. Measurements of the CMB from WMAP put tight constraints on many of these parameters; however it is important to confirm and further constrain them using an independent method with different systematics.

The BAO signal is a standard ruler such that the length of the sound horizon can be measured as a function of cosmic time.[3] This measures two cosmological distances: the Hubble parameter, $H(z)$, and the angular diameter distance, $d_A(z)$, as a function of redshift (z) .[24] By measuring the subtended angle, $\Delta\theta$, of the ruler of length $\Delta\chi$, these parameters are determined as follows:[24]

$$\Delta\theta = \frac{\Delta\chi}{d_A(z)}$$

$$d_A(z) \propto \int_0^z \frac{dz'}{H(z')}$$

the redshift interval, Δz , can be measured from the data and thus determining the Hubble parameter as a function of redshift:

$$c\Delta z = H(z)\Delta\chi$$

Therefore, the BAO technique helps constrain cosmological parameters and provide further insight into the nature of dark energy.

26.7 See also

- Baryon Oscillation Spectroscopic Survey

26.8 References

[1] Perlmutter, S.; et al. (1999). "Measurements of Ω and Λ from 42 High-Redshift Supernovae". *The Astrophysical Journal.* **517** (2): 565. Bibcode:1999ApJ...517..565P. arXiv:astro-ph/9812133. doi:10.1086/307221.

[2] Eisenstein, D. J. (2005). "Dark energy and cosmic sound". *New Astronomy Reviews.* **49** (7–9): 360. Bibcode:2005NewAR..49..360E. doi:10.1016/j.newar.2005.08.005.

[3] Eisenstein, D. J.; et al. (2005). "Detection of the Baryon Acoustic Peak in the Large-Scale Correlation Function of SDSS Luminous Red Galaxies". *The Astrophysical Journal.* **633** (2): 560. Bibcode:2005ApJ...633..560E. arXiv:astro-ph/0501171. doi:10.1086/466512.

[4] Dodelson, S. (2003). *Modern Cosmology*. Academic Press. ISBN 978-0122191411.

[5] Gannon, M. (December 21, 2012). "New 'Baby Picture' of Universe Unveiled". Space.com. Retrieved December 21, 2012.

[6] Bennett, C. L.; et al. (2012). "Nine-Year Wilkinson Microwave Anisotropy Probe (WMAP) Observations: Final Maps and Results". arXiv:1212.5225 [astro-ph.CO].

[7] Hinshaw, G.; et al. (2009). "Five-year Wilkinson Microwave Anisotropy Probe observations: Data processing, sky maps, and basic results" (PDF). *The Astrophysical Journal Supplement Series.* **180** (2): 225. Bibcode:2009ApJS..180..225H. arXiv:0803.0732. doi:10.1088/0067-0049/180/2/225.

[8] Sunyaev, R.; Zeldovich, Ya. B. (1970). "Small-Scale Fluctuations of Relic Radiation". *Astrophysics and Space Science.* **7** (1): 3. Bibcode:1970Ap&SS...7....3S. doi:10.1007/BF00653471.

[9] Peebles, P. J. E.; Yu, J. T. (1970). "Primeval Adiabatic Perturbation in an Expanding Universe". *The Astrophysical Journal.* **162**: 815. Bibcode:1970ApJ...162..815P. doi:10.1086/150713.

[10] See http://www.cfa.harvard.edu/~{}deisenst/acousticpeak/anim.gif

[11] See http://www.cfa.harvard.edu/~{}deisenst/acousticpeak/anim_many.gif

[12] Landy, S. D.; Szalay, A. S. (1993). "Bias and variance of angular correlation functions". *The Astrophysical Journal.* **412**: 64. Bibcode:1993ApJ...412...64L. doi:10.1086/172900.

[13] Peebles, P. J. E. (1980). *The large-scale structure of the universe.* Princeton University Press. Bibcode:1980lssu.book.....P. ISBN 978-0-691-08240-0.

[14] http://www.sdss.org/news/releases/20050111.yardstick.html

[15] Cole, S.; et al. (2005). "The 2dF Galaxy Redshift Survey: Power-spectrum analysis of the final data set and cosmological implications". *Monthly Notices of the Royal Astronomical Society.* **362** (2): 505. Bibcode:2005MNRAS.362..505C. arXiv:astro-ph/0501174. doi:10.1111/j.1365-2966.2005.09318.x.

[16] Shaw Prize 2014

[17] Beutler, F.; et al. (2011). "The 6dF Galaxy Survey: Baryon acoustic oscillations and the local Hubble constant". *Monthly Notices of the Royal Astronomical Society.* **416** (4): 3017B. Bibcode:2011MNRAS.416.3017B. arXiv:1106.3366. doi:10.1111/j.1365-2966.2011.19250.x.

[18] Blake, C.; et al. (2011). "The WiggleZ Dark Energy Survey: Mapping the distance-redshift relation with baryon acoustic oscillations". *Monthly Notices of the Royal Astronomical Society.* **418** (3): 1707. Bibcode:2011MNRAS.418.1707B. arXiv:1108.2635. doi:10.1111/j.1365-2966.2011.19592.x.

[19] Anderson, L.; et al. (2012). "The clustering of galaxies in the SDSS-III Baryon Oscillation Spectroscopic Survey: Baryon acoustic oscillations in the Data Release 9 spectroscopic galaxy sample". *Monthly Notices of the Royal Astronomical Society.* **427** (4): 3435. Bibcode:2012MNRAS.427.3435A. arXiv:1203.6594. doi:10.1111/j.1365-2966.2012.22066.x.

[20] Chevallier, M; Polarski, D. (2001). "Accelerating Universes with Scaling Dark Matter". *International Journal of Modern Physics D.* **10**: 213–224. Bibcode:2001IJMPD..10..213C. arXiv:gr-qc/0009008. doi:10.1142/S0218271801000822.

[21] Barbosa Jr., E. M.; Alcaniz, J. S. (2008). "A parametric model for dark energy". *Physics Letters B.* **666** (5): 415–419. Bibcode:2008PhLB..666..415B. arXiv:0805.1713. doi:10.1016/j.physletb.2008.08.012.

[22] Shi, K.; Yong, H.; Lu, T. (2011). "The effects of parametrization of the dark energy equation of state". *Research in Astronomy and Astrophysics*. **11** (12): 1403–1412. Bibcode:2011RAA....11.1403S. doi:10.1088/1674-4527/11/12/003.

[23] Albrecht, A.; et al. (2006). "Report of the Dark Energy Task Force". arXiv:astro-ph/0609591 [astro-ph].

[24] White, M. (2007). "The Echo of Einstein's Greatest Blunder" (PDF). *Santa Fe Cosmology Workshop*.

26.9 External links

- Martin White's Baryon Acoustic Oscillations and Dark Energy Web Page
- Daniel Eisenstein's Detection of the Baryon Acoustic Peak Web Page
- Review of Baryon Acoustic Oscillations
- SDSS BAO Press Release

Chapter 27

Recombination (cosmology)

In cosmology, **recombination** refers to the epoch at which charged electrons and protons first became bound to form electrically neutral hydrogen atoms.[nb 1] Recombination occurred about 378,000 years after the Big Bang (at a redshift of $z = 1100$). The word "recombination" is misleading, since the big bang theory doesn't posit that protons and electrons had been combined before, but the name exists for historical reasons since it was named before the Big Bang hypothesis became the primary theory of the creation of the universe.

Immediately after the Big Bang, the universe was a hot, dense plasma of photons, electrons, and quarks: the Quark epoch. At .000001 seconds, the Universe had expanded and cooled sufficiently to allow for the formation of protons: the Hadron epoch. This plasma was effectively opaque to electromagnetic radiation due to Thomson scattering by free electrons, as the mean free path each photon could travel before encountering an electron was very short. This is the current state of the interior of the Sun. As the universe expanded, it also cooled. Eventually, the universe cooled to the point that the formation of neutral hydrogen was energetically favored, and the fraction of free electrons and protons as compared to neutral hydrogen decreased to a few parts in 10,000.

Shortly after, photons decoupled from matter in the universe, which leads to recombination sometimes being called **photon decoupling**, but recombination and photon decoupling are distinct events. Once photons decoupled from matter, they traveled freely through the universe without interacting with matter and constitute what is observed today as cosmic microwave background radiation (in that sense, the cosmic background radiation is infrared black-body radiation emitted when the universe was at a temperature of some 4000 K, redshifted by a factor of 1100 from the visible spectrum to the microwave spectrum).

27.1 The recombination history of hydrogen

The cosmic ionization history is generally described in terms of the free electron fraction x_e as a function of redshift. It is the ratio of the abundance of free electrons to the total abundance of hydrogen (both neutral and ionized). Denoting by n_e the number density of free electrons, nH that of atomic hydrogen and n_p that of ionized hydrogen (i.e. protons), x_e is defined as

$$x_{\rm e} = \frac{n_{\rm e}}{n_{\rm p} + n_{\rm H}}.$$

Since hydrogen only recombines once helium is fully neutral, charge neutrality implies $n_e = n_p$, i.e. x_e is also the fraction of ionized hydrogen.

27.1.1 Rough estimate from equilibrium theory

It is possible to find a rough estimate of the redshift of the recombination epoch assuming the recombination reaction $p + e^- \longleftrightarrow H + \gamma$ is fast enough that it proceeds near thermal equilibrium. The relative abundance of free electrons, protons and neutral hydrogen is then given by the Saha equation:

$$\frac{n_{\rm p} n_{\rm e}}{n_{\rm H}} = \left(\frac{m_{\rm e} k_{\rm B} T}{2\pi\hbar^2}\right)^{3/2} \exp\left(-\frac{E_{\rm I}}{k_{\rm B} T}\right),$$

where m_e is the mass of the electron, kB is Boltzmann's constant, T is the temperature, $\hbar$ is the reduced Planck's constant, and EI = 13.6 eV is the ionization energy of hydrogen.[1] Charge neutrality requires $n_e = n_p$, and the Saha equation can be rewritten in terms of the free electron fraction x_e:

$$\frac{x_e^2}{1-x_e} = (n_H + n_p)^{-1} \left(\frac{m_e k_B T}{2\pi\hbar^2}\right)^{3/2} \exp\left(-\frac{E_I}{k_B T}\right).$$

All quantities in the right-hand side are known functions of redshift: the temperature is given by $T = 2.728\,(1 + z)$ K,[2] and the total density of hydrogen (neutral and ionized) is given by $n_p + nH = 1.6\,(1{+}z)^3$ m^{-3}.

Solving this equation for a 50 percent ionization fraction yields a recombination temperature of roughly 4000 K, corresponding to redshift $z = 1500$.

27.1.2 The effective three-level atom

In 1968, physicists Jim Peebles[3] in the US and Yakov Borisovich Zel'dovich and collaborators[4] in the USSR independently computed the non-equilibrium recombination history of hydrogen. The basic elements of the model are the following.

- Direct recombinations to the ground state of hydrogen are very inefficient: each such event leads to a photon with energy greater than 13.6 eV, which almost immediately re-ionizes a neighboring hydrogen atom.
- Electrons therefore only efficiently recombine to the excited states of hydrogen, from which they cascade very quickly down to the first excited state, with principal quantum number $n = 2$.
- From the first excited state, electrons can reach the ground state $n = 1$ through two pathways:
    - Decay from the *2p* state by emitting a Lyman-α photon. This photon will almost always be reabsorbed by another hydrogen atom in its ground state. However, cosmological redshifting systematically decreases the photon frequency, and there is a small chance that it escapes reabsorption if it gets redshifted far enough from the Lyman-α line resonant frequency before encountering another hydrogen atom.
    - Decay from the *2s* state by emitting two photons. This two-photon decay process is very slow, with a rate[5] of 8.22 s^{-1}. It is however competitive with the slow rate of Lyman-α escape in producing ground-state hydrogen.
- Atoms in the first excited state may also be re-ionized by the ambient CMB photons before they reach the ground state. When this is the case, it is as if the recombination to the excited state did not happen in the first place. To account for this possibility, Peebles defines the factor C as the probability that an atom in the first excited state reaches the ground state through either of the two pathways described above before being photoionized.

This model is usually described as an "effective three-level atom" as it requires keeping track of hydrogen under three forms: in its ground state, in its first excited state (assuming all the higher excited states are in Boltzmann equilibrium with it), and ionized.

Accounting for these processes, the recombination history is then described by the differential equation

$$\frac{dx_e}{dt} = -C\left(\alpha_B(T) n_p x_e - 4(1-x_e)\beta_B(T) e^{-E_{21}/T}\right),$$

where αB is the "case B" recombination coefficient to the excited states of hydrogen, βB is the corresponding photoionization rate and $E_{21} = 10.2$ eV is the energy of the first excited state. Note that the second term in the right-hand side of the above equation can be obtained by a detailed balance argument. The equilibrium result given in the previous section would be recovered by setting the left-hand side to zero, i.e. assuming that the net rates of recombination and photoionization are large in comparison to the Hubble expansion rate, which sets the overall evolution timescale for the temperature and density. However, $C\,\alpha B\, n_p$ is comparable to the Hubble expansion rate, and even gets significantly lower at low redshifts, leading to an evolution of the free electron fraction much slower than what one would obtain from the Saha equilibrium calculation. With modern values of cosmological parameters, one finds that the universe is 90% neutral at $z \approx 1070$.

27.1.3 Modern developments

The simple effective three-level atom model described above accounts for the most important physical processes. However it does rely on approximations which lead to errors on the predicted recombination history at the level of 10% or so. Due to the importance of recombination for the precise prediction of cosmic microwave background anisotropies,[6] several research groups have revisited the details of this picture over the last two decades.

The refinements to the theory can be divided into two categories:

- Accounting for the non-equilibrium populations of the highly excited states of hydrogen. This effectively amounts to modifying the recombination coefficient αB.
- Accurately computing the rate of Lyman-α escape and the effect of these photons on the *2s-1s* transition. This requires solving a time-dependent radiative transfer equation. In addition, one needs to account for

higher-order Lyman transitions. These refinements effectively amount to a modification of Peebles' C factor.

Modern recombination theory is believed to be accurate at the level of 0.1%, and is implemented in publicly available fast recombination codes.[7][8]

27.2 Primordial helium recombination

Helium nuclei are produced during Big Bang nucleosynthesis, and make up about 24% of the total mass of baryonic matter. The ionization energy of helium is larger than that of hydrogen and it therefore recombines earlier. Because neutral helium carries two electrons, its recombination proceeds in two steps. The first recombination, $He^{++} + e^{-} \longrightarrow He^{+} + \gamma$ proceeds near Saha equilibrium and takes place around redshift $z \approx 6000$.[9] The second recombination, $He^{+} + e^{-} \longrightarrow He + \gamma$, is slower than what would be predicted from Saha equilibrium and takes place around redshift $z \approx 2000$.[10] The details of helium recombination are less critical than those of hydrogen recombination for the prediction of cosmic microwave background anisotropies, since the universe is still very optically thick after helium has recombined and before hydrogen has started its recombination.

27.3 Primordial light barrier

Prior to recombination, photons were not able to freely travel through the universe, as they constantly scattered off the free electrons and protons. This scattering causes a loss of information, and "there is therefore a photon barrier at a redshift" near that of recombination that prevents us from using photons directly to learn about the universe at larger redshifts.[11] Once recombination had occurred, however, the mean free path of photons greatly increased due to the lower number of free electrons. Shortly after recombination, the photon mean free path became larger than the Hubble length, and photons traveled freely without interacting with matter.[12] For this reason, recombination is closely associated with the last scattering surface, which is the name for the last time at which the photons in the cosmic microwave background interacted with matter.[13] However, these two events are distinct, and in a universe with different values for the baryon-to-photon ratio and matter density, recombination and photon decoupling need not have occurred at the same epoch.[12]

27.4 See also

- Timeline of the Big Bang or Chronology of the universe
- Age of the universe
- Big Bang

27.5 Notes

[1] The term recombination is actually a misnomer since it represents the *first* time that electrically neutral hydrogen formed.

27.6 References

[1] Ryden (2003), p. 157.

[2] Longair (2006), p. 32.

[3] Peebles, P. J. E., "Recombination of the Primeval Plasma", Astrophysical Journal, vol. 153, p.1, 1968

[4] Zeldovich, Y. B.; Kurt, V. G.; Syunyaev, R. A., "Recombination of Hydrogen in the Hot Model of the Universe", Zhurnal Eksperimental'noi i Teoreticheskoi Fiziki, V.55, N.1, P. 278-286, 1968

[5] Nussbaumer, H. and Schmutz, W., "The hydrogenic 2s-1s two-photon emission", Astronomy and Astrophysics, vol. 138, no. 2, Sept. 1984, p. 495

[6] Hu, W.; Scott, D.; Sugiyama, N.; White, M., "Effect of physical assumptions on the calculation of microwave background anisotropies", Physical Review D, Volume 52, Issue 10, 15 November 1995, p.5498

[7] Cosmorec: http://www.cita.utoronto.ca/~{}jchluba/Science_Jens/Recombination/CosmoRec.html

[8] Hyrec: http://www.sns.ias.edu/~{}yacine/hyrec/hyrec.html

[9] Switzer, E. R. & Hirata, C. M., "Primordial helium recombination. III. Thomson scattering, isotope shifts, and cumulative results", Physical Review D, Volume 77, Issue 8, p. 083008, 2008

[10] Switzer, E. R. & Hirata, C. M., "Primordial helium recombination. I. Feedback, line transfer, and continuum opacity", Physical Review D, Volume 77, Issue 8, p. 083006, 2008

[11] Longair (2006), p. 280.

[12] Padmanabhan (1993), p. 115.

[13] Longair (2006), p. 281.

27.7 Bibliography

- Peebles, P. J. E. (1968). "Recombination of the Primeval Plasma". *Astrophysical Journal*. **153**: 1. Bibcode:1968ApJ...153....1P. doi:10.1086/149628.
- Zeldovich, Y. B.; Kurt, V. G.; Syunyaev, R. A. (1968). "Recombination of Hydrogen in the Hot Model of the Universe". *Zhurnal Eksperimental'noi i Teoreticheskoi Fiziki*. **55**: 278. Bibcode:1968ZhETF..55..278Z.
- Longair, Malcolm (2006). *Galaxy Formation*. Springer. ISBN 978-3-540-73477-2.
- Padmanabhan, Thanu (1993). *Structure formation in the universe*. Cambridge University Press. ISBN 0-521-42486-0.
- Ryden, Barbara (2003). *Introduction to Cosmology*. Addison-Wesley. ISBN 0-8053-8912-1.

Chapter 28

Decoupling (cosmology)

In cosmology, **decoupling** refers to a period in the development of the universe when different types of particles fall out of thermal equilibrium with each other. This occurs as a result of the expansion of the universe, as their interaction rates decrease (and mean free paths increase) up to this critical point. The two verified instances of decoupling since the Big Bang which are most often discussed are photon decoupling and neutrino decoupling, as these led to the cosmic microwave background and cosmic neutrino background, respectively.

28.1 Photon decoupling

Main article: Recombination (cosmology)

Photon decoupling occurred during the epoch known as the recombination. During this time, electrons combined with protons to form hydrogen atoms, resulting in a sudden drop in free electron density. Decoupling occurred abruptly when the rate of Compton scattering of photons Γ was approximately equal to the rate of expansion of the universe H , or alternatively when the mean free path of the photons λ was approximately equal to the horizon size of the universe H^{-1} . After this photons were able to stream freely, producing the cosmic microwave background as we know it, and the universe became transparent.[1]

The interaction rate of the photons is given by

$$\Gamma = \frac{c}{\lambda} = n_e \sigma_e c$$

where n_e is the electron number density, σ_e is the electron cross sectional area, and c is the speed of light.

In the matter-dominated era (when recombination takes place),

$$H \propto a^{-3/2}$$

where a is the cosmic scale factor. Γ also decreases as a more complicated function of a , at a faster rate than H .[2] By working out the precise dependence of H and Γ on the scale factor and equating $\Gamma = H$, it is possible to show that photon decoupling occurred approximately 380,000 years after the Big Bang, at a redshift of $z = 1100$ [3] when the universe was at a temperature around 3000 K.

28.2 Neutrino decoupling

Main article: Neutrino decoupling

Another example is the neutrino decoupling which occurred within one second of the Big Bang.[4] Analogous to the decoupling of photons, neutrinos decoupled when the rate of weak interactions between neutrinos and other forms of matter dropped below the rate of expansion of the universe, which produced a cosmic neutrino background of freely streaming neutrinos. An important consequence of neutrino decoupling is that the temperature of this neutrino background is lower than the temperature of the cosmic microwave background.

28.3 WIMPs: non-relativistic decoupling

Decoupling may also have occurred for the dark matter candidate, WIMPs. These are known as “cold relics”, meaning they decoupled after they became non-relativistic (by comparison, photons and neutrinos decoupled while still relativistic and are known as “hot relics”). By calculating the hypothetical time and temperature of decoupling for non-relativistic WIMPs of a particular mass, it is possible to find their density.[5] Comparing this to the measured density parameter of cold dark matter today of 0.1198 ± 0.0027 [6] it is possible to rule out WIMPs of certain masses as reasonable dark matter candidates.[7]

28.4 See also

- Recombination
- Chronology of the universe
- reference needed

28.5 References

[1] Ryden, Barbara Sue (2003). *Introduction to cosmology*. San Francisco: Addison-Wesley.

[2] Kolb, Edward; Turner, Michael (1994). *The Early Universe*. New York: Westview Press.

[3] Hinshaw, G.; Weiland, J. L.; Hill, R. S.; Odegard, N.; Larson, D.; Bennett, C. L.; Dunkley, J.; Gold, B.; Greason, M. R.; Jarosik, N. (1 February 2009). "Five-Year Wilkinson Microwave Anisotropy Probe (WMAP) Observations: Data Processing, Sky Maps, and Basic Results". *The Astrophysical Journal Supplement Series*. **180** (2): 225–245. Bibcode:2009ApJS..180..225H. arXiv:0803.0732. doi:10.1088/0067-0049/180/2/225.

[4] Longair, M.S. (2008). *Galaxy formation* (2nd ed.). Berlin: Springer.

[5] Bringmann, Torsten; Hofmann, Stefan (23 April 2007). "Thermal decoupling of WIMPs from first principles". *Journal of Cosmology and Astroparticle Physics*. **2007** (04): 016–016. Bibcode:2007JCAP...04..016B. arXiv:hep-ph/0612238. doi:10.1088/1475-7516/2007/04/016.

[6] Jarosik, N. (4 December 2010). "Seven-Year Wilkinson Microwave Anisotropy Probe (WMAP) Observations: Sky Maps, Systematic Errors, and Basic Results. Table 8.". *Astrophysical Journal Supplement Series*. **192**: 14. Bibcode:2011ApJS..192...14J. arXiv:1001.4744. doi:10.1088/0067-0049/192/2/14.

[7] Weinheimer, C. "Dark Matter Results from 100 Live Days of XENON100 Data". *Physical Review Letters*. **107** (13): 131302. Bibcode:2011PhRvL.107m1302A. PMID 22026838. arXiv:1104.2549. doi:10.1103/physrevlett.107.131302.

Chapter 29

Structure formation

In physical cosmology, **structure formation** refers to the formation of galaxies, galaxy clusters and larger structures from small early density fluctuations. The universe, as is now known from observations of the cosmic microwave background radiation, began in a hot, dense, nearly uniform state approximately 13.8 billion years ago.[1] However, looking in the sky today, we see structures on all scales, from stars and planets to galaxies and, on still larger scales still, galaxy clusters and sheet-like structures of galaxies separated by enormous voids containing few galaxies. Structure formation attempts to model how these structures formed by gravitational instability of small early density ripples.[2][3][4][5]

The modern Lambda-CDM model is successful at predicting the observed large-scale distribution of galaxies, clusters and voids; but on the scale of individual galaxies there are many complications due to highly nonlinear processes involving baryonic physics, gas heating and cooling, star formation and feedback. Understanding the processes of galaxy formation is a major topic of modern cosmology research, both via observations such as the Hubble Ultra-Deep Field and via large computer simulations.

29.1 Overview

Under present models, the structure of the visible universe was formed in the following stages:

29.1.1 Very early universe

In this stage, some mechanism, such as cosmic inflation, was responsible for establishing the initial conditions of the universe: homogeneity, isotropy, and flatness.[3][6] Cosmic inflation also would have amplified minute quantum fluctuations (pre-inflation) into slight density ripples of overdensity and underdensity (post-inflation).

29.1.2 Growth of structure

The early universe was dominated by radiation; in this case density fluctuations larger than the cosmic horizon grow proportional to the scale factor, as the gravitational potential fluctuations remain constant. Structures smaller than the horizon remained essentially frozen due to radiation domination impeding growth. As the universe expanded, the density of radiation drops faster than matter (due to redshifting of photon energy); this led to a crossover called matter-radiation equality at ~ 50,000 years after the Big Bang. After this all dark matter ripples could grow freely, forming seeds into which the baryons could later fall. The size of the universe at this epoch forms a turnover in the matter power spectrum which can be measured in large redshift surveys.

29.1.3 Recombination

The universe was dominated by radiation for most of this stage, and due to the intense heat and radiation, the primordial hydrogen and helium were fully ionized into nuclei and free electrons. In this hot and dense situation, the radiation (photons) could not travel far before Thomson scattering off an electron. The universe was very hot and dense, but expanding rapidly and therefore cooling. Finally, at a little less than 400,000 years after the 'bang', it become cool enough (around 3000 K) for the protons to capture negatively charged electrons, forming neutral hydrogen atoms. (Helium atoms formed somewhat earlier due to their larger binding energy). Once nearly all the charged particles were bound in neutral atoms, the photons no longer interacted with them and were free to propagate for the next 13.8 billion years; we currently detect those photons redshifted by a factor 1090 down to 2.725 K as the Cosmic Microwave Background Radiation (CMB) filling today's universe. Several remarkable space-based missions (COBE, WMAP, Planck), have detected very slight variations in the density and temperature of the CMB. These variations were subtle, and the CMB appears very nearly uniformly the same in every direction. However, the slight temperature variations of order a few parts in 100,000 are of enormous importance, for they essentially were early "seeds" from which all subsequent complex structures in the universe ultimately developed.

The theory of what happened after the universe's first 400,000 years is one of hierarchical structure formation: the smaller gravitationally bound structures such as matter peaks containing the first stars and stellar clusters formed first, and these subsequently merged with gas and dark matter to form galaxies, followed by groups, clusters and superclusters of galaxies.

29.2 Very early universe

The very early universe is still a poorly understood epoch, from the viewpoint of fundamental physics. The prevailing theory, cosmic inflation, does a good job explaining the observed flatness, homogeneity and isotropy of the universe, as well as the absence of exotic relic particles (such as magnetic monopoles). Another prediction borne out by observation is that tiny perturbations in the primordial universe seed the later formation of structure. These fluctuations, while they form the foundation for all structure, appear most clearly as tiny temperature fluctuations at one part in 100,000. (To put this in perspective, the same level of fluctuations on a topographic map of the United States would show no feature taller than a few centimeters.) These fluctuations are critical, because they provide the seeds from which the largest structures can grow and eventually collapse to form galaxies and stars. COBE (Cosmic Background Explorer) provided the first detection of the intrinsic fluctuations in the cosmic microwave background radiation in the 1990s.

These perturbations are thought to have a very specific character: they form a Gaussian random field whose covariance function is diagonal and nearly scale-invariant. Observed fluctuations appear to have exactly this form, and in addition the *spectral index* measured by WMAP—the spectral index measures the deviation from a scale-invariant (or Harrison-Zel'dovich) spectrum—is very nearly the value predicted by the simplest and most robust models of inflation. Another important property of the primordial perturbations, that they are adiabatic (or isentropic between the various kinds of matter that compose the universe), is predicted by cosmic inflation and has been confirmed by observations.

Other theories of the very early universe have been proposed that are claimed to make similar predictions, such as the brane gas cosmology, cyclic model, pre-big bang model and holographic universe, but they remain nascent and are not widely accepted. Some theories, such as cosmic strings, have largely been refuted by increasingly precise data.

29.2.1 The horizon problem

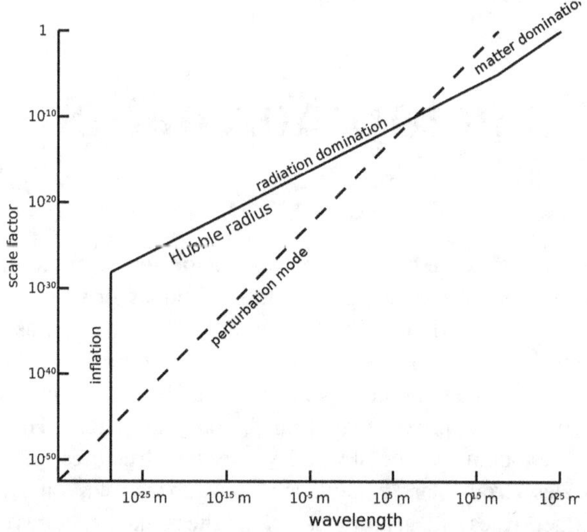

The physical size of the Hubble radius (solid line) as a function of the scale factor of the universe. The physical wavelength of a perturbation mode (dashed line) is shown as well. The plot illustrates how the perturbation mode exits the horizon during cosmic inflation in order to reenter during radiation domination. If cosmic inflation never happened, and radiation domination continued back until a gravitational singularity, then the mode would never have exited the horizon in the very early universe.

An important concept in structure formation is the notion of the Hubble radius, often called simply the *horizon,* as it is closely related to the particle horizon. The Hubble radius, which is related to the Hubble parameter H as $R = c/H$, where c is the speed of light, defines, roughly speaking, the volume of the nearby universe that has recently (in the last expansion time) been in causal contact with an observer. Since the universe is continually expanding, its energy density is continually decreasing (in the absence of truly exotic matter such as phantom energy). The Friedmann equation relates the energy density of the universe to the Hubble parameter and shows that the Hubble radius is continually increasing.

The horizon problem of big bang cosmology says that, without inflation, perturbations were never in causal contact before they entered the horizon and thus the homogeneity and isotropy of, for example, the large scale galaxy distributions cannot be explained. This is because, in an ordinary Friedmann–Lemaître–Robertson–Walker cosmology, the Hubble radius increases more rapidly than space expands, so perturbations only enter the Hubble radius, and are not pushed out by the expansion. This paradox is resolved by cosmic inflation, which suggests that during a phase of rapid expansion in the early universe the Hubble radius was nearly constant. Thus, large scale isotropy is due to quantum fluctuations produced during cosmic inflation that are pushed outside the horizon.

29.3 Primordial plasma

The end of inflation is called reheating, when the inflation particles decay into a hot, thermal plasma of other particles. In this epoch, the energy content of the universe is entirely radiation, with standard model particles having relativistic velocities. As the plasma cools, baryogenesis and leptogenesis are thought to occur, as the quark–gluon plasma cools, electroweak symmetry breaking occurs and the universe becomes principally composed of ordinary protons, neutrons and electrons. As the universe cools further, big bang nucleosynthesis occurs and small quantities of deuterium, helium and lithium nuclei are created. As the universe cools and expands, the energy in photons begins to redshift away, particles become non-relativistic and ordinary matter begins to dominate the universe. Eventually, atoms begin to form as free electrons bind to nuclei. This suppresses Thomson scattering of photons. Combined with the rarefaction of the universe (and consequent increase in the mean free path of photons), this makes the universe transparent and the cosmic microwave background is emitted at recombination (the *surface of last scattering*).

29.3.1 Acoustic oscillations

Main article: baryon acoustic oscillations

The primordial plasma would have had very slight overdensities of matter, thought to have derived from the enlargement of quantum fluctuations during inflation. Whatever the source, these overdensities gravitationally attract matter. But the intense heat of the near constant photon-matter interactions of this epoch rather forcefully seeks thermal equilibrium, which creates a large amount of outward pressure. These counteracting forces of gravity and pressure create oscillations, analogous to sound waves created in air by pressure differences.

These perturbations are important, as they are responsible for the subtle physics that result in the cosmic microwave background anisotropy. In this epoch, the amplitude of perturbations that enter the horizon oscillate sinusoidally, with dense regions becoming more rarefied and then becoming dense again, with a frequency which is related to the size of the perturbation. If the perturbation oscillates an integral or half-integral number of times between coming into the horizon and recombination, it appears as an acoustic peak of the cosmic microwave background anisotropy. (A half-oscillation, in which a dense region becomes a rarefied region or vice versa, appears as a peak because the anisotropy is displayed as a *power spectrum*, so underdensities contribute to the power just as much as overdensities.) The physics that determines the detailed peak structure of the microwave background is complicated, but these oscillations provide the essence.[7][8][9][10][11]

29.4 Linear structure

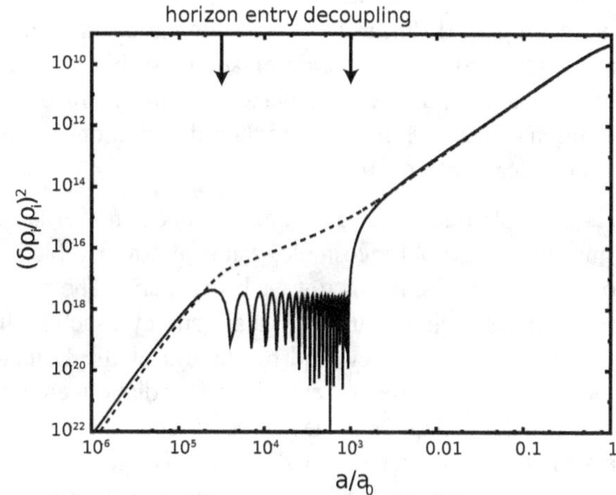

Evolution of two perturbations to the ΛCDM homogeneous big bang model. Between entering the horizon and decoupling, the dark matter perturbation (dashed line) grows logarithmically, before the growth accelerates in matter domination. On the other hand, between entering the horizon and decoupling, the perturbation in the baryon-photon fluid (solid line) oscillates rapidly. After decoupling, it grows rapidly to match the dominant matter perturbation, the dark matter mode.

One of the key realizations made by cosmologists in the 1970s and 1980s was that the majority of the matter content of the universe was composed not of atoms, but rather a mysterious form of matter known as dark matter. Dark matter interacts through the force of gravity, but it is not composed of baryons and it is known with very high accuracy that it does not emit or absorb radiation. It may be composed of particles that interact through the weak interaction, such as neutrinos, but it cannot be composed entirely of the three known kinds of neutrinos (although some have suggested it is a sterile neutrino). Recent evidence suggests that there is about five times as much dark matter as baryonic matter, and thus the dynamics of the universe in this epoch are dominated by dark matter.

Dark matter plays a key role in structure formation because it feels only the force of gravity: the gravitational Jeans instability which allows compact structures to form is not opposed by any force, such as radiation pressure. As a result, dark matter begins to collapse into a complex network of dark matter halos well before ordinary matter, which is impeded by pressure forces. Without dark matter, the epoch of galaxy formation would occur substantially later in the universe than is observed.

The physics of structure formation in this epoch is particularly simple, as dark matter perturbations with different wavelengths evolve independently. As the Hubble radius grows in the expanding universe, it encompasses larger and larger perturbations. During matter domination, all causal dark matter perturbations grow through gravitational clustering. However, the shorter-wavelength perturbations that are encompassed during radiation domination have their growth retarded until matter domination. At this stage, luminous, baryonic matter is expected to simply mirror the evolution of the dark matter, and their distributions should closely trace one another.

It is a simple matter to calculate this "linear power spectrum" and, as a tool for cosmology, it is of comparable importance to the cosmic microwave background. The power spectrum has been measured by galaxy surveys, such as the Sloan Digital Sky Survey, and by surveys of the Lyman-α forest. Since these surveys observe radiation emitted from galaxies and quasars, they do not directly measure the dark matter, but the large scale distribution of galaxies (and of absorption lines in the Lyman-α forest) is expected to closely mirror the distribution of dark matter. This depends on the fact that galaxies will be larger and more numerous in denser parts of the universe, whereas they will be comparatively scarce in rarefied regions.

29.5 Nonlinear structure

When the perturbations have grown sufficiently, a small region might become substantially denser than the mean density of the universe. At this point, the physics involved becomes substantially more complicated. When the deviations from homogeneity are small, the dark matter may be treated as a pressureless fluid and evolves by very simple equations. In regions which are significantly denser than the background, the full Newtonian theory of gravity must be included. (The Newtonian theory is appropriate because the masses involved are much less than those required to form a black hole, and the speed of gravity may be ignored as the light-crossing time for the structure is still smaller than the characteristic dynamical time.) One sign that the linear and fluid approximations become invalid is that dark matter starts to form caustics in which the trajectories of adjacent particles cross, or particles start to form orbits. These dynamics are generally best understood using N-body simulations (although a variety of semi-analytic schemes, such as the Press–Schechter formalism, can be used in some cases). While in principle these simulations are quite simple, in practice they are very difficult to implement, as they require simulating millions or even billions of particles. Moreover, despite the large number of particles, each particle typically weighs 10^9 solar masses and discretization effects may become significant. The largest such simulation as of 2005 is the Millennium simulation.[12]

The result of N-body simulations suggests that the universe is composed largely of voids, whose densities might be as low as one tenth the cosmological mean. The matter condenses in large filaments and haloes which have an intricate web like structure. These form galaxy groups, clusters and superclusters. While the simulations appear to agree broadly with observations, their interpretation is complicated by the understanding of how dense accumulations of dark matter spur galaxy formation. In particular, many more small haloes form than we see in astronomical observations as dwarf galaxies and globular clusters. This is known as the galaxy bias problem, and a variety of explanations have been proposed. Most account for it as an effect in the complicated physics of galaxy formation, but some have suggested that it is a problem with our model of dark matter and that some effect, such as warm dark matter, prevents the formation of the smallest haloes.

29.6 Gas evolution

See also: galaxy formation and evolution and stellar evolution

The final stage in evolution comes when baryons condense in the centres of galaxy haloes to form galaxies, stars and quasars. A paradoxical aspect of structure formation is that while dark matter greatly accelerates the formation of dense haloes, because dark matter does not have radiation pressure, the formation of smaller structures from dark matter is impossible because dark matter cannot dissipate angular momentum, whereas ordinary baryonic matter can collapse to form dense objects by dissipating angular momentum through radiative cooling. Understanding these processes is an enormously difficult computational problem, because they can involve the physics of gravity, magnetohydrodynamics, atomic physics, nuclear reactions, turbulence and even general relativity. In most cases, it is not yet possible to perform simulations that can be compared quantitatively with observations, and the best that can be achieved are approximate simulations that illustrate the main qualitative features of a process such as star formation.

29.7 Modelling structure formation

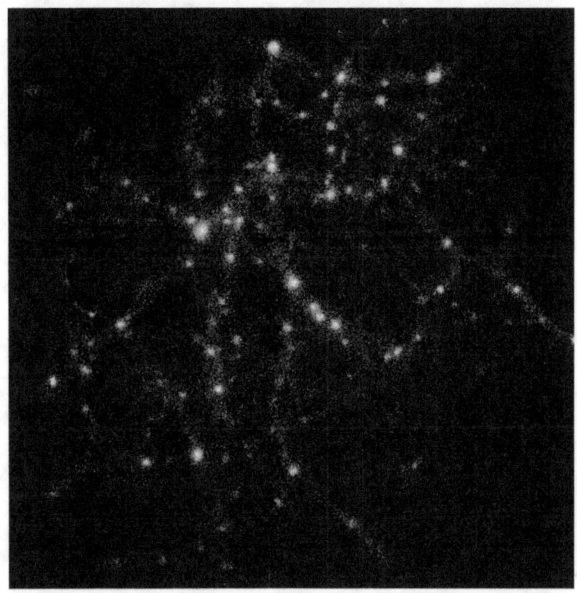

Snapshot from a computer simulation of large scale structure formation in a Lambda-CDM universe.

29.7.1 Cosmological perturbations

Main article: cosmological perturbation theory

Much of the difficulty, and many of the disputes, in understanding the large-scale structure of the universe can be resolved by better understanding the choice of gauge in general relativity. By the scalar-vector-tensor decomposition, the metric includes four scalar perturbations, two vector perturbations, and one tensor perturbation. Only the scalar perturbations are significant: the vectors are exponentially suppressed in the early universe, and the tensor mode makes only a small (but important) contribution in the form of primordial gravitational radiation and the B-modes of the cosmic microwave background polarization. Two of the four scalar modes may be removed by a physically meaningless coordinate transformation. Which modes are eliminated determine the infinite number of possible gauge fixings. The most popular gauge is Newtonian gauge (and the closely related conformal Newtonian gauge), in which the retained scalars are the Newtonian potentials Φ and Ψ, which correspond exactly to the Newtonian potential energy from Newtonian gravity. Many other gauges are used, including synchronous gauge, which can be an efficient gauge for numerical computation (it is used by CMBFAST). Each gauge still includes some unphysical degrees of freedom. There is a so-called gauge-invariant formalism, in which only gauge invariant combinations of variables are considered.

29.7.2 Inflation and initial conditions

The initial conditions for the universe are thought to arise from the scale invariant quantum mechanical fluctuations of cosmic inflation. The perturbation of the background energy density at a given point $\rho(\mathbf{x},t)$ in space is then given by an isotropic, homogeneous Gaussian random field of mean zero. This means that the spatial Fourier transform of ρ – $\hat{\rho}(\mathbf{k},t)$ has the following correlation functions

$$\langle\hat{\rho}(\mathbf{k},t)\hat{\rho}(\mathbf{k}',t)\rangle = f(k)\delta^{(3)}(\mathbf{k}-\mathbf{k}')$$

where $\delta^{(3)}$ is the three-dimensional Dirac delta function and $k = |\mathbf{k}|$ is the length of $\mathbf{k}$. Moreover, the spectrum predicted by inflation is nearly scale invariant, which means

$$\langle\hat{\rho}(\mathbf{k},t)\hat{\rho}(\mathbf{k}',t)\rangle = k^{n_s-1}\delta^{(3)}(\mathbf{k}-\mathbf{k}')$$

where $n_s - 1$ is a small number. Finally, the initial conditions are adiabatic or isentropic, which means that the fractional perturbation in the entropy of each species of particle is equal.

29.8 See also

- Big Bang
- Chronology of the universe
- Galaxy formation and evolution
- Illustris project
- Stellar evolution
- Timeline of the Big Bang

29.9 References

[1] "Cosmic Detectives". The European Space Agency (ESA). 2013-04-02. Retrieved 2013-04-15.

[2] Dodelson, Scott (2003). *Modern Cosmology*. Academic Press. ISBN 0-12-219141-2.

[3] Liddle, Andrew; David Lyth (2000). *Cosmological Inflation and Large-Scale Structure*. Cambridge. ISBN 0-521-57598-2.

[4] Padmanabhan, T. (1993). *Structure formation in the universe.* Cambridge University Press. ISBN 0-521-42486-0.

[5] Peebles, P. J. E. (1980). *The Large-Scale Structure of the Universe.* Princeton University Press. ISBN 0-691-08240-5.

[6] Kolb, Edward; Michael Turner (1988). *The Early Universe.* Addison-Wesley. ISBN 0-201-11604-9.

[7] Harrison, E. R. (1970). "Fluctuations at the threshold of classical cosmology". *Phys. Rev.* **D1** (10): 2726. Bibcode:1970PhRvD...1.2726H. doi:10.1103/PhysRevD.1.2726.

[8] Peebles, P. J. E.; Yu, J. T. (1970). "Primeval adiabatic perturbation in an expanding universe". *Astrophysical Journal.* **162**: 815. Bibcode:1970ApJ...162..815P. doi:10.1086/150713.

[9] Zel'dovich, Yaa B. (1972). "A hypothesis, unifying the structure and entropy of the Universe". *Monthly Notices of the Royal Astronomical Society.* **160**: 1P. Bibcode:1972MNRAS.160P...1Z. doi:10.1093/mnras/160.1.1p.

[10] R. A. Sunyaev, "Fluctuations of the microwave background radiation", in *Large Scale Structure of the Universe* ed. M. S. Longair and J. Einasto, 393. Dordrecht: Reidel 1978.

[11] U. Seljak & M. Zaldarriaga (1996). "A line-of-sight integration approach to cosmic microwave background anisotropies". *Astrophys. J.* **469**: 437–444. Bibcode:1996ApJ...469..437S. arXiv:astro-ph/9603033. doi:10.1086/177793.

[12] Springel, V.; et al. (2005). "Simulations of the formation, evolution and clustering of galaxies and quasars". *Nature.* **435** (7042): 629–636. Bibcode:2005Natur.435..629S. PMID 15931216. arXiv:astro-ph/0504097. doi:10.1038/nature03597.

Chapter 30

Cosmic microwave background

"CMB" redirects here. For other uses, see CMB (disambiguation).

The **cosmic microwave background** (**CMB**) is electromagnetic radiation left over from an early stage of the universe in Big Bang cosmology. In older literature, the CMB is also variously known as cosmic microwave background radiation (CMBR) or "relic radiation". The CMB is a faint cosmic background radiation filling all space that is an important source of data on the early universe because it is the oldest light in the universe, dating to the epoch of recombination. With a traditional optical telescope, the space between stars and galaxies (the *background*) is completely dark. However, a sufficiently sensitive radio telescope shows a faint background noise, or glow, almost isotropic, that is not associated with any star, galaxy, or other object. This glow is strongest in the microwave region of the radio spectrum. The accidental discovery of the CMB in 1964 by American radio astronomers Arno Penzias and Robert Wilson[1][2] was the culmination of work initiated in the 1940s, and earned the discoverers the 1978 Nobel Prize in Physics.

The discovery of CMB is landmark evidence of the Big Bang origin of the universe. When the universe was young, before the formation of stars and planets, it was denser, much hotter, and filled with a uniform glow from a white-hot fog of hydrogen plasma. As the universe expanded, both the plasma and the radiation filling it grew cooler. When the universe cooled enough, protons and electrons combined to form neutral hydrogen atoms. These atoms could no longer absorb the thermal radiation, and so the universe became transparent instead of being an opaque fog.[3] Cosmologists refer to the time period when neutral atoms first formed as the *recombination epoch*, and the event shortly afterwards when photons started to travel freely through space rather than constantly being scattered by electrons and protons in plasma is referred to as photon decoupling. The photons that existed at the time of photon decoupling have been propagating ever since, though growing fainter and less energetic, since the expansion of space causes their wavelength to increase over time (and wavelength is inversely proportional to energy according to Planck's relation). This is the source of the alternative term *relic radiation*. The *surface of last scattering* refers to the set of points in space at the right distance from us so that we are now receiving photons originally emitted from those points at the time of photon decoupling.

Precise measurements of the CMB are critical to cosmology, since any proposed model of the universe must explain this radiation. The CMB has a thermal black body spectrum at a temperature of 2.72548±0.00057 K.[4] The spectral radiance $dE_\nu/d\nu$ peaks at 160.23 GHz, in the microwave range of frequencies. The photon energy of CMB photons is about 6.626534×10^{-4} eV. Alternatively, if spectral radiance is defined as $dE_\lambda/d\lambda$, then the peak wavelength is 1.063 mm. The glow is very nearly uniform in all directions, but the tiny residual variations show a very specific pattern, the same as that expected of a fairly uniformly distributed hot gas that has expanded to the current size of the universe. In particular, the spectral radiance at different angles of observation in the sky contains small anisotropies, or irregularities, which vary with the size of the region examined. They have been measured in detail, and match what would be expected if small thermal variations, generated by quantum fluctuations of matter in a very tiny space, had expanded to the size of the observable universe we see today. This is a very active field of study, with scientists seeking both better data (for example, the Planck spacecraft) and better interpretations of the initial conditions of expansion. Although many different processes might produce the general form of a black body spectrum, no model other than the Big Bang has yet explained the fluctuations. As a result, most cosmologists consider the Big Bang model of the universe to be the best explanation for the CMB.

The high degree of uniformity throughout the observable universe and its faint but measured anisotropy lend strong support for the Big Bang model in general and the ΛCDM ("Lambda Cold Dark Matter") model in particular. Moreover, the fluctuations are coherent on angular scales that are larger than the apparent cosmological horizon at recombination. Either such coherence is acausally fine-tuned, or cosmic inflation occurred.[5][6]

30.1 Features

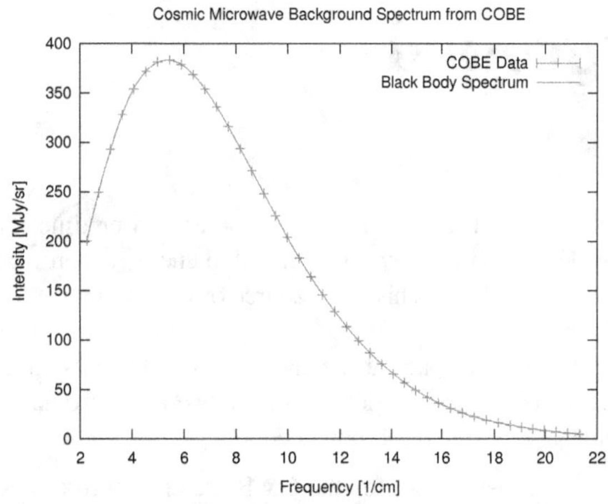

Graph of cosmic microwave background spectrum measured by the FIRAS instrument on the COBE, the most precisely measured black body spectrum in nature.[7] The error bars are too small to be seen even in an enlarged image, and it is impossible to distinguish the observed data from the theoretical curve.

The cosmic microwave background radiation is an emission of uniform, black body thermal energy coming from all parts of the sky. The radiation is isotropic to roughly one part in 100,000: the root mean square variations are only 18 μK,[8] after subtracting out a dipole anisotropy from the Doppler shift of the background radiation. The latter is caused by the peculiar velocity of the Earth relative to the comoving cosmic rest frame as the planet moves at some 371 km/s towards the constellation Leo. The CMB dipole as well as aberration at higher multipoles have been measured, consistent with galactic motion.[9]

In the Big Bang model for the formation of the universe, Inflationary Cosmology predicts that after about 10^{-37} seconds[10] the nascent universe underwent exponential growth that smoothed out nearly all irregularities. The remaining irregularities were caused by quantum fluctuations in the inflaton field that caused the inflation event.[11] Before the formation of stars and planets (after 10^{-6} seconds), the early universe was smaller, much hotter, and filled with a uniform glow from its white-hot fog of interacting plasma of photons, electrons, and baryons.

As the universe expanded, adiabatic cooling caused the energy density of the plasma to decrease until it became favorable for electrons to combine with protons, forming hydrogen atoms. This recombination event happened when the temperature was around 3000 K or when the universe was approximately 379,000 years old.[12] As photons did not interact with these electrically neutral atoms, the former began to travel freely through space, resulting in the decoupling of matter and radiation.[13]

The color temperature of the ensemble of decoupled photons has continued to diminish ever since; now down to 2.7260±0.0013 K,[4] it will continue to drop as the universe expands. The intensity of the radiation also corresponds to black-body radiation at 2.726 K because red-shifted black-body radiation is just like black-body radiation at a lower temperature. According to the Big Bang model, the radiation from the sky we measure today comes from a spherical surface called *the surface of last scattering*. This represents the set of locations in space at which the decoupling event is estimated to have occurred[14] and at a point in time such that the photons from that distance have just reached observers. Most of the radiation energy in the universe is in the cosmic microwave background,[15] making up a fraction of roughly 6×10^{-5} of the total density of the universe.[16]

Two of the greatest successes of the Big Bang theory are its prediction of the almost perfect black body spectrum and its detailed prediction of the anisotropies in the cosmic microwave background. The CMB spectrum has become the most precisely measured black body spectrum in nature.[7]

Density of energy for CMB is 0.25 eV/cm^3[17] (4.005×10^{-14} J/m^3) or (400–500 photons/cm^3[18]).

30.2 History

See also: Discovery of cosmic microwave background radiation

The cosmic microwave background was first predicted in 1948 by Ralph Alpher and Robert Herman.[19][20][21] Alpher and Herman were able to estimate the temperature of the cosmic microwave background to be 5 K, though two years later they re-estimated it at 28 K. This high estimate was due to a mis-estimate of the Hubble constant by Alfred Behr, which could not be replicated and was later abandoned for the earlier estimate. Although there were several previous estimates of the temperature of space, these suffered from two flaws. First, they were measurements of the *effective* temperature of space and did not suggest that space was filled with a thermal Planck spectrum. Next, they depend on our being at a special spot at the edge of the Milky Way galaxy and they did not suggest the radiation is isotropic. The estimates would yield very different predictions if Earth happened to be located elsewhere in the universe.[22]

The Holmdel Horn Antenna on which Penzias and Wilson discovered the cosmic microwave background

The 1948 results of Alpher and Herman were discussed in many physics settings through about 1955, when both left the Applied Physics Laboratory at Johns Hopkins University. The mainstream astronomical community, however, was not intrigued at the time by cosmology. Alpher and Herman's prediction was rediscovered by Yakov Zel'dovich in the early 1960s, and independently predicted by Robert Dicke at the same time. The first published recognition of the CMB radiation as a detectable phenomenon appeared in a brief paper by Soviet astrophysicists A. G. Doroshkevich and Igor Novikov, in the spring of 1964.[23] In 1964, David Todd Wilkinson and Peter Roll, Dicke's colleagues at Princeton University, began constructing a Dicke radiometer to measure the cosmic microwave background.[24] In 1964, Arno Penzias and Robert Woodrow Wilson at the Crawford Hill location of Bell Telephone Laboratories in nearby Holmdel Township, New Jersey had built a Dicke radiometer that they intended to use for radio astronomy and satellite communication experiments. On 20 May 1964 they made their first measurement clearly showing the presence of the microwave background,[25] with their instrument having an excess 4.2K antenna temperature which they could not account for. After receiving a telephone call from Crawford Hill, Dicke said "Boys, we've been scooped."[1][26][27] A meeting between the Princeton and Crawford Hill groups determined that the antenna temperature was indeed due to the microwave background. Penzias and Wilson received the 1978 Nobel Prize in Physics for their discovery.[28]

The interpretation of the cosmic microwave background was a controversial issue in the 1960s with some proponents of the steady state theory arguing that the microwave background was the result of scattered starlight from distant galaxies.[29] Using this model, and based on the study of narrow absorption line features in the spectra of stars, the astronomer Andrew McKellar wrote in 1941: "It can be calculated that the 'rotational temperature' of interstellar space is 2 K."[30] However, during the 1970s the consensus was established that the cosmic microwave background is a remnant of the big bang. This was largely because new measurements at a range of frequencies showed that the spectrum was a thermal, black body spectrum, a result that the steady state model was unable to reproduce.[31]

Harrison, Peebles, Yu and Zel'dovich realized that the early universe would have to have inhomogeneities at the level of 10^{-4} or 10^{-5}.[32][33][34] Rashid Sunyaev later calculated the observable imprint that these inhomogeneities would have on the cosmic microwave background.[35] Increasingly stringent limits on the anisotropy of the cosmic microwave background were set by ground based experiments during the 1980s. RELIKT-1, a Soviet cosmic microwave background anisotropy experiment on board the Prognoz 9 satellite (launched 1 July 1983) gave upper limits on the large-scale anisotropy. The NASA COBE mission clearly confirmed the primary anisotropy with the Differential Microwave Radiometer instrument, publishing their findings in 1992.[36][37] The team received the Nobel Prize in physics for 2006 for this discovery.

Inspired by the COBE results, a series of ground and balloon-based experiments measured cosmic microwave background anisotropies on smaller angular scales over the next decade. The primary goal of these experiments was to measure the scale of the first acoustic peak, which COBE did not have sufficient resolution to resolve. This peak corresponds to large scale density variations in the early universe that are created by gravitational instabilities, resulting in acoustical oscillations in the plasma.[38] The first peak in the anisotropy was tentatively detected by the Toco experiment and the result was confirmed by the BOOMERanG and MAXIMA experiments.[39][40][41] These measurements demonstrated that the geometry of the universe is approximately flat, rather than curved.[42] They ruled out cosmic strings as a major component of cosmic structure formation and suggested cosmic inflation was the right theory of structure formation.[43]

The second peak was tentatively detected by several experiments before being definitively detected by WMAP, which has also tentatively detected the third peak.[44] As of 2010, several experiments to improve measurements of the polarization and the microwave background on small angular scales are ongoing. These include DASI, WMAP, BOOMERanG, QUaD, Planck spacecraft, Atacama Cosmology Telescope, South Pole Telescope and the QUIET telescope.

30.3

Relationship to the Big Bang

←

The cosmic microwave background radiation and the cosmological redshift-distance relation are together regarded as the best available evidence for the Big Bang theory. Measurements of the CMB have made the inflationary Big Bang theory the Standard Model of Cosmology.[45] The discovery of the CMB in the mid-1960s curtailed interest in alternatives such as the steady state theory.[46]

The CMB essentially confirms the Big Bang theory. In the late 1940s Alpher and Herman reasoned that if there was a big bang, the expansion of the universe would have stretched and cooled the high-energy radiation of the very early universe into the microwave region of the electromagnetic spectrum, and down to a temperature of about 5 K. They were slightly off with their estimate, but they had exactly the right idea. They predicted the CMB. It took another 15 years for Penzias and Wilson to stumble into discovering that the microwave background was actually there.[47]

The CMB gives a snapshot of the universe when, according to standard cosmology, the temperature dropped enough to allow electrons and protons to form hydrogen atoms, thereby making the universe nearly transparent to radiation because light was no longer being scattered off free electrons. When it originated some 380,000 years after the Big Bang—this time is generally known as the "time of last scattering" or the period of recombination or decoupling—the temperature of the universe was about 3000 K. This corresponds to an energy of about 0.25 eV, which is much less than the 13.6 eV ionization energy of hydrogen.[48]

Since decoupling, the temperature of the background radiation has dropped by a factor of roughly 1,100[49] due to the expansion of the universe. As the universe expands, the CMB photons are redshifted, causing them to decrease in energy. The temperature of this radiation stays inversely proportional to a parameter that describes the relative expansion of the universe over time, known as the scale length. The temperature T_r of the CMB as a function of redshift, z, can be shown to be proportional to the temperature of the CMB as observed in the present day (2.725 K or 0.235 meV):[50]

$$T_r = 2.725(1 + z)$$

For details about the reasoning that the radiation is evidence for the Big Bang, see Cosmic background radiation of the Big Bang.

30.3.1 Primary anisotropy

The anisotropy, or directional dependency, of the cosmic microwave background is divided into two types: primary anisotropy, due to effects that occur at the last scattering surface and before; and secondary anisotropy, due to effects such as interactions of the background radiation with hot gas or gravitational potentials, which occur between the last scattering surface and the observer.

The structure of the cosmic microwave background anisotropies is principally determined by two effects: acoustic oscillations and diffusion damping (also called collisionless damping or Silk damping). The acoustic oscillations arise because of a conflict in the photon–baryon plasma in the early universe. The pressure of the photons tends to erase anisotropies, whereas the gravitational attraction of the baryons, moving at speeds much slower

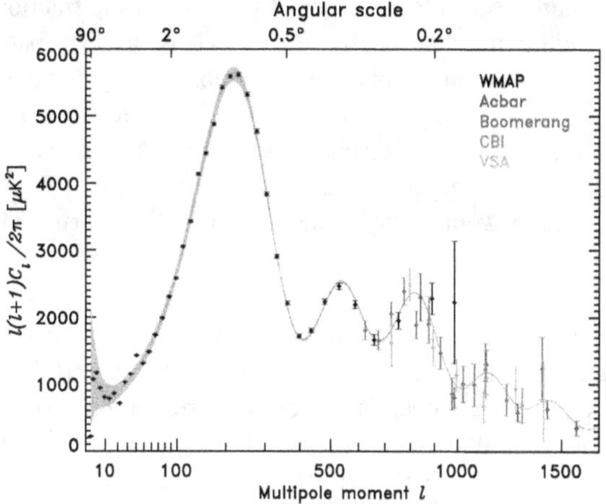

The power spectrum of the cosmic microwave background radiation temperature anisotropy in terms of the angular scale (or multipole moment). The data shown comes from the WMAP (2006), Acbar (2004) Boomerang (2005), CBI (2004), and VSA (2004) instruments. Also shown is a theoretical model (solid line).

than light, makes them tend to collapse to form overdensities. These two effects compete to create acoustic oscillations, which give the microwave background its characteristic peak structure. The peaks correspond, roughly, to resonances in which the photons decouple when a particular mode is at its peak amplitude.

The peaks contain interesting physical signatures. The angular scale of the first peak determines the curvature of the universe (but not the topology of the universe). The next peak—ratio of the odd peaks to the even peaks—determines the reduced baryon density.[51] The third peak can be used to get information about the dark-matter density.[52]

The locations of the peaks also give important information about the nature of the primordial density perturbations. There are two fundamental types of density perturbations called *adiabatic* and *isocurvature*. A general density perturbation is a mixture of both, and different theories that purport to explain the primordial density perturbation spectrum predict different mixtures.

Adiabatic density perturbations The fractional additional number density of each type of particle (baryons, photons ...) is the same. That is, if at one place there is a 1% higher number density of baryons than average, then at that place there is also a 1% higher number density of photons (and a 1% higher number density in neutrinos) than average. Cosmic inflation predicts that the primordial perturbations are adiabatic.

Isocurvature density perturbations In each place the

sum (over different types of particle) of the fractional additional densities is zero. That is, a perturbation where at some spot there is 1% more energy in baryons than average, 1% more energy in photons than average, and 2% *less* energy in neutrinos than average, would be a pure isocurvature perturbation. Cosmic strings would produce mostly isocurvature primordial perturbations.

The CMB spectrum can distinguish between these two because these two types of perturbations produce different peak locations. Isocurvature density perturbations produce a series of peaks whose angular scales (*l* values of the peaks) are roughly in the ratio 1:3:5:..., while adiabatic density perturbations produce peaks whose locations are in the ratio 1:2:3:...[53] Observations are consistent with the primordial density perturbations being entirely adiabatic, providing key support for inflation, and ruling out many models of structure formation involving, for example, cosmic strings.

Collisionless damping is caused by two effects, when the treatment of the primordial plasma as fluid begins to break down:

- the increasing mean free path of the photons as the primordial plasma becomes increasingly rarefied in an expanding universe,
- the finite depth of the last scattering surface (LSS), which causes the mean free path to increase rapidly during decoupling, even while some Compton scattering is still occurring.

These effects contribute about equally to the suppression of anisotropies at small scales and give rise to the characteristic exponential damping tail seen in the very small angular scale anisotropies.

The depth of the LSS refers to the fact that the decoupling of the photons and baryons does not happen instantaneously, but instead requires an appreciable fraction of the age of the universe up to that era. One method of quantifying how long this process took uses the *photon visibility function* (PVF). This function is defined so that, denoting the PVF by $P(t)$, the probability that a CMB photon last scattered between time t and $t + dt$ is given by $P(t)\ dt$.

The maximum of the PVF (the time when it is most likely that a given CMB photon last scattered) is known quite precisely. The first-year WMAP results put the time at which $P(t)$ has a maximum as 372,000 years.[54] This is often taken as the "time" at which the CMB formed. However, to figure out how *long* it took the photons and baryons to decouple, we need a measure of the width of the PVF. The WMAP team finds that the PVF is greater than half of its maximal value (the "full width at half maximum", or FWHM) over an interval of 115,000 years. By this measure, decoupling took place over roughly 115,000 years, and when it was complete, the universe was roughly 487,000 years old.

30.3.2 Late time anisotropy

Since the CMB came into existence, it has apparently been modified by several subsequent physical processes, which are collectively referred to as late-time anisotropy, or secondary anisotropy. When the CMB photons became free to travel unimpeded, ordinary matter in the universe was mostly in the form of neutral hydrogen and helium atoms. However, observations of galaxies today seem to indicate that most of the volume of the intergalactic medium (IGM) consists of ionized material (since there are few absorption lines due to hydrogen atoms). This implies a period of reionization during which some of the material of the universe was broken into hydrogen ions.

The CMB photons are scattered by free charges such as electrons that are not bound in atoms. In an ionized universe, such charged particles have been liberated from neutral atoms by ionizing (ultraviolet) radiation. Today these free charges are at sufficiently low density in most of the volume of the universe that they do not measurably affect the CMB. However, if the IGM was ionized at very early times when the universe was still denser, then there are two main effects on the CMB:

1. Small scale anisotropies are erased. (Just as when looking at an object through fog, details of the object appear fuzzy.)
2. The physics of how photons are scattered by free electrons (Thomson scattering) induces polarization anisotropies on large angular scales. This broad angle polarization is correlated with the broad angle temperature perturbation.

Both of these effects have been observed by the WMAP spacecraft, providing evidence that the universe was ionized at very early times, at a redshift more than 17. The detailed provenance of this early ionizing radiation is still a matter of scientific debate. It may have included starlight from the very first population of stars (population III stars), supernovae when these first stars reached the end of their lives, or the ionizing radiation produced by the accretion disks of massive black holes.

The time following the emission of the cosmic microwave background—and before the observation of the first stars—is semi-humorously referred to by cosmologists as the dark age, and is a period which is under intense study by astronomers (see 21 centimeter radiation).

Two other effects which occurred between reionization and our observations of the cosmic microwave background, and which appear to cause anisotropies, are the Sunyaev–Zel'dovich effect, where a cloud of high-energy electrons scatters the radiation, transferring some of its energy to the CMB photons, and the Sachs–Wolfe effect, which causes photons from the Cosmic Microwave Background to be gravitationally redshifted or blueshifted due to changing gravitational fields.

30.4 Polarization

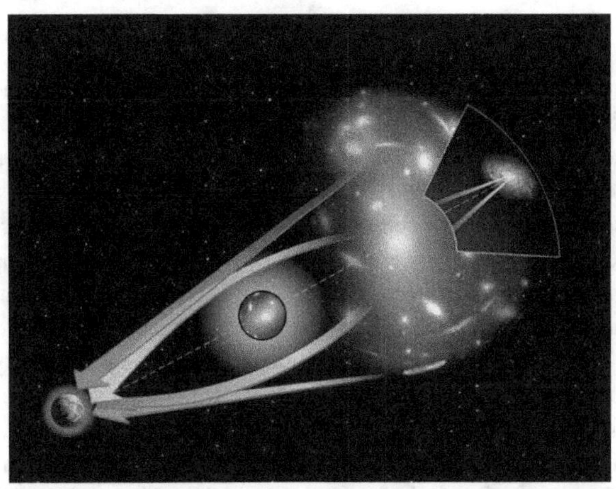

This artist's impression shows how light from the early universe is deflected by the gravitational lensing effect of massive cosmic structures forming B-modes as it travels across the universe.

The cosmic microwave background is polarized at the level of a few microkelvin. There are two types of polarization, called E-modes and B-modes. This is in analogy to electrostatics, in which the electric field (E-field) has a vanishing curl and the magnetic field (B-field) has a vanishing divergence. The E-modes arise naturally from Thomson scattering in a heterogeneous plasma. The B-modes are not produced by standard scalar type perturbations. Instead they can be created by two mechanisms: the first one is by gravitational lensing of E-modes, which has been measured by the South Pole Telescope in 2013;[55] the second one is from gravitational waves arising from cosmic inflation. Detecting the B-modes is extremely difficult, particularly as the degree of foreground contamination is unknown, and the weak gravitational lensing signal mixes the relatively strong E-mode signal with the B-mode signal.[56]

30.4.1 E-modes

E-modes were first seen in 2002 by the Degree Angular Scale Interferometer (DASI).

30.4.2 B-modes

Cosmologists predict two types of B-modes, the first generated during cosmic inflation shortly after the big bang,[57][58][59] and the second generated by gravitational lensing at later times.[60]

Primordial gravitational waves

Primordial gravitational waves are gravitational waves that could be observed in the polarisation of the cosmic microwave background and having their origin in the early universe. Models of cosmic inflation predict that such gravitational waves should appear; thus, their detection supports the theory of inflation, and their strength can confirm and exclude different models of inflation. It is the result of three things: inflationary expansion of space itself, reheating after inflation, and turbulent fluid mixing of matter and radiation. [61]

On 17 March 2014 it was announced that the BICEP2 instrument had detected the first type of B-modes, consistent with inflation and gravitational waves in the early universe at the level of r = 0.20+0.07
−0.05, which is the amount of power present in gravitational waves compared to the amount of power present in other scalar density perturbations in the very early universe. Had this been confirmed it would have provided strong evidence of cosmic inflation and the Big Bang,[62][63] [64][65] [66][67][68] but on 19 June 2014, considerably lowered confidence in confirming the findings was reported[67][69][70] and on 19 September 2014 new results of the Planck experiment reported that the results of BICEP2 can be fully attributed to cosmic dust.[71][72]

Gravitational lensing

The second type of B-modes was discovered in 2013 using the South Pole Telescope with help from the Herschel Space Observatory.[73] This discovery may help test theories on the origin of the universe. Scientists are using data from the Planck mission by the European Space Agency, to gain a better understanding of these waves.[74][75][76]

In October 2014, a measurement of the B-mode polarization at 150 GHz was published by the POLARBEAR experiment.[77] Compared to BICEP2, POLARBEAR focuses on a smaller patch of the sky and is less susceptible to dust effects. The team reported that POLARBEAR's measured B-mode polarization was of cosmological origin (and not just due to dust) at a 97.2% confidence level.[78]

30.5 Microwave background observations

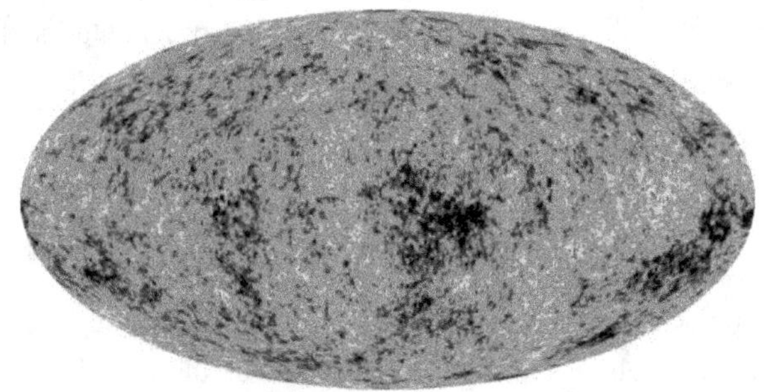

All-sky map of the CMB, created from 9 years of WMAP data

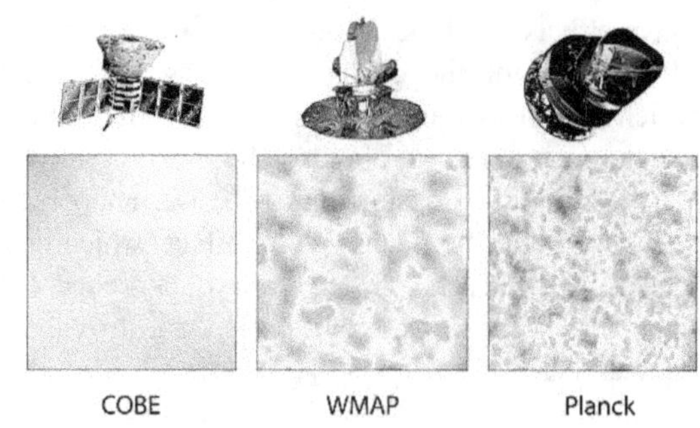

Comparison of CMB results from COBE, WMAP and Planck (March 21, 2013)

Main article: List of cosmic microwave background experiments

Subsequent to the discovery of the CMB, hundreds of cosmic microwave background experiments have been conducted to measure and characterize the signatures of the radiation. The most famous experiment is probably the NASA Cosmic Background Explorer (COBE) satellite that orbited in 1989–1996 and which detected and quantified the large scale anisotropies at the limit of its detection capabilities. Inspired by the initial COBE results of an extremely isotropic and homogeneous background, a series of ground- and balloon-based experiments quantified CMB anisotropies on smaller angular scales over the next decade. The primary goal of these experiments was to measure the angular scale of the first acoustic peak, for which COBE did not have sufficient resolution. These measurements were able to rule out cosmic strings as the leading theory of cosmic structure formation, and suggested cosmic inflation was the right theory. During the 1990s, the first peak was measured with increasing sensitivity and by 2000 the BOOMERanG experiment reported that the highest power fluctuations occur at scales of approximately one degree. Together with other cosmological data, these results implied that the geometry of the universe is flat. A number of ground-based interferometers provided measurements of the fluctuations with higher accuracy over the next three years, including the Very Small Array, Degree Angular Scale Interferometer (DASI), and the Cosmic Background Imager (CBI). DASI made the first detection of the polarization of the CMB and the CBI provided the first E-mode polarization spectrum with compelling evidence that it is out of phase with the T-mode spectrum.

In June 2001, NASA launched a second CMB space mission, WMAP, to make much more precise measurements of the large scale anisotropies over the full sky. WMAP used symmetric, rapid-multi-modulated scanning, rapid switching radiometers to minimize non-sky signal noise.[49] The first results from this mission, disclosed in 2003, were detailed measurements of the angular power spectrum at a scale of less than one degree, tightly constraining various cosmological parameters. The results are broadly consistent with those expected from cosmic inflation as well as various other competing theories, and are available in detail at NASA's data bank for Cosmic Microwave Background (CMB) (see links below). Although WMAP provided very accurate measurements of the large scale angular fluctuations in the CMB (structures about as broad in the sky as the moon), it did not have the angular resolution to measure the smaller scale fluctuations which had been observed by former ground-based interferometers.

A third space mission, the ESA (European Space Agency) Planck Surveyor, was launched in May 2009 and performed an even more detailed investigation until it was shut down in October 2013. Planck employs both HEMT radiometers and bolometer technology and will measure the CMB at a smaller scale than WMAP. Its detectors were trialled in the Antarctic Viper telescope as ACBAR (Arcminute Cosmology Bolometer Array Receiver) experiment—which has produced the most precise measurements at small angular scales to date—and in the Archeops balloon telescope.

On 21 March 2013, the European-led research team behind the Planck cosmology probe released the mission's all-sky map (565x318 jpeg, 3600x1800 jpeg) of the cosmic microwave background.[79][80] The map suggests the universe is slightly older than researchers thought. According to the map, subtle fluctuations in temperature were imprinted on the deep sky when the cosmos was about 370,000 years old. The imprint reflects ripples that arose as early, as the existence of the universe, as the first nonillionth of a second. Apparently, these ripples gave rise to the present vast cosmic web of galaxy clusters and dark matter. Based on the 2013 data, the universe contains 4.9% ordinary matter, 26.8% dark matter and 68.3% dark energy. On 5 February 2015, new data was released by the Planck mission, according to which the age of the universe is 13.799 ± 0.021 billion years old and the Hubble constant was measured to be 67.74 ± 0.46 (km/s)/Mpc.[81]

Additional ground-based instruments such as the South Pole Telescope in Antarctica and the proposed Clover Project, Atacama Cosmology Telescope and the QUIET telescope in Chile will provide additional data not available from satellite observations, possibly including the B-mode polarization.

30.6 Data reduction and analysis

Raw CMBR data from the space vehicle (i.e. WMAP) contain foreground effects that completely obscure the fine-scale structure of the cosmic microwave background. The fine-scale structure is superimposed on the raw CMBR data but is too small to be seen at the scale of the raw data. The most prominent of the foreground effects is the dipole anisotropy caused by the Sun's motion relative to the CMBR background. The dipole anisotropy and others due to Earth's annual motion relative to the Sun and numerous microwave sources in the galactic plane and elsewhere must be subtracted out to reveal the extremely tiny variations characterizing the fine-scale structure of the CMBR background.

The detailed analysis of CMBR data to produce maps, an angular power spectrum, and ultimately cosmological parameters is a complicated, computationally difficult problem. Although computing a power spectrum from a map is in principle a simple Fourier transform, decomposing the map of the sky into spherical harmonics, in practice it is hard to take the effects of noise and foreground sources into account. In particular, these foregrounds are dominated by galactic emissions such as Bremsstrahlung, synchrotron, and dust that emit in the microwave band; in practice, the galaxy has to be removed, resulting in a CMB map that is not a full-sky map. In addition, point sources like galaxies and clusters represent another source of foreground which must be removed so as not to distort the short scale structure of the CMB power spectrum.

Constraints on many cosmological parameters can be obtained from their effects on the power spectrum, and results are often calculated using Markov Chain Monte Carlo sampling techniques.

30.6.1 CMBR dipole anisotropy

From the CMB data it is seen that the Local Group (the galaxy group that includes the Milky Way galaxy) appears to be moving at 627±22 km/s relative to the reference frame of the CMB (also called the CMB rest frame, or the frame of reference in which there is no motion through the CMB) in the direction of galactic longitude l = 276°±3°, b = 30°±3°.[82][83] This motion results in an anisotropy of the data (CMB appearing slightly warmer in the direction of movement than in the opposite direction).[84] From a theoretical point of view, the existence of a CMB rest frame breaks Lorentz invariance even in empty space far away from any galaxy.[85] The standard interpretation of this temperature variation is a simple velocity red shift and blue shift due to motion relative to the CMB, but alternative cosmological models can explain some fraction of the observed dipole temperature distribution in the CMB.[86]

30.6.2 Low multipoles and other anomalies

With the increasingly precise data provided by WMAP, there have been a number of claims that the CMB exhibits anomalies, such as very large scale anisotropies, anomalous alignments, and non-Gaussian distributions.[87][88][89] The most longstanding of these is the low-l multipole controversy. Even in the COBE map, it was observed that the quadrupole (l = 2, spherical harmonic) has a low amplitude compared to the predictions of the Big Bang. In particular, the quadrupole and octupole (l = 3) modes appear to have an unexplained alignment with each other and with both the ecliptic plane and equinoxes,[90][91][92] A number of groups have suggested that this could be the signature of new physics at the greatest observable scales; other groups suspect systematic errors in the data.[93][94][95] Ultimately, due to the foregrounds and the cosmic variance problem, the greatest modes will never be as well measured as the small angular scale modes. The analyses were performed on two maps that have had the foregrounds removed as far as possible: the "internal linear combination" map of the WMAP collaboration and a similar map prepared by Max Tegmark and others.[44][49][96] Later analyses have pointed out that these are the modes most susceptible to foreground contamination from synchrotron, dust, and Bremsstrahlung emission, and from experimental uncertainty in the monopole and dipole. A full Bayesian analysis of the WMAP power spectrum demonstrates that the quadrupole prediction of Lambda-CDM cosmology is consistent with the data at the 10% level and that the observed octupole is not remarkable.[97] Carefully accounting for the procedure used to remove the foregrounds from the full sky map further reduces the significance of the alignment by ~5%.[98][99][100][101] Recent observations with the Planck telescope, which is very much more sensitive than WMAP and has a larger angular resolution, record the same anomaly, and so instrumental error (but not foreground contamination) appears to be ruled out.[102] Coincidence is a possible explanation, chief scientist from WMAP, Charles L. Bennett suggested coincidence and human psychology were involved, *"I do think there is a bit of a psychological effect; people want to find unusual things."* [103]

See also: Cosmological principle, Axis of evil (cosmology), and CMB cold spot

30.7 Future evolution

Assuming the universe keeps expanding and it does not suffer a Big Crunch, a Big Rip, or another similar fate, the cosmic microwave background will continue redshifting until it will no longer be detectable,[104] and will be overtaken first by the one produced by starlight, and later by the background radiation fields of processes that are assumed will take place in the far future of the universe.[105], §VD.

30.8 Timeline of prediction, discovery and interpretation

30.8.1 Thermal (non-microwave background) temperature predictions

- 1896 – Charles Édouard Guillaume estimates the "radiation of the stars" to be 5.6K.[106]
- 1926 – Sir Arthur Eddington estimates the non-thermal radiation of starlight in the galaxy "... by the formula $E = \sigma T^4$ the effective temperature corresponding to this density is 3.18° absolute ... black body"[107]
- 1930s – Cosmologist Erich Regener calculates that the non-thermal spectrum of cosmic rays in the galaxy has an effective temperature of 2.8 K
- 1931 – Term *microwave* first used in print: "When trials with wavelengths as low as 18 cm. were made known, there was undisguised surprise+that the problem of the micro-wave had been solved so soon." *Telegraph & Telephone Journal* XVII. 179/1
- 1934 – Richard Tolman shows that black-body radiation in an expanding universe cools but remains thermal
- 1938 – Nobel Prize winner (1920) Walther Nernst reestimates the cosmic ray temperature as 0.75K
- 1941 – Andrew McKellar was attempting to measure the average temperature of the interstellar medium, and used the excitation of CN doublet lines to measure that the "effective temperature of space" (the average bolometric temperature) is about 2.3 K[30][108]
- 1946 – Robert Dicke predicts "... radiation from cosmic matter" at <20 K, but did not refer to background radiation [109]
- 1946 – George Gamow calculates a temperature of 50 K (assuming a 3-billion year old universe),[110] commenting it "... is in reasonable agreement with the actual temperature of interstellar space", but does not mention background radiation.[111]
- 1953 – Erwin Finlay-Freundlich in support of his tired light theory, derives a blackbody temperature for intergalactic space of 2.3K [112] with comment from Max Born suggesting radio astronomy as the arbitrator between expanding and infinite cosmologies.

30.8.2 Microwave background radiation predictions and measurements

- 1946 – George Gamow calculates a temperature of 50 K (assuming a 3-billion year old universe),[110] commenting it "... is in reasonable agreement with the actual temperature of interstellar space", but does not mention background radiation.
- 1948 – Ralph Alpher and Robert Herman estimate "the temperature in the universe" at 5 K. Although they do not specifically mention microwave background radiation, it may be inferred.[113]
- 1949 – Ralph Alpher and Robert Herman re-re-estimate the temperature at 28 K.
- 1953 – George Gamow estimates 7 K.[109]
- 1956 – George Gamow estimates 6 K.[109]
- 1955 – Émile Le Roux of the Nançay Radio Observatory, in a sky survey at $\lambda = 33$ cm, reported a near-isotropic background radiation of 3 kelvins, plus or minus 2.[109]
- 1957 – Tigran Shmaonov reports that "the absolute effective temperature of the radioemission background ... is 4±3 K".[114] It is noted that the "measurements showed that radiation intensity was independent of either time or direction of observation ... it is now clear that Shmaonov did observe the cosmic microwave background at a wavelength of 3.2 cm"[115][116]
- 1960s – Robert Dicke re-estimates a microwave background radiation temperature of 40 K[109][117]
- 1964 – A. G. Doroshkevich and Igor Dmitrievich Novikov publish a brief paper suggesting microwave searches for the black-body radiation predicted by Gamow, Alpher, and Herman, where they name the CMB radiation phenomenon as detectable.[118]

- 1964–65 – Arno Penzias and Robert Woodrow Wilson measure the temperature to be approximately 3 K. Robert Dicke, James Peebles, P. G. Roll, and D. T. Wilkinson interpret this radiation as a signature of the big bang.
- 1966 – Rainer K. Sachs and Arthur M. Wolfe theoretically predict microwave background fluctuation amplitudes created by gravitational potential variations between observers and the last scattering surface (see Sachs-Wolfe effect)
- 1968 – Martin Rees and Dennis Sciama theoretically predict microwave background fluctuation amplitudes created by photons traversing time-dependent potential wells
- 1969 – R. A. Sunyaev and Yakov Zel'dovich study the inverse Compton scattering of microwave background photons by hot electrons (see Sunyaev-Zel'dovich effect)
- 1983 – Researchers from the Cambridge Radio Astronomy Group and the Owens Valley Radio Observatory first detect the Sunyaev-Zel'dovich effect from clusters of galaxies
- 1983 – RELIKT-1 Soviet CMB anisotropy experiment was launched.
- 1990 – FIRAS on the Cosmic Background Explorer (COBE) satellite measures the black body form of the CMB spectrum with exquisite precision, and shows that the microwave background has a nearly perfect black-body spectrum and thereby strongly constrains the density of the intergalactic medium.
- January 1992 – Scientists that analysed data from the RELIKT-1 report the discovery of anisotropy in the cosmic microwave background at the Moscow astrophysical seminar.[119]
- 1992 – Scientists that analysed data from COBE DMR report the discovery of anisotropy in the cosmic microwave background.[120]
- 1995 – The Cosmic Anisotropy Telescope performs the first high resolution observations of the cosmic microwave background.
- 1999 – First measurements of acoustic oscillations in the CMB anisotropy angular power spectrum from the TOCO, BOOMERANG, and Maxima Experiments. The BOOMERanG experiment makes higher quality maps at intermediate resolution, and confirms that the universe is “flat”.
- 2002 – Polarization discovered by DASI.[121]
- 2003 – E-mode polarization spectrum obtained by the CBI.[122] The CBI and the Very Small Array produces yet higher quality maps at high resolution (covering small areas of the sky).
- 2003 – The WMAP spacecraft produces an even higher quality map at low and intermediate resolution of the whole sky (WMAP provides *no* high-resolution data, but improves on the intermediate resolution maps from BOOMERanG).
- 2004 – E-mode polarization spectrum obtained by the CBI.[123]
- 2004 – The Arcminute Cosmology Bolometer Array Receiver produces a higher quality map of the high resolution structure not mapped by WMAP.
- 2005 – The Arcminute Microkelvin Imager and the Sunyaev-Zel'dovich Array begin the first surveys for very high redshift clusters of galaxies using the Sunyaev-Zel'dovich effect.
- 2005 – Ralph A. Alpher is awarded the National Medal of Science for his groundbreaking work in nucleosynthesis and prediction that the universe expansion leaves behind background radiation, thus providing a model for the Big Bang theory.
- 2006 – The long-awaited three-year WMAP results are released, confirming previous analysis, correcting several points, and including polarization data.
- 2006 – Two of COBE's principal investigators, George Smoot and John Mather, received the Nobel Prize in Physics in 2006 for their work on precision measurement of the CMBR.
- 2006-2011 – Improved measurements from WMAP, new supernova surveys ESSENCE and SNLS, and baryon acoustic oscillations from SDSS and WiggleZ, continue to be consistent with the standard Lambda-CDM model.
- 2010 – The first all-sky map from the Planck telescope is released.
- 2013 – An improved all-sky map from the Planck telescope is released, improving the measurements of WMAP and extending them to much smaller scales.
- 2014 – On March 17, 2014, astrophysicists of the BICEP2 collaboration announced the detection of inflationary gravitational waves in the B-mode power spectrum, which if confirmed, would provide clear experimental evidence for the theory of inflation.[62][63][64][65][67][124] However, on 19 June 2014, lowered confidence in confirming the cosmic inflation findings was reported.[67][69][70]

- 2015 – On January 30, 2015, the same team of astronomers from BICEP2 withdrew the claim made on the previous year. Based on the combined data of BICEP2 and Planck, the European Space Agency announced that the signal can be entirely attributed to dust in the Milky Way.[125]

30.9 In popular culture

- In the *Stargate Universe* TV series, an Ancient spaceship, *Destiny*, was built to study patterns in the CMBR which indicate that the universe as we know it might have been created by some form of sentient intelligence.[126]
- In *Wheelers*, a novel by Ian Stewart & Jack Cohen, CMBR is explained as the encrypted transmissions of an ancient civilization. This allows the Jovian "blimps" to have a society older than the currently-observed age of the universe.
- In *The Three-Body Problem*, a novel by Liu Cixin, CMBR becomes observable to the naked eye due to interference from an alien civilization.

30.10 See also

- Computational packages for Cosmologists
- Cosmic neutrino background
- Cosmic gravitational wave background
- Cosmological perturbation theory
- Axis of evil (cosmology)
- Gravitational wave background
- Heat death of the universe
- Lambda-CDM model
- Observational cosmology
- Observation history of galaxies
- Physical cosmology

30.11 References

[1] Penzias, A. A.; Wilson, R. W. (1965). "A Measurement of Excess Antenna Temperature at 4080 Mc/s". *The Astrophysical Journal*. **142** (1): 419–421. Bibcode:1965ApJ...142..419P. doi:10.1086/148307.

[2] Smoot Group (28 March 1996). "The Cosmic Microwave Background Radiation". Lawrence Berkeley Lab. Retrieved 2008-12-11.

[3] Kaku, M. (2014). "First Second of the Big Bang". *How the Universe Works*. Discovery Science.

[4] Fixsen, D. J. (2009). "The Temperature of the Cosmic Microwave Background". *The Astrophysical Journal*. **707** (2): 916–920. Bibcode:2009ApJ...707..916F. arXiv:0911.1955. doi:10.1088/0004-637X/707/2/916.

[5] Dodelson, S. (2003). "Coherent Phase Argument for Inflation". *AIP Conference Proceedings*. **689**: 184–196. Bibcode:2003AIPC..689..184D. arXiv:hep-ph/0309057. doi:10.1063/1.1627736.

[6] Baumann, D. (2011). "The Physics of Inflation" (PDF). University of Cambridge. Retrieved 2015-05-09.

[7] White, M. (1999). "Anisotropies in the CMB". *Proceedings of the Los Angeles Meeting, DPF 99*. UCLA. Bibcode:1999dpf..conf.....W. arXiv:astro-ph/9903232.

[8] Wright, E.L. (2004). "Theoretical Overview of Cosmic Microwave Background Anisotropy". In W. L. Freedman. *Measuring and Modeling the Universe*. Carnegie Observatories Astrophysics Series. Cambridge University Press. p. 291. ISBN 0-521-75576-X. arXiv:astro-ph/0305591.

[9] The Planck Collaboration, "Planck 2013 results. XXVII. Doppler boosting of the CMB: Eppur si muove", *Astronomy*, **571**: A27, Bibcode:2014A&A...571A..27P, arXiv:1303.5087, doi:10.1051/0004-6361/201321556

[10] Guth, A. H. (1998). *The Inflationary Universe: The Quest for a New Theory of Cosmic Origins*. Basic Books. p. 186. ISBN 978-0201328400. OCLC 35701222.

[11] Cirigliano, D.; de Vega, H.J.; Sanchez, N. G. (2005). "Clarifying inflation models: The precise inflationary potential from effective field theory and the WMAP data". *Physical Review D*. **71** (10): 77–115. Bibcode:2005PhRvD..71j3518C. arXiv:astro-ph/0412634. doi:10.1103/PhysRevD.71.103518.

[12] Abbott, B. (2007). "Microwave (WMAP) All-Sky Survey". Hayden Planetarium. Archived from the original on 2013-02-13. Retrieved 2008-01-13.

[13] Gawiser, E.; Silk, J. (2000). "The cosmic microwave background radiation". *Physics Reports*. 333–334: 245–267. Bibcode:2000PhR...333..245G. arXiv:astro-ph/0002044. doi:10.1016/S0370-1573(00)00025-9.

[14] Smoot, G. F. (2006). "Cosmic Microwave Background Radiation Anisotropies: Their Discovery and Utilization". *Nobel Lecture*. Nobel Foundation. Retrieved 2008-12-22.

[15] Hobson, M.P.; Efstathiou, G.; Lasenby, A.N. (2006). *General Relativity: An Introduction for Physicists*. Cambridge University Press. p. 388. ISBN 0-521-82951-8.

[16] Unsöld, A.; Bodo, B. (2002). *The New Cosmos, An Introduction to Astronomy and Astrophysics* (5th ed.). Springer-Verlag. p. 485. ISBN 3-540-67877-8.

[17] Confrontation of Cosmological Theories with Observational Data, M. S. Longair, page 144

[18] Cosmology II: The thermal history of the Universe, Ruth Durrer

[19] Gamow, G. (1948). "The Origin of Elements and the Separation of Galaxies". *Physical Review.* **74** (4): 505–506. Bibcode:1948PhRv...74..505G. doi:10.1103/PhysRev.74.505.2.

[20] Gamow, G. (1948). "The evolution of the universe". *Nature.* **162** (4122): 680–682. Bibcode:1948Natur.162..680G. PMID 18893719. doi:10.1038/162680a0.

[21] Alpher, R. A.; Herman, R. C. (1948). "On the Relative Abundance of the Elements". *Physical Review.* **74** (12): 1737–1742. Bibcode:1948PhRv...74.1737A. doi:10.1103/PhysRev.74.1737.

[22] Assis, A. K. T.; Neves, M. C. D. (1995). "History of the 2.7 K Temperature Prior to Penzias and Wilson" (PDF) (3): 79–87. but see also Wright, E. L. (2006). "Eddington's Temperature of Space". UCLA. Retrieved 2008-12-11.

[23] Penzias, A. A. (2006). "The origin of elements" (PDF). *Nobel lecture.* Nobel Foundation. Retrieved 2006-10-04.

[24] Dicke, R. H. (1946). "The Measurement of Thermal Radiation at Microwave Frequencies". *Review of Scientific Instruments.* **17** (7): 268–275. Bibcode:1946RScI...17..268D. PMID 20991753. doi:10.1063/1.1770483. This basic design for a radiometer has been used in most subsequent cosmic microwave background experiments.

[25] The Cosmic Microwave Background Radiation (Nobel Lecture) by Robert Wilson 8 Dec 1978, p. 474

[26] Dicke, R. H.; et al. (1965). "Cosmic Black-Body Radiation". *Astrophysical Journal.* **142**: 414–419. Bibcode:1965ApJ...142..414D. doi:10.1086/148306.

[27] The history is given in Peebles, P. J. E (1993). *Principles of Physical Cosmology.* Princeton University Press. pp. 139–148. ISBN 0-691-01933-9.

[28] "The Nobel Prize in Physics 1978". Nobel Foundation. 1978. Retrieved 2009-01-08.

[29] Narlikar, J. V.; Wickramasinghe, N. C. (1967). "Microwave Background in a Steady State Universe". *Nature.* **216** (5110): 43–44. Bibcode:1967Natur.216...43N. doi:10.1038/216043a0.

[30] McKellar, A.; Kan-Mitchell, June; Conti, Peter S. (1941). "Molecular Lines from the Lowest States of Diatomic Molecules Composed of Atoms Probably Present in Interstellar Space". *Publications of the Dominion Astrophysical Observatory (Victoria, BC).* **7** (6): 251–272. Bibcode:1941PDAO....7..251P.

[31] Peebles, P. J. E.; et al. (1991). "The case for the relativistic hot big bang cosmology". *Nature.* **352** (6338): 769–776. Bibcode:1991Natur.352..769P. doi:10.1038/352769a0.

[32] Harrison, E. R. (1970). "Fluctuations at the threshold of classical cosmology". *Physical Review D.* **1** (10): 2726–2730. Bibcode:1970PhRvD...1.2726H. doi:10.1103/PhysRevD.1.2726.

[33] Peebles, P. J. E.; Yu, J. T. (1970). "Primeval Adiabatic Perturbation in an Expanding Universe". *Astrophysical Journal.* **162**: 815–836. Bibcode:1970ApJ...162..815P. doi:10.1086/150713.

[34] Zeldovich, Y. B. (1972). "A hypothesis, unifying the structure and the entropy of the Universe". *Monthly Notices of the Royal Astronomical Society.* **160** (7–8): 1P–4P. doi:10.1016/S0026-0576(07)80178-4.

[35] Doroshkevich, A. G.; Zel'Dovich, Y. B.; Syunyaev, R. A. (1978) [12–16 September 1977]. "Fluctuations of the microwave background radiation in the adiabatic and entropic theories of galaxy formation". In Longair, M. S.; Einasto, J. *The large scale structure of the universe; Proceedings of the Symposium.* Tallinn, Estonian SSR: Dordrecht, D. Reidel Publishing Co. pp. 393–404. Bibcode:1978IAUS...79..393S. While this is the first paper to discuss the detailed observational imprint of density inhomogeneities as anisotropies in the cosmic microwave background, some of the groundwork was laid in Peebles and Yu, above.

[36] Smooth, G. F.; et al. (1992). "Structure in the COBE differential microwave radiometer first-year maps". *Astrophysical Journal Letters.* **396** (1): L1–L5. Bibcode:1992ApJ...396L...1S. doi:10.1086/186504.

[37] Bennett, C.L.; et al. (1996). "Four-Year COBE DMR Cosmic Microwave Background Observations: Maps and Basic Results". *Astrophysical Journal Letters.* **464**: L1–L4. Bibcode:1996ApJ...464L...1B. arXiv:astro-ph/9601067. doi:10.1086/310075.

[38] Grupen, C.; et al. (2005). *Astroparticle Physics.* Springer. pp. 240–241. ISBN 3-540-25312-2.

[39] Miller, A. D.; et al. (1999). "A Measurement of the Angular Power Spectrum of the Microwave Background Made from the High Chilean Andes". *Astrophysical Journal.* **521** (2): L79–L82. Bibcode:1999ApJ...521L..79T. arXiv:astro-ph/9905100. doi:10.1086/312197.

[40] Melchiorri, A.; et al. (2000). "A Measurement of Ω from the North American Test Flight of Boomerang". *The Astrophysical Journal Letters.* **536** (2): L63–L66. Bibcode:2000ApJ...536L..63M. arXiv:astro-ph/9911445. doi:10.1086/312744.

[41] Hanany, S.; et al. (2000). "MAXIMA-1: A Measurement of the Cosmic Microwave Background Anisotropy on Angular Scales of 10'–5°". *Astrophysical Journal.* **545** (1): L5–L9.

Bibcode:2000ApJ...545L...5H. arXiv:astro-ph/0005123. doi:10.1086/317322.

[42] de Bernardis, P.; et al. (2000). "A flat Universe from high-resolution maps of the cosmic microwave background radiation". *Nature*. **404** (6781): 955–959. Bibcode:2000Natur.404..955D. PMID 10801117. arXiv:astro-ph/0004404. doi:10.1038/35010035.

[43] Pogosian, L.; et al. (2003). "Observational constraints on cosmic string production during brane inflation". *Physical Review D*. **68** (2): 023506. Bibcode:2003PhRvD..68b3506P. arXiv:hep-th/0304188. doi:10.1103/PhysRevD.68.023506.

[44] Hinshaw, G.; (WMAP collaboration); Bennett, C. L.; Bean, R.; Doré, O.; Greason, M. R.; Halpern, M.; Hill, R. S.; Jarosik, N.; Kogut, A.; Komatsu, E.; Limon, M.; Odegard, N.; Meyer, S. S.; Page, L.; Peiris, H. V.; Spergel, D. N.; Tucker, G. S.; Verde, L.; Weiland, J. L.; Wollack, E.; Wright, E. L.; et al. (2007). "Three-year Wilkinson Microwave Anisotropy Probe (WMAP) observations: temperature analysis". *Astrophysical Journal Supplement Series*. **170** (2): 288–334. Bibcode:2007ApJS..170..288H. arXiv:astro-ph/0603451. doi:10.1086/513698.

[45] Scott, D. (2005). "The Standard Cosmological Model". arXiv:astro-ph/0510731 [astro-ph].

[46] Durham, Frank; Purrington, Robert D. (1983). *Frame of the universe: a history of physical cosmology*. Columbia University Press. pp. 193–209. ISBN 0-231-05393-2.

[47] Assis, A. K. T.; Paulo, São; Neves, M. C. D. (July 1995). "History of the 2.7 K Temperature Prior to Penzias and Wilson" (PDF). *Apeiron*. **2** (3): 79–87.

[48] Brandenberger, Robert H. (1995). "Formation of Structure in the Universe": 8159. Bibcode:1995astro.ph..8159B. arXiv:astro-ph/9508159.

[49] Bennett, C. L.; (WMAP collaboration); Hinshaw, G.; Jarosik, N.; Kogut, A.; Limon, M.; Meyer, S. S.; Page, L.; Spergel, D. N.; Tucker, G. S.; Wollack, E.; Wright, E. L.; Barnes, C.; Greason, M. R.; Hill, R. S.; Komatsu, E.; Nolta, M. R.; Odegard, N.; Peiris, H. V.; Verde, L.; Weiland, J. L.; et al. (2003). "First-year Wilkinson Microwave Anisotropy Probe (WMAP) observations: preliminary maps and basic results". *Astrophysical Journal Supplement Series*. **148**: 1–27. Bibcode:2003ApJS..148....1B. arXiv:astro-ph/0302207. doi:10.1086/377253. This paper warns, "the statistics of this internal linear combination map are complex and inappropriate for most CMB analyses."

[50] Noterdaeme, P.; Petitjean, P.; Srianand, R.; Ledoux, C.; López, S. (February 2011). "The evolution of the cosmic microwave background temperature. Measurements of TCMB at high redshift from carbon monoxide excitation". *Astronomy and Astrophysics*. **526**: L7. Bibcode:2011A&A...526L...7N. arXiv:1012.3164. doi:10.1051/0004-6361/201016140.

[51] Wayne Hu. "Baryons and Inertia".

[52] Wayne Hu. "Radiation Driving Force".

[53] Hu, W.; White, M. (1996). "Acoustic Signatures in the Cosmic Microwave Background". *Astrophysical Journal*. **471**: 30–51. Bibcode:1996ApJ...471...30H. arXiv:astro-ph/9602019. doi:10.1086/177951.

[54] WMAP Collaboration; Verde, L.; Peiris, H. V.; Komatsu, E.; Nolta, M. R.; Bennett, C. L.; Halpern, M.; Hinshaw, G.; et al. (2003). "First-Year Wilkinson Microwave Anisotropy Probe (WMAP) Observations: Determination of Cosmological Parameters". *Astrophysical Journal Supplement Series*. **148** (1): 175–194. Bibcode:2003ApJS..148..175S. arXiv:astro-ph/0302209. doi:10.1086/377226.

[55] Hanson, D.; et al. (2013). "Detection of B-mode polarization in the Cosmic Microwave Background with data from the South Pole Telescope". *Physical Review Letters*. **111** (14). Bibcode:2013PhRvL.111n1301H. arXiv:1307.5830. doi:10.1103/PhysRevLett.111.141301.

[56] Lewis, A.; Challinor, A. (2006). "Weak gravitational lensing of the CMB". *Physics Reports*. **429**: 1–65. Bibcode:2006PhR...429....1L. arXiv:astro-ph/0601594. doi:10.1016/j.physrep.2006.03.002.

[57] Seljak, U. (June 1997). "Measuring Polarization in the Cosmic Microwave Background". *Astrophysical Journal*. **482**: 6–16. Bibcode:1997ApJ...482....6S. arXiv:astro-ph/9608131. doi:10.1086/304123.

[58] Seljak, U.; Zaldarriaga M. (March 17, 1997). "Signature of Gravity Waves in the Polarization of the Microwave Background". *Phys. Rev. Lett.* **78** (11): 2054–2057. Bibcode:1997PhRvL..78.2054S. arXiv:astro-ph/9609169. doi:10.1103/PhysRevLett.78.2054.

[59] Kamionkowski, M.; Kosowsky A. & Stebbins A. (March 17, 1997). "A Probe of Primordial Gravity Waves and Vorticity". *Phys. Rev. Lett.* **78** (11): 2058–2061. Bibcode:1997PhRvL..78.2058K. arXiv:astro-ph/9609132. doi:10.1103/PhysRevLett.78.2058.

[60] Zaldarriaga, M.; Seljak U. (July 15, 1998). "Gravitational lensing effect on cosmic microwave background polarization". *Physical Review D*. 2. **58**. Bibcode:1998PhRvD..58b3003Z. arXiv:astro-ph/9803150. doi:10.1103/PhysRevD.58.023003.

[61] "Scientists Report Evidence for Gravitational Waves in Early Universe". Retrieved 2007-06-20.

[62] Staff (17 March 2014). "BICEP2 2014 Results Release". *National Science Foundation*. Retrieved 18 March 2014.

[63] Clavin, Whitney (March 17, 2014). "NASA Technology Views Birth of the Universe". *NASA*. Retrieved March 17, 2014.

[64] Overbye, Dennis (March 17, 2014). "Space Ripples Reveal Big Bang's Smoking Gun". *The New York Times*. Retrieved March 17, 2014.

[65] Overbye, Dennis (March 24, 2014). "Ripples From the Big Bang". *New York Times*. Retrieved March 24, 2014.

[66] "Gravitational waves: have US scientists heard echoes of the big bang?". The Guardian. 2014-03-14. Retrieved 2014-03-14.

[67] Ade, P.A.R. (BICEP2 Collaboration) (19 June 2014). "Detection of B-Mode Polarization at Degree Angular Scales by BICEP2" (PDF). *Physical Review Letters*. **112** (24): 241101. Bibcode:2014PhRvL.112x1101B. PMID 24996078. arXiv:1403.3985 ⓐ. doi:10.1103/PhysRevLett.112.241101. Retrieved 20 June 2014.

[68] "Space Ripples Reveal Big Bang's Smoking Gun". March 17, 2014.

[69] Overbye, Dennis (June 19, 2014). "Astronomers Hedge on Big Bang Detection Claim". *New York Times*. Retrieved June 20, 2014.

[70] Amos, Jonathan (June 19, 2014). "Cosmic inflation: Confidence lowered for Big Bang signal". *BBC News*. Retrieved June 20, 2014.

[71] Planck Collaboration Team (9 February 2016). "Planck intermediate results. XXX. The angular power spectrum of polarized dust emission at intermediate and high Galactic latitudes". *Astronomy & Astrophysics*. **586**: A133. Bibcode:2016A&A...586A.133P. arXiv:1409.5738 ⓐ. doi:10.1051/0004-6361/201425034.

[72] Overbye, Dennis (22 September 2014). "Study Confirms Criticism of Big Bang Finding". *New York Times*. Retrieved 22 September 2014.

[73] "Polarization detected in Big Bang's echo". *Nature News & Comment*.

[74] ESA Planck (Oct 22, 2013). "Planck Space Mission". Retrieved Oct 23, 2013.

[75] NASA/Jet Propulsion Laboratory (October 22, 2013). "Long-sought pattern of ancient light detected". *ScienceDaily*. Retrieved October 23, 2013.

[76] Hanson, D.; et al. (Sep 30, 2013). "Detection of B-Mode Polarization in the Cosmic Microwave Background with Data from the South Pole Telescope". *Physical Review Letters*. 14. **111**. Bibcode:2013PhRvL.111n1301H. arXiv:1307.5830 ⓐ. doi:10.1103/PhysRevLett.111.141301.

[77] The Polarbear Collaboration (October 2014). "A Measurement of the Cosmic Microwave Background B-Mode Polarization Power Spectrum at Sub-Degree Scales with POLARBEAR" (PDF). *The Astrophysical Journal*. **794**: 171. Bibcode:2014ApJ...794..171T. arXiv:1403.2369 ⓐ. doi:10.1088/0004-637X/794/2/171. Retrieved November 16, 2014.

[78] "POLARBEAR project offers clues about origin of universe's cosmic growth spurt". *Christian Science Monitor*. October 21, 2014.

[79] Clavin, Whitney; Harrington, J.D. (21 March 2013). "Planck Mission Brings Universe Into Sharp Focus". *NASA*. Retrieved 21 March 2013.

[80] Staff (21 March 2013). "Mapping the Early Universe". *New York Times*. Retrieved 23 March 2013.

[81] Planck Collaboration (2015). "Planck 2015 results. XIII. Cosmological parameters (See Table 4 on page 31 of pfd).". *Astronomy & Astrophysics*. **594**: A13. Bibcode:2016A&A...594A..13P. arXiv:1502.01589 ⓐ. doi:10.1051/0004-6361/201525830.

[82] Kogut, A.; Lineweaver, C.; Smoot, G. F.; Bennett, C. L.; Banday, A.; Boggess, N. W.; Cheng, E. S.; De Amici, G.; Fixsen, D. J.; Hinshaw, G.; Jackson, P. D.; Janssen, M.; Keegstra, P.; Loewenstein, K.; Lubin, P.; Mather, J. C.; Tenorio, L.; Weiss, R.; Wilkinson, D. T.; Wright, E. L. (1993). "Dipole Anisotropy in the COBE Differential Microwave Radiometers First-Year Sky Maps". *Astrophysical Journal*. **419**: 1–6. Bibcode:1993ApJ...419....1K. arXiv:astro-ph/9312056 ⓐ. doi:10.1086/173453.

[83] Aghanim, N.; Armitage-Caplan, C.; et al. (2013). "Planck 2013 results. XXVII. Doppler boosting of the CMB: Eppur si muove". *Astronomy & Astrophysics*. **571** (27): A27. Bibcode:2014A&A...571A..27P. arXiv:1303.5087 ⓐ. doi:10.1051/0004-6361/201321556.

[84] http://antwrp.gsfc.nasa.gov/apod/ap090906.html

[85] http://iopscience.iop.org/1126-6708/2005/07/029/

[86] Inoue, K. T.; Silk, J. (2007). "Local Voids as the Origin of Large-Angle Cosmic Microwave Background Anomalies: The Effect of a Cosmological Constant". *Astrophysical Journal*. **664** (2): 650–659. Bibcode:2007ApJ...664..650I. arXiv:astro-ph/0612347 ⓐ. doi:10.1086/517603.

[87] Rossmanith, G.; Räth, C.; Banday, A. J.; Morfill, G. (2009). "Non-Gaussian Signatures in the five-year WMAP data as identified with isotropic scaling indices". *Monthly Notices of the Royal Astronomical Society*. **399** (4): 1921–1933. Bibcode:2009MNRAS.399.1921R. arXiv:0905.2854 ⓐ. doi:10.1111/j.1365-2966.2009.15421.x.

[88] Bernui, A.; Mota, B.; Rebouças, M. J.; Tavakol, R. (2005). "Mapping the large-scale anisotropy in the WMAP data". *Astronomy and Astrophysics*. **464** (2): 479–485. Bibcode:2007A&A...464..479B. arXiv:astro-ph/0511666 ⓐ. doi:10.1051/0004-6361:20065585.

[89] Jaffe, T.R.; Banday, A. J.; Eriksen, H. K.; Górski, K. M.; Hansen, F. K. (2005). "Evidence of vorticity and shear at large angular scales in the WMAP data: a violation of cosmological isotropy?". *The Astrophysical Journal*. **629**: L1–L4. Bibcode:2005ApJ...629L...1J. arXiv:astro-ph/0503213. doi:10.1086/444454.

[90] de Oliveira-Costa, A.; Tegmark, Max; Zaldarriaga, Matias; Hamilton, Andrew (2004). "The significance of the largest scale CMB fluctuations in WMAP". *Physical Review D*. **69** (6): 063516. Bibcode:2004PhRvD..69f3516D. arXiv:astro-ph/0307282. doi:10.1103/PhysRevD.69.063516.

[91] Schwarz, D. J.; Starkman, Glenn D.; et al. (2004). "Is the low-*l* microwave background cosmic?". *Physical Review Letters*. **93** (22): 221301. Bibcode:2004PhRvL..93v1301S. arXiv:astro-ph/0403353. doi:10.1103/PhysRevLett.93.221301.

[92] Bielewicz, P.; Gorski, K. M.; Banday, A. J. (2004). "Low-order multipole maps of CMB anisotropy derived from WMAP". *Monthly Notices of the Royal Astronomical Society*. **355** (4): 1283–1302. Bibcode:2004MNRAS.355.1283B. arXiv:astro-ph/0405007. doi:10.1111/j.1365-2966.2004.08405.x.

[93] Liu, Hao; Li, Ti-Pei (2009). "Improved CMB Map from WMAP Data". arXiv:0907.2731v3 [astro-ph].

[94] Sawangwit, Utane; Shanks, Tom (2010). "Lambda-CDM and the WMAP Power Spectrum Beam Profile Sensitivity". arXiv:1006.1270v1 [astro-ph].

[95] Liu, Hao; et al. (2010). "Diagnosing Timing Error in WMAP Data". arXiv:1009.2701v1 [astro-ph].

[96] Tegmark, M.; de Oliveira-Costa, A.; Hamilton, A. (2003). "A high resolution foreground cleaned CMB map from WMAP". *Physical Review D*. **68** (12): 123523. Bibcode:2003PhRvD..68l3523T. arXiv:astro-ph/0302496. doi:10.1103/PhysRevD.68.123523. This paper states, "Not surprisingly, the two most contaminated multipoles are [the quadrupole and octupole], which most closely trace the galactic plane morphology."

[97] O'Dwyer, I.; Eriksen, H. K.; Wandelt, B. D.; Jewell, J. B.; Larson, D. L.; Górski, K. M.; Banday, A. J.; Levin, S.; Lilje, P. B. (2004). "Bayesian Power Spectrum Analysis of the First-Year Wilkinson Microwave Anisotropy Probe Data". *Astrophysical Journal Letters*. **617** (2): L99–L102. Bibcode:2004ApJ...617L..99O. arXiv:astro-ph/0407027. doi:10.1086/427386.

[98] Slosar, A.; Seljak, U. (2004). "Assessing the effects of foregrounds and sky removal in WMAP". *Physical Review D*. **70** (8): 083002. Bibcode:2004PhRvD..70h3002S. arXiv:astro-ph/0404567. doi:10.1103/PhysRevD.70.083002.

[99] Bielewicz, P.; Eriksen, H. K.; Banday, A. J.; Górski, K. M.; Lilje, P. B. (2005). "Multipole vector anomalies in the first-year WMAP data: a cut-sky analysis". *Astrophysical Journal*. **635** (2): 750–60. Bibcode:2005ApJ...635..750B. arXiv:astro-ph/0507186. doi:10.1086/497263.

[100] Copi, C.J.; Huterer, Dragan; Schwarz, D. J.; Starkman, G. D. (2006). "On the large-angle anomalies of the microwave sky". *Monthly Notices of the Royal Astronomical Society*. **367**: 79–102. Bibcode:2006MNRAS.367...79C. arXiv:astro-ph/0508047. doi:10.1111/j.1365-2966.2005.09980.x.

[101] de Oliveira-Costa, A.; Tegmark, M. (2006). "CMB multipole measurements in the presence of foregrounds". *Physical Review D*. **74** (2): 023005. Bibcode:2006PhRvD..74b3005D. arXiv:astro-ph/0603369. doi:10.1103/PhysRevD.74.023005.

[102] Planck shows almost perfect cosmos – plus axis of evil

[103] Found: Hawking's initials written into the universe

[104] Krauss, Lawrence M.; Scherrer, Robert J. (2007). "The return of a static universe and the end of cosmology". *General Relativity and Gravitation*. **39** (10): 1545–1550. Bibcode:2007GReGr..39.1545K. arXiv:0704.0221. doi:10.1007/s10714-007-0472-9.

[105] Adams, Fred C.; Laughlin, Gregory (1997). "A dying universe: The long-term fate and evolution of astrophysical objects". *Reviews of Modern Physics*. **69** (2): 337–372. Bibcode:1997RvMP...69..337A. arXiv:astro-ph/9701131. doi:10.1103/RevModPhys.69.337.

[106] Guillaume, C.-É., 1896, *La Nature* 24, series 2, p. 234, cited in "History of the 2.7 K Temperature Prior to Penzias and Wilson" (PDF)

[107] Eddington, A., The Internal Constitution of the Stars, cited in "History of the 2.7 K Temperature Prior to Penzias and Wilson" (PDF)

[108] Weinberg, S. (1972). *Oxford Astronomy Encyclopedia*. John Wiley & Sons. p. 514. ISBN 0-471-92567-5.

[109] Kragh, H. (1999). *Cosmology and Controversy: The Historical Development of Two Theories of the Universe*. ISBN 0-691-00546-X. "In 1946, Robert Dicke and coworkers at MIT tested equipment that could test a cosmic microwave background of intensity corresponding to about 20K in the microwave region. However, they did not refer to such a background, but only to 'radiation from cosmic matter'. Also, this work was unrelated to cosmology and is only mentioned because it suggests that by 1950, detection of the background radiation might have been technically possible, and also because of Dicke's later role in the discovery". See also Dicke, R. H.; et al. (1946). "Atmospheric Absorption Measurements with a Microwave Radiometer". *Physical Review*. **70** (5–6): 340–348. Bibcode:1946PhRv...70..340D. doi:10.1103/PhysRev.70.340.

[110] George Gamow, *The Creation Of The Universe* p.50 (Dover reprint of revised 1961 edition) ISBN 0-486-43868-6

[111] Gamow, G. (2004) [1961]. *Cosmology and Controversy: The Historical Development of Two Theories of the Universe*. Courier Dover Publications. p. 40. ISBN 978-0-486-43868-9.

[112] Erwin Finlay-Freundlich, "Ueber die Rotverschiebung der Spektrallinien" (1953) *Contributions from the Observatory, University of St. Andrews* ; no. 4, p. 96–102. Finlay-Freundlich also gave two extreme values of 1.9K and 6.0K in Finlay-Freundlich, E.: 1954, “Red shifts in the spectra of celestial bodies”, Phil. Mag., Vol. 45, pp. 303–319.

[113] Helge Kragh, Cosmology and Controversy: The Historical Development of Two Theories of the Universe (1999) ISBN 0-691-00546-X. “Alpher and Herman first calculated the present temperature of the decoupled primordial radiation in 1948, when they reported a value of 5 K. Although it was not mentioned either then or in later publications that the radiation is in the microwave region, this follows immediately from the temperature ... Alpher and Herman made it clear that what they had called “the temperature in the univerese” the previous year referred to a blackbody distributed background radiation quite different from sunliight”.

[114] Shmaonov, T. A. (1957). “Commentary”. *Pribory i Tekhnika Experimenta* (in Russian). **1**: 83. doi:10.1016/S0890-5096(06)60772-3.

[115] It is noted that the “measurements showed that radiation intensity was independent of either time or direction of observation ... it is now clear that Shmaonov did observe the cosmic microwave background at a wavelength of 3.2cm”

[116] Naselsky, P. D.; Novikov, D.I.; Novikov, I. D. (2006). *The Physics of the Cosmic Microwave Background*. ISBN 0-521-85550-0.

[117] Helge Kragh, Cosmology and Controversy: The Historical Development of Two Theories of the Universe

[118] Doroshkevich, A. G.; Novikov, I.D. (1964). “Mean Density of Radiation in the Metagalaxy and Certain Problems in Relativistic Cosmology”. *Soviet Physics Doklady*. **9** (23): 4292–4298. Bibcode:1999EnST...33.4292W. doi:10.1021/es990537g.

[119] *Nobel Prize In Physics: Russia's Missed Opportunities*, RIA Novosti, Nov 21, 2006

[120] Sanders, R.; Kahn, J. (13 October 2006). “UC Berkeley, LBNL cosmologist George F. Smoot awarded 2006 Nobel Prize in Physics”. UC Berkeley News. Retrieved 2008-12-11.

[121] Kovac, J.M.; et al. (2002). “Detection of polarization in the cosmic microwave background using DASI”. *Nature*. **420** (6917): 772–787. Bibcode:2002Natur.420..772K. PMID 12490941. arXiv:astro-ph/0209478 . doi:10.1038/nature01269.

[122] Readhead, A. C. S.; et al. (2004). “Polarization Observations with the Cosmic Background Imager”. *Science*. **306** (5697): 836–844. Bibcode:2004Sci...306..836R. PMID 15472038. arXiv:astro-ph/0409569 . doi:10.1126/science.1105598.

[123] A. Readhead et al., “Polarization observations with the Cosmic Background Imager”, Science 306, 836-844 (2004).

[124] http://www.math.columbia.edu/~{ }woit/wordpress/?p=6865

[125] Cowen, Ron (2015-01-30). “Gravitational waves discovery now officially dead”. *nature*. doi:10.1038/nature.2015.16830.

[126] Cosmic Rebirth Encoded in Background Radiation?

30.12 Further reading

- Balbi, Amedeo (2008). *The music of the big bang : the cosmic microwave background and the new cosmology*. Berlin: Springer. ISBN 3540787267.
- Evans, Rhodri (2015). *The Cosmic Microwave Background: How It Changed Our Understanding of the Universe*. Springer. ISBN 9783319099279.

30.13 External links

- Student Friendly Intro to the CMB A pedagogic, step-by-step introduction to the cosmic microwave background power spectrum analysis suitable for those with an undergraduate physics background. More in depth than typical online sites. Less dense than cosmology texts.
- CMBR Theme on arxiv.org
- Audio: Fraser Cain and Dr. Pamela Gay – Astronomy Cast. The Big Bang and Cosmic Microwave Background – October 2006
- Visualization of the CMB data from the Planck mission
- Copeland, Ed. “CMBR: Cosmic Microwave Background Radiation”. *Sixty Symbols*. Brady Haran for the University of Nottingham.

Chapter 31

Angular diameter distance

The **angular diameter distance** is a distance measure used in astronomy. It is defined in terms of an object's physical size, x , and θ the angular size of the object as viewed from earth.

$d_A = \frac{x}{\theta}$ The angular diameter distance depends on the assumed cosmology of the universe. The angular diameter distance to an object at redshift, z , is expressed in terms of the comoving distance, r as:

$d_A = \frac{S_k(r)}{1+z}$ Where $S_k(r)$ is the FLRW coordinate defined as:

$$S_k(r) = \begin{cases} \sin\left(\sqrt{-\Omega_k} H_0 r\right) / \left(H_0 \sqrt{|\Omega_k|}\right) & \Omega_k < 0 \\ r & \Omega_k = 0 \\ \sinh\left(\sqrt{\Omega_k} H_0 r\right) / \left(H_0 \sqrt{|\Omega_k|}\right) & \Omega_k > 0 \end{cases}$$

Where Ω_k is the curvature density and H_0 is the value of the Hubble parameter today.

In the currently favoured geometric model of our Universe, the "angular diameter distance" of an object is a good approximation to the "real distance", i.e. the proper distance when the light left the object. Note that beyond a certain redshift, the angular diameter distance gets smaller with increasing redshift. In other words, an object "behind" another of the same size, beyond a certain redshift (roughly z=1.5), appears larger on the sky, and would therefore have a *smaller* "angular diameter distance".

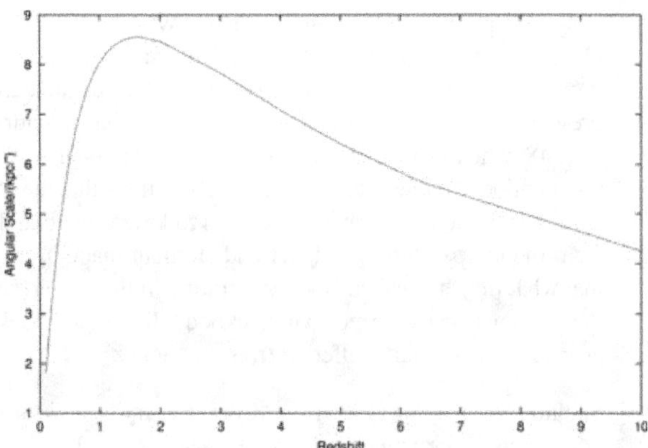

The angular size redshift relation for a Lambda cosmology, with on the vertical scale kiloparsecs per arcsecond.

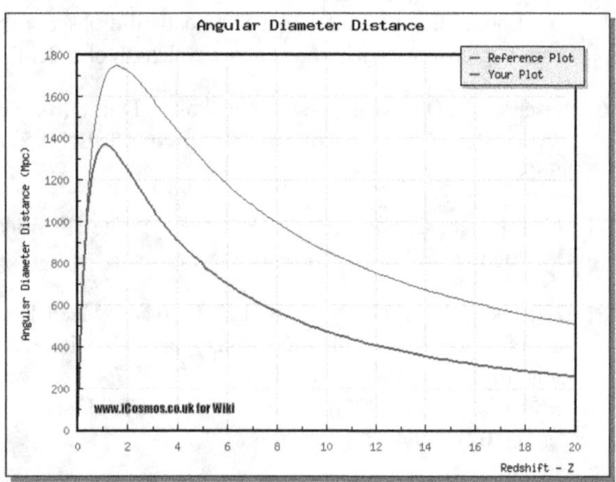

The angular size redshift relation for a Lambda cosmology, with on the vertical scale megaparsecs.

31.1 Angular size redshift relation

The **angular size redshift relation** describes the relation between the angular size observed on the sky of an object of given physical size, and the objects redshift from Earth (which is related to its distance, d , from Earth). In a Euclidean geometry the relation between size on the sky and distance from Earth would simply be given by the equation:

$\tan(\theta) = \frac{x}{d}$

where θ is the angular size of the object on the sky, x is the size of the object and d is the distance to the object. Where θ is small this approximates to:

$\theta \approx \frac{x}{d}$.

However, in the ΛCDM model (the currently favored cosmology), the relation is more complicated. In this model, objects at redshifts greater than about 1.5 appear larger on the sky with increasing redshift.

This is related to the angular diameter distance, which is the distance an object is calculated to be at from θ and x , assuming the Universe is Euclidean.

The actual relation between the angular-diameter distance, d_A , and redshift is given below. q_0 is called the deceleration parameter and measures the deceleration of the expansion rate of the Universe; in the simplest models, $q_0 < 0.5$ corresponds to the case where the Universe will expand for ever, $q_0 > 0.5$ to closed models which will ultimately stop expanding and contract $q_0 = 0.5$ corresponds to the critical case – Universes which will just be able to expand to infinity without re-contracting.

$$d_A = \frac{c}{H_0 q_0^2} \frac{(z q_0 + (q_0 - 1)(\sqrt{2 q_0 z + 1} - 1))}{(1+z)^2}$$

The Mattig relation yields the angular-diameter distance as a function of redshift for a universe with ΩΛ = 0.[1]

31.2 See also

- Distance measures (cosmology)
- Standard ruler

31.3 References

[1] An introduction to the science of cosmology, Chapter 6:2 by Derek J. Raine & Edwin George Thomas (2001)

31.4 External links

- iCosmos: Cosmology Calculator (With Graph Generation)

Chapter 32

Galaxy cluster

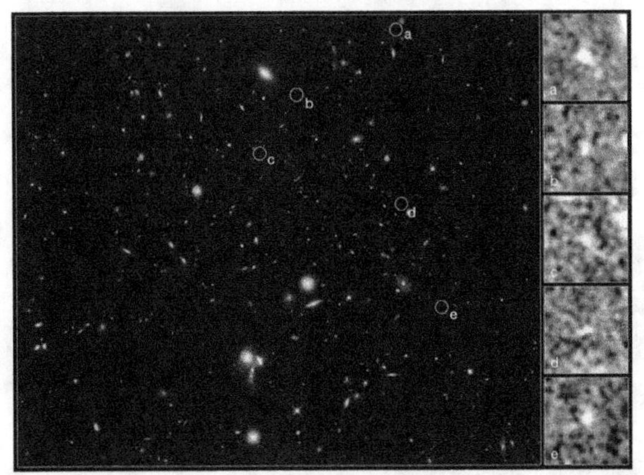

Composite image of five galaxies clustered together just 600 million years after the Universe's birth[1]

A **galaxy cluster**, or **cluster of galaxies**, is a structure that consists of anywhere from hundreds to thousands of galaxies that are bound together by gravity[1] with typical masses ranging from 10^{14}–10^{15} solar masses. They are the largest known gravitationally bound structures in the universe and were believed to be the largest known structures in the universe until the 1980s, when superclusters were discovered.[2] One of the key features of clusters is the intracluster medium (ICM). The ICM consists of heated gas between the galaxies and has a peak temperature between 2–15 keV that is dependent on the total mass of the cluster. Galaxy clusters should not be confused with star clusters, such as open clusters, which are structures of stars *within* galaxies, or with globular clusters, which typically orbit galaxies. Small aggregates of galaxies are referred to as groups of galaxies rather than clusters of galaxies. The groups and clusters can themselves cluster together to form superclusters.

Notable galaxy clusters in the relatively nearby Universe include the Virgo Cluster, Fornax Cluster, Hercules Cluster, and the Coma Cluster. A very large aggregation of galaxies known as the Great Attractor, dominated by the Norma Cluster, is massive enough to affect the local expansion of the Universe. Notable galaxy clusters in the distant, high-redshift Universe include SPT-CL J0546-5345 and SPT-CL J2106-5844, the most massive galaxy clusters found in the early Universe. In the last few decades, they are also found to be relevant sites of particle acceleration, a feature that has been discovered by observing non-thermal diffuse radio emissions, such as radio halos and radio relics. Using the Chandra X-ray Observatory, structures such as cold fronts and shock waves have also been found in many galaxy clusters.

32.1 Basic properties

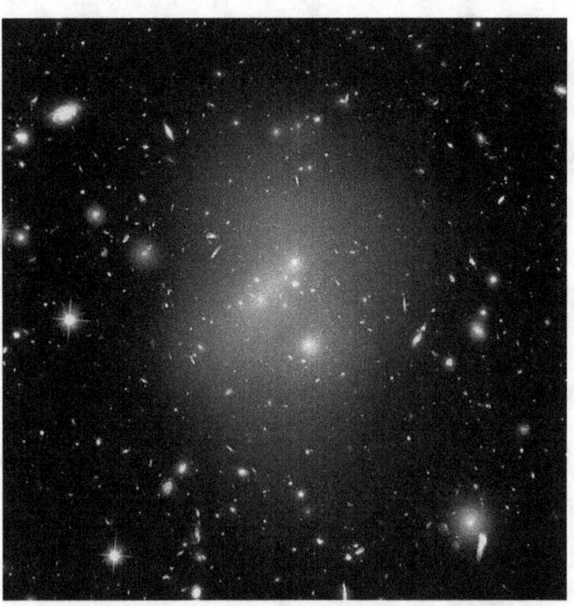

Galaxy cluster IDCS J1426 is located 10 billion light-years from Earth and weighs almost 500 trillion suns.[3]

Galaxy clusters typically have the following properties:

- They contain 100 to 1,000 galaxies, hot X-ray- emitting gas and large amounts of dark matter.[4] Details are described in the “Composition” section.

- The distribution of the three components is approximately the same in the cluster.
- They have total masses of 10^{14} to 10^{15} solar masses.
- They typically have a diameter from 2 to 10 Mpc (see 10^{23} m for distance comparisons).
- The spread of velocities for the individual galaxies is about 800–1000 km/s.

32.2 Composition

There are three main components of a galaxy cluster. They are tabulated below:

32.3 Classification

See also: Bautz-Morgan classification

Stars, Star clusters, Galaxies, Galaxy clusters, Super clusters

32.4 List

Main article: List of galaxy groups and clusters

32.5 Gallery

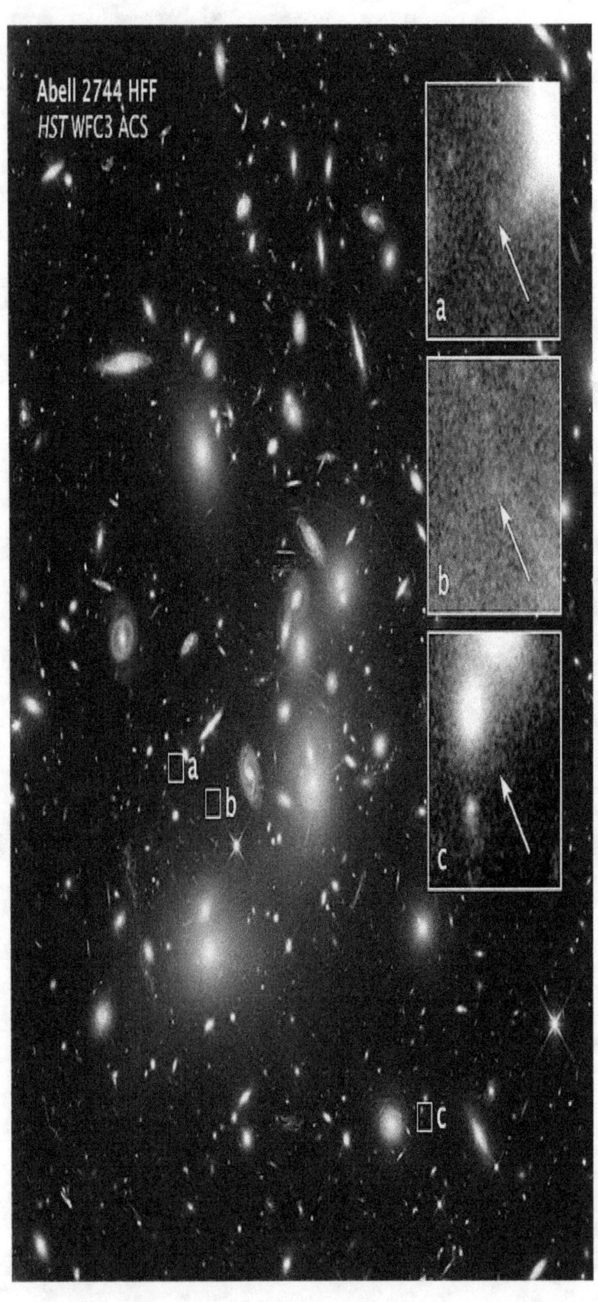

2744 galaxy cluster - extremely distant galaxies revealed by gravitational lensing (16 October 2014).[5][6]

32.5.1 Images

- "Smiley" image - galaxy cluster (SDSS J1038+4849) & gravitational lensing (an Einstein ring) (HST).[7]

- 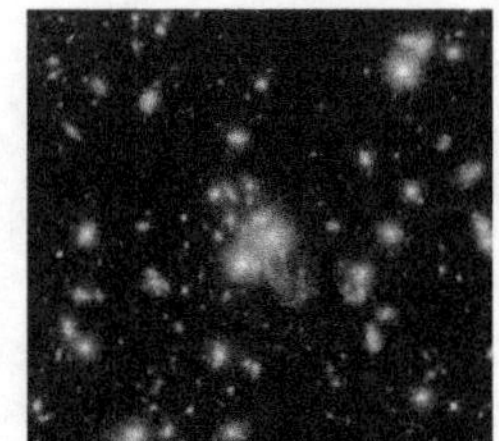Galaxy cluster SpARCS1049 taken by Spitzer and the Hubble Space Telescope.[8]

- Galaxy cluster MOO J1142+1527 discovered by the MaDCoWS survey

- Abell 2744 galaxy cluster (HST).[5]

- 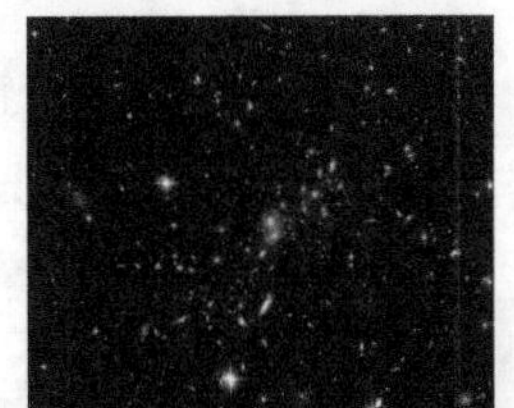Magnifying the distant universe through MACS J0454.1-0300.[9]

- Turbulence may prevent galaxy clusters from cooling ; illustrated: Perseus Cluster and Virgo Cluster (Chandra X-ray).

- 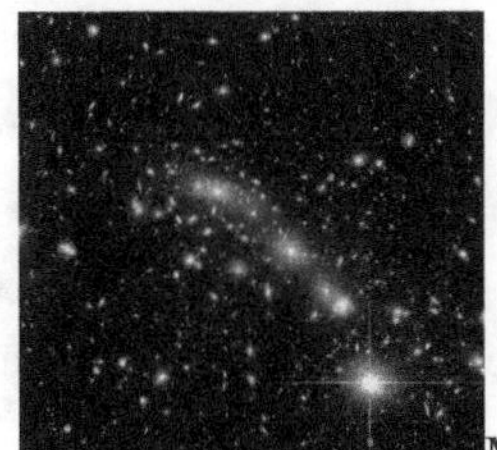MACS0416.1-2403 imaged by the HST

32.5.2 Videos

- Video: Formation of galaxy cluster MRC 1138-262 (artist's concept).

32.6 See also

- Abell catalogue
- Intracluster medium
- List of Abell clusters

32.7 References

[1] "Hubble Pinpoints Furthest Protocluster of Galaxies Ever Seen". *ESA/Hubble Press Release*. Retrieved 13 January 2012.

[2] Kravtsov, A. V.; Borgani, S. (2012). "Formation of Galaxy Clusters". *Annual Review of Astronomy and Astrophysics*. **50**: 353. Bibcode:2012ARA&A..50..353K. arXiv:1205.5556 . doi:10.1146/annurev-astro-081811-125502.

[3] "Galaxy cluster IDCS J1426". Retrieved 11 January 2016.

[4] http://chandra.harvard.edu/xray_sources/galaxy_clusters.html

[5] Clavin, Whitney; Jenkins, Ann; Villard, Ray (7 January 2014). "NASA's Hubble and Spitzer Team up to Probe Faraway Galaxies". *NASA*. Retrieved 8 January 2014.

[6] Chou, Felecia; Weaver, Donna (16 October 2014). "RELEASE 14-283 - NASA's Hubble Finds Extremely Distant Galaxy through Cosmic Magnifying Glass". *NASA*. Retrieved 17 October 2014.

[7] Loff, Sarah; Dunbar, Brian (10 February 2015). "Hubble Sees A Smiling Lens". *NASA*. Retrieved 10 February 2015.

[8] "Image of the galaxy cluster SpARCS1049". Retrieved 11 September 2015.

[9] "Magnifying the distant Universe". *ESA/Hubble Picture of the Week*. Retrieved 10 April 2014.

Chapter 33

Number density

In physics, astronomy, chemistry, biology and geography, **number density** (symbol: n or ϱN) is an intensive quantity used to describe the degree of concentration of countable objects (particles, molecules, phonons, cells, galaxies, etc.) in physical space: three-dimensional volume number density, two-dimensional area number density, or one-dimensional line number density. Population density is an example of areal number density. The term **number concentration** (symbol: C, to avoid confusion with amount of substance n) is sometimes used in chemistry for the same quantity, particularly when comparing with other concentrations.

33.1 Definition

Volume number density is the number of specified objects per unit volume:[1]

$$n = \frac{N}{V},$$

where N is the total number of objects in a volume V.

Here it is assumed[2] that N is large enough that rounding of the count to the nearest integer does not introduce much of an error, however V is chosen to be small enough that the resulting n does not depend much on the size or shape of the volume V.

33.2 Units

In SI units, number density is measured in m^{-3}, although cm^{-3} is often used. However, these units are not quite practical when dealing with atoms or molecules of gases, liquids or solids at room temperature and atmospheric pressure, because the resulting numbers are extremely large (on the order of 10^{20}). Using the number density of an ideal gas at 0 °C and 1 atm as a yardstick: n_0 = 1 amg = 2.686,777,4 × 10^{25} m^{-3} is often introduced as a unit of number density, for any substances at any conditions (not necessarily limited to an ideal gas at 0 °C and 1 atm).[3]

33.3 Usage

Using the number density as a function of spatial coordinates, the total number of objects N in the entire volume V can be calculated as

$$N = \iiint_V n(x,\, y,\, z)\, \mathrm{d}V,$$

where $\mathrm{d}V = \mathrm{d}x\ \mathrm{d}y\ \mathrm{d}z$ is a volume element. If each object possesses the same mass m_0, the total mass m of all the objects in the volume V can be expressed as

$$m = \iiint_V m_0 n(x,\, y,\, z)\, \mathrm{d}V.$$

Similar expressions are valid for electric charge or any other extensive quantity associated with countable objects. For example, replacing m with q (total charge) and m_0 with q_0 (charge of each object) in the above equation will lead to a correct expression for charge.

The number density of solute molecules in a solvent is sometimes called concentration, although usually concentration is expressed as a number of moles per unit volume (and thus called molar concentration).

33.4 Relation to other quantities

33.4.1 Molar concentration

For any substance, the number density can be expressed in terms of its amount concentration c (in mol/m^3) as

$$n = \mathrm{N_A} c,$$

where NA is the Avogadro constant. This is still true if the spatial dimension unit, metre, in both *n* and *c* is consistently replaced by any other spatial dimension unit, e.g. if *n* is in cm^{-3} and *c* is in mol/cm^3, or if *n* is in L^{-1} and *c* is in mol/L, etc.

33.4.2 Mass density

For atoms or molecules of a well-defined molar mass *M* (in kg/mol), the number density can be expressed in terms of their mass density ϱ_m (in kg/m^3) as

$$n = \frac{\mathrm{N_A}}{M} \rho_\mathrm{m}.$$

Note that the ratio *M*/NA is the mass of a single atom or molecule in kg.

33.5 Examples

The following table lists common examples of number densities at 1 atm and 20 °C, unless otherwise noted.

33.6 See also

- Columnar number density

33.7 References and notes

[1] IUPAC, *Compendium of Chemical Terminology*, 2nd ed. (the "Gold Book") (1997). Online corrected version: (2006–) "number concentration".

[2] Clayton T. Crowe; Martin Sommerfeld; Yutaka Tsuji (1998), *Multiphase flows with droplets and particles: allelochemical interactions*, CRC Press, p. 18, ISBN 0-8493-9469-4

[3] Joseph Kestin (1979), *A Course in Thermodynamics*, **2**, Taylor & Francis, p. 230, ISBN 0-89116-641-6

[4] For elemental substances, atomic densities/concentrations are used

Chapter 34

Phantom energy

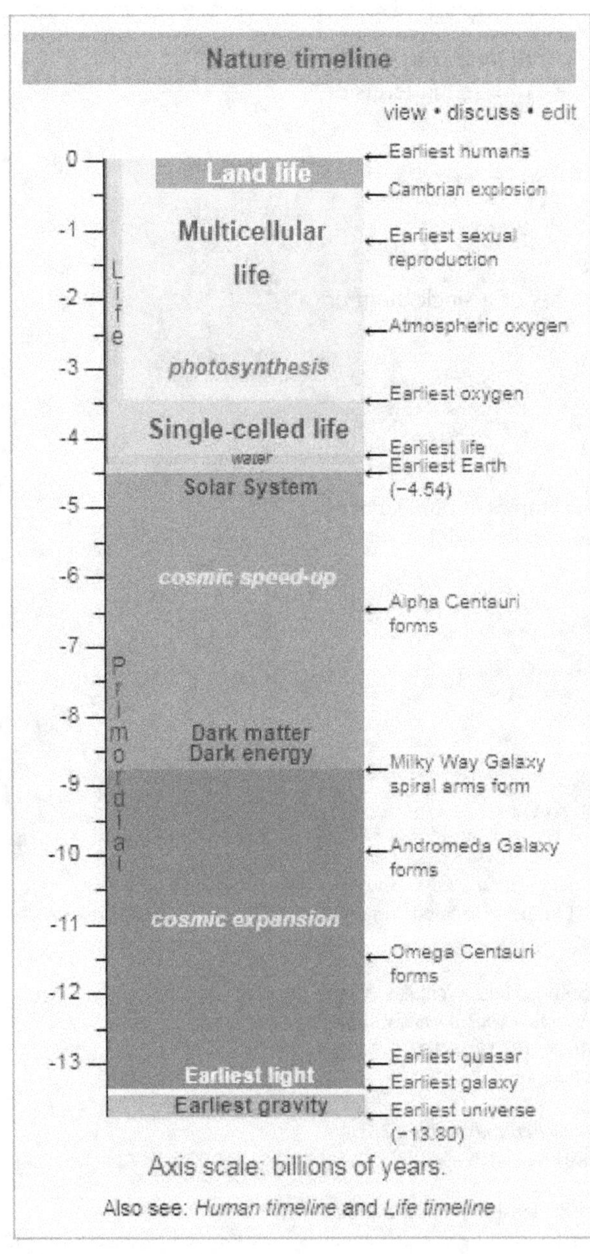

Phantom energy is a hypothetical form of dark energy satisfying the equation of state with $w < -1$. It possesses negative kinetic energy, and predicts expansion of the universe in excess of that predicted by a cosmological constant, which leads to a Big Rip.

34.1 Consequences

The existence of phantom energy could cause the expansion of the universe to accelerate so quickly that a scenario known as the Big Rip, a possible end to the universe.

34.1.1 Big Rip mechanism

Main article: Big Rip

The expansion of the universe reaches an infinite degree in finite time, causing expansion to accelerate without bounds. This acceleration necessarily passes the speed of light (since it involves expansion of the universe itself, not particles moving within it), causing more and more objects to leave our observable universe faster than its expansion, as light and information emitted from distant stars and other cosmic sources cannot "catch up" with the expansion. As the observable universe expands, objects will be unable to interact with each other via fundamental forces, and eventually the expansion will prevent any action of forces between any particles, even within atoms, "ripping apart" the universe.

One application of phantom energy in 2007 was to a cyclic model of the universe.[1]

34.2 References

[1] Lauris Baum and Paul Frampton (2007). "Turnaround In Cyclic Cosmology". *Phys. Rev. Lett.* **98** (7): 071301. Bibcode:2007PhRvL..98g1301B. PMID 17359014. arXiv:hep-th/0610213. doi:10.1103/PhysRevLett.98.071301.

34.3 Further reading

- Robert R. Caldwell et al.: Phantom Energy and Cosmic Doomsday

Chapter 35

Quintessence (physics)

For other forms of quintessence, see Quintessence (disambiguation). For theories and defunct or classical concepts named after the synonym "Aether", see Aether (disambiguation).

In physics, **quintessence** is a hypothetical form of dark energy, more precisely a scalar field, postulated as an explanation of the observation of an accelerating rate of expansion of the universe, rather than due to a true cosmological constant. The first example of this scenario was proposed by Ratra and Peebles (1988).[1] The concept was expanded to more general types of time-varying dark energy and the term "quintessence" was first introduced in a paper by Robert R. Caldwell, Rahul Dave and Paul Steinhardt.[2] It has been proposed by some physicists to be a fifth fundamental force. Quintessence differs from the cosmological constant explanation of dark energy in that it is dynamic; that is, it changes over time, unlike the cosmological constant which, by definition, does not change. It is suggested that quintessence can be either attractive or repulsive depending on the ratio of its kinetic and potential energy. Those working with this postulate believe that quintessence became repulsive about ten billion years ago, about 3.5 billion years after the Big Bang.[3]

35.1 Scalar field

Quintessence is a scalar field with an equation of state where wq, the ratio of pressure pq and density $\rho\, q$, is given by the potential energy $V(Q)$ and a kinetic term:

$$w_q = p_q/\rho_q = \frac{\frac{1}{2}\dot{Q}^2 - V(Q)}{\frac{1}{2}\dot{Q}^2 + V(Q)}$$

Hence, quintessence is dynamic, and generally has a density and wq parameter that varies with time. By contrast, a cosmological constant is static, with a fixed energy density and $wq = -1$.

35.2 Tracker behavior

Many models of quintessence have a *tracker* behavior, which according to Ratra and Peebles (1988) and Paul Steinhardt *et al.* (1999) partly solves the cosmological constant problem.[4] In these models, the quintessence field has a density which closely tracks (but is less than) the radiation density until matter-radiation equality, which triggers quintessence to start having characteristics similar to dark energy, eventually dominating the universe. This naturally sets the low scale of the dark energy.[5] When comparing the predicted expansion rate of the universe as given by the tracker solutions with cosmological data, a main feature of tracker solutions is that one needs four parameters to properly describe the behavior of their equation of state,[6][7] whereas it has been shown that at most a two-parameter model can optimally be constrained by mid-term future data (horizon 2015-2020).[8]

35.3 Specific models

Some special cases of quintessence are phantom energy, in which $wq < -1$,[9] and k-essence (short for kinetic quintessence), which has a non-standard form of kinetic energy. If this type of energy were to exist, it would cause a big rip[10] in the universe due to the growing energy density of dark energy which would cause the expansion of the universe to increase at a faster-than-exponential rate.

35.3.1 Holographic Dark Energy

Holographic Dark Energy models compared to Cosmological Constant models, imply a high degeneracy.[11] It has been suggested that dark energy might originate from quantum fluctuations of spacetime, and are limited by the event horizon of the universe.[12]

Studies with quintessence dark energy found that it dominates gravitational collapse in a spacetime simulation, based on the holographic thermalization. These results show that the smaller the state parameter of quintessence is, the harder it is for the plasma to thermalize.[13]

35.4 Quintom scenario

In 2004, when scientists fitted the evolution of dark energy with the cosmological data, they found that the equation of state had possibly crossed the cosmological constant boundary (w = –1) from above to below. A proven no-go theorem indicates this situation, called the Quintom scenario, requires at least two degrees of freedom for dark energy models.[14]

35.5 Terminology

The name comes from the classical elements in ancient Greece. The aether, a pure "fifth element" (*quinta essentia* in Latin), was thought to fill the universe beyond Earth. Similarly, modern quintessence would be the fifth known contribution to the overall mass-energy content of the universe. (The other four in the modern interpretation, different from the ancient ideas, are: baryonic matter; radiation – photons and the highly relativistic neutrinos, which may be considered hot dark matter; cold dark matter; and the term due to spatial curvature – loosely, gravitational self-energy.)

35.6 See also

- Dark-energy-dominated era

35.7 References

[1] Ratra, P.; Peebles, L. (1988). "Cosmological consequences of a rolling homogeneous scalar field". *Physical Review D*. **37** (12): 3406. Bibcode:1988PhRvD..37.3406R. doi:10.1103/PhysRevD.37.3406.

[2] Caldwell, R.R.; Dave, R.; Steinhardt, P.J. (1998). "Cosmological Imprint of an Energy Component with General Equation-of-State". *Phys. Rev. Lett.* **80** (8): 1582–1585. Bibcode:1998PhRvL..80.1582C. arXiv:astro-ph/9708069. doi:10.1103/PhysRevLett.80.1582.

[3] Christopher Wanjek; "Quintessence, accelerating the Universe?"; http://www.astronomytoday.com/cosmology/quintessence.html

[4] Zlatev, I.; Wang, L.; Steinhardt, P. (1999). "Quintessence, Cosmic Coincidence, and the Cosmological Constant". *Physical Review Letters*. **82** (5): 896–899. Bibcode:1999PhRvL..82..896Z. arXiv:astro-ph/9807002. doi:10.1103/PhysRevLett.82.896.

[5] Steinhardt, P.; Wang, L.; Zlatev, I. (1999). "Cosmological tracking solutions". *Physical Review D*. **59** (12): 123504. Bibcode:1999PhRvD..5913504S. arXiv:astro-ph/9812313. doi:10.1103/PhysRevD.59.123504.

[6] Linden, Sebastian; Virey, Jean-Marc (2008). "Test of the Chevallier-Polarski-Linder parametrization for rapid dark energy equation of state transitions". *Physical Review D*. **78** (2): 023526. Bibcode:2008PhRvD..78b3526L. arXiv:0804.0389. doi:10.1103/PhysRevD.78.023526.

[7] Ferramacho, L.; Blanchard, A.; Zolnierowsky, Y.; Riazuelo, A. (2010). "Constraints on dark energy evolution". *Astronomy & Astrophysics*. **514**: A20. Bibcode:2010A&A...514A..20F. arXiv:0909.1703. doi:10.1051/0004-6361/200913271.

[8] Linder, Eric V.; Huterer, Dragan (2005). "How many cosmological parameters". *Physical Review D*. **72** (4): 043509. Bibcode:2005PhRvD..72d3509L. arXiv:astro-ph/0505330. doi:10.1103/PhysRevD.72.043509.

[9] Caldwell, R. R. (2002). "A phantom menace? Cosmological consequences of a dark energy component with super-negative equation of state". *Physics Letters B*. **545** (1–2): 23–29. Bibcode:2002PhLB..545...23C. arXiv:astro-ph/9908168. doi:10.1016/S0370-2693(02)02589-3.

[10] Antoniou, Ioannis; Perivolaropoulos, Leandros (2016). "Geodesics of McVittie Spacetime with a Phantom Cosmological Background". *Phys. Rev. D*. **93** (12): 123520. Bibcode:2016PhRvD..9313520A. arXiv:1603.02569. doi:10.1103/PhysRevD.93.123520.

[11] Hu, Yazhou; Li, Miao; Li, Nan; Zhang, Zhenhui (2015). "Holographic Dark Energy with Cosmological Constant". *Journal of Cosmology and Astroparticle Physics*. **2015** (8): 012. Bibcode:2015JCAP...08..012H. arXiv:1502.01156. doi:10.1088/1475-7516/2015/08/012.

[12] Shan Gao (2013). "Explaining Holographic Dark Energy". *Galaxies*. **1** (3): 180. Bibcode:2013Galax...1..180G. doi:10.3390/galaxies1030180.

[13] Zeng, Xiao-Xiong; Chen, De-You; Li, Li-Fang (2014). "Holographic thermalization and gravitational collapse in the spacetime dominated by quintessence dark energy". *Physical Review D*. **91** (4): 046005. Bibcode:2015PhRvD..91d6005Z. arXiv:1408.6632. doi:10.1103/PhysRevD.91.046005.

[14] Hu, Wayne (2005). "Crossing the phantom divide: Dark energy internal degrees of freedom". *Physical Review D*. **71** (4): 047301. Bibcode:2005PhRvD..71d7301H. arXiv:astro-ph/0410680. doi:10.1103/PhysRevD.71.047301.

35.8 Further reading

- Ostriker JP; Steinhardt P (January 2001). "The Quintessential Universe". *Scientific American*. **284** (1): 46–53. PMID 11132422. doi:10.1038/scientificamerican0101-46.
- Lawrence M. Krauss (2000). *Quintessence: The Search for Missing Mass in the Universe*. Basic Books. ISBN 978-0465037414.

Chapter 36

Gravitational interaction of antimatter

The **gravitational interaction of antimatter** with matter or antimatter has not been conclusively observed by physicists. While the consensus among physicists is that gravity will attract both matter and antimatter at the same rate that matter attracts matter, there is a strong desire to confirm this experimentally.

Antimatter's rarity and tendency to annihilate when brought into contact with matter makes its study a technically demanding task. Most methods for the creation of antimatter (specifically antihydrogen) result in high-energy particles and atoms of high kinetic energy, which are unsuitable for gravity-related study. In recent years, first ALPHA [1][2] and then ATRAP [3] have trapped antihydrogen atoms at CERN; in 2012 ALPHA used such atoms to set the first free-fall loose bounds on the gravitational interaction of antimatter with matter, measured to within ±7500% of ordinary gravity[4], not enough for a clear scientific statement about the sign of gravity acting on antimatter. Future experiments need to be performed with higher precision, either with beams of antihydrogen (AEGIS or GBAR) or with trapped antihydrogen (ALPHA).

36.1 Three hypotheses

Thus far, there are three hypotheses about how *antimatter* gravitationally interacts *with normal matter*:

- **Normal gravity**: The standard assumption is that gravitational interactions of matter and antimatter are identical.
- **Antigravity**: Some authors argue that antimatter repels matter with the same magnitude as matter attracts itself. (see below).
- **Gravivector and graviscalar**: Later difficulties in creating quantum gravity theories have led to the idea that antimatter may react with a slightly different magnitude.[5]

36.2 Experiments

36.2.1 Supernova 1987A

One source of experimental evidence in favor of normal gravity was the observation of neutrinos from Supernova 1987A. In 1987, three neutrino detectors around the world simultaneously observed a cascade of neutrinos emanating from a supernova in the Large Magellanic Cloud. Although the supernova happened about 164,000 light years away, both neutrinos and antineutrinos may have been detected virtually simultaneously. If both were actually observed, then any difference in the gravitational interaction would have to be very small. However, neutrino detectors cannot distinguish perfectly between neutrinos and antineutrinos; in fact, the two may be identical. Some physicists conservatively estimate that there is less than a 10% chance that no regular neutrinos were observed at all. Others estimate even lower probabilities, some as low as 1%.[6] Unfortunately, this accuracy is unlikely to be improved by duplicating the experiment any time soon. The last known supernova to occur at such a close range prior to Supernova 1987A was around 1867.[7]

36.2.2 Fairbank's experiments

Physicist William Fairbank attempted a laboratory experiment to directly measure the gravitational acceleration of both electrons and positrons. However, their charge-to-mass ratio is so large that electromagnetic effects overwhelmed the experiment.

It is difficult to directly observe gravitational forces at the particle level. For charged particles, the electromagnetic force overwhelms the much weaker gravitational interaction. Even antiparticles in neutral antimatter, such as antihydrogen, must be kept separate from their counterparts in the matter that forms the experimental equipment, which requires strong electromagnetic fields. These fields, e.g. in the form of atomic traps, exert forces on these antiparticles which easily overwhelm the gravitational force of Earth and nearby test masses. Since all production methods for antiparticles result in high-energy antimatter particles, the necessary cooling for observation of gravitational effects in a laboratory environment requires very elaborate experimental techniques and very careful control of the trapping fields.

36.2.3 Cold neutral antihydrogen experiments

Since 2010 the production of cold antihydrogen has become possible at the Antiproton Decelerator at CERN. Antihydrogen, which is electrically neutral, should make it possible to directly measure the gravitational attraction of antimatter particles to the matter Earth. In 2013, experiments on antihydrogen atoms released from the ALPHA trap set direct, i.e. freefall, coarse limits on antimatter gravity.[4] These limits were coarse, with a relative precision of ± 100%, thus far from a clear statement even for the sign of gravity acting on antimatter. Future experiments at CERN with beams of antihydrogen, such as AEGIS and GBAR, or with trapped antihydrogen such as ALPHA, have to improve the sensitivity to make a clear, scientific statement about gravity on antimatter.[8]

36.3 Arguments against a gravitational repulsion of matter and antimatter

When antimatter was first discovered in 1932, physicists wondered about how it would react to gravity. Initial analysis focused on whether antimatter should react the same as matter or react oppositely. Several theoretical arguments arose which convinced physicists that antimatter would react exactly the same as normal matter. They inferred that a gravitational repulsion between matter and antimatter was implausible as it would violate CPT invariance, conservation of energy, result in vacuum instability, and result in CP violation. It was also theorized that it would be inconsistent with the results of the Eötvös test of the weak equivalence principle. Many of these early theoretical objections were later overturned.[9]

36.3.1 The equivalence principle

The equivalence principle predicts that the gravitational acceleration of antimatter is the same as that of ordinary matter. A matter-antimatter gravitational repulsion is thus excluded from this point of view. Furthermore, photons, which are their own antiparticles in the framework of the Standard Model, have in a large number of astronomical tests (gravitational redshift and gravitational lensing, for example) been observed to interact with the gravitational field of ordinary matter exactly as predicted by the general theory of relativity. This is a feature that has to be explained by any theory predicting that matter and antimatter repel.

36.3.2 CPT theorem

The CPT theorem implies that the difference between the properties of a matter particle and those of its antimatter counterpart is *completely* described by C-inversion. Since this C-inversion doesn't affect gravitational mass, the CPT theorem predicts that the gravitational mass of antimatter is the same as that of ordinary matter.[10] A repulsive gravity is then excluded, since that would imply a difference in sign between the observable gravitational mass of matter and antimatter.

36.3.3 Morrison's argument

In 1958, Philip Morrison argued that antigravity would violate conservation of energy. If matter and antimatter responded oppositely to a gravitational field, then it would take no energy to change the height of a particle-antiparticle pair. However, when moving through a gravitational potential, the frequency and energy of light is shifted. Morrison argued that energy would be created by producing matter and antimatter at one height and then annihilating it higher up, since the photons used in production would have less energy than the photons yielded from annihilation.[11] However, it was later found that antigravity would still not violate the second law of thermodynamics.[12]

36.3.4 Schiff's argument

Later in 1958, L. Schiff used quantum field theory to argue that antigravity would be inconsistent with the results of the Eötvös experiment.[13] However, the renormalization technique used in Schiff's analysis is heavily criticized, and his work is seen as inconclusive.[9] In 2014 the argument was redone by Cabbolet, who concluded however that it merely demonstrates the incompatibility of the Standard Model and gravitational repulsion.[14]

36.3.5 Good's argument

In 1961, Myron L. Good argued that antigravity would result in the observation of an unacceptably high amount of CP violation in the anomalous regeneration of kaons.[15] At the time, CP violation had not yet been observed. However, Good's argument is criticized for being expressed in terms of absolute potentials. By rephrasing the argument in terms of relative potentials, Gabriel Chardin found that it resulted in an amount of kaon regeneration which agrees with observation.[16] He argues that antigravity is in fact a potential explanation for CP violation based on his models on K mesons. His results date back to 1992. Since then however, studies on CP violation mechanisms in the B mesons systems have fundamentally invalidated these explanations.

36.3.6 Gerard 't Hooft's argument

According to Gerard 't Hooft, every physicist recognizes immediately what is wrong with the idea of gravitational repulsion: if a ball is thrown high up in the air so that it falls back, then its motion is symmetric under time-reversal; and therefore, the ball falls also down in opposite time-direction.[17] Since a matter particle in opposite time-direction is an antiparticle, this proves according to 't Hooft that antimatter falls down on earth just like "normal" matter. However, Cabbolet replied that 't Hooft's argument is false, and only proves that an anti-ball falls down on an anti-earth - which is not disputed.[18]

36.4 Theories of gravitational repulsion

As long as repulsive gravity has not been refuted experimentally, one can speculate about physical principles that would bring about such a repulsion. Thus far, three radically different theories have been published:

- The first theory of repulsive gravity was a quantum theory published by Kowitt.[19] In this modified Dirac theory, Kowitt postulated that the positron is not a hole in the sea of electrons-with-negative-energy as in usual Dirac hole theory, but instead is a hole in the sea of electrons-with-negative-energy-and-positive-gravitational-mass: this yields a modified C-inversion, by which the positron has positive energy but negative gravitational mass. Repulsive gravity is then described by adding extra terms ($m_g\Phi_g$ and m_gA_g) to the wave equation. The idea is that the wave function of a positron moving in the gravitational field of a matter particle evolves such that in time it becomes more probable to find the positron further away from the matter particle.

- Classical theories of repulsive gravity have been published by Santilli and Villata.[20][21][22][23] Both theories are extensions of General Relativity, and are experimentally indistinguishable. The general idea remains that gravity is the deflection of a continuous particle trajectory due to the curvature of spacetime, but antiparticles now 'live' in an inverted spacetime. The equation of motion for antiparticles is then obtained from the equation of motion of ordinary particles by applying the C, P, and T-operators (Villata) or by applying *isodual maps* (Santilli), which amounts to the same thing: the equation of motion for antiparticles then predicts a repulsion of matter and antimatter. It has to be taken that the *observed* trajectories of antiparticles are projections on *our* spacetime of the true trajectories in the inverted spacetime. However, it has been argued on methodological and ontological grounds that the area of application of Villata's theory cannot be extended to include the microcosmos.[24] These objections were subsequently dismissed by Villata.[25]

- The first non-classical, non-quantum physical principles underlying a matter-antimatter gravitational repulsion have been published by Cabbolet.[10][26] He introduces the Elementary Process Theory, which uses a new language for physics, i.e. a new mathematical formalism and new physical concepts, and which is incompatible with both quantum mechanics and general relativity. The core idea is that nonzero rest mass particles such as electrons, protons, neutrons and their antimatter counterparts exhibit stepwise motion as they alternate between a particlelike state of rest and a wavelike state of motion. Gravitation then takes place in a wavelike state, and the theory allows, for example, that the wavelike states of protons and antiprotons interact differently with the earth's gravitational field.

Further authors[27][28][29] have used a matter-antimatter gravitational repulsion to explain cosmological observations, but these publications do not address the physical principles of gravitational repulsion.

36.5 See also

- AEgIS
- Dark energy
- Dark matter
- General relativity (where gravity is a curvature of spacetime caused by matter and energy)

36.6 References

[1] Andresen, G. B.; Ashkezari, M. D.; Baquero-Ruiz, M.; Bertsche, W.; Bowe, P. D.; Butler, E.; Cesar, C. L.; Chapman, S.; Charlton, M.; Deller, A.; Eriksson, S.; Fajans, J.; Friesen, T.; Fujiwara, M. C.; Gill, D. R.; Gutierrez, A.; Hangst, J. S.; Hardy, W. N.; Hayden, M. E.; Humphries, A. J.; Hydomako, R.; Jenkins, M. J.; Jonsell, S.; Jørgensen, L. V.; Kurchaninov, L.; Madsen, N.; Menary, S.; Nolan, P.; Olchanski, K.; Olin, A. (2010). "Trapped antihydrogen". *Nature*. **468** (7324): 673–676. Bibcode:2010Natur.468..673A. PMID 21085118. doi:10.1038/nature09610.

[2] Andresen, G. B.; Ashkezari, M. D.; Baquero-Ruiz, M.; Bertsche, W.; Bowe, P. D.; Butler, E.; Cesar, C. L.; Charlton, M.; Deller, A.; Eriksson, S.; Fajans, J.; Friesen, T.; Fujiwara, M. C.; Gill, D. R.; Gutierrez, A.; Hangst, J. S.; Hardy, W. N.; Hayano, R. S.; Hayden, M. E.; Humphries, A. J.; Hydomako, R.; Jonsell, S.; Kemp, S. L.; Kurchaninov, L.; Madsen, N.; Menary, S.; Nolan, P.; Olchanski, K.; Olin, A.; et al. (2011). "Confinement of antihydrogen for 1,000 seconds". *Nature Physics*. **7** (7): 558–564. Bibcode:2011NatPh...7..558A. arXiv:1104.4982. doi:10.1038/NPHYS2025.

[3] Gabrielse, G.; Kalra, R.; Kolthammer, W. S.; McConnell, R.; Richerme, P.; Grzonka, D.; Oelert, W.; Sefzick, T.; Zielinski, M.; Fitzakerley, D. W.; George, M. C.; Hessels, E. A.; Storry, C. H.; Weel, M.; Müllers, A.; Walz, J. (2012). "Trapped Antihydrogen in Its Ground State". *Physical Review Letters*. **108** (11): 113002. Bibcode:2012PhRvL.108k3002G. PMID 22540471. arXiv:1201.2717. doi:10.1103/PhysRevLett.108.113002.

[4] Amole, C.; Ashkezari, M. D.; Baquero-Ruiz, M.; Bertsche, W.; Butler, E.; Capra, A.; Cesar, C. L.; Charlton, M.; Eriksson, S.; Fajans, J.; Friesen, T.; Fujiwara, M. C.; Gill, D. R.; Gutierrez, A.; Hangst, J. S.; Hardy, W. N.; Hayden, M. E.; Isaac, C. A.; Jonsell, S.; Kurchaninov, L.; Little, A.; Madsen, N.; McKenna, J. T. K.; Menary, S.; Napoli, S. C.; Nolan, P.; Olin, A.; Pusa, P.; Rasmussen, C. Ø; Robicheaux, F.; Sarid, E.; Silveira, D. M.; So, C.; Thompson, R. I.; van der Werf, D. P.; Wurtele, J. S.; Zhmoginov, A. I.; Charman, A. E. (2013). "Description and first application of a new technique to measure the gravitational mass of antihydrogen". *Nature Communications*. **4**: 1785. Bibcode:2013NatCo...4E1785A. PMC 3644108. PMID 23653197. doi:10.1038/ncomms2787.

[5] Nieto, M. M.; Hughes, R. J.; Goldman, T. (March 1988). "Gravity and Antimatter". *Scientific American*. Retrieved December 21, 2016. (Subscription required (help)).

[6] Pakvasa, S.; Simmons, W. A.; Weiler, T. J. (1989). "Test of equivalence principle for neutrinos and antineutrinos". *Physical Review D*. **39** (6): 1761–1763. doi:10.1103/PhysRevD.39.1761.

[7] Reynolds, S. P.; Borkowski, K. J.; Green, D. A.; Hwang, U.; Harrus, I.; Petre, R. (2008). "The Youngest Galactic Supernova Remnant: G1.9+0.3". *The Astrophysical Journal*. **680** (1): L41–L44. Bibcode:2008ApJ...680L..41R. doi:10.1086/589570.

[8] Amos, J. (2011-06-06). "Antimatter atoms are corralled even longer". BBC News Online. Retrieved 2013-09-03.

[9] Nieto, M. M.; Goldman, T. (1991). "The arguments against 'antigravity' and the gravitational acceleration of antimatter". *Physics Reports*. **205** (5): 221–281. doi:10.1016/0370-1573(91)90138-C. Note: errata issued in 1992 in volume 216.

[10] Cabbolet, M. J. T. F. (2010). "Elementary Process Theory: a formal axiomatic system with a potential application as a foundational framework for physics supporting gravitational repulsion of matter and antimatter". *Annalen der Physik*. **522** (10): 699–738. doi:10.1002/andp.201000063.

[11] Morrison, P. (1958). "Approximate Nature of Physical Symmetries". *American Journal of Physics*. **26** (6): 358–368. doi:10.1119/1.1996159.

[12] Chardin, G. (1993). "CP violation and antigravity (revisited)". *Nuclear Physics A*. **558**: 477–495. doi:10.1016/0375-9474(93)90415-T.

[13] Schiff, L. I. (1958). "Sign of the Gravitational Mass of a Positron". *Physical Review Letters*. **1** (7): 254–255. Bibcode:1958PhRvL...1..254S. doi:10.1103/PhysRevLett.1.254.

[14] Cabbolet, M. J. T. F. (2014). "Incompatibility of QED/QCD and repulsive gravity, and implications for some recent approaches to dark energy". *Astrophysics and Space Science*. **350** (2): 777–780. doi:10.1007/s10509-014-1791-4.

[15] Good, M. L. (1961). "$K_2{}^0$ and the Equivalence Principle". *Physical Review*. **121** (1): 311–313. doi:10.1103/PhysRev.121.311.

[16] Chardin, G.; Rax, J.-M. (1992). "*CP* violation. A matter of (anti)gravity?". *Physics Letters B*. **282** (1–2): 256–262. doi:10.1016/0370-2693(92)90510-B.

[17] G. 't Hooft, Spookrijders in de wetenschap (in Dutch), DUB (2014)

[18] M.J.T.F. Cabbolet, 't Hooft slaat plank mis over spookrijders (in Dutch), DUB (2014)

[19] Kowitt, M. (1996). "Gravitational repulsion and Dirac antimatter". *International Journal of Theoretical Physics*. **35** (3): 605–631. doi:10.1007/BF02082828.

[20] Santilli, R.M. (1999). "A classical isodual theory of antimatter and its prediction of antigravity". *International Journal of Modern Physics A*. **14** (14): 2205–2238. doi:10.1142/S0217751X99001111.

[21] Villata, M. (2011). "CPT symmetry and antimatter gravity in general relativity". *EPL*. **94** (2): 20001. doi:10.1209/0295-5075/94/20001.

[22] Villata, M. (2013). "On the nature of dark energy: the lattice Universe". *Astrophysics and Space Science*. **345** (1): 1–9. doi:10.1007/s10509-013-1388-3.

[23] Villata, M. (2015). "The matter-antimatter interpretation of Kerr spacetime". *Annalen der Physik*. **527** (7–8): 507–512. doi:10.1002/andp.201500154.

[24] Cabbolet, M. J. T. F. (2011). "Comment to a paper of M. Villata on antigravity". *Astrophysics and Space Science*. **337** (1): 5–7. doi:10.1007/s10509-011-0939-8.

[25] Villata, M. (2011). "Reply to 'Comment to a paper of M. Villata on antigravity'". *Astrophysics and Space Science*. **337** (1): 15–17. doi:10.1007/s10509-011-0940-2.

[26] Cabbolet, M. J. T. F. (2011). "Addendum to the Elementary Process Theory". *Annalen der Physik*. **523** (12): 990–994. doi:10.1002/andp.201100194.

[27] Blanchet, L.; Le Tiec, A. (2008). "Model of dark matter and dark energy based on gravitational polarization". *Physical Review D*. **78** (2). doi:10.1103/PhysRevD.78.024031.

[28] Hajdukovic, D. S. (2011). "Is dark matter an illusion created by the gravitational polarization of the quantum vacuum?". *Astrophysics and Space Science*. **334** (2): 215–218. doi:10.1007/s10509-011-0744-4.

[29] Benoit-Lévy, A.; Chardin, G. (2012). "Introducing the Dirac-Milne universe". *Astronomy and Astrophysics*. **537**: A78. Bibcode:2012A&A...537A..78B. doi:10.1051/0004-6361/201016103.

Chapter 37

Future of an expanding universe

"Big Freeze" redirects here. For other uses, see Big Freeze (disambiguation).

Observations suggest that the expansion of the universe will continue forever. If so, then a popular theory is that the universe will cool as it expands, eventually becoming too cold to sustain life. For this reason, this future scenario is popularly called the *Big Freeze* or *Heat Death*.[1]

If dark energy—represented by the cosmological constant, a *constant* energy density filling space homogeneously,[2] or scalar fields, such as quintessence or moduli, *dynamic* quantities whose energy density can vary in time and space—accelerates the expansion of the universe, then the space between clusters of galaxies will grow at an increasing rate. Redshift will stretch ancient, incoming photons (even gamma rays) to undetectably long wavelengths and low energies.[3] Stars are expected to form normally for 10^{12} to 10^{14} (1–100 trillion) years, but eventually the supply of gas needed for star formation will be exhausted. As existing stars run out of fuel and cease to shine, the universe will slowly and inexorably grow darker, one star at a time.[4][5] According to theories that predict proton decay, the stellar remnants left behind will disappear, leaving behind only black holes, which themselves eventually disappear as they emit Hawking radiation.[6] Ultimately, if the universe reaches a state in which the temperature approaches a uniform value, no further work will be possible, resulting in a final heat death of the universe.[7]

37.1 Cosmology

L
i
f
e

Infinite expansion does not determine the spatial curvature of the universe. It can be open (with negative spatial curvature), flat, (positive spatial curvature), although if it is closed, sufficient dark energy must be present to counteract the gravitational forces. Open and flat universes will expand forever even in the absence of dark energy.[8]

Observations of the cosmic background radiation by the Wilkinson Microwave Anisotropy Probe and the Planck mission suggest that the universe is spatially flat and has a significant amount of dark energy.[9][10] In this case, the universe should continue to expand at an accelerating rate. The acceleration of the universe's expansion has also been confirmed by observations of distant supernovae.[8] If, as in the concordance model of physical cosmology (Lambda-cold dark matter or ΛCDM), the dark energy is in the form of a cosmological constant, the expansion will eventually become exponential, with the size of the universe doubling at a constant rate.

If the theory of inflation is true, the universe went through an episode dominated by a different form of dark energy in the first moments of the Big Bang; but inflation ended, indicating an equation of state much more complicated than those assumed so far for present-day dark energy. It is possible that the dark energy equation of state could change again resulting in an event that would have consequences which are extremely difficult to parametrize or predict.

37.2 Future history

In the 1970s, the future of an expanding universe was studied by the astrophysicist Jamal Islam[11] and the physicist Freeman Dyson.[12] Then, in their 1999 book The Five Ages of the Universe, the astrophysicists Fred Adams and Gregory Laughlin have divided the past and future history of an expanding universe into five eras. The first, the *Primordial Era*, is the time in the past just after the Big Bang when stars had not yet formed. The second, the *Stelliferous Era*, includes the present day and all of the stars and galaxies we see. It is the time during which stars form from collapsing clouds of gas. In the subsequent *Degenerate Era*, the stars will have burnt out, leaving all stellar-mass objects as stellar remnants—white dwarfs, neutron stars, and black holes. In the *Black Hole Era*, white dwarfs, neutron stars, and other smaller astronomical objects have been destroyed by proton decay, leaving only black holes. Finally, in the *Dark Era*, even black holes have disappeared, leaving only a dilute gas of photons and leptons.[13]

This future history and the timeline below assume the continued expansion of the universe. If the universe begins to recontract, subsequent events in the timeline may not occur because the Big Crunch, the recontraction of the universe into a hot, dense state similar to that after the Big Bang, will supervene.[13][14]

37.3 Timeline

For the past, including the Primordial Era, see Chronology of the universe.

37.3.1 Stelliferous Era

The observable universe is currently 1.38×10^{10} (13.8 billion) years old.[15] This time is in the Stelliferous Era. About 155 million years after the Big Bang, the first star formed. Since then, stars have formed by the collapse of small, dense core regions in large, cold molecular clouds of hydrogen gas. At first, this produces a protostar, which is hot and bright because of energy generated by gravitational contraction. After the protostar contracts for a while, its center will become hot enough to fuse hydrogen and its lifetime as a star will properly begin.[13]

Stars of very low mass will eventually exhaust all their fusible hydrogen and then become helium white dwarfs.[16] Stars of low to medium mass, such as our own sun, will expel some of their mass as a planetary nebula and eventually become white dwarfs; more massive stars will explode in a core-collapse supernova, leaving behind neutron stars or black holes.[17] In any case, although some of the star's matter may be returned to the interstellar medium, a degenerate remnant will be left behind whose mass is not returned to the interstellar medium. Therefore, the supply of gas available for star formation is steadily being exhausted.

Milky Way Galaxy and the Andromeda Galaxy merge into one

4–8 billion years from now (17.7 – 21.7 billion years after the Big Bang)

Main article: Andromeda–Milky Way collision

The Andromeda Galaxy is currently approximately 2.5 million light years away from our galaxy, the Milky Way Galaxy, and they are moving towards each other at approximately 300 kilometers (186 miles) per second. Approximately five billion years from now, or 19 billion years after the Big Bang, the Milky Way and the Andromeda Galaxy will collide with one another and merge into one large galaxy based on current evidence. Up until 2012, there was no way to know whether the possible collision was definitely going to happen or not.[18] In 2012, researchers came to the conclusion that the collision is definite after using the Hubble Space Telescope between 2002 and 2010 to track the motion of Andromeda.[19]

Coalescence of Local Group and galaxies outside the Local Group are no longer accessible

10^{11} (100 billion) to 10^{12} (1 trillion) years

The galaxies in the Local Group, the cluster of galaxies which includes the Milky Way and the Andromeda Galaxy, are gravitationally bound to each other. It is expected that between 10^{11} (100 billion) and 10^{12} (1 trillion) years from now, their orbits will decay and the entire Local Group will merge into one large galaxy.[4]

Assuming that dark energy continues to make the universe expand at an accelerating rate, in about 150 billion years all galaxies outside the Local Group will pass behind the cosmological horizon. It will then be impossible for events in the Local Group to affect other galaxies. Similarly it will be impossible for events after 150 billion years, as seen by observers in distant galaxies, to affect events in the Local Group.[3] However, an observer in the Local Group will continue to see distant galaxies, but events they observe will become exponentially more time dilated (and red shifted[3]) as the galaxy approaches the horizon until time in the distant galaxy seems to stop. The observer in the Local Group never actually sees the distant galaxy pass beyond the horizon and never observes events after 150 billion years in their local time. Therefore, after 150 billion years intergalactic transportation and communication beyond the Local Group becomes causally impossible.

Luminosities of galaxies begin to diminish

$8x10^{11}$ (800 billion) years

$8x10^{11}$ (800 billion) years from now, the luminosities of the different galaxies, approximately similar until then to the current ones thanks to the increasing luminosity of the remaining stars as they age, will start to decrease, as the less massive red dwarf stars begin to die as white dwarfs.[20]

Galaxies outside the Local Supercluster are no longer detectable

2×10^{12} (2 trillion) years

2×10^{12} (2 trillion) years from now, all galaxies outside the Local Supercluster will be red-shifted to such an extent that even gamma rays they emit will have wavelengths longer than the size of the observable universe of the time. Therefore, these galaxies will no longer be detectable in any way.[3]

37.3.2 Degenerate Era

From 10^{14} (100 trillion) to 10^{40} (10 duodecillion) years

By 10^{14} (100 trillion) years from now, star formation will end,[4] leaving all stellar objects in the form of degenerate remnants. If protons do not decay, stellar-mass objects will disappear more slowly, making this era last longer.

Star formation ceases

10^{14} (100 trillion) years

By 10^{14} (100 trillion) years from now, star formation will end. This period, known as the Degenerate Era, will last until the degenerate remnants finally decay.[21] The least massive stars take the longest to exhaust their hydrogen fuel (see stellar evolution). Thus, the longest living stars in the universe are low-mass red dwarfs, with a mass of about 0.08 solar masses ($M\odot$), which have a lifetime of order 10^{13} (10 trillion) years.[22] Coincidentally, this is comparable to the length of time over which star formation takes place.[4] Once star formation ends and the least massive red dwarfs exhaust their fuel, nuclear fusion will cease. The low-mass red dwarfs will cool and become black dwarfs.[16] The only objects remaining with more than planetary mass will be brown dwarfs, with mass less than 0.08 $M\odot$, and degenerate remnants; white dwarfs, produced by stars with initial masses between about 0.08 and 8 solar masses; and neutron stars and black holes, produced by stars with initial masses over 8 $M\odot$. Most of the mass of this collection, approximately 90%, will be in the form of white dwarfs.[5] In the absence of any energy source, all of these formerly luminous bodies will cool and become faint.

The universe will become extremely dark after the last star burns out. Even so, there can still be occasional light in the universe. One of the ways the universe can be illuminated is if two carbon–oxygen white dwarfs with a combined mass of more than the Chandrasekhar limit of about 1.4 solar masses happen to merge. The resulting object will then undergo runaway thermonuclear fusion, producing a Type Ia supernova and dispelling the darkness of the Degenerate Era for a few weeks.[23][24] If the combined mass is not above the Chandrasekhar limit but is larger than the minimum mass to fuse carbon (about 0.9 $M\odot$), a carbon star could be produced, with a lifetime of around 10^6 (1 million) years.[13] Also, if two helium white dwarfs with a combined mass of at least 0.3 $M\odot$ collide, a helium star may be produced, with a lifetime of a few hundred million years.[13] Finally brown dwarfs can form new stars colliding with each other to form a red dwarf star, that can survive for 10^{13} (10 trillion) years,[22][23] or accreting gas at very slow rates from the remaining interstellar medium until they have enough mass to start hydrogen burning as red dwarfs too. This process, at least on white dwarfs, could induce Type Ia supernovae too.[25]

Planets fall or are flung from orbits by a close encounter with another star

10^{15} (1 quadrillion) years

Over time, the orbits of planets will decay due to gravitational radiation, or planets will be ejected from their local systems by gravitational perturbations caused by encounters with another stellar remnant.[26]

Stellar remnants escape galaxies or fall into black holes

10^{19} to 10^{20} (10 to 100 quintillion) years

Over time, objects in a galaxy exchange kinetic energy in a process called dynamical relaxation, making their velocity distribution approach the Maxwell–Boltzmann distribution.[27] Dynamical relaxation can proceed either by close encounters of two stars or by less violent but more frequent distant encounters.[28] In the case of a close encounter, two brown dwarfs or stellar remnants will pass close to each other. When this happens, the trajectories of the objects involved in the close encounter change slightly. After a large number of encounters, lighter objects tend to gain kinetic energy while the heavier objects lose it.[13]

Because of dynamical relaxation, some objects will gain enough energy to reach galactic escape velocity and depart the galaxy, leaving behind a smaller, denser galaxy. Since encounters are more frequent in the denser galaxy, the process then accelerates. The end result is that most objects (90% to 99%) are ejected from the galaxy, leaving a small fraction (maybe 1% to 10%) which fall into the central supermassive black hole.[4][13] It has been suggested that the matter of the fallen remnants will form an accretion disk around it that will create a quasar, as long as enough matter is present there.[29]

Nucleons start to decay

See also: Nucleon
Chance: 10^{34} (10 decillion) < 10^{39} years (1 duodecillion)

The subsequent evolution of the universe depends on the possibility and rate of proton decay. Experimental evidence shows that if the proton is unstable, it has a half-life of at least 10^{34} years.[30] Some of the Grand Unified theories

(GUTs) predict long-term proton instability between 10^{31} and 10^{36} years, with the upper bound on standard (non-SUSY) proton decay at 1.4 x 10^{36} years and an overall upper limit maximum for any proton decay (including SUSY models) at 6 x 10^{39} years.[31][32] Recent research showing proton lifetime (if unstable) at or exceeding $10^{34}-10^{35}$ year range rules out simpler GUTs and most non-SUSY models.

Neutrons bound into nuclei are also expected to decay with a half-life comparable to that of protons. Planets (substellar objects) would decay in a simple cascade process from heavier elements to pure hydrogen while radiating energy.[33]

In the event that the proton does not decay at all, stellar objects would still disappear, but more slowly. See Future without proton decay below.

Shorter or longer proton half-lives will accelerate or decelerate the process. This means that after 10^{37} years (the maximum proton half-life used by Adams & Laughlin (1997)), one-half of all baryonic matter will have been converted into gamma ray photons and leptons through proton decay.

All nucleons decay

10^{40} (10 duodecillion) years

Given our assumed half-life of the proton, nucleons (protons and bound neutrons) will have undergone roughly 1,000 half-lives by the time the universe is 10^{40} years old. To put this into perspective, there are an estimated 10^{80} protons currently in the universe.[34] This means that the number of nucleons will be slashed in half 1,000 times by the time the universe is 10^{40} years old. Hence, there will be roughly $½^{1,000}$ (approximately 10^{-301}) as many nucleons remaining as there are today; that is, *zero* nucleons remaining in the universe at the end of the Degenerate Age. Effectively, all baryonic matter will have been changed into photons and leptons. Some models predict the formation of stable positronium atoms with a greater diameter than the observable universe's current diameter in 10^{85} years, and that these will in turn decay to gamma radiation in 10^{141} years.[4][5]

If protons decay on higher order nuclear processes

Chance: 10^{100} years to 10^{200} years

In the event that the proton does not decay according to the GUT theories above, the Degenerate Era will last longer, and will overlap or surpass the Black Hole Era. However,

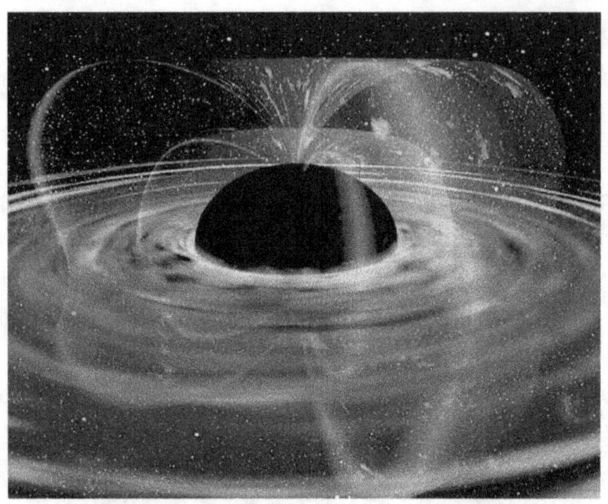

The supermassive black holes are all that remains of galaxies once all protons decay, but even these giants are not immortal.

degenerate stellar objects can still experience proton decay, for example via processes involving virtual black hole, or higher-dimension supersymmetry with a half-life of under 10^{200} years.[4]

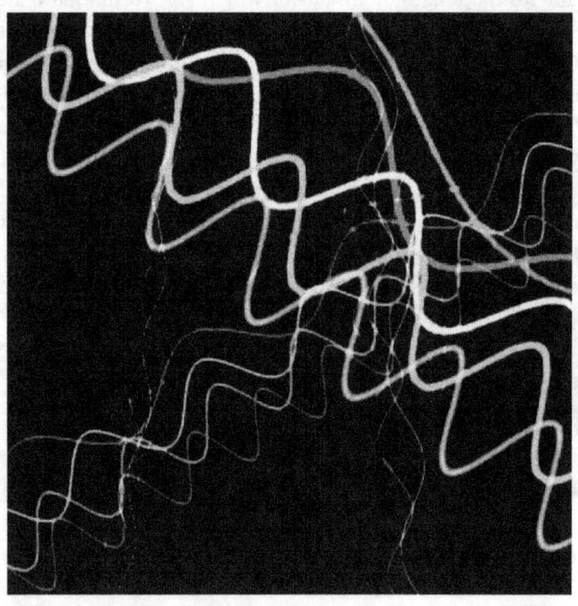

The photon is now the king of the universe as the last of the supermassive black holes evaporates.

37.3.3 Black Hole Era

10^{40} (10 duodecillion) years to 10^{100} (1 googol) years

After 10^{40} years, black holes will dominate the universe. They will slowly evaporate via Hawking radiation.[4] A black hole with a mass of around 1 $M_\odot$ will vanish in around 2×10^{66} years. As the lifetime of a black hole is proportional to the cube of its mass, more massive black holes take longer to decay. A supermassive black hole with a mass of 10^{11} (100 billion) $M_\odot$ will evaporate in around 2×10^{99} years.[35]

Hawking radiation has a thermal spectrum. During most of a black hole's lifetime, the radiation has a low temperature and is mainly in the form of massless particles such as photons and hypothetical gravitons. As the black hole's mass decreases, its temperature increases, becoming comparable to the Sun's by the time the black hole mass has decreased to 10^{19} kilograms. The hole then provides a temporary source of light during the general darkness of the Black Hole Era. During the last stages of its evaporation, a black hole will emit not only massless particles, but also heavier particles, such as electrons, positrons, protons, and antiprotons.[13]

37.3.4 Dark Era and Photon Age

From 10^{100} years (10 duotrigintillion years)

After all the black holes have evaporated (and after all the ordinary matter made of protons has disintegrated, if protons are unstable), the universe will be nearly empty. Photons, neutrinos, electrons, and positrons will fly from place to place, hardly ever encountering each other. Gravitationally, the universe will be dominated by dark matter, electrons, and positrons (not protons).[36]

By this era, with only very diffuse matter remaining, activity in the universe will have tailed off dramatically (compared with previous eras), with very low energy levels and very large time scales. Electrons and positrons drifting through space will encounter one another and occasionally form positronium atoms. These structures are unstable, however, and their constituent particles must eventually annihilate.[37] Other low-level annihilation events will also take place, albeit very slowly. The universe now reaches an extremely low-energy state.

37.3.5 Beyond

Beyond 10^{2500} years to the infinite future

What happens after this is speculative. It is possible that a Big Rip or a Big Freeze event may occur far off into the future.[38][39] The former singularity takes place at a finite scale factor while the latter occurs at an infinitely large radius. Also, the universe may enter a second inflationary epoch, or, assuming that the current vacuum state is a false vacuum, the vacuum may decay into a lower-energy state.[40]

Presumably, extreme low-energy states imply that localized quantum events become major macroscopic phenomena rather than negligible microscopic events because the smallest perturbations make the biggest difference in this era, so there is no telling what may happen to space or time. It is perceived that the laws of "macro-physics" will break down, and the laws of "quantum-physics" will prevail.[7]

The universe could possibly avoid eternal heat death through random quantum tunnelling and quantum fluctuations, given the non-zero probability of producing a new Big Bang in roughly $10^{10^{10^{56}}}$ years.[41]

Over an infinite time there could be a spontaneous entropy decrease, by a Poincaré recurrence or through thermal fluctuations (see also fluctuation theorem).[42][43][44][45]

37.4 Future without proton decay

If the protons do not decay, stellar-mass objects will still become black holes, but more slowly. The following timeline assumes that proton decay does not take place.

37.4.1 Degenerate Era

Possible ionization of matter

>10^{23} years from now

In an expanding universe with decreasing density and nonzero cosmological constant, matter density would reach zero, resulting in all matter including stellar objects and planets ionizing and dissipating at thermal equilibrium.[46]

Sphaleron transitions and possible baryon violation

>10^{150} years from now

Although protons are stable in standard model physics, a quantum anomaly may exist on the electroweak level, which can cause groups of baryons (protons and neutrons) to annihilate into antileptons via the sphaleron transition.[47] Such baryon/lepton violations have a number of 3 and can only occur in multiples or groups of three baryons, which can restrict or prohibit such events. No experimental evidence of sphalerons has yet been observed at low energy levels, though they are believed to occur regularly at high energies and temperatures.

Matter decays into iron

10^{1500} years from now

In 10^{1500} years, cold fusion occurring via quantum tunnelling should make the light nuclei in ordinary matter fuse into iron-56 nuclei (see isotopes of iron). Fission and alpha particle emission should make heavy nuclei also decay to iron, leaving stellar-mass objects as cold spheres of iron, called iron stars.[12]

37.4.2 Black Hole Era

Collapse of iron star to black hole

$10^{(10^{26})}$ to $10^{(10^{76})}$ years from now

Quantum tunnelling should also turn large objects into black holes. Depending on the assumptions made, the time this takes to happen can be calculated as from $10^{(10^{26})}$ years to $10^{(10^{76})}$ years. Quantum tunnelling may also make iron stars collapse into neutron stars in around $10^{(10^{76})}$ years.[12]

37.5 Graphical timeline

Main article: Graphical timeline from Big Bang to Heat Death
See also: Graphical timeline of the universe and Graphical timeline of the Big Bang

37.6 Route diagram styled timeline

For use of this RDT-styled timeline, see Wikipedia:Route diagram template.

37.7 See also

- Big Rip
- Big Crunch
- Big Bounce
- Big Bang
- Chronology of the universe
- Cyclic model
- Dyson's eternal intelligence
- Entropy (arrow of time)
- Final anthropic principle
- Graphical timeline of the Stelliferous Era
- Graphical timeline of the Big Bang
- Graphical timeline from Big Bang to Heat Death. This timeline uses the double-logarithmic scale for comparison with the graphical timeline included in this article.
- Graphical timeline of the universe. This timeline uses the more intuitive linear time, for comparison with this article.
- Heat death of the universe
- Timeline of the Big Bang
- Timeline of the far future
- The Last Question, a short story by Isaac Asimov which considers the inevitable oncome of heat death in the universe and how it may be reversed.
- Ultimate fate of the universe

37.8 References

[1] WMAP – Fate of the Universe, *WMAP's Universe*, NASA. Accessed on line July 17, 2008.

[2] Sean Carroll (2001). "The cosmological constant". *Living Reviews in Relativity*. **4**. Bibcode:2001LRR.....4....1C. arXiv:astro-ph/0004075 . doi:10.12942/lrr-2001-1. Retrieved 2006-09-28.

[3] Krauss, Lawrence M.; Starkman, Glenn D. (2000). "Life, the Universe, and Nothing: Life and Death in an Ever-expanding Universe". *Astrophysical Journal*. **531**: 22–30. Bibcode:2000ApJ...531...22K. arXiv:astro-ph/9902189 . doi:10.1086/308434.

[4] Adams, Fred C.; Laughlin, Gregory (1997). "A dying universe: the long-term fate and evolution of astrophysical objects". *Reviews of Modern Physics*. **69**: 337–372. Bibcode:1997RvMP...69..337A. arXiv:astro-ph/9701131 . doi:10.1103/RevModPhys.69.337.

[5] Adams & Laughlin (1997), §IIE.

[6] Adams & Laughlin (1997), §IV.

[7] Adams & Laughlin (1997), §VID

[8] Chapter 7, *Calibrating the Cosmos*, Frank Levin, New York: Springer, 2006, ISBN 0-387-30778-8.

[9] Five-Year Wilkinson Microwave Anisotropy Probe (WMAP) Observations: Data Processing, Sky Maps, and Basic Results, G. Hinshaw et al., *The Astrophysical Journal Supplement Series* (2008), submitted, arXiv:0803.0732, Bibcode: 2008arXiv0803.0732H.

[10] Planck 2015 results. XIII. Cosmological parameters arXiv:1502.01589

[11] Possible Ultimate Fate of the Universe, Jamal N. Islam, *Quarterly Journal of the Royal Astronomical Society* **18** (March 1977), pp. 3–8, Bibcode: 1977QJRAS..18....3I

[12] Dyson, Freeman J. (1979). "Time without end: Physics and biology in an open universe". *Reviews of Modern Physics*. **51**: 447–460. Bibcode:1979RvMP...51..447D. doi:10.1103/RevModPhys.51.447.

[13] *The Five Ages of the Universe*, Fred Adams and Greg Laughlin, New York: The Free Press, 1999, ISBN 0-684-85422-8.

[14] Adams & Laughlin (1997), §VA

[15] Planck collaboration (2013). "Planck 2013 results. XVI. Cosmological parameters". *Astronomy & Astrophysics*. **571**: A16. Bibcode:2014A&A...571A..16P. arXiv:1303.5076. doi:10.1051/0004-6361/201321591.

[16] Laughlin, Gregory; Bodenheimer, Peter; Adams, Fred C. (1997). "The End of the Main Sequence". *The Astrophysical Journal*. **482**: 420–432. Bibcode:1997ApJ...482..420L. doi:10.1086/304125.

[17] Heger, A.; Fryer, C. L.; Woosley, S. E.; Langer, N.; Hartmann, D. H. (2003). "How Massive Single Stars End Their Life". *Astrophysical Journal*. **591**: 288–300. Bibcode:2003ApJ...591..288H. arXiv:astro-ph/0212469. doi:10.1086/375341.

[18] van der Marel, G.; et al. (2012). "The M31 Velocity Vector. III. Future Milky Way M31-M33 Orbital Evolution, Merging, and Fate of the Sun". *The Astrophysical Journal*. **753**: 9. Bibcode:2012ApJ...753....9V. arXiv:1205.6865. doi:10.1088/0004-637X/753/1/9.

[19] Cowen, R. (31 May 2012). "Andromeda on collision course with the Milky Way". *Nature*. doi:10.1038/nature.2012.10765.

[20] Adams, F. C.; Graves, G. J. M.; Laughlin, G. (December 2004). García-Segura, G.; Tenorio-Tagle, G.; Franco, J.; Yorke, H. W., eds. "Gravitational Collapse: From Massive Stars to Planets. / First Astrophysics meeting of the Observatorio Astronomico Nacional. / A meeting to celebrate Peter Bodenheimer for his outstanding contributions to Astrophysics: Red Dwarfs and the End of the Main Sequence". *Revista Mexicana de Astronomía y Astrofísica (Serie de Conferencias)*. **22**: 46–49. Bibcode:2004RMxAC..22...46A. See Fig. 3.

[21] Adams & Laughlin (1997), § III–IV.

[22] Adams & Laughlin (1997), §IIA and Figure 1.

[23] Adams & Laughlin (1997), §IIIC.

[24] The Future of the Universe, M. Richmond, lecture notes, "Physics 240", Rochester Institute of Technology. Accessed on line July 8, 2008.

[25] Brown Dwarf Accretion: Nonconventional Star Formation over Very Long Timescales, Cirkovic, M. M., *Serbian Astronomical Journal* **171**, (December 2005), pp. 11–17. Bibcode: 2005SerAJ.171...11C

[26] Adams & Laughlin (1997), §IIIF, Table I.

[27] p. 428, A deep focus on NGC 1883, A. L. Tadross, *Bulletin of the Astronomical Society of India* **33**, #4 (December 2005), pp. 421–431, Bibcode: 2005BASI...33..421T.

[28] Reading notes, Liliya L. R. Williams, Astrophysics II: Galactic and Extragalactic Astronomy, University of Minnesota, accessed on line July 20, 2008.

[29] *Deep Time*, David J. Darling, New York: Delacorte Press, 1989, ISBN 978-0-38529-757-8.

[30] G Senjanovic *Proton decay and grand unification*, Dec 2009

[31] "Upper Bound on the Proton Lifetime and the Minimal Non-SUSY Grand Unified Theory", Pavel Fileviez Perez, Max Planck Institute for Nuclear Physics, June 2006. doi:10.1063/1.2735205 https://www.researchgate.net/publication/2020161_Upper_Bound_on_the_Proton_Lifetime_and_the_Minimal_Non-SUSY_Grand_Unified_Theory

[32] Pran Nath and Pavel Fileviez Perez, "Proton Stability in Grand Unified Theories, in Strings and in Branes", Appendix H; 23 April 2007. arXiv:hep-ph/0601023 http://arxiv.org/abs/hep-ph/0601023

[33] Adams & Laughlin (1997), §IV-H.

[34] Solution, exercise 17, *One Universe: At Home in the Cosmos*, Neil de Grasse Tyson, Charles Tsun-Chu Liu, and Robert Irion, Washington, D.C.: Joseph Henry Press, 2000. ISBN 0-309-06488-0.

[35] Particle emission rates from a black hole: Massless particles from an uncharged, nonrotating hole, Don N. Page, *Physical Review D* **13** (1976), pp. 198–206. doi:10.1103/PhysRevD.13.198. See in particular equation (27).

[36] Adams & Laughlin (1997), §VD.

[37] Adams & Laughlin (1997), §VF3.

[38] Caldwell, Robert R.; Kamionkowski, Marc; and Weinberg, Nevin N. (2003). "Phantom energy and cosmic doomsday". arXiv:astro-ph/0302506.

[39] Bohmadi-Lopez, Mariam; Gonzalez-Diaz, Pedro F.; and Martin-Moruno, Prado (2008). “Worse than a big rip?". arXiv:gr-qc/0612135.

[40] Adams & Laughlin (1997), §VE.

[41] Carroll, Sean M. and Chen, Jennifer (2004). “Spontaneous Inflation and Origin of the Arrow of Time”. arXiv:hep-th/0410270.

[42] Tegmark, Max (2003) “Parallel Universes”. arXiv:astro-ph/0302131.

[43] Werlang, T., Ribeiro, G. A. P. and Rigolin, Gustavo (2012) “Interplay between quantum phase transitions and the behavior of quantum correlations at finite temperatures”. arXiv:1205.1046.

[44] Xing, Xiu-San (2007) “Spontaneous entropy decrease and its statistical formula”. arXiv:0710.4624.

[45] Linde, Andrei (2007) “Sinks in the Landscape, Boltzmann Brains, and the Cosmological Constant Problem”. arXiv:hep-th/0611043.

[46] John Baez, University of California-Riverside (Department of Mathematics), “The End of the Universe” 7 Feb 2016 http://math.ucr.edu/home/baez/end.html

[47] G. 't Hooft, “Symmetry breaking through Bell-Jackiw anomalies”. Phys. Rev. Lett. 37 (1976) 8

Chapter 38

De Sitter space

In mathematics and physics, a **de Sitter space** is the analog in Minkowski space, or spacetime, of a sphere in ordinary, Euclidean space. The n-dimensional de Sitter space, denoted dSn, is the Lorentzian manifold analog of an n-sphere (with its canonical Riemannian metric); it is maximally symmetric, has constant positive curvature, and is simply connected for n at least 3. De Sitter space and anti-de Sitter space are named after Willem de Sitter (1872–1934), professor of astronomy at Leiden University and director of the Leiden Observatory. Willem de Sitter and Albert Einstein worked in the 1920s in Leiden closely together on the spacetime structure of our universe.

In the language of general relativity, de Sitter space is the maximally symmetric vacuum solution of Einstein's field equations with a positive cosmological constant Λ (corresponding to a positive vacuum energy density and negative pressure). When $n = 4$ (3 space dimensions plus time), it is a cosmological model for the physical universe; see de Sitter universe.

De Sitter space[1][2] was also discovered, independently, and about the same time, by Tullio Levi-Civita.[3]

More recently it has been considered as the setting for special relativity rather than using Minkowski space, since a group contraction reduces the isometry group of de Sitter space to the Poincaré group, allowing a unification of the spacetime translation subgroup and Lorentz transformation subgroup of the Poincaré group into a simple group rather than a semi-simple group. This alternate formulation of special relativity is called de Sitter relativity.

38.1 Definition

De Sitter space can be defined as a submanifold of a generalized Minkowski space of one higher dimension. Take Minkowski space $\mathbf{R}^{1,n}$ with the standard metric:

$$ds^2 = -dx_0^2 + \sum_{i=1}^{n} dx_i^2.$$

De Sitter space is the submanifold described by the hyperboloid of one sheet

$$-x_0^2 + \sum_{i=1}^{n} x_i^2 = \alpha^2$$

where α is some nonzero constant with dimensions of length. The metric on de Sitter space is the metric induced from the ambient Minkowski metric. The induced metric is nondegenerate and has Lorentzian signature. (Note that if one replaces α^2 with $-\alpha^2$ in the above definition, one obtains a hyperboloid of two sheets. The induced metric in this case is positive-definite, and each sheet is a copy of hyperbolic n-space. For a detailed proof, see geometry of Minkowski space.)

De Sitter space can also be defined as the quotient $O(1, n) / O(1, n-1)$ of two indefinite orthogonal groups, which shows that it is a non-Riemannian symmetric space.

Topologically, de Sitter space is $\mathbf{R} \times S^{n-1}$ (so that if $n \geq 3$ then de Sitter space is simply connected).

38.2 Properties

The isometry group of de Sitter space is the Lorentz group $O(1, n)$. The metric therefore then has $n(n + 1)/2$ independent Killing vector fields and is maximally symmetric. Every maximally symmetric space has constant curvature. The Riemann curvature tensor of de Sitter is given by

$$R_{\rho\sigma\mu\nu} = \frac{1}{\alpha^2}(g_{\rho\mu}g_{\sigma\nu} - g_{\rho\nu}g_{\sigma\mu}).$$

De Sitter space is an Einstein manifold since the Ricci tensor is proportional to the metric:

$$R_{\mu\nu} = \frac{n-1}{\alpha^2} g_{\mu\nu}$$

This means de Sitter space is a vacuum solution of Einstein's equation with cosmological constant given by

$$\Lambda = \frac{(n-1)(n-2)}{2\alpha^2}.$$

The scalar curvature of de Sitter space is given by

$$R = \frac{n(n-1)}{\alpha^2} = \frac{2n}{n-2}\Lambda.$$

For the case $n = 4$, we have $\Lambda = 3/\alpha^2$ and $R = 4\Lambda = 12/\alpha^2$.

38.3 Static coordinates

We can introduce static coordinates $(t, r, \ldots)$ for de Sitter as follows:

$$x_0 = \sqrt{\alpha^2 - r^2}\sinh(t/\alpha)$$
$$x_1 = \sqrt{\alpha^2 - r^2}\cosh(t/\alpha)$$
$$x_i = r z_i \qquad 2 \le i \le n.$$

where z_i gives the standard embedding the $(n-2)$-sphere in $\mathbf{R}^{n-1}$. In these coordinates the de Sitter metric takes the form:

$$ds^2 = -\left(1 - \frac{r^2}{\alpha^2}\right)dt^2 + \left(1 - \frac{r^2}{\alpha^2}\right)^{-1} dr^2 + r^2 d\Omega^2_{n-2}.$$

Note that there is a cosmological horizon at $r = \alpha$.

38.4 Flat slicing

Let

$$x_0 = \alpha\sinh(t/\alpha) + r^2 e^{t/\alpha}/2\alpha,$$
$$x_1 = \alpha\cosh(t/\alpha) - r^2 e^{t/\alpha}/2\alpha,$$
$$x_i = e^{t/\alpha} y_i, \qquad 2 \le i \le n$$

where $r^2 = \sum_i y_i^2$. Then in the (t, y_i) coordinates metric reads:

$$ds^2 = -dt^2 + e^{2t/\alpha} dy^2$$

where $dy^2 = \sum_i dy_i^2$ is the flat metric on y_i 's.

38.5 Open slicing

Let

$$x_0 = \alpha\sinh(t/\alpha)\cosh\xi,$$
$$x_1 = \alpha\cosh(t/\alpha),$$
$$x_i = \alpha z_i \sinh(t/\alpha)\sinh\xi, \qquad 2 \le i \le n$$

where $\sum_i z_i^2 = 1$ forming a S^{n-2} with the standard metric $\sum_i dz_i^2 = d\Omega^2_{n-2}$. Then the metric of the de Sitter space reads

$$ds^2 = -dt^2 + \alpha^2\sinh^2(t/\alpha) dH^2_{n-1},$$

where

$$dH^2_{n-1} = d\xi^2 + \sinh^2(\xi) d\Omega^2_{n-2}$$

is the metric of a Euclidean hyperbolic space.

38.6 Closed slicing

Let

$$x_0 = \alpha\sinh(t/\alpha),$$
$$x_i = \alpha\cosh(t/\alpha) z_i, \qquad 1 \le i \le n$$

where z_i s describe a S^{n-1} . Then the metric reads:

$$ds^2 = -dt^2 + \alpha^2\cosh^2(t/\alpha) d\Omega^2_{n-1}.$$

Changing the time variable to the conformal time via $\tan(\eta/2) = \tanh(t/2\alpha)$ we obtain a metric conformally equivalent to Einstein static universe:

$$ds^2 = \frac{\alpha^2}{\cos^2\eta}(-d\eta^2 + d\Omega^2_{n-1}).$$

This serves to find the Penrose diagram of de Sitter space.

38.7 dS slicing

Let

$$x_0 = \alpha \sin(\chi/\alpha) \sinh(t/\alpha) \cosh \xi,$$

$$x_1 = \alpha \cos(\chi/\alpha),$$

$$x_2 = \alpha \sin(\chi/\alpha) \cosh(t/\alpha),$$

$$x_i = \alpha z_i \sin(\chi/\alpha) \sinh(t/\alpha) \sinh \xi, \qquad 3 \leq i \leq n$$

where z_i s describe a S^{n-3} . Then the metric reads:

$$ds^2 = d\chi^2 + \sin^2(\chi/\alpha) ds^2_{dS,\alpha,n-1},$$

where

$$ds^2_{dS,\alpha,n-1} = -dt^2 + \alpha^2 \sinh^2(t/\alpha) dH^2_{n-2}$$

is the metric of an $n-1$ dimensional de Sitter space with radius of curvature α in open slicing coordinates. The hyperbolic metric is given by:

$$dH^2_{n-2} = d\xi^2 + \sinh^2 \xi d\Omega^2_{n-3}.$$

This is the analytic continuation of the open slicing coordinates under $(t, \xi, \theta, \phi_1, \phi_2, \cdots, \phi_{n-3}) \to (i\chi, \xi, it, \theta, \phi_1, \cdots, \phi_{n-4})$ and also switching x_0 and x_2 because they change their timelike/spacelike nature.

38.8 See also

- Anti-de Sitter space
- de Sitter universe
- AdS/CFT correspondence
- De Sitter–Schwarzschild metric

38.9 Notes

[1] de Sitter, W. (1917), "On the relativity of inertia: Remarks concerning Einstein's latest hypothesis", *Proc. Kon. Ned. Acad. Wet.*, **19**: 1217–1225

[2] de Sitter, W. (1917), "On the curvature of space", *Proc. Kon. Ned. Acad. Wet.*, **20**: 229–243

[3] Levi-Civita, Tullio (1917), "Realtà fisica di alcuni spazî normali del Bianchi", *Rendiconti, Reale Accademia Dei Lincei*, **26**: 519–31

38.10 References

- Qingming Cheng (2001), "De Sitter space", in Hazewinkel, Michiel, *Encyclopedia of Mathematics*, Springer, ISBN 978-1-55608-010-4
- Nomizu, Katsumi (1982), "The Lorentz–Poincaré metric on the upper half-space and its extension", *Hokkaido Mathematical Journal*, **11** (3): 253–261
- Coxeter, H. S. M. (1943), "A geometrical background for de Sitter's world", *American Mathematical Monthly*, Mathematical Association of America, **50** (4): 217–228, JSTOR 2303924, doi:10.2307/2303924
- Susskind, L.; Lindesay, J. (2005), *An Introduction to Black Holes, Information and the String Theory Revolution:The Holographic Universe*, p. 119(11.5.25)

Chapter 39

Ultimate fate of the universe

"End of the Universe" redirects here. For the physical location, see Shape of the universe. For the TV series episode, see End of the Universe (LEXX episode).

The **ultimate fate of the universe** is a topic in physical cosmology, whose theoretical restrictions can usefully and scientifically predict the future behaviour of the universe as it ages. Based on available observational evidence, deciding the fate and evolution of the universe now have become valid cosmological questions, being beyond the mostly untestable constraints of mythological or theological beliefs. Many possible futures have been predicted by rival scientific hypotheses, including that the universe might have existed for a finite and infinite duration, or towards explaining how and in what circumstances it was created.

Observations made by Edwin Hubble during the 1920s-1930s found that most galaxies appeared to be moving away from each other, leading to current accepted Big Bang Theory. This suggests that the universe began in the far distant past about 13.8 billion years ago and ever since, continues to expand.[1] Confirmation of the Big Bang mostly depends on knowing the rate of expansion, average density of matter, and the physical properties of the mass/energy in the universe.

There is a strong consensus among cosmologists that the universe is flat and will continue to expand forever.[2][3] Yet many other factors may influence the universe's origin and final destiny, including, for example: the average motions of galaxies, the shape and structure of the universe, or the amount of dark matter and dark energy the universe contains.

39.1 Emerging scientific basis

39.1.1 Theory

The theoretical scientific exploration of the ultimate fate of the universe became possible with Albert Einstein's 1916

Alexander Friedmann

theory of general relativity. General relativity can be employed to describe the universe on the largest possible scale. There are many possible solutions to the equations of general relativity, and each solution implies a possible ultimate fate of the universe.

Alexander Friedmann proposed several solutions in 1922, as did Georges Lemaître in 1927.[4] In some of these solutions, the universe has been expanding from an initial singularity which was, essentially, the Big Bang.

39.1.2 Observation

In 1931, Edwin Hubble published his conclusion, based on his observations of Cepheid variable stars in distant galaxies, that the universe was expanding. From then on, the *beginning* of the universe and its possible *end* have been the subjects of serious scientific investigation.

The Reverend Monsignor Georges Lemaître

39.1.3 Big Bang and Steady State theories

In 1927, Georges Lemaître set out a theory that has since come to be called the Big Bang theory of the origin of the universe.[4] In 1948, Fred Hoyle set out his opposing Steady State theory in which the universe continually expanded but remained statistically unchanged as new matter is constantly created. These two theories were active contenders until the 1965 discovery, by Arno Penzias and Robert Wilson, of the cosmic microwave background radiation, a fact that is a straightforward prediction of the Big Bang theory, and one that the original Steady State theory could not account for. As a result, The Big Bang theory quickly became the most widely held view of the origin of the universe.

39.1.4 Cosmological constant

When Einstein formulated general relativity, he and his contemporaries believed in a static universe. When Einstein found that his equations could easily be solved in such a way as to allow the universe to be expanding now, and to contract in the far future, he added to those equations what he called a cosmological constant, essentially a constant energy density unaffected by any expansion or contraction, whose role was to offset the effect of gravity on the universe as a whole in such a way that the universe would remain static. After Hubble announced his conclusion that the universe was expanding, Einstein wrote that his cosmological constant was "the greatest blunder of my life".[5]

39.1.5 Density parameter

An important parameter in fate of the universe theory is the density parameter, Omega (Ω), defined as the average matter density of the universe divided by a critical value of that density. This selects one of three possible geometries depending on whether Ω is equal to, less than, or greater than 1. These are called, respectively, the flat, open and closed universes. These three adjectives refer to the overall geometry of the universe, and not to the local curving of spacetime caused by smaller clumps of mass (for example, galaxies and stars). If the primary content of the universe is inert matter, as in the dust models popular for much of the 20th century, there is a particular fate corresponding to each geometry. Hence cosmologists aimed to determine the fate of the universe by measuring Ω, or equivalently the rate at which the expansion was decelerating.

39.1.6 Repulsive force

Starting in 1998, observations of supernovas in distant galaxies have been interpreted as consistent with a universe whose expansion is *accelerating*. Subsequent cosmological theorizing has been designed so as to allow for this possible acceleration, nearly always by invoking dark energy, which in its simplest form is just a positive cosmological constant. In general, dark energy is a catch-all term for any hypothesised field with negative pressure, usually with a density that changes as the universe expands.

39.2 Role of the shape of the universe

See also: Shape of the universe

The current scientific consensus of most cosmologists is that the ultimate fate of the universe depends on its overall shape, how much dark energy it contains, and on the equation of state which determines how the dark energy density responds to the expansion of the universe.[3] Recent observations conclude, from 7.5 billion years after the Big Bang, that the expansion rate of the universe has likely been increasing, commensurate with the Open Universe theory.[6] However, other recent measurements by

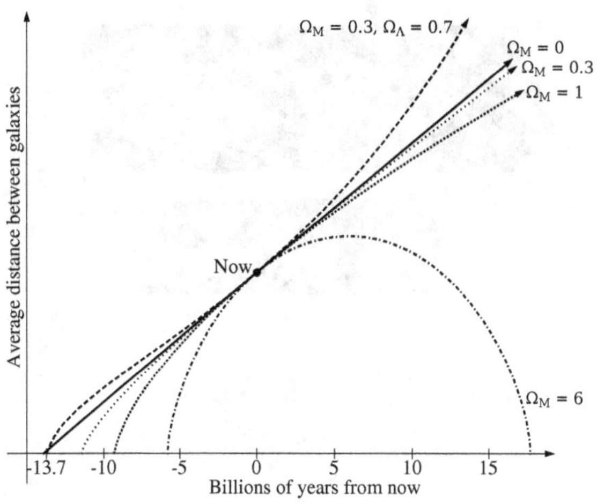

The ultimate fate of an expanding universe depends on the matter density ΩM and the dark energy density ΩΛ.

Wilkinson Microwave Anisotropy Probe suggest that the universe is either flat or very close to flat.[2]

39.2.1 Closed universe

If $\Omega > 1$, then the geometry of space is closed like the surface of a sphere. The sum of the angles of a triangle exceeds 180 degrees and there are no parallel lines; all lines eventually meet. The geometry of the universe is, at least on a very large scale, elliptic.

In a closed universe, gravity eventually stops the expansion of the universe, after which it starts to contract until all matter in the universe collapses to a point, a final singularity termed the "Big Crunch", the opposite of the Big Bang. Some new modern theories assume the universe may have a significant amount of dark energy, whose repulsive force may be sufficient to cause the expansion of the universe to continue forever—even if $\Omega > 1$.[7]

39.2.2 Open universe

If $\Omega < 1$, the geometry of space is open, i.e., negatively curved like the surface of a saddle. The angles of a triangle sum to less than 180 degrees, and lines that do not meet are never equidistant; they have a point of least distance and otherwise grow apart. The geometry of such a universe is hyperbolic.

Even without dark energy, a negatively curved universe expands forever, with gravity negligibly slowing the rate of expansion. With dark energy, the expansion not only continues but accelerates. The ultimate fate of an open universe is either universal heat death, the "Big Freeze", or the "Big Rip", where the acceleration caused by dark energy eventually becomes so strong that it completely overwhelms the effects of the gravitational, electromagnetic and strong binding forces.

Conversely, a *negative* cosmological constant, which would correspond to a negative energy density and positive pressure, would cause even an open universe to re-collapse to a big crunch. This option has been ruled out by observations.

39.2.3 Flat universe

If the average density of the universe exactly equals the critical density so that $\Omega = 1$, then the geometry of the universe is flat: as in Euclidean geometry, the sum of the angles of a triangle is 180 degrees and parallel lines continuously maintain the same distance. Measurements from Wilkinson Microwave Anisotropy Probe have confirmed the universe is flat with only a 0.4% margin of error.[2]

In absence of dark energy, a flat universe expands forever but at a continually decelerating rate, with expansion asymptotically approaching zero. With dark energy, the expansion rate of the universe initially slows down, due to the effect of gravity, but eventually increases. The ultimate fate of the universe is the same as an open universe.

39.3 Theories about the end of the universe

The fate of the universe is determined by the density of the universe. The preponderance of evidence to date, based on measurements of the rate of expansion and the mass density, favors a universe that will continue to expand indefinitely, resulting in the “big freeze” scenario below.[8] However, observations are not conclusive, and alternative models are still possible.[9]

39.3.1 Big Freeze or heat death

Main articles: Future of an expanding universe and Heat death of the universe

The Big Freeze is a scenario under which continued expansion results in a universe that asymptotically approaches absolute zero temperature.[10] This scenario, in combination with the Big Rip scenario, is currently gaining ground as the most important hypothesis.[11] It could, in the absence of dark energy, occur only under a flat or hyperbolic geometry. With a positive cosmological constant, it could also occur in a closed universe. In this scenario, stars are expected to form normally for 10^{12} to 10^{14} (1–100 trillion) years, but eventually the supply of gas needed for star formation will be exhausted. As existing stars run out of fuel and cease to shine, the universe will slowly and inexorably grow darker. Eventually black holes will dominate the universe, which themselves will disappear over time as they emit Hawking radiation.[12] Over infinite time, there would be a spontaneous entropy decrease by the Poincaré recurrence theorem, thermal fluctuations,[13][14] and the fluctuation theorem.[15][16]

A related scenario is heat death, which states that the universe goes to a state of maximum entropy in which everything is evenly distributed and there are no gradients—which are needed to sustain information processing, one form of which is life. The heat death scenario is compatible with any of the three spatial models, but requires that the universe reach an eventual temperature minimum.[17]

39.3.2 Big Rip

Main article: Big Rip

In the special case of phantom dark energy, which has even more negative pressure than a simple cosmological constant, the density of dark energy increases with time, causing the *rate* of acceleration to increase, leading to a steady increase in the Hubble constant. As a result, all material objects in the universe, starting with galaxies and eventually (in a finite time) all forms, no matter how small, will disintegrate into unbound elementary particles and radiation, ripped apart by the phantom energy force and shooting apart from each other. The end state of the universe is a singularity, as the dark energy density and expansion rate becomes infinite.

39.3.3 Big Crunch

Main article: Big Crunch

The Big Crunch hypothesis is a symmetric view of the ultimate fate of the universe. Just as the Big Bang started a cosmological expansion, this theory assumes that the average density of the universe is enough to stop its expansion and begin contracting. The end result is unknown; a simple estimation would have all the matter and space-time in the universe collapse into a dimensionless singularity, but at these scales unknown quantum effects need to be considered (see Quantum gravity). Recent evidence suggests that this scenario is not likely but it has not been ruled out as measurements are only available over a short period of time and could reverse in the future.[11]

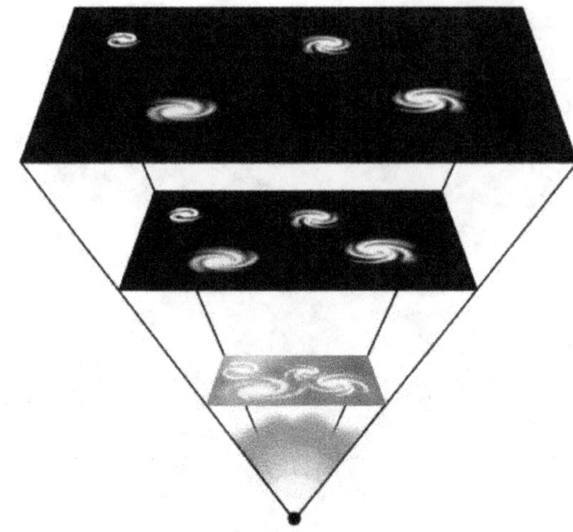

The Big Crunch. The vertical axis can be considered as either plus or minus time.

This scenario allows the Big Bang to occur immediately after the Big Crunch of a preceding universe. If this happens repeatedly, it creates a cyclic model, which is also known as an oscillatory universe. The universe could then consist of an infinite sequence of finite universes, with each finite universe ending with a Big Crunch that is also the Big Bang of the next universe. Theoretically, the cyclic universe could not be reconciled with the second law of thermodynamics: entropy would build up from oscillation to oscillation and cause heat death. Current evidence also indicates the universe is not closed. This has caused cosmologists to abandon the oscillating universe model. A somewhat similar idea is embraced by the cyclic model, but this idea evades heat death because of an expansion of the branes that dilutes entropy accumulated in the previous cycle.

39.3.4 Big Bounce

Main article: Big Bounce

The Big Bounce is a theorized scientific model related to the beginning of the known universe. It derives from the oscillatory universe or cyclic repetition interpretation of the Big Bang where the first cosmological event was the result of the collapse of a previous universe.

According to one version of the Big Bang theory of cosmology, in the beginning the universe was infinitely dense. Such a description seems to be at odds with everything else in physics, and especially quantum mechanics and its uncertainty principle. It is not surprising, therefore, that quantum mechanics has given rise to an alternative version of the Big Bang theory. Also, if the universe is closed, this theory would predict that once this universe collapses it will spawn another universe in an event similar to the Big Bang after a universal singularity is reached or a repulsive quantum force causes re-expansion.

In simple terms, this theory states that the universe will continuously repeat the cycle of a Big Bang, followed up with a Big Crunch.

39.3.5 False vacuum

Main article: False vacuum

In order to best understand the false vacuum collapse theory, one must first understand the Higgs field which permeates the universe. Much like an electromagnetic field it varies in strength, based upon its potential. A true vacuum exists so long as the universe exists in its lowest energy state, in which case the false vacuum theory is irrelevant. However, if the vacuum is not in its lowest energy state (a false vacuum), it could tunnel into a lower energy state.[18] This is called the vacuum metastability event. This has the potential to fundamentally alter our universe; in more audacious scenarios even the various physical constants could have different values, severely affecting the foundations of matter, energy, and spacetime. It is also possible that all structures will be destroyed instantaneously, without any forewarning.[19] Studies of a particle similar to the Higgs boson support the theory of a false vacuum collapse billions of years from now.[20]

39.3.6 Cosmic uncertainty

Each possibility described so far is based on a very simple form for the dark energy equation of state. But as the name is meant to imply, very little is currently known about the physics of the dark energy. If the theory of inflation is true, the universe went through an episode dominated by a different form of dark energy in the first moments of the Big Bang; but inflation ended, indicating an equation of state far more complex than those assumed so far for present-day dark energy. It is possible that the dark energy equation of state could change again resulting in an event that would have consequences which are extremely difficult to predict or parametrize. As the nature of dark energy and dark matter remain enigmatic, even hypothetical, the possibilities surrounding their coming role in the universe are currently unknown.

39.4 Observational constraints on theories

Choosing among these rival scenarios is done by 'weighing' the universe, for example, measuring the relative contributions of matter, radiation, dark matter, and dark energy to the critical density. More concretely, competing scenarios are evaluated against data on galaxy clustering and distant supernovae, and on the anisotropies in the cosmic microwave background.

39.5 See also

- Alan Guth
- Andrei Linde
- Anthropic principle
- Arrow of time
- Cosmological horizon
- Cyclic model
- Freeman Dyson
- General relativity
- John D. Barrow
- Kardashev scale
- Multiverse
- Shape of the universe
- Timeline of the far future
- Zero-energy universe

39.6 References

[1] Wollack, Edward J. (10 December 2010). "Cosmology: The Study of the Universe". *Universe 101: Big Bang Theory*. NASA. Retrieved 27 April 2011.

[2] Will the Universe expand forever?

[3] What is the Ultimate Fate of the Universe?

[4] Lemaître, Georges (1927). "Un univers homogène de masse constante et de rayon croissant rendant compte de la vitesse radiale des nébuleuses extra-galactiques". *Annales de la Société Scientifique de Bruxelles*. **A47**: 49–56.

Bibcode:1927ASSB...47...49L *translated by A. S. Eddington*: Lemaître, Georges (1931). "Expansion of the universe, A homogeneous universe of constant mass and increasing radius accounting for the radial velocity of extra-galactic nebulæ". *Monthly Notices of the Royal Astronomical Society*. **91**: 483–490. Bibcode:1931MNRAS..91..483L. doi:10.1093/mnras/91.5.483

[5] Did Einstein Predict Dark Energy?, hubblesite.org

[6] Dark Energy, Dark Matter

[7] Ryden, Barbara. *Introduction to Cosmology*. The Ohio State University. p. 56.

[8] WMAP - Fate of the Universe, *WMAP's Universe*, NASA. Accessed online July 17, 2008.

[9] "The Return of the Phoenix Universe", Princeton Center For Theoretical Science. Accessed online April 15, 2009.

[10] Glanz, James (1998). "Breakthrough of the year 1998. Astronomy: Cosmic Motion Revealed". *Science*. **282** (5397): 2156–2157. Bibcode:1998Sci...282.2156G. doi:10.1126/science.282.5397.2156a.

[11] Y Wang, J M Kratochvil, A Linde, and M Shmakova, *Current Observational Constraints on Cosmic Doomsday*. JCAP 0412 (2004) 006, astro-ph/0409264

[12] Adams, Fred C.; Laughlin, Gregory (1997). "A dying universe: the long-term fate and evolution of astrophysical objects". *Reviews of Modern Physics*. **69** (2): 337–372. Bibcode:1997RvMP...69..337A. arXiv:astro-ph/9701131. doi:10.1103/RevModPhys.69.337.

[13] "Parallel Universes". *Scientific American*. **288**: 40–51. May 2003. Bibcode:2003SciAm.288e..40T. PMID 12701329. arXiv:astro-ph/0302131. doi:10.1038/scientificamerican0503-40.

[14] Werlang, T.; Ribeiro, G. A. P.; Rigolin, Gustavo (2013). "Interplay Between Quantum Phase Transitions and the Behavior of Quantum Correlations at Finite Temperatures". *International Journal of Modern Physics B*. **27**: 1345032. Bibcode:2012IJMPB..2745032W. arXiv:1205.1046. doi:10.1142/S021797921345032X.

[15] Spontaneous entropy decrease and its statistical formula

[16] Linde, Andrei (2007). "Sinks in the landscape, Boltzmann brains and the cosmological constant problem". *Journal of Cosmology and Astroparticle Physics*. **2007**: 022. Bibcode:2007JCAP...01..022L. arXiv:hep-th/0611043. doi:10.1088/1475-7516/2007/01/022.

[17] Yurov, A. V.; Astashenok, A. V.; González-Díaz, P. F. (2008). "Astronomical bounds on a future Big Freeze singularity". *Gravitation and Cosmology*. **14** (3): 205–212. Bibcode:2008GrCo...14..205Y. arXiv:0705.4108. doi:10.1134/S0202289308030018.

[18]
- M. Stone (1976). "Lifetime and decay of excited vacuum states". *Phys. Rev. D*. **14** (12): 3568–3573. Bibcode:1976PhRvD..14.3568S. doi:10.1103/PhysRevD.14.3568.
- P.H. Frampton (1976). "Vacuum Instability and Higgs Scalar Mass". *Phys. Rev. Lett.* **37** (21): 1378–1380. Bibcode:1976PhRvL..37.1378F. doi:10.1103/PhysRevLett.37.1378.
- M. Stone (1977). "Semiclassical methods for unstable states". *Phys. Lett. B*. **67** (2): 186–183. Bibcode:1977PhLB...67..186S. doi:10.1016/0370-2693(77)90099-5.
- P.H. Frampton (1977). "Consequences of Vacuum Instability in Quantum Field Theory". *Phys. Rev.* **D15** (10): 2922–28. Bibcode:1977PhRvD..15.2922F. doi:10.1103/PhysRevD.15.2922.
- S. Coleman (1977). "Fate of the false vacuum: Semiclassical theory". *Phys. Rev.* **D15**: 2929–36. Bibcode:1977PhRvD..15.2929C. doi:10.1103/physrevd.15.2929.
- C. Callan & S. Coleman (1977). "Fate of the false vacuum. II. First quantum corrections". *Phys. Rev.* **D16**: 1762–68. Bibcode:1977PhRvD..16.1762C. doi:10.1103/physrevd.16.1762.

[19] S. W. Hawking & I. G. Moss (1982). "Supercooled phase transitions in the very early universe". *Phys. Lett.* **B110**: 35–8. Bibcode:1982PhLB..110...35H. doi:10.1016/0370-2693(82)90946-7.

[20] "Will our universe end in a 'big slurp'? Higgs-like particle suggests it might". NBC News. 18 February 2013. Retrieved 19 February 2013.

39.7 Further reading

- Adams, Fred; Gregory Laughlin (2000). *The Five Ages of the Universe: Inside the Physics of Eternity*. Simon & Schuster Australia. ISBN 0-684-86576-9.
- Chaisson, Eric (2001). *Cosmic Evolution: The Rise of Complexity in Nature*. Harvard University Press. ISBN 0-674-00342-X.
- Dyson, Freeman (2004). *Infinite in All Directions (the 1985 Gifford Lectures)*. Harper Perennial. ISBN 0-06-039081-6.
- Harrison, Edward (2003). *Masks of the Universe: Changing Ideas on the Nature of the Cosmos*. Cambridge University Press. ISBN 0-521-77351-2.
- Penrose, Roger (2004). *The Road to Reality*. Alfred A. Knopf. ISBN 0-679-45443-8.

- Prigogine, Ilya (2003). *Is Future Given?*. World Scientific Publishing. ISBN 981-238-508-8.
- Smolin, Lee (2001). *Three Roads to Quantum Gravity: A New Understanding of Space, Time and the Universe*. Phoenix. ISBN 0-7538-1261-4.

39.8 External links

- Baez, J., 2004, "The End of the Universe".
- Caldwell, R. R.; Kamionski, M.; Weinberg, N. N. (2003). "Phantom Energy and Cosmic Doomsday". *Physical Review Letters*. **91** (7): 071301. Bibcode:2003PhRvL..91g1301C. PMID 12935004. arXiv:astro-ph/0302506 . doi:10.1103/physrevlett.91.071301.
- Hjalmarsdotter, Linnea, 2005, "Cosmological parameters."
- George Musser (2010). "Could Time End?". *Scientific American*. **303** (3): 84–91. PMID 20812485. doi:10.1038/scientificamerican0910-84.
- Vaas, R., 2006, "Dark Energy and Life's Ultimate Future," in Burdyuzha, V. (ed.) *The Future of Life and the Future of our Civilization*. Springer: 231–247.
- A Brief History of the End of Everything, a BBC Radio 4 series.
- Cosmology at Caltech.

Chapter 40

Big Rip

In physical cosmology, the **Big Rip** is a hypothetical cosmological model concerning the ultimate fate of the universe, in which the matter of the universe, from stars and galaxies to atoms and subatomic particles, and even spacetime itself, is progressively torn apart by the expansion of the universe at a certain time in the future. According to the hypothesis, first published in 2003, the scale factor of the universe and with it all distances in the universe will become infinite at a finite time in the future. The possibility of sudden singularities and crunch or rip singularities at late times occur only for hypothetical matter with implausible physical properties.[1]

40.1 Overview

The hypothesis relies crucially on the type of dark energy in the universe. The key value is the equation of state parameter w , the ratio between the dark energy pressure and its energy density. If $w < -1$, this dynamical vacuum energy is known as phantom energy, an extreme form of quintessence.

40.1.1 Expansion

Main articles: Future of an expanding universe, Heat death of the universe, Timeline of the far future, and Ultimate fate of the universe

A universe dominated by phantom energy is an accelerating universe, expanding at an ever-increasing rate. However, this implies that the size of the observable universe is continually shrinking; the distance to the edge of the observable universe which is moving away at the speed of light from any point moves ever closer. When the size of the observable universe becomes smaller than any particular structure, no interaction by any of the fundamental forces can occur between the most remote parts of the structure. When these interactions become impossible, the structure is "ripped apart". The model implies that after a finite time there will be a final singularity, called the "Big Rip", in which all distances diverge to infinite values.

The authors of this hypothesis, led by Robert R. Caldwell of Dartmouth College, calculate the time from the present to the end of the universe as we know it for this form of energy to be

$$t_{rip} - t_0 \approx \frac{2}{3|1+w|H_0\sqrt{1-\Omega_m}}$$

where w is defined above, H_0 is Hubble's constant and Ω_m is the present value of the density of all the matter in the universe.

40.2 Author's example

In their paper, the authors consider a hypothetical example with $w = -1.5$, $H_0 = 70$ km/s/Mpc, and $\Omega_m = 0.3$, in which case the Big Rip will happen approximately 22 billion years from the present.

For $w = -1.5$, the galaxies would first be separated from each other. About 60 million years before the Big Rip, gravity would be too weak to hold the Milky Way and other individual galaxies together. Approximately three months before the Big Rip, the Solar System (or systems similar to our own at this time, as the fate of the Solar System 22 billion years in the future is questionable) would be gravitationally unbound. In the last minutes, stars and planets would be torn apart, and an extremely short amount of time before the Big Rip, atoms would be destroyed. At the time the Big Rip occurs, the scale factor will be infinity.[2]

40.3 Observed universe

Evidence indicates w to be very close to -1 in our universe, which makes w the dominating term in the equation. The closer that w is to -1, the closer the denominator is to zero and the further the Big Rip is in the future. If w were exactly equal to -1, the Big Rip could not happen, regardless of the values of H_0 or Ω_m.

According to the latest cosmological data available, the uncertainties are still too large to discriminate among the three cases $w < -1$, $w = -1$, and $w > -1$.[3][4]

40.4 See also

- Big Bounce
- Big Crunch
- Big Freeze
- False vacuum
- Entropy (arrow of time)

40.5 References

[1] Ellis, George F. R., R. Maartens, and M. A. H. MacCallum. Relativistic Cosmology. Cambridge: Cambridge UP, 2012. 146-47. Print.

[2] Caldwell, Robert R.; Kamionkowski, Marc; Weinberg, Nevin N. (2003). "Phantom Energy and Cosmic Doomsday". *Physical Review Letters*. **91** (7): 071301. Bibcode:2003PhRvL..91g1301C. PMID 12935004. arXiv:astro-ph/0302506. doi:10.1103/PhysRevLett.91.071301.

[3] http://wmap.gsfc.nasa.gov/news/

[4] Allen, S. W.; Rapetti, D. A.; Schmidt, R. W.; Ebeling, H.; Morris, R. G.; Fabian, A. C. (2008). "Improved constraints on dark energy from Chandra X-ray observations of the largest relaxed galaxy clusters". *Monthly Notices of the Royal Astronomical Society*. **383** (3): 879. Bibcode:2008MNRAS.383..879A. arXiv:0706.0033. doi:10.1111/j.1365-2966.2007.12610.x.

40.6 External links

- New York Times article
-
-

Chapter 41

Big Bounce

For other uses, see The Big Bounce (disambiguation).

The **Big Bounce** is a hypothetical cosmological model for the origin of the known universe. It was originally suggested as a phase of the *cyclic model* or *oscillatory universe* interpretation of the Big Bang, where the first cosmological event was the result of the collapse of a previous universe. It receded from serious consideration in the early 1980s after inflation theory emerged as a solution to the horizon problem, which had arisen from advances in observations revealing the large-scale structure of the universe. In the early 2000s, inflation was found by some theorists to be problematic and unfalsifiable in that its various parameters could be adjusted to fit any observations, so that the properties of the observable universe are a matter of chance. An alternative picture including a Big Bounce was conceived as a predictive and falsifiable possible solution to the horizon problem, and is under active investigation as of 2017.[1]

41.1 Expansion and contraction

The concept of the Big Bounce envisions the Big Bang as the beginning of a period of expansion that followed a period of contraction. In this view, one could talk of a *Big Crunch* followed by a *Big Bang*, or more simply, a *Big Bounce*. This suggests that we could be living at any point in an infinite sequence of universes, or conversely the current universe could be the very first iteration. However, if the condition of the interval phase "between bounces", considered the 'hypothesis of the primeval atom', is taken into full contingency such enumeration may be meaningless because that condition could represent a singularity in time at each instance, if such perpetual return was absolute and undifferentiated.

The main idea behind the quantum theory of a Big Bounce is that, as density approaches infinity, the behavior of the *quantum foam* changes. All the so-called fundamental physical constants, including the speed of light in a vacuum, need not remain constant during a Big Crunch, especially in the time interval smaller than that in which measurement may never be possible (one unit of Planck time, roughly 10^{-43} seconds) spanning or bracketing the point of inflection.

If the fundamental physical constants were determined in a quantum-mechanical manner during the Big Crunch, then their apparently inexplicable values in this universe would not be so surprising, it being understood here that a *universe* is that which exists between a Big Bang and its Big Crunch.

The Big Bounce Models, however do not explain much about that how the currently expanding universe will manage to contract. This constant and steady expansion is explained by NASA through the metric expansion of space.

41.2 History

Big bounce models have a venerable history and were endorsed on largely aesthetic grounds by cosmologists including Willem de Sitter, Carl Friedrich von Weizsäcker, George McVittie and George Gamow (who stressed that "from the physical point of view we must forget entirely about the precollapse period").[2]

By the early 1980s, the advancing precision and scope of observational cosmology had revealed that the large-scale structure of the universe is flat, homogenous and isotropic, a finding later accepted as the Cosmological Principle to apply at scales beyond roughly 300 million light-years. It was recognized that it was necessary to find an explanation for how distant regions of the universe could have essentially identical properties without ever having been in light-like communication. A solution was proposed to be a period of exponential expansion of space in the early universe, as a basis for what became known as Inflation theory. Following the brief inflationary period, the universe continues to expand, but at a less rapid rate.

Various formulations of inflation theory and their detailed implications became the subject of intense theoretical study. In the absence of a compelling alternative, inflation became the leading solution to the horizon problem. In the early 2000s, inflation was found by some theorists to be problematic and unfalsifiable in that its various parameters could be adjusted to fit any observations, a situation known as a fine-tuning problem. Furthermore, inflation was found to be inevitably eternal, creating an infinity of different universes with typically different properties, so that the properties of the observable universe are a matter of chance.[3] An alternative concept including a Big Bounce was conceived as a predictive and falsifiable possible solution to the horizon problem,[4] and is under active investigation as of 2017.[5][1]

The phrase "Big Bounce" appeared in the scientific literature in 1987, when it was first used in the title of a pair of articles (in German) in *Stern und Weltraum* by Wolfgang Priester and Hans-Joachim Blome.[6] It reappeared in 1988 in Iosif Rozental's *Big Bang, Big Bounce*, a revised English-language translation of a Russian-language book (by a different title), and in a 1991 article (in English) by Priester and Blome in *Astronomy and Astrophysics*. (The phrase apparently originated as the title of a novel by Elmore Leonard in 1969, shortly after increased public awareness of the Big Bang model with of the discovery of the cosmic microwave background by Penzias and Wilson in 1965.)

Martin Bojowald, an assistant professor of physics at Pennsylvania State University, published a study in July 2007 detailing work somewhat related to loop quantum gravity that claimed to mathematically solve the time before the Big Bang, which would give new weight to the oscillatory universe and Big Bounce theories.[7]

One of the main problems with the Big Bang theory is that at the moment of the Big Bang, there is a singularity of zero volume and infinite energy. This is normally interpreted as the end of the physics as we know it; in this case, of the theory of general relativity. This is why one expects quantum effects to become important and avoid the singularity.

However, research in loop quantum cosmology purported to show that a previously existing universe collapsed, not to the point of singularity, but to a point before that where the quantum effects of gravity become so strongly repulsive that the universe rebounds back out, forming a new branch. Throughout this collapse and bounce, the evolution is unitary.

Bojowald also claims that some properties of the universe that collapsed to form ours can also be determined. Some properties of the prior universe are not determinable however due to some kind of uncertainty principle.

This work is still in its early stages and very speculative. Some extensions by further scientists have been published in *Physical Review* Letters.[8]

In 2003, Peter Lynds has put forward a new cosmology model in which time is cyclic. In his theory our Universe will eventually stop expanding and then contract. Before becoming a singularity, as one would expect from Hawking's black hole theory, the universe would bounce. Lynds claims that a singularity would violate the second law of thermodynamics and this stops the universe from being bounded by singularities. The Big Crunch would be avoided with a new Big Bang. Lynds suggests the exact history of the universe would be repeated in each cycle in an eternal recurrence. Some critics argue that while the universe may be cyclic, the histories would all be variants. Lynds' theory has been dismissed by mainstream physicists for the lack of a mathematical model behind its philosophical considerations.[9]

In 2006, it was proposed that the application of loop quantum gravity techniques to Big Bang cosmology can lead to a bounce that need not be cyclic.[10]

In 2011, Nikodem Popławski showed that a nonsingular Big Bounce appears naturally in the Einstein-Cartan-Sciama-Kibble theory of gravity.[11] This theory extends general relativity by removing a constraint of the symmetry of the affine connection and regarding its antisymmetric part, the torsion tensor, as a dynamical variable. The minimal coupling between torsion and Dirac spinors generates a spin-spin interaction which is significant in fermionic matter at extremely high densities. Such an interaction averts the unphysical Big Bang singularity, replacing it with a cusp-like bounce at a finite minimum scale factor, before which the universe was contracting. This scenario also explains why the present Universe at largest scales appears spatially flat, homogeneous and isotropic, providing a physical alternative to cosmic inflation.

In 2012, a new theory of nonsingular big bounce was successfully constructed within the frame of standard Einstein gravity.[12] This theory combines the benefits of matter bounce and Ekpyrotic cosmology. Particularly, the famous BKL instability, that the homogeneous and isotropic background cosmological solution is unstable to the growth of anisotropic stress, is resolved in this theory. Moreover, curvature perturbations seeded in matter contraction are able to form a nearly scale-invariant primordial power spectrum and thus provides a consistent mechanism to explain the cosmic microwave background (CMB) observations.

41.3 See also

- Abhay Ashtekar
- Anthropic principle
- Big Crunch
- Big Rip
- Big Freeze
- False vacuum
- Eternal return
- John Archibald Wheeler
- Loop quantum cosmology
- Loop quantum gravity
- Supernova

41.4 References

[1] Brandenberger, Robert; Peter, Patrick (2017). "Bouncing Cosmologies: Progress and Problems" (PDF). *Foundations of Physics*. ISSN 0015-9018. doi:10.1007/s10701-016-0057-0.

[2] Kragh, Helge (1996). *Cosmology*. Princeton, NJ, USA: Princeton University Press. ISBN 0-691-00546-X.

[3] McKee, Maggie (25 September 2014). "Ingenious: Paul J. Steinhardt — The Princeton physicist on what's wrong with inflation theory and his view of the Big Bang". *Nautilus* (017). NautilusThink Inc. Retrieved 31 March 2017.

[4] Steinhardt, Paul J.; Turok, Neil (2005). "The cyclic model simplified". *New Astronomy Reviews*. **49** (2-6): 43–57. ISSN 1387-6473. doi:10.1016/j.newar.2005.01.003.

[5] Lehners, Jean-Luc; Steinhardt, Paul J. (2013). "Planck 2013 results support the cyclic universe" (PDF). *Physical Review D*. **87** (12). ISSN 1550-7998. doi:10.1103/PhysRevD.87.123533.

[6] Overduin, James; Hans-Joachim Blome; Josef Hoell (June 2007). "Wolfgang Priester: from the big bounce to the Λ-dominated universe". *Naturwissenschaften*. **94** (6): 417–429. Bibcode:2007NW.....94..417O. PMID 17146687. arXiv:astro-ph/0608644. doi:10.1007/s00114-006-0187-x.

[7] Bojowald, Martin (2007). "What happened before the Big Bang?". *Nature Physics*. **3** (8): 523–525. Bibcode:2007NatPh...3..523B. doi:10.1038/nphys654.

[8] Ashtekar, Abhay; Corichi, Alejandro; Singh, Parampreet (2008). "Robustness of key features of loop quantum cosmology". *Physical Review D*. **77** (2): 024046. Bibcode:2008PhRvD..77b4046A. arXiv:0710.3565. doi:10.1103/PhysRevD.77.024046.

[9] David Adam (14 August 2003). "The Strange story of Peter Lynds". *The Guardian*.

[10] "Penn State Researchers Look Beyond The Birth Of The Universe". *Science Daily*. May 17, 2006. Referring to Ashtekar, Abhay; Pawlowski, Tomasz; Singh, Parmpreet (2006). "Quantum Nature of the Big Bang". *Physical Review Letters*. **96** (14): 141301. Bibcode:2006PhRvL..96n1301A. PMID 16712061. arXiv:gr-qc/0602086. doi:10.1103/PhysRevLett.96.141301.

[11] Poplawski, N. J. (2012). "Nonsingular, big-bounce cosmology from spinor-torsion coupling". *Physical Review D*. **85**: 107502. Bibcode:2012PhRvD..85j7502P. arXiv:1111.4595. doi:10.1103/PhysRevD.85.107502.

[12] Cai, Yi-Fu; Damien Easson; Robert Brandenberger (2012). "Towards a Nonsingular Bouncing Cosmology". *Journal of Cosmology and Astroparticle Physics*. **08**: 020. Bibcode:2012JCAP...08..020C. arXiv:1206.2382. doi:10.1088/1475-7516/2012/08/020.

41.5 Further reading

- Angha, Nader (2001). *Expansion & Contraction Within Being (Dahm)*. Riverside, CA: M.T.O Shahmaghsoudi Publications. ISBN 0-910735-61-1.
- Bojowald, Martin (2008). "Follow the Bouncing Universe". *Scientific American*. **299** (October 2008): 44–51. PMID 18847084. doi:10.1038/scientificamerican1008-44.
- Magueijo, João (2003). *Faster than the Speed of Light: the Story of a Scientific Speculation*. Cambridge, MA: Perseus Publishing. ISBN 0-7382-0525-7.

41.6 External links

- Wolfgang Priester: from the big bounce to the Lambda-dominated universe, James Overduin, 2006
- Dark Matter, Antimatter, and Time-Symmetry, Trevor Pitts, 1999
- Penn State Researchers Look Beyond The Birth Of The Universe (Penn State) May 12, 2006
- What Happened Before the Big Bang? (Penn State) July 1, 2007
- From big bang to big bounce (Pen State) NewScientist December 13, 2008
- SpringerLink - Gravitation and Cosmology, Volume 16, Number 4

Chapter 42

Big Crunch

An animation of the supposed behaviour of a Big Crunch

The **Big Crunch** is one possible scenario for the ultimate fate of the universe, in which the metric expansion of space eventually reverses and the universe recollapses, ultimately causing the cosmic scale factor to reach zero or causing a reformation of the universe starting with another Big Bang. Sudden singularities and crunch or rip singularities at late times occur only for hypothetical matter with implausible physical properties.[1]

42.1 Overview

If the universe's expansion speed does not exceed the escape velocity, then the mutual gravitational attraction of all its matter will eventually cause it to contract. If entropy continues to increase in the contracting phase (see Ergodic hypothesis), the contraction would appear very different from the time reversal of the expansion. While the early universe was highly uniform, a contracting universe would become increasingly clumped.[2] Eventually all matter would collapse into black holes, which would then coalesce producing a unified black hole or Big Crunch singularity.

The exact details of the events that would take place before such final collapse depend on the length of both the expansion phase as well as the previous contraction phase; the longer both lasted, the more events expected to take place in an ever-expanding universe would happen; nonetheless it's expected that the contraction phase would not immediately be noticed by hypothetical observers because of the delay caused by the speed of light, that the temperature of the cosmic microwave background would rise during contraction symmetrically compared to the previous expansion phase, and that the events that took place during the Big Bang would take in opposite order.[3] For a contracting Universe similar to ours in composition it's expected that superclusters would merge among themselves followed by galaxy clusters and later galaxies. By the time stars were so close together that collisions among them were frequent, the temperature of the cosmic microwave background would have increased so much that stars would be unable to expel their internal heat, slowly *cooking* until they exploded leaving behind a hot and highly heterogeneus gas, whose atoms would break down in their constituent subatomic particles because of the increasing temperature, that would be absorbed by the already coalescing black holes before the Big Crunch itself.[3]

The Hubble Constant measures the current state of expansion in the universe, and the strength of the gravitational force depends on the density and pressure of matter in the universe, or in other words, the critical density of the universe. If the density of the universe is greater than the critical density, then the strength of the gravitational force will stop the universe from expanding and the universe will collapse back on itself[2]—assuming that there is no repulsive force such as a cosmological constant. Conversely, if the density of the universe is less than the critical density, the universe will continue to expand and the gravitational pull will not be enough to stop the universe from expanding. This scenario would result in the Big Freeze, where the universe cools as it expands and reaches a state of entropy.[4] One theory proposes that the universe could collapse to the state where it began and then initiate another Big Bang,[2] so in this way the universe would last forever, but would pass through phases of expansion (Big Bang) and contraction (Big Crunch).[5] Another scenario results in a flat universe which occurs when the critical density is just right. In this state the universe would always be slowing down, and eventually come to a stop in an interminable amount of time. Although, it is now understood that the critical density has been measured and determined to be a flat universe.[6]

Recent experimental evidence (namely the observation of distant supernovae as standard candles, and the well-resolved mapping of the cosmic microwave background) has led to speculation that the expansion of the universe is not being slowed down by gravity but rather accelerating. However, since the nature of the dark energy that is postulated to drive the acceleration is unknown, it is still possible (though not observationally supported as of today) that it might eventually reverse its developmental path and cause a collapse.[7] [8]

42.2 See also

- Space portal
- Arrow of time
- Bentley's paradox
- Big Bounce
- Big Rip
- Chronology of the universe
- Cyclic model
- Entropy (arrow of time)
- Eternal return
- FRW model
- Gravitational collapse
- Heat death of the universe
- Timeline of the far future
- Timeline of the formation of the Universe

42.3 References

[1] Ellis, George F. R., R. Maartens, and M. A. H. MacCallum. Relativistic Cosmology. Cambridge: Cambridge UP, 2012. 146–47. Print.

[2] *How the Universe Works 3*. End of the Universe. Discovery Channel. 2014.

[3] Davies, Paul (January 9, 1997). *The Last Three Minutes: Conjectures About The Ultimate Fate Of The Universe*. Basic Books. ISBN 978-0-465-03851-0.

[4] Dr. Gary F. Hinshaw, *WMAP Introduction to Cosmology*. NASA (2008)

[5] Jennifer Bergman, *The Big Crunch*, Windows to the Universe (2003)

[6] Fraser Cain (2013-10-17), *How Will The Universe End?*, retrieved 2016-06-13

[7] Y Wang, J M Kratochvil, A Linde, and M Shmakova, *Current Observational Constraints on Cosmic Doomsday*. JCAP 0412 (2004) 006, astro-ph/0409264

[8] McSween, Stephen A. "Dark Energy and the Red Shift in a Contracting Universe."

Chapter 43

Proton decay

This article is about decay of protons into subatomic particles. For the type of radioactive decay in which a nucleus ejects a proton, see Proton emission.

In particle physics, **proton decay** is a hypothetical form

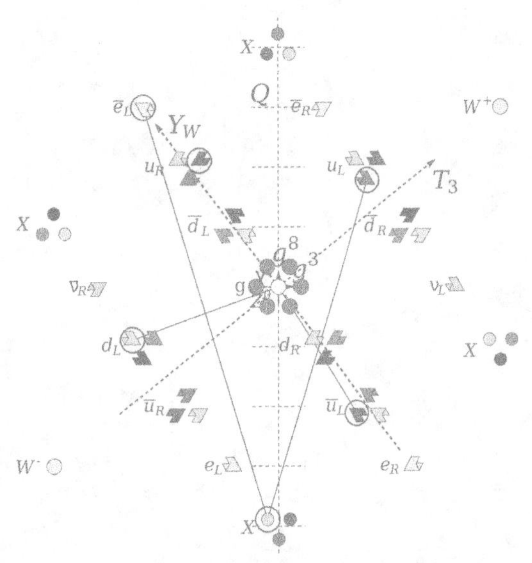

The pattern of weak isospins, weak hypercharges, and color charges for particles in the Georgi–Glashow model. Here, a proton, consisting of two up quarks and a down, decays into a pion, consisting of an up and anti-up, and a positron, via an X boson with electric charge −4/3.

of radioactive decay in which the proton decays into lighter subatomic particles, such as a neutral pion and a positron.[1] There is currently no experimental evidence that proton decay occurs.

According to the Standard Model, protons, a type of baryon, are stable because baryon number (quark number) is conserved (under normal circumstances; see chiral anomaly for exception). Therefore, protons will not decay into other particles on their own, because they are the lightest (and therefore least energetic) baryon. Positron emission, a form of radioactive decay which sees a proton become a neutron, is not proton decay, since the proton interacts with other particles within the atom.

Some beyond-the-Standard Model grand unified theories (GUTs) explicitly break the baryon number symmetry, allowing protons to decay via the Higgs particle, magnetic monopoles or new X bosons with a half-life of 10^{31} to 10^{36} years. To date, all attempts to observe new phenomena predicted by GUTs (like proton decay or the existence of magnetic monopoles) have failed.

Quantum gravity (via virtual black holes) may also provide a venue of proton decay at magnitudes or lifetimes well beyond the GUT scale decay range above, as well as extra dimensions in supersymmetry.

There are other theoretical methods of baryon violation other than proton decay including interactions with changes of baryon and/or lepton number other than 1 (as required in proton decay). These included B and/or L violations of 2, 3 or other numbers, or B-L violation. Such examples include neutron oscillations and the electroweak sphaleron anomaly at high energies and temperatures that can result between the collision of protons into antileptons[2] or vice versa (a key factor in leptogenesis and non-GUT baryogenesis).

43.1 Baryogenesis

Main article: Baryogenesis

One of the outstanding problems in modern physics is the predominance of matter over antimatter in the universe. The universe, as a whole, seems to have a nonzero positive baryon number density — that is, matter exists. Since it is assumed in cosmology that the particles we see were created using the same physics we measure today, it would normally be expected that the overall baryon number should be zero, as matter and antimatter should have been created in equal amounts. This has led to a number of proposed mechanisms for symmetry breaking that favour the creation of normal matter (as opposed to antimatter) under certain conditions. This imbalance would have been exceptionally small, on the order of 1 in every 10000000000 (10^{10}) particles a small fraction of a second after the Big Bang, but after most of the matter and antimatter annihilated, what was left over was all the baryonic matter in the current universe, along with a much greater number of bosons. Experiments reported in 2010 at Fermilab, however, seem to show that this imbalance is much greater than previously assumed. In an experiment involving a series of particle collisions, the amount of generated matter was approximately 1% larger than the amount of generated antimatter. The reason for this discrepancy is yet unknown.[3]

Most grand unified theories explicitly break the baryon number symmetry, which would account for this discrepancy, typically invoking reactions mediated by very massive X bosons (X)or massiveHiggs bosons (H0). The rate at which these events occur is governed largely by the mass of the intermediate X or H0 particles, so by assuming these reactions are responsible for the majority of the baryon number seen today, a maximum mass can be calculated above which the rate would be too slow to explain the presence of matter today. These estimates predict that a large volume of material will occasionally exhibit a spontaneous proton decay.

43.2 Experimental evidence

Proton decay is one of the key predictions of the various grand unified theories (GUTs) proposed in the 1970s, another major one being the existence of magnetic monopoles. Both concepts have been the focus of major experimental physics efforts since the early 1980s. To date, all attempts to observe these events have failed. Best results come from the super-Kamiokande water Cherenkov radiation detector in Japan. 2015 analysis gave half-life higher than 1.67×10^{34} years via positron decay[4] and 2012 analysis gave 1.08×10^{34} years via antimuon decay,[5] close to a supersymmetry (SUSY) prediction of 10^{34}–10^{36} yr.[6] An upgraded version, Hyper-Kamiokande, probably will have sensitivity 5–10 times better than super-Kamiokande.[4]

43.3 Theoretical motivation

Despite the lack of observational evidence for proton decay, some grand unification theories, such as the SU(5) Georgi–Glashow model and SO(10), along with their supersymmetric variants, require it. According to such theories, the proton has a half-life of about 10^{31} to 10^{36} years and decays into a positron and a neutral pion that itself immediately decays into 2 gamma ray photons:

Since a positron is an antilepton this decay preserves B-L number, which is conserved in most GUTs.

Additional decay modes are available (e.g.:
p+ → μ+ + π0),[5]
both directly and when catalyzed via interaction with GUT-predicted magnetic monopoles.[7] Though this process has not been observed experimentally, it is within the realm of experimental testability for future planned very large-scale detectors on the megaton scale. Such detectors include the Hyper-Kamiokande.

Early grand unification theories (GUTs) such as the Georgi–Glashow model, which were the first consistent theories to suggest proton decay, postulated that the proton's half-life would be at least 10^{31} years. As further experiments and calculations were performed in the 1990s, it became clear that the proton half-life could not lie below 10^{32} years. Many books from that period refer to this figure for the possible decay time for baryonic matter. More recent findings have pushed the minimum proton half-life to at least $10^{34}-10^{35}$ years, ruling out the simpler GUTs (including minimal SU(5)/Georgi–Glashow) and most non-SUSY models. The maximum upper limit on proton lifetime (if unstable), is calculated at 6×10^{39} years, a bound applicable to SUSY models,[8] with a maximum for (minimal) non-SUSY GUTs at 1.4×10^{36} years.[9]

Although the phenomenon is referred to as "proton decay", the effect would also be seen in neutrons bound inside atomic nuclei. Free neutrons—those not inside an atomic nucleus—are already known to decay into protons (and an electron and an antineutrino) in a process called beta decay. Free neutrons have a half-life of about 10 minutes (610.2±0.8 s)[10] due to the weak interaction. Neutrons bound inside a nucleus have an immensely longer half-life—apparently as great as that of the proton.

43.4 Projected proton lifetimes

43.5 Decay operators

43.5.1 Dimension-6 proton decay operators

The dimension−6 proton decay operators are $\frac{qqql}{\Lambda^2}$, $\frac{d^c u^c u^c e^c}{\Lambda^2}$, $\frac{\overline{e^c u^c q q}}{\Lambda^2}$ and $\frac{\overline{d^c u^c q l}}{\Lambda^2}$ where Λ is the cutoff scale for the Standard Model. All of these operators violate both baryon number (*B*) and lepton number (*L*) conservation but not the combination *B* − *L*.

In GUT models, the exchange of an X or Y boson with the mass ΛGUT can lead to the last two operators suppressed by $\frac{1}{\Lambda_{GUT}^2}$. The exchange of a triplet Higgs with mass *M* can lead to all of the operators suppressed by $1/M^2$. See doublet–triplet splitting problem.

- Proton decay. These graphics refer to the X bosons and Higgs bosons.

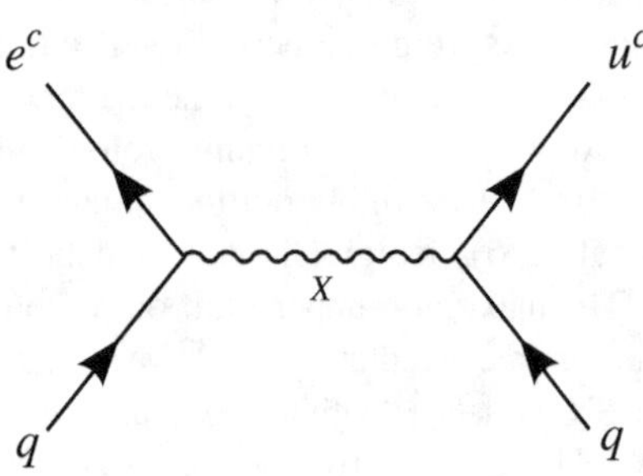

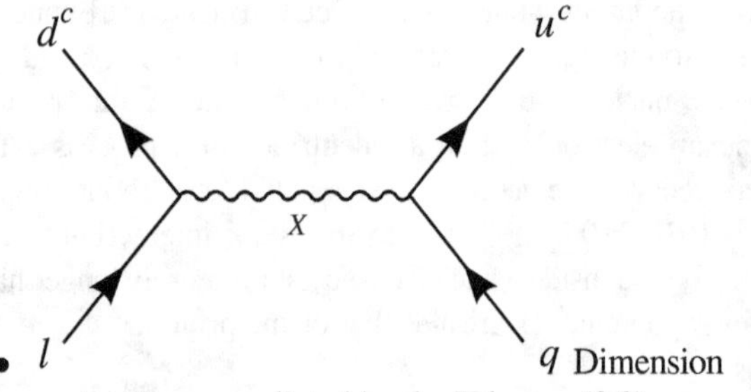

- Dimension 6 proton decay mediated by the X boson (3,2) − ⅚ in SU(5) GUT

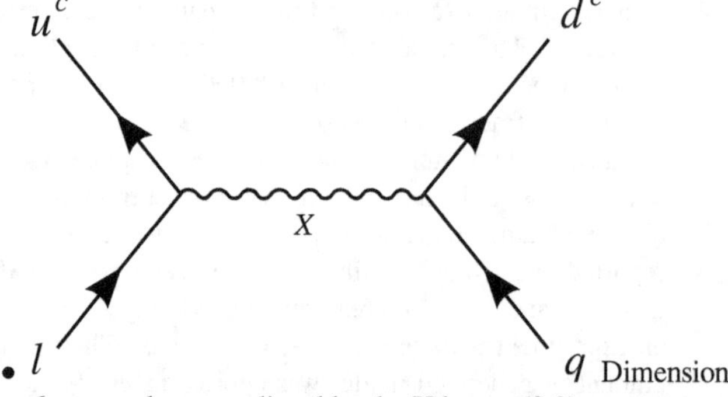

- Dimension 6 proton decay mediated by the X boson (3,2) ⅙ in flipped SU(5) GUT

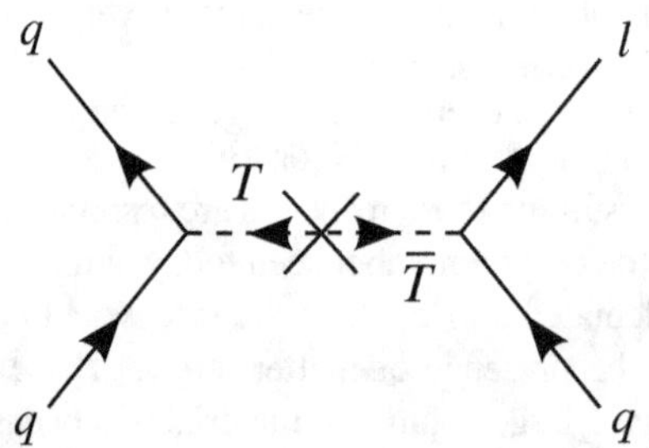

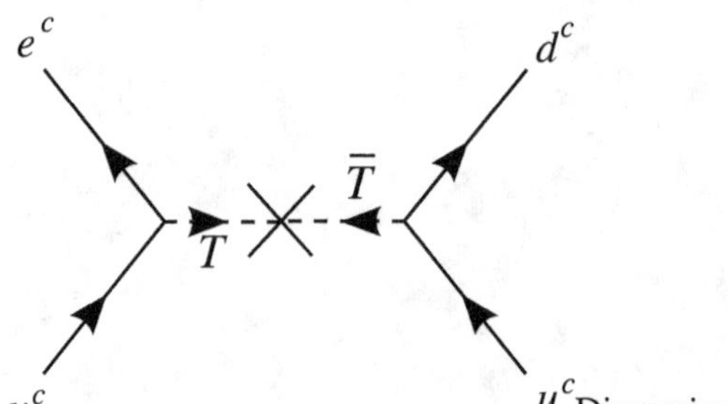

- Dimension 6 proton decay mediated by the triplet Higgs T (3,1) − ⅓ and the anti-triplet Higgs T (3,1) ⅓ in SU(5) GUT

43.5.2 Dimension-5 proton decay operators

In supersymmetric extensions (such as the MSSM), we can also have dimension-5 operators involving two fermions and two sfermions caused by the exchange of a tripletino of mass *M*. The sfermions will then exchange a gaugino or Higgsino or gravitino leaving two fermions. The overall Feynman diagram has a loop (and other complications due to strong interaction physics). This decay rate is suppressed by $\frac{1}{MM_{SUSY}}$ where *M*SUSY is the mass scale of the superpartners.

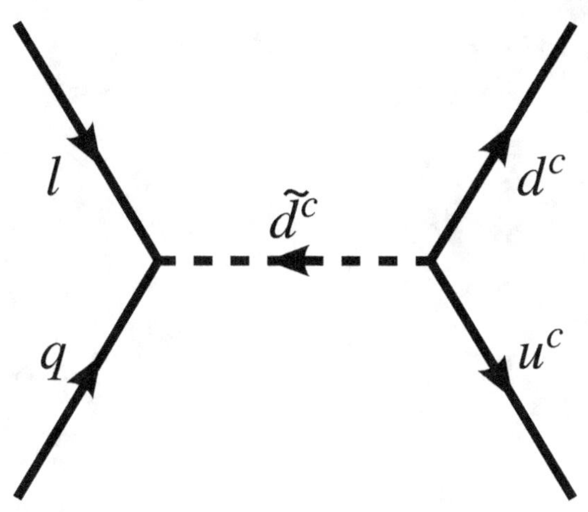

43.5.3 Dimension-4 proton decay operators

In the absence of matter parity, supersymmetric extensions of the Standard Model can give rise to the last operator suppressed by the inverse square of sdown quark mass. This is due to the dimension-4 operators
q l đ c

and c

u c d đ c

.

The proton decay rate is only suppressed by

$$\frac{1}{M_{SUSY}^2} \text{ which}$$

is far too fast unless the couplings are very small.

43.6 Proton decay in media

When Woody Allen in his 1980 film *Stardust Memories* launches a depressive soliloquy with the quote, "Did anybody read on the front page of *The Times* that matter is decaying?", this was almost certainly a reference to the GUTs models of the 1970s, given the film's period, their importance at the time and the many contemporary layperson articles in general media about some of the model's most striking consequences, particularly its mechanism for proton decay.

43.7 See also

- Virtual black hole
- Weak hypercharge
- B − L
- X and Y bosons

43.8 References

[1] Radioactive decays by Protons. Myth or reality?, Ishfaq Ahmad, The Nucleus, 1969. pp 69–70

[2] "Bloch Wave Function for the Periodic Sphaleron Potential and Unsuppressed Baryon and Lepton Number Violating Processes", S.H. Henry Tyne & Sam S.C. Wong, Phys.Rev. D92 (2015) no.4, 045005 (2015-08-05) DOI: 10.1103/PhysRevD.92.045005

[3] V.M. Abazov; et al. (2010). "Evidence for an anomalous like-sign dimuon charge asymmetry". *Physical Review D*. **82** (3). arXiv:1005.2757. doi:10.1103/PhysRevD.82.032001.

[4] Bajc, Borut; Hisano, Junji; Kuwahara, Takumi; Omura, Yuji (2016). "Threshold corrections to dimension-six proton decay operators in non-minimal SUSY SU(5) GUTs". *Nuclear Physics B*. **910**: 1. Bibcode:2016NuPhB.910....1B. arXiv:1603.03568. doi:10.1016/j.nuclphysb.2016.06.017.

[5] H. Nishino; Super-K Collaboration (2012). "Search for Proton Decay via
p+ → e+ π0 and p+ → μ+ π0
in a Large Water Cherenkov Detector". *Physical Review Letters*. **102** (14): 141801. Bibcode:2009PhRvL.102n1801N. PMID 19392425. doi:10.1103/PhysRevLett.102.141801.

[6] "Proton lifetime is longer than 10^{34} years". *www-sk.icrr.u-tokyo.ac.jp*. 25 November 2009.

[7] B. V. Sreekantan (1984). "Searches for Proton Decay and Superheavy Magnetic Monopoles" (PDF). *Journal of Astrophysics and Astronomy*. **5** (3): 251–271. Bibcode:1984JApA....5..251S. doi:10.1007/BF02714542.

[8] Nath, Pran; Fileviez Pérez, Pavel (2007). "Proton stability in grand unified theories, in strings and in branes". *Physics Reports*. **441** (5–6): 191.

Bibcode:2007PhR...441..191N. arXiv:hep-ph/0601023. doi:10.1016/j.physrep.2007.02.010.

[9] Nath and Perez, 2007, part 5.6

[10] K.A. Olive; et al. (2014). "Review of Particle Physics – *N* Baryons" (PDF). *Chinese Physics C*. **38** (9): 090001. Bibcode:2006JPhG...33....1Y. arXiv:astro-ph/0601168. doi:10.1088/1674-1137/38/9/090001.

[11] "Grand Unified Theories and Proton Decay", Ed Kearns, Boston University, 2009, page 15. http://physics.bu.edu/NEPPSR/TALKS-2009/Kearns_GUTs_ProtonDecay.pdf

43.9 Further reading

- C. Amsler; Particle Data Group (2008). "Review of Particle Physics – *N* Baryons" (PDF). *Physics Letters B*. **667**: 1–6. Bibcode:2008PhLB..667....1A. doi:10.1016/j.physletb.2008.07.018.
- K. Hagiwara; Particle Data Group (2002). "Review of Particle Physics – *N* Baryons" (PDF). *Physical Review D*. **66**: 010001. Bibcode:2002PhRvD..66a0001H. doi:10.1103/PhysRevD.66.010001.
- F. Adams; G. Laughlin. *The Five Ages of the Universe : Inside the Physics of Eternity*. ISBN 978-0-684-86576-8.
- L.M. Krauss. *Atom : An Odyssey from the Big Bang to Life on Earth*. ISBN 0-316-49946-3.
- D.-D. Wu; T.-Z. Li (1985). "Proton decay, annihilation or fusion?". *Zeitschrift für Physik C*. **27** (2): 321–323. Bibcode:1985ZPhyC..27..321W. doi:10.1007/BF01556623.
- P. Nath; P. Fileviez Perez (2007). "Proton stability in grand unified theories, in strings and in branes". *Physics Reports*. **441** (5–6): 191–317. Bibcode:2007PhR...441..191N. arXiv:hep-ph/0601023. doi:10.1016/j.physrep.2007.02.010.

43.10 External links

- Proton decay at Super-Kamiokande
- Pictorial history of the IMB experiment
- Luciano Maiani (8 February 2006). *The problem of proton decay* (PDF). Third NO-VE International Workshop on Neutrino Oscillations in Venice. Venice.

43.11 Text and image sources, contributors, and licenses

43.11.1 Text

- **Accelerating expansion of the universe** *Source:* https://en.wikipedia.org/wiki/Accelerating_expansion_of_the_universe?oldid=787262338 *Contributors:* AxelBoldt, Bth, Dbundy, Tim Starling, EddEdmondson, Alfio, Looxix~enwiki, William M. Connolley, Julesd, Evercat, Hashar, Timwi, Dysprosia, Mw66, The Anomebot, DW40, Dragons flight, Phys, Peak, Lowellian, Rursus, JerryFriedman, Giftlite, Graeme Bartlett, DavidCary, Barbara Shack, Joe Kress, ConradPino, Bbbl67, Burschik, Eep², JimJast, Guanabot, Vsmith, LeeHunter, Dbachmann, Bender235, AdamSolomon, RJHall, Mr. Billion, Sietse Snel, Bobo192, Foobaz, 9SGjOSfyHJaQVsEmy9NS, Hackwrench, Eric Kvaalen, Vuo, Falcorian, Bobrayner, Zanaq, Richard Arthur Norton (1958-), Woohookitty, OCNative, Joke137, Wisq, Driftwoodzebulin, Drbogdan, Rjwilmsi, Seandop, Kolbasz, ThunderPeel2001, Marcperkel, Chase me ladies, I'm the Cavalry, Closedmouth, Dutch-Bostonian, SmackBot, Melchoir, Vald, Nickst, AnOddName, Dreadstar, Pulu, Wkerney, Ckatz, Hypnosifl, Xxxiv34, Geral Corasjo, CRGreathouse, Kjknohw, Meno25, Michael C Price, Doug Weller, DumbBOT, Peter Gulutzan, Dawnseeker2000, KrakatoaKatie, Seaphoto, Obeattie, Steelpillow, Arch dude, Bpmullins, Cardamon, Cgingold, Robin S, NatureA16, Hedwig in Washington, McSly, Mannhoodd, Tarotcards, VolkovBot, Jackfork, UnitedStatesian, Insanity Incarnate, Nihil novi, Puzhok, Lightmouse, Coldcreation, Gevgiorbran, Aggel000, Dr. Leif Rongved, Dmyersturnbull, Kentgen1, El bot de la dieta, DanielPharos, DumZiBoT, XLinkBot, Ladsgroup, Ost316, Addbot, Ridgepg, Simonm223, Iliketitz93, AkhtaBot, Marx01, Verbal, Legobot, Cosmos72, Luckas-bot, Yobot, Systemizer, Fraggle81, Amirobot, Amble, Synchronism, AnomieBOT, Jim1138, Citation bot, Xqbot, Gap9551, GrouchoBot, RibotBOT, Waleswatcher, Bellerophon, Michael93555, DivineAlpha, Citation bot 4, Jonesey95, Tom.Reding, Issuesixty soulsgreat, Trappist the monk, Lotje, RjwilmsiBot, John of Reading, Primefac, Tinss, Vanjka-ivanych, Solomonfromfinland, ZéroBot, Brandmeister, ChuispastonBot, MrChandmari, Mechachomp, ClueBot NG, Astrocog, Bibcode Bot, BG19bot, Yizlpku, Anukool.rajoriya, Minsbot, Pheng13, RiseUpAgain, Mediran, Kozmokonstans, Makecat-bot, Wjs64, Illuusio, Rfassbind, Yheyma, Blackbombchu, The Herald, BDwinds, Monkbot, Leegrc, Jsaur, Igby Kollektiv, Garfield Garfield, HannahFord428, Tetra quark, Isambard Kingdom, Absolutelypuremilk, MusikBot, Sir Cumference, Youknowwhatimsayin, ICameHereToEditNotToFeel, The Cube Root Of Infinity, DatGuy, Fitindia, Xx Cool Guy7202 xX, Bender the Bot, Penskins, YuRi YuZi and Anonymous: 106
- **Supernova Cosmology Project** *Source:* https://en.wikipedia.org/wiki/Supernova_Cosmology_Project?oldid=731445833 *Contributors:* Giftlite, RJHall, Falcorian, Frankie1969, Malcolma, Alain r, RG2, SmackBot, Kurtan~enwiki, Alaibot, Thijs!bot, Headbomb, Rosarinagazo, TAnthony, WolfmanSF, Quantling, VolkovBot, TXiKiBoT, UnitedStatesian, MystBot, Addbot, FrescoBot, Jonesey95, Plucas58, Tom.Reding, Nacen, Earthandmoon, Solarflare100, ZéroBot, Sgerbic, FConway-Derley, Bibcode Bot, Giambrox, Hmainsbot1, Stamptrader, Leegrc, Tetra quark, Sro23 and Anonymous: 4
- **Shape of the universe** *Source:* https://en.wikipedia.org/wiki/Shape_of_the_universe?oldid=788219961 *Contributors:* Bryan Derksen, Timo Honkasalo, The Anome, JeLuF, Olivier, Boud, Michael Hardy, Oliver Pereira, Menchi, Dcljr, Looxix~enwiki, Darkwind, Mark Foskey, Cyan, AugPi, Timwi, Reddi, Furrykef, Phys, Jusjih, BenRG, Fredrik, Romanm, Stewartadcock, Sverdrup, Rursus, Hadal, UtherSRG, Smjg, Graeme Bartlett, Just Another Dan, Isidore, LucasVB, Erikp, Clemwang, Mike Rosoft, Rfl, JTN, Discospinster, Pjacobi, Ascánder, SpookyMulder, Bender235, El C, Art LaPella, Bobo192, Smalljim, C S, 9SGjOSfyHJaQVsEmy9NS, Franl, Wayfarer, Schaefer, Wtmitchell, Eddie Dealtry, DV8 2XL, Gene Nygaard, Oleg Alexandrov, Philthecow, Jersyko, Jeff3000, GregorB, Joke137, BD2412, Qwertyus, Drbogdan, Rjwilmsi, Mike Peel, DVdm, Wavelength, Gaius Cornelius, ONEder Boy, Justin Eiler, Froth, Leptictidium, 2over0, Caco de vidro, KnightRider~enwiki, SmackBot, Incnis Mrsi, Auctoris, Gilliam, Bluebot, PrimeHunter, Nbarth, Hve, Bear Eagleson, Jbergquist, Captainbeefart, Lambiam, J 1982, Disavian, AstroChemist, JorisvS, 041744, Ckatz, Macellarius, DabMachine, Richard Nowell, Twas Now, Amakuru, Blehfu, Jonathan W, CRGreathouse, CmdrObot, Laplacian, JohnCD, Linus M., Jsd, Shirulashem, DarkLink, Repliedthemockturtle, Kablammo, Headbomb, I do not exist, Peter Gulutzan, Rriegs, Porqin, AntiVandalBot, Sluzzelin, MER-C, .anacondabot, TassadarAlpha, VoABot II, Tripbeetle, Fusionmix, Tonicthebrown, CommonsDelinker, Alexrussell101, Maurice Carbonaro, OttoMäkelä, Sarregouset, Sheliak, Taiwania Justo, VolkovBot, John Darrow, Red Act, Qxz, Manchurian candidate, UnitedStatesian, BotKung, Wiae, SwordSmurf, BOTijo, AlleborgoBot, Drschawrz, SieBot, Flyer22 Reborn, Shahidur Rahman, Sunrise, Laurentseries, Contestcen, Sphilbrick, ClueBot, Masamafo, Cliff, LittleDevil0071, Margit rudy, ChandlerMapBot, DumZiBoT, Nathan Johnson, John318, ErgoSum88, Richard-of-Earth, Addbot, Roentgenium111, Willking1979, Ersik, Redheylin, Debresser, Prim Ethics, Lightbot, Yobot, Aldebaran66, AnomieBOT, Gallaiis, Materialscientist, Citation bot, Brightgalrs, TechBot, Gap9551, Srich32977, Omnipaedista, Point-set topologist, RibotBOT, MeDrewNotYou, Peter470, Scientia-knowledge-knowing, FrescoBot, Paine Ellsworth, Majopius, Sae1962, Citation bot 1, Pinethicket, Jonesey95, Tom.Reding, Night Jaguar, OxEdit, Mathwordedit, Orenburg1, Mono, FuzzyBS, Spacestationinfo, Aoidh, Earthandmoon, Turnley, WildBot, DASHBot, EmausBot, Takahiro4, Lunaibis, The Mysterious El Willstro, Dcirovic, Cmck1980, AsceticRose, Solomonfromfinland, Thecheesykid, Traxs7, JoeSperrazza, ClueBot NG, SpikeTorontoRCP, Wcherowi, Gilderien, J kay831, LMFIV, Scottip3, Helpful Pixie Bot, Bibcode Bot, BG19bot, Mn-imhotep, Urltom, Unnown1, BattyBot, StarryGrandma, U-95, Dexbot, Caroline1981, CuriousMind01, LTWoods, Dan23234, Cadillac000, Me, Myself, and I are Here, Epicgenius, DASL51984, HFEO, Defiantdeacon, Evonj, Kaulfuss, Jadhachem, Prokaryotes, Empresschild, Kdmeaney, Christer Berner, ErinPancakes, Anonymous-232, Npmitchell, Yoshi24517, Monkbot, Maxwell Verbeek, AHusain3141, Eteethan, SpiritUVTruth, Bapehu, Tetra quark, Isambard Kingdom, Tony North, Jmc76, Israel Folau, Kitcher45, L3erdnik, John "Hannibal" Smith, W33dscoper, Tomaz 2 and Anonymous: 183
- **High-Z Supernova Search Team** *Source:* https://en.wikipedia.org/wiki/High-Z_Supernova_Search_Team?oldid=732165080 *Contributors:* Bearcat, RJHall, Nobi, 9SGjOSfyHJaQVsEmy9NS, Falcorian, Pol098, Frankie1969, Mandarax, DuoDeathscyther 02, Tim Pierce, Joseph Solis in Australia, Alaibot, Headbomb, Brianpschmidt, Doesper, Lbeaumont, Quantling, Nsuntzeff, Kyle the bot, UnitedStatesian, Puzhok, Addbot, Backpacker314159, AnomieBOT, ClarketheK, Anna Frodesiak, FrescoBot, Nacen, Earthandmoon, ZéroBot, Techmaniak, Cgt, ClueBot NG, BG19bot, HiddenValley123, Astrodud, Stridgway, ChrisGualtieri, BerdanII, Fy8F2fbP8, Tetra quark, Astro4686, OmniBot and Anonymous: 20
- **Type Ia supernova** *Source:* https://en.wikipedia.org/wiki/Type_Ia_supernova?oldid=783984761 *Contributors:* Roadrunner, Nv8200pa, Topbanana, Rursus, Melikamp, Cyclopia, Neko-chan, RJHall, El C, Shenme, Quaoar, Alansohn, Ceyockey, Falcorian, Oleg Alexandrov, Etacar11, K Lepo, Chrkl, Drbogdan, Rjwilmsi, Mike Peel, Hairy Dude, Hellbus, Bjf, Długosz, Ilmari Karonen, Algae, Attilios, SmackBot, Gilliam, Modest Genius, J 1982, JorisvS, IronGargoyle, Soulkeeper, George The Dragon, Rock4arolla, Nehrams2020, Newone, Nutster, Robertinventor, Keraunos, Headbomb, Nick Number, Orionus, Igodard, WolfmanSF, R'n'B, CommonsDelinker, Fconaway, Ihutchesson, Tambora1815, KylieTastic, DorganBot, Idioma-bot, Vestboy Myst, Rei-bot, UnitedStatesian, BotKung, AlleborgoBot, KGyST, LeadSongDog, Reuqr, Lightmouse,

Nergaal, Fedhere, ClueBot, GambitNC, StigBot, Noca2plus, Sun Creator, Crowsnest, Mchaddock, Drewtaylor1978, SkyLined, Hobbema, Addbot, DOI bot, Sn2007Cb, Proxima Centauri, LinkFA-Bot, Somerandomjackass, Wikbot, Wisconsinsurfer, Yobot, Piano non troppo, Citation bot, Maxis ftw, Xqbot, WingedSkiCap, Lithopsian, ChristopherKingChemist, Mnmngb, Fotaun, FrescoBot, LucienBOT, Paine Ellsworth, Originalwana, JohnMenninghaus7, Citation bot 1, Emaus, Jonesey95, Tom.Reding, Jim Fitzgerald, Fartherred, IVAN3MAN, Trappist the monk, Ukmaxi, JLincoln, Louiselives, TGCP, EmausBot, WikitanvirBot, Primefac, GoingBatty, Sp33dyphil, Jmencisom, Luiscalcada, EMurciano, Solarflare100, ZéroBot, Quondum, Daniel.kassl, ChuispastonBot, Mikhail Ryazanov, ClueBot NG, Phidica, Helpful Pixie Bot, Calabe1992, Bibcode Bot, BG19bot, Narayan89, Vagobot, AvocatoBot, Mark Arsten, Gallina3795, MythosMagic, Zedshort, Wer900, BattyBot, Ecegdagdeviren, Khazar2, Forty Seven Nine, Dexbot, Rfassbind, Devious3144, YiFeiBot, Kogge, Snkoysean, TomLoredo, Craneis, Monkbot and Anonymous: 58

- **Cosmic distance ladder** *Source:* https://en.wikipedia.org/wiki/Cosmic_distance_ladder?oldid=783590000 *Contributors:* Roadrunner, Patrick, Michael Hardy, Alfio, Chris 73, Babbage, Inkling, Foot, Mperrin, Radius, Peter Ellis, HorsePunchKid, Beland, WhiteDragon, Urhixidur, Tsemii, Julianonions, Rich Farmbrough, Pie4all88, Night Gyr, Bender235, RJHall, Pt, Lycurgus, 9SGjOSfyHJaQVsEmy9NS, Vuo, WilliamKF, Xover, StradivariusTV, BD2412, Rjwilmsi, Quiddity, Nihiltres, Gurch, Mathrick, DVdm, Wavelength, Ospalh, Mtu, SmackBot, Ashill, Onebravemonkey, Bluebot, Bidgee, Rick7425, ALK, JH-man, JorisvS, Physis, Onionmon, JForget, Markjoseph125, CRGreathouse, MaxEnt, Thijs!bot, Oerjan, Publicola, Headbomb, Dawnseeker2000, Felix116, Magioladitis, VoABot II, BSVulturis, Galukalock, Jim.henderson, PrestonH, Hans Dunkelberg, Maurice Carbonaro, TomS TDotO, BobEnyart, It Is Me Here, Bengandesbery, Eragontherider123456789, TXiKiBoT, Oshwah, Wingedsubmariner, FKmailliW, Glob.au, Coronellian~enwiki, Macdonald-ross, SieBot, ClueBot, Agge1000, Brews ohare, MilesAgain, RexxS, Jhorthos, Hobbema, Addbot, Mortense, Jncraton, Sn2007Cb, Chzz, 84user, Lightbot, Luckas-bot, AnomieBOT, Yunus.sendag, Citation bot, ArthurBot, GreaterPoobah6, Gap9551, Ataleh, Lhuntr, MeDrewNotYou, A. di M., Fotaun, Constructive editor, Kxx, FrescoBot, Citation bot 1, Jonesey95, Tom.Reding, IVAN3MAN, Trappist the monk, Tbhotch, NotKlayman, Octaazacubane, Quantanew, Iamchenzetian, Jmencisom, Josve05a, Dondervogel 2, Ὁ οἶστρος, Suslindisambiguator, Kristijh, ClueBot NG, P0lise, Frietjes, Braincricket, Doctore, Helpful Pixie Bot, Gob Lofa, Bibcode Bot, BG19bot, Vagobot, ElphiBot, BML0309, CitationCleanerBot, Zedshort, The1337gamer, Khazar2, Avneref, Lugia2453, Wywin, DavidLeighEllis, Kogge, Dough34, Coreyemotela, Tushar Shrotriya, Monkbot, Sangdeboeuf, Tetra quark, KasparBot, Pulsarwind, Magic links bot and Anonymous: 78
- **Scale factor (cosmology)** *Source:* https://en.wikipedia.org/wiki/Scale_factor_(cosmology)?oldid=781018046 *Contributors:* XJaM, Patrick, Fropuff, Dbachmann, El C, Chrisvls, 9SGjOSfyHJaQVsEmy9NS, Keenan Pepper, Ketiltrout, Sjakkalle, Rjwilmsi, TiagoTiago, Geoffrey.landis, Caco de vidro, SmackBot, Snori, Jbergquist, Lambiam, JHunterJ, Hypnosifl, Stanlekub, Stebbins, Michael C Price, Peter Gulutzan, Neilljones, Hamiltondaniel, Vreezkid, Brews ohare, Panos84, Addbot, Yobot, Ht686rg90, AnomieBOT, ??, Tom.Reding, Helpful Pixie Bot, Bibcode Bot, Cyberbot II, YiFeiBot, Monkbot, Tetra quark, GreenC bot, Bender the Bot and Anonymous: 30
- **Vacuum energy** *Source:* https://en.wikipedia.org/wiki/Vacuum_energy?oldid=784933143 *Contributors:* Bryan Derksen, The Anome, Roadrunner, Mjb, SebastianHelm, Looxix~enwiki, Ahoerstemeier, William M. Connolley, Cyan, Evercat, Reddi, Selket, Asser hassanain, Phys, Ann O'nyme, Fairandbalanced, Hemanshu, BovineBeast, Nagelfar, Jyril, Herbee, Markus Kuhn, Unconcerned, OldakQuill, Utcursch, Superborsuk, Icairns, Haisch, Pjacobi, Pavel Vozenilek, Houston~enwiki, Nabla, Smalljim, La goutte de pluie, Mc6809e, ShardPhoenix, Kdau, Drbreznjev, Axeman89, Kazvorpal, Dan100, Linas, Timharwoodx, Lofor, Mpatel, Arzachel, GregorB, Georgelazenby, Dwype, Hunterd, Bambaiah, Hillman, RussBot, Ytrottier, Bill52270, Salsb, Zunaid, Georgewilliamherbert, Enormousdude, Xaxafrad, Petri Krohn, Moonsleeper7, KasugaHuang, SmackBot, Master Jay, Jjalexand, Rrburke, Radagast83, Ne0Freedom, Yevgeny Kats, JorisvS, Mgiganteus1, Cerowyn, Ckatz, Theflyingman, Tawkerbot2, Chetvorno, Patrickwooldridge, CRGreathouse, Geremia, Vyznev Xnebara, Colombiano21, Burkedavis, Arnavion, Hamish Ross!!, Cydebot, Dr.enh, Mbell, Kathovo, Hcobb, Nick Number, Dawnseeker2000, Cstreet, Alphachimpbot, JAnDbot, VoABot II, Nyq, SHCarter, Email4mobile, GeorgeDishman, R'n'B, Leyo, Natsirtguy, ColdCase, Dextrose, Irwin McLean, Liometopum, UnitedStatesian, Enviroboy, Paradoctor, Mverleg, Arjen Dijksman, Henry Delforn (old), BartekChom, Dlrohrer2003, Jonathanstray, ClueBot, CohesionBot, Estirabot, Brews ohare, DCCougar, Imagine Reason, Addbot, Gravitophoton, Ntesla66, Deamon138, WikiDreamer Bot, Quantumobserver, Fushigi-kun, Yobot, Aldebaran66, IW.HG, Magog the Ogre, AnomieBOT, Jim1138, Materialscientist, Omnipaedista, FrescoBot, Paine Ellsworth, Merongb10, Tom.Reding, Johann137, Netheril96, Brandmeister, ClueBot NG, Harizotoh9, Physicsch, Ugurbost, YFdyh-bot, Ginsuloft, Kamil.schumann, Jiale8331, Haxxorz596, AHusain3141, TheCoffeeAddict, CaptainSmegma, CLCStudent, Sparkyscience and Anonymous: 110
- **Dark energy** *Source:* https://en.wikipedia.org/wiki/Dark_energy?oldid=786425853 *Contributors:* The Anome, Dachshund, Roadrunner, Schewek, Stevertigo, Thesteve, Nealmcb, Michael Hardy, Tim Starling, FrankH, Bobby D. Bryant, SebastianHelm, Ahoerstemeier, Glenn, Tristanb, Reddi, Wik, DW40, Dragons flight, Anupamsr, Pierre Boreal, BenRG, Jeffq, Donarreiskoffer, Robbot, Zandperl, Korath, Scott McNay, Vespristiano, Peak, Gandalf61, Rursus, Mlaine, UtherSRG, SC, Mattflaschen, Acm, Ancheta Wis, Giftlite, Graeme Bartlett, Awolf002, Jyril, Art Carlson, Herbee, Perl, Curps, Henry Flower, Gzornenplatz, Manuel Anastácio, Andycjp, BruceR, LucasVB, Antandrus, Beland, Karol Langner, Kevin B12, Bbbl67, Urvabara, JimJast, Discospinster, Rich Farmbrough, Pjacobi, Vsmith, D-Notice, Dbachmann, Bender235, Eric Forste, RJHall, JustinWick, Omnibus, El C, Lycurgus, Jomel, Kwamikagami, Frankenschulz, RoyBoy, Stesmo, Reuben, Russ3Z, 9SGjOSfyHJaQVsEmy9NS, Diego Moya, Keenan Pepper, Slugmaster, Axl, Benna, Wtmitchell, RainbowOfLight, Mikeo, Vuo, Freyr, DV8 2XL, Kazvorpal, Falcorian, Velho, Batintherain, Hottscubbard, OwenX, Mindmatrix, FeanorStar7, Velvetsmog, Uncle G, Netdragon, Jeff3000, GregorB, Isnow, SDC, ??????, Joke137, Abd, Christopher Thomas, Sneakums, Dysepsion, BD2412, Doc Savage, Malangthon, RadioActive~enwiki, Drbogdan, Loris Bennett, Rjwilmsi, Strait, TheRingess, Salleman, HappyCamper, Sohmc, Ems57fcva, Afterwriting, DonJuan~enwiki, BitterMan, Tomer Ish Shalom, Srleffler, Smithbrenon, CJLL Wright, Chobot, DVdm, Wavelength, RobotE, SamuelR, Diliff, Bhny, Stephenb, CambridgeBayWeather, Merick, NawlinWiki, Msikma, FFLaguna, LiamE, SCZenz, FoolsWar, Bota47, Rwxrwxrwx, Daniel C, Enormousdude, 2over0, Helge Rosé, Pb30, Dr.alf, Joedixon, Rlove, Geoffrey.landis, Ilmari Karonen, Moonsleeper7, Kungfuadam, Bernd in Japan, GrinBot~enwiki, Treesmill, Attilios, SmackBot, Ashill, Saravask, Bayardo, Tom Lougheed, InverseHypercube, KnowledgeOfSelf, Melchoir, J.Sarfatti, Nickst, Silverhand, Edgar181, Vixus, Gilliam, Skizzik, Jlsilva, Andy M. Wang, Tyciol, Sirex98, Oli Filth, DHN-bot~enwiki, Sbharris, Colonies Chris, Jdthood, Hgrosser, Can't sleep, clown will eat me, ThePromenader, PoiZaN, Chlewbot, Joema, Cybercobra, Lpgeffen, Rpf, Kendrick7, Cchambers, Byelf2007, Rory096, Boradis, Titus III, Richard L. Peterson, Xerxesx18, Writtenonsand, JorisvS, Mgiganteus1, Ckatz, Hypnosifl, Megane~enwiki, Ryulong, Quaeler, Dan Gluck, Spebudmak, Paul venter, Cxat, UncleDouggie, Courcelles, Tawkerbot2, JRSpriggs, Atomobot, Trevor.tombe, JForget, CRGreathouse, Ale jrb, Lavateraguy, Banedon, Nadyes, Mlsmith10, Arnavion, Logical2u, Rob Maguire, Cydebot, Stebbins, Gmusser, 879(CoDe), Rracecarr, Soetermans, Michael C Price, Chrislk02, Kozuch, Landroo, Thijs!bot, Headbomb, Marek69, Electron9, Second Quantization, Chris goulet, Davidhorman, Turelli, Dawnseeker2000, AntiVandalBot, Orionus, Gnixon, Fayenatic london,

Tim Shuba, Empyrius, Archmagusrm, AstroPaul, Bagster, JAnDbot, Carl1011, Davewho2, MER-C, CosineKitty, Rkomatsu, Michael Wood-Vasey, Felix116, Acroterion, WolfmanSF, Bongwarrior, VoABot II, Tripbeetle, LordCémOnur, Seleucus, Kevinwiatrowski, Ours18, DerHexer, Nevit, Robin S, Simplizissimus, NatureA16, Johann1870, Jimmilu, ARCG, Niclisp, TechnoFaye, Christian424, Thirdright, J.delanoy, Trusilver, Maurice Carbonaro, Natty4bumpo, Komowkwa, OttoMäkelä, Jlechem, Tsuite, SJP, Videokunst~enwiki, Malerin, Jorfer, Potatoswatter, Cmichael, KylieTastic, DorganBot, Jcmargeson, Ross Fraser, Ja 62, JHussein, Jjabellar, Sheliak, Johnassassin, Caribbean H.Q., VolkovBot, ColdCase, JohnBlackburne, D A Patriarche, AlnoktaBOT, Fences and windows, Philip Trueman, Darren22, HowardFrampton, TXiKiBoT, Oshwah, Dwight666, Matthias Buchmeier, Zanardm, Someguy1221, Oxfordwang, Jackfork, UnitedStatesian, Mazarin07, Venny85, Goaliemaster121, SwordSmurf, Lamro, RayNorris, Fourthark, Wanchung Hu, Obsidianmile, Radical Robert, Noncompliant one, Donauland~enwiki, PlanetStar, TrulyBlue, Murad.Shibli, Likebox, Flyer22 Reborn, Hotdiggity, TimothyFreeman, Avidallred, Faradayplank, Poindexter Propellerhead, IdreamofJeanie, OKBot, Aquijex, Loren.wilton, Martarius, BillWilliam, ClueBot, Dead10ck, The Thing That Should Not Be, Rodhullandemu, SuperHamster, Andwor, Tms9, Jusdafax, Da rulz07, Barbarinaz, Kentgen1, Razorflame, Stevecrye, AC+79 3888, Pillar of Babel, TimothyRias, Gwark, Ost316, PL290, MikeSmith10, Parejkoj, Andreaprins, Dgirl1723, HexaChord, D.M. from Ukraine, Addbot, Gravitophoton, Uruk2008, DOI bot, Nernom, LaaknorBot, Adfellin, Glane23, Delaszk, ChenzwBot, Jasper Deng, Sophia8891, Combatman~enwiki, Craigsjones, Arbitrarily0, Gurusoft2, Cosmos72, Luckas-bot, Yobot, Cosoce, Systemizer, Aldebaran66, Fulcanelli, Amble, AnomieBOT, Iluziat, Materialscientist, Citation bot, Icosmology, ArthurBot, Xqbot, S h i v a (Visnu), Sionus, Drilnoth, Wperdue, Tomwsulcer, Gap9551, BLP-outrageous move logs, ProtectionTaggingBot, Mathonius, Shadowjams, Finncarey, PrimeMatter, FrescoBot, Paine Ellsworth, Tobby72, Sławomir Biały, Zero Thrust, Kvgyarmati, Woodingdean, Alpha plus (a+), Citation bot 1, Redrose64, Pinethicket, I dream of horses, Jonesey95, Three887, Tom.Reding, Shahidur Rahman Sikder, Efficiency1101e, Casimir9999, Aknochel, IVAN3MAN, Meier99, BradTheBadWiki, Trappist the monk, TADEET, Jordgette, Heurisko, Michael9422, Adi4094, Earthandmoon, Wellsmax, RjwilmsiBot, Alph Bot, EmausBot, Mmpcq, Grrow, Quantanew, RA0808, Slightsmile, Dcirovic, Italia2006, NicatronTg, H3llBot, Suslindisambiguator, Paulstarpaulstar, Frigotoni, Colin.campbell.27, Iiar, HCPotter, Tunborough, RockMagnetist, Herk1955, Deathglass, DASHBotAV, Fire Vortex, Mjbmrbot, Yceren Loq, ClueBot NG, Ccalen, Chester Markel, Matias Pocobi, Jj1236, Frietjes, Habil zare, Helvitica Bold, Curb Chain, Bibcode Bot, BG19bot, Gordonben, Cheeseray1, FiveColourMap, Hippokrateszholdacskai, Yizlpku, Snow Blizzard, Gerhardtschmerhardt, Migrainus, Mcspaans, Szczureq, Unclejoe0306, Guanghuilin, Akshay Lattimardi, Dexbot, CityOfUr, CuriousMind01, Wjs64, JustAMuggle, WorldWideJuan, Epicgenius, Yheyma, MiceEater, LindaYeah, DavidLeighEllis, Federicoturner, Babitaarora, Isateach, Onecreation, Prokaryotes, Christophe1946, BerdanII, Anrnusna, Stamptrader, Suelru, 22merlin, Monkbot, Mlsmith55, Haxxorz596, Ceosad, THemanRE$%S23, Jnojha007, Richard.drapeau, UrDreamViola, MF22, ChamithN, Larsyxa, EpicLX, Tibenas, Mediavalia, ScrapIronIV, 39Debangshu, Iazyges, Anunaki truth, Tetra quark, Isambard Kingdom, Absolutelypuremilk, Anand2202, GeneralizationsAreBad, Jman135, Jtrrs0, Grammarian3.14159265359, KasparBot, ShankZeTank, Tgorewic, Sir Cumference, ShiningSword, Esadri21, Phseek, Srisri19962003, TychosElk, Buckbill10, Farank olamaeian, Alopresti777, Themalina, Sire-Wonton, Adithya2804, The Cube Root Of Infinity, Khrpr, Kigaei, Gerald wish, Latex-yow, Dr. Hung M. Choi, GreenC bot, Soopdish, John "Hannibal" Smith, Aaron.iji3, Bender the Bot, Bobsala, Sparkyscience, MinervasOwl, DbDotNet1, Cleko and Anonymous: 573

- **Dark matter** *Source:* https://en.wikipedia.org/wiki/Dark_matter?oldid=788281335 *Contributors:* AxelBoldt, Chenyu, Derek Ross, CYD, BF, Bryan Derksen, The Anome, Tarquin, Taw, XJaM, Arvindn, William Avery, Roadrunner, Mintguy, Bth, Stevertigo, Edward, Nealmcb, Boud, FrankH, Cprompt, DopefishJustin, Bobby D. Bryant, Ixfd64, SebastianHelm, Alfio, CesarB, Looxix~enwiki, Mkweise, William M. Connolley, JWSchmidt, Glenn, Mxn, Charles Matthews, Timwi, Fuzheado, Rednblu, Haukurth, DW40, Dragons flight, Furrykef, Saltine, Dogface, Populus, Jusjih, Finlay McWalter, Bearcat, Robbot, Zandperl, Korath, Nurg, Naddy, Arkuat, Gandalf61, Pingveno, Rursus, Rtfisher, Wereon, Diberri, Adam78, Aasim75, Marc Venot, Ancheta Wis, Giftlite, Graeme Bartlett, Laudaka, Barbara Shack, Herbee, Fropuff, Xerxes314, Dratman, Curps, Joconnor, Jdavidb, Unconcerned, Eequor, Bobblewik, Andycjp, Alexf, Geni, Antandrus, HorsePunchKid, Melikamp, PDH, Rdsmith4, Anythingyouwant, Bosmon, Bbbl67, Icairns, Sam Hocevar, Cynical, Lumidek, Iantresman, Burschik, Joyous!, Adashiel, Urvabara, Discospinster, Rich Farmbrough, Oliver Lineham, Vsmith, Jpk, ArnoldReinhold, Murtasa, D-Notice, JPX7, KaiSeun, SpookyMulder, Bender235, Kjoonlee, Kaisershatner, Pk2000, PsychoDave, RJHall, Mr. Billion, El C, Huntster, Bletch, PhilHibbs, Shanes, Frankenschulz, Art LaPella, RoyBoy, Themusicgod1, Bobo192, Smalljim, Shenme, Cmdrjameson, Reuben, Kmaguire, 9SGjOSfyHJaQVsEmy9NS, Zelda~enwiki, Mr. Brownstone, E is for Ian, Jumbuck, Storm Rider, Alansohn, Gary, Anthony Appleyard, Guy Harris, Eric Kvaalen, Arthena, Keenan Pepper, Kocio, Bart133, RPellessier, Benna, Wtmitchell, ClockworkSoul, Cal 1234, Count Iblis, Guthrie, H2g2bob, Bsadowski1, GabrielF, Pauli133, Leondz, DV8 2XL, Gene Nygaard, Feline1, Oleg Alexandrov, Brookie, Natalya, Flying fish, WilliamKF, Yeastbeast, Mindmatrix, RHaworth, Plek, BillC, JPFlip, Benbest, JFG, ?, WadeSimMiser, Gxojo, MONGO, Jwanders, Torqueing, ??????, Joke137, Wisq, Christopher Thomas, Palica, Mandarax, RedBLACKandBURN, Aarghdvaark, RichardWeiss, Ashmoo, Graham87, Seb-Gibbs, Malangthon, Mamling, Jclemens, Drbogdan, Loris Bennett, Rjwilmsi, Lars T., Strait, Patrick Gill, Tangotango, Tawker, Smithfarm, Stevenscollege, Mike Peel, HappyCamper, SeanMack, ScottJ, Bubba73, Krash, Dermeister, Rangek, Madcat87, FlaBot, Naraht, Ian Pitchford, PlatypeanArchcow, A scientist, Margosbot~enwiki, Gark, Nivix, Gparker, Pathoschild, Gurch, Stevenfruitsmaak, Goudzovski, Tomer Ish Shalom, Smithbrenon, Chobot, Moocha, DVdm, Bgwhite, Gwernol, The Rambling Man, YurikBot, Wavelength, RobotE, Koveras, Hairy Dude, Huw Powell, Phmer, Hillman, RussBot, Michael Slone, Ohwilleke, Bhny, JabberWok, GLaDOS, DanMS, Zelmerszoetrop, Stephenb, Eleassar, Merick, Big Brother 1984, NawlinWiki, Alpertron, Długosz, Schlafly, FFLaguna, BlackAndy, Dbmag9, SCZenz, Haoie, Raven4x4x, Ospalh, Durval, Bota47, Supspirit, Pegship, Noosfractal, Charlie Wiederhold, WAS 4.250, Smoggyrob, Reyk, Tvaughan, Joedixon, Eric TF Bat, Emc2, Ilmari Karonen, Allens, Bernd in Japan, InsayneWrapper, Bclayabt, Attilios, MacsBug, SmackBot, Cubs Fan, Ashill, IddoGenuth, Tomer yaffe, Stellea, InverseHypercube, KnowledgeOfSelf, Allixpeeke, Clpo13, Nickst, RedSpruce, Nightbat, Doc Strange, Herbm, Edgar181, HalfShadow, Flux.books, Dheerajkakar, Yamaguchi??, Richmeister, Gilliam, Folajimi, The Gnome, Oscarthecat, Skizzik, Kmarinas86, Chris the speller, SuperBuuBuu, Quinsareth, Persian Poet Gal, Sirex98, MalafayaBot, Silly rabbit, Sangrolu, Villarinho, DHN-bot~enwiki, Sbharris, Hongooi, Jdthood, CheerLeone, Modest Genius, Gtkysor, Can't sleep, clown will eat me, Nick Levine, Tamfang, Kelvin Case, V1adis1av, Vanished User 0001, Rrburke, Jgoulden, Auvii, Krich, Wen D House, Radagast83, Engwar, Nakon, VegaDark, John D. Croft, Alexander110, KimO, Adrigon, SpiderJon, Ultraexactzz, Zadignose, Tesseran, Byelf2007, L337p4wn, K7lim, SashatoBot, Mchavez, Swatjester, Leftydan6, Minaker, Attys, Brillow, John, Ashoat, Scientizzle, J 1982, Acitrano, Gobonobo, Linnell, JoshuaZ, James.S, JorisvS, Coredesat, Goodnightmush, ICBB, Plunge, JHunterJ, Hypnosifl, Silverthorn, Descubes, Freederick, Dr.K., Vanished user, Iridescent, Darkerprojects, Astrobayes, Newone, MOBle, Igoldste, CapitalR, AGK, Courcelles, Tawkerbot2, Dlohcierekim, Chetvorno, Hammer Raccoon, Owen214, Eastlaw, Peledre, Pukkie, Anakata, Banedon, Runningonbrains, DKOH, NickW557, Gregbard, MikeWren, Phatom87, Vttoth, Necessary Evil, Ryan, Viciouspiggy, Gogo Dodo, Anonymi, Xxanthippe, A Softer Answer, Odie5533, Tawkerbot4, Doug Weller, DumbBOT, Robertinventor, Kozuch, Mtpaley, Philza85, Starship Trooper, UberScienceNerd, Crum375, Thijs!bot, Epbr123, Astroceltica, Passaggio, Barbarina, Mbell, Eugenespeed, N5iln, Mojo Hand, Carlif, Headbomb, Tonyle, Marek69, RickinBaltimore, Lars Lindberg Christensen, OtterSmith, SusanLesch, Dawnseeker2000, Mmortal03, Hmrox, Hires an editor, AntiVandalBot, Seaphoto, Orionus, Opelio, Shirt58, Rehnn83,

Joehodge, AaronY, Jj137, TTN, Dylan Lake, Chill doubt, Spencer, Yellowdesk, Sniktaw, Lfstevens, CPitt76, Gökhan, Jcarter1, ClassicSC, JAnDbot, Leuko, Husond, MER-C, CosineKitty, Plantsurfer, Mcorazao, Therealintellectual, Folkform, Balbers, 100110100, Autotheist, Wasell, Magioladitis, Bongwarrior, VoABot II, Timothy McVeigh, Charlesrkiss, AuburnPilot, Krkaiser, Mbarbier, Kaivosukeltaja, Foroa, Swpb, Stigmj, T a y l o s, Ekantik, Brusegadi, Bubba hotep, Fabrictramp, Catgut, Lilian.Kaufmann, Zhanghia, Acornwithwings, BatteryIncluded, Vssun, LtHija, Whisky5, Mcfar54, DerHexer, Prisca6023, PeteSF, KenyaSong, Rickard Vogelberg, NatureA16, DancingPenguin, MartinBot, Schmloof, STBot, Pagw, Fs644, Niclisp, Drewmutt, Anaxial, CommonsDelinker, Jean-Pierre Petit~enwiki, Gah4, PrestonH, WelshMatt, Dudley Miles, Chrishy man, Thirdright, J.delanoy, Pharaoh of the Wizards, Trusilver, Adavidb, Kpvats, Kudpung, Rod57, Arion 3x3, PedEye1, McSly, Tarotcards, Davy p, HiLo48, NewEnglandYankee, Ohms law, Jorfer, Blckavnger, Potatoswatter, KylieTastic, Joshua Issac, Infiniteglitch, Remember the dot, Pitpif, Vanished user 39948282, Neekap, Natl1, Ldebain, BernardZ, SoCalSuperEagle, Squids and Chips, Borat fan, CardinalDan, Idioma-bot, Sheliak, Funandtrvl, Lights, Hammersoft, VolkovBot, Craigheinke, Itsfullofstars, ColdCase, Jeff G., JohnBlackburne, Mocirne, AlnoktaBOT, Scikid, Grammarmonger, Hootis, Leojohns, Larry R. Holmgren, Philip Trueman, TXiKiBoT, Oshwah, Docanton, Authorized User, Theophilus reed, Drestros power, Strichek, MarekMahut, Monkey Bounce, Lradrama, Sintaku, Carillonatreides, Martin451, Broadbot, Wiae, Mazarin07, Inductiveload, Knightshield, Telecineguy, Spiral5800, Kurowoofwoof111, Greswik, RobertFritzius, SwordSmurf, Falcon8765, Hellothere17, Enviroboy, Littlehollah, Wanchung Hu, Dpleibovitz, Illumini85, SonOfMog Worf, Jazzman123, PGWG, 19merlin69, FlyingLeopard2014, Neparis, Bfpage, S-n-ushakov, SieBot, Calliopejen1, Tresiden, Wibubba48, Tachyonics, Pallab1234, Paradoctor, KGyST, Bentogoa, Jimlester51, Battlepace, Oda Mari, Aaarnooo, Suomichris, Crowstar, PromX1, Lightmouse, Tombomp, Cyberplasm, Diego Grez-Cañete, Spartan-James, Thinghy, Mygerardromance, Hamiltondaniel, Superbeecat, Denisarona, JL-Bot, Escape Orbit, Starcluster, Troy 07, Atif.t2, ArepoEn, Ak47gforce, Ratemonth, Sfan00 IMG, ClueBot, Phoenix-wiki, GorillaWarfare, The Thing That Should Not Be, ArdClose, Rodhullandemu, Cptmurdok, Drravisharma, Drmies, Frmorrison, Uncle Milty, Iuhkjhk87y678, Niceguyedc, MrBosnia, Bhaskarns, Andwor, Ktr101, Excirial, Dombom12, Cromescythe, Barbarinaz, FOARP, Brews ohare, Jotterbot, Iohannes Animosus, R.Andrae, Kentgen1, Ordovico, Mastertek, Rgoogin, Thehelpfulone, DanielPharos, 1ForTheMoney, Versus22, Palmer666palmer, PCHS-NJROTC, Burner0718, Pillar of Babel, SoxBot III, Erodium, Vanished user uih38riiw4hjlsd, 1ofhissheep, TimothyRias, Arianewiki1, XLinkBot, DCCougar, Oldnoah, Rror, Gwark, Sakura Cartelet, Ost316, Avoided, Webmaster369, Gthomson, Tugrul irmak, Noctibus, Ploversegg, ZooFari, Parejkoj, Tayste, Addbot, Xp54321, Grayfell, Experimental Hobo Infiltration Droid, Willking1979, Some jerk on the Internet, Uruk2008, 04aeverington, DOI bot, Tcncv, Non-dropframe, Nohomers48, CharlesChandler, Gmeyerowitz, Haasfelix, Download, Proxima Centauri, Ashirgo, RTG, Redheylin, Luke Maurer, Glane23, Darkmatter654, SamatBot, Nanzilla, Lzkelley, Clone 209, Tassedethe, Numbo3-bot, Peridon, Chinchinthehun, Evildeathmath, Tide rolls, Lightbot, OlEnglish, Qemist, Gail, North Polaris, Legobot, Artichoke-Boy, Luckas-bot, Yobot, WikiDan61, Cosoce, Dov Henis, Aldebaran66, KillYourLove, CzechFalcon, Amble, Mmxx, CinchBug, Perusnarpk, IW.HG, Einstein vs Dark energys, Eric-Wester, Tempodivalse, Synchronism, AnomieBOT, Letuño, Girl Scout cookie, IRP, JackieBot, RBM 72, AdjustShift, Nicolaas Vroom, Henrykandrup, Iluziat, Materialscientist, Dendlai, ImperatorExercitus, The High Fin Sperm Whale, Citation bot, Ternity0127, Maxis ftw, Frankenpuppy, Quebec99, LilHelpa, Aksel89, Xqbot, Stlwebs, Random astronomer, Sionus, Cureden, Jradis1337, Capricorn42, Wperdue, Julianhyde, Deleance, Raspw, Tomwsulcer, Magicxcian, Gap9551, AbigailAbernathy, Srich32977, NOrbeck, Artemis6234, Almabot, Abell 1367, Feldhaus, False vacuum, Omnipaedista, RibotBOT, Waleswatcher, Mikedr, Kongkokhaw, Rvnieuwe, Shadowjams, MeDrewNotYou, A. di M., Peter470, Sageman7, ??, Luminique, Captain-n00dle, Imyfujita, FrescoBot, Andyradke0, Ag allstar, Paine Ellsworth, Originalwana, Styxpaint, Mark Renier, VS6507, PhysicsExplorer, Dbirkhofer, Steve Quinn, Nestlefolife, Adrian Akau, 1414rwbt, SF88, Citation bot 1, Redrose64, DUUJEEGWEEM, Tyler6298, Pinethicket, I dream of horses, Grammarspellchecker, Danlof, 10metreh, Jonesey95, Tom.Reding, Pmokeefe, A8UDI, Brian Everlasting, For.a.limited.time.only, Elentirno, TedderBot, Aknochel, SkyMachine, IVAN3MAN, Kgrad, Nieuwenh, Trappist the monk, Puzl bustr, Fama Clamosa, Domeinthebumhole, Michael9422, UrukHaiLoR, Allen4names, JLincoln, Jeffrd10, Lovemybluetooth, Diannaa, Fastilysock (usurped), Innotata, DrCrisp, Whisky drinker, One15969, RjwilmsiBot, 5mgoblue5, Blakelewis122, Þorri, Mathewsyriac, Leandro.lelas, Mserard313, Mdznr, Sbugnon, Smd75jr, Ultima821, EmausBot, Francophile124, Grrow, Super48paul, GoingBatty, RA0808, Gimmetoo, Solarra, Jmencisom, Slightsmile, Tommy2010, Winner 42, SusanaMultidark, Gocows2, Wikipelli, Dcirovic, Serketan, Krifferjel, Zurich Astro, Hhhippo, Mz7, Mhatthei, Svolin, Micahqgecko, JSquish, Josve05a, Trojanmice, MithrandirAgain, Edwinkaren, Devilaza, Arbnos, Oraclan, Suslindisambiguator, SporkBot, AlbertusmagnusOP, Mcmatter, Tolly4bolly, L1A1 FAL, Ancient Anomaly, Rcsprinter123, Brandmeister, L Kensington, Maj den, Corabilek, Donner60, Aldnonymous, Ihardlythinkso, RockMagnetist, Terra Novus, TYelliot, DASHBotAV, Kroupap, D Phoesheezey, Mannix Chan, Travies10, Jxraynor, TheTimesAreAChanging, ClueBot NG, Rich Smith, Gareth Griffith-Jones, Afjvanraan, Lyla1205, Crystal7878, Catinthehat93, Bped1985, Infinifold, Wiggit002, Jj1236, PapaMike, MonEyshOt42069210, Muon, Esdacosta, Asukite, Masssly, Ph.d Carl edenburgh, Widr, Gavin.perch, Helpful Pixie Bot, Curb Chain, Calabe1992, Bibcode Bot, BG19bot, Dualus, Kishanparekh, Stevenwilkins, NacowY, Cheeseray1, Cyberguy5, Darkmatter adam, Yomomma8102, Hza a 9, Rarelight, Cyberpower678, Cosmologist77, MusikAnimal, தென்காசி சுப்பிரமணியன், Dahliamtl, Dodshe, Mark Arsten, Darkmatterotheruniverses, Samcstewart, Cadiomals, Trevayne08, Zedshort, Achowat, Rolandwilliamson, A2Die, Clint55555, Mgka79, NotEither, BattyBot, Millennium bug, Ronin712, Babymushrooms, Davidmexican, Drphilmarshall, Dilaton, Quin71901, U-95, ChrisGualtieri, Npmay, Kvark92, Lukasz.astrus, Ducknish, JYBot, Davidlwinkler, IjonTichyIjonTichy, Astrohap, Hunterf12, Dexbot, Caroline1981, BinaryPhoton, Gravityking100, Junavia, Fredrikdn, CuriousMind01, TowardsTheLight, Lugia2453, Wjs64, Andwor42, Frosty, Honneydewp243, Junjunone, DrHowzer73, JustAMuggle, Me, Myself, and I are Here, WadiElNatrun, Reatlas, Leprof 7272, Rfassbind, Acetotyce, I am One of Many, DirkXcal, Melonkelon, Ybidzian, Gig9876, M.ashrafinia, Trolololman12, Ilikedeletingstufffromhere, DavidLeighEllis, Onecreation, LahmacunKebab, The Herald, Zenibus, Jernahthern, JustBerry, Hipposaregrey, Frinthruit, Derekdoth, Stamptrader, Cyberalchemyst, Aaronknowsitall, VicLucas, FelixRosch, Chadchumley, Darkmer, Doubleknockout, Monkbot, Yikkayaya, Wardinstrument, Leegrc, Vikas Rauniyar, Apipia, Kinetic37, Upsalla, Jkvaternik, Lol kaptyn troll, HMSLavender, A Great Catholic Person, Mohammedshukoor, MRD2014, Deadwiper, Callum92, Stefania.deluca, Jiten Dhandha, Ashweigh, Waters.Justin, Mrpiecjohnson, Oldstone James, Astezar, 39Debangshu, YoYoDude012, Anunaki truth, Pyrotle, Tetra quark, Carazmatic, God of matterrr, Silversparkcontributions, Isambard Kingdom, VexorAbVikipædia, Rizi0909, Absolutelypuremilk, Tralala0, Anand2202, Kbap2002, Kb2002, DN-boards1, Yohoona, Denniscabrams, Omanyd, KasparBot, I love trains sooo much, Ira Leviton, Sir Cumference, Ricardo A. Olea, Id6040, MHolland85, TychosElk, Boowiebear, Stephane Le Corre, Mustachman71, Abdelrahmam shawky, Huritisho, Jeffman257, The Cube Root Of Infinity, Outedexits, Reg7d88, Maka Tree, Eslam nsr, Incendiary Iconoclasm, GSS-1987, Boobety boop boop boop, Atharv4321, Zhakhan9er, XABHISHEK23x, Elaysungur, Not a creative person, The Voidwalker, Shane ducharme, Tyler kwejew, Igojamon, Iamcoolswag, Riki71144, Kdkskdfj, Soopdish, SkyWarrior, Mr ST 142, Portal da Ciência, WikiEditor79, Wiki62838351936, MoneyMonkey112, Aidos the great, Trefoily+, ZeroDaVinci, FabulousFerd, EeeveeeFrost, Kuffey, RichardLewis41, Bender the Bot, J.Kaesser, Grigorii-Torkel, Davidfpaul, George Granville, PowerLien, Stevejb904, Magicmesh48, Goodboi123, Zingvin, Classicalphysics, JonHMDavis, Hyperbolick, Gimtired, Home Lander, Tompholmen, Telly1414, MinervasOwl, Karlpoppery, DbDotNet1, Alvateroks, Ugc454, Al Legorhythm and Anonymous: 1413

- **Dark fluid** *Source:* https://en.wikipedia.org/wiki/Dark_fluid?oldid=780797386 *Contributors:* Hike395, Dratman, Bbbl67, Urvabara, Discospinster, Rich Farmbrough, Pauli133, Kazvorpal, Camw, Uncle G, Rjwilmsi, Bhny, Limulus, Dan Harkless, Sbyrnes321, Tom Morris, OrganicMan, SmackBot, Nickst, Derek farn, John, Alaibot, Ebichu63, Tim Shuba, Lenticel, JohnBlackburne, Oshwah, Suraj.kapil.singh, Scog, Ost316, Addbot, Ashton1983, Lzkelley, Yobot, AnomieBOT, Westerness, Citation bot, StrontiumDogs, LilHelpa, Paine Ellsworth, Sae1962, Cogiati, Suslindisambiguator, RockMagnetist, ClueBot NG, Fergusnoble, Bibcode Bot, Markwilliamlee, Junjunone, Saectar, Olidog, Sizeofint, Tetra quark, Sir Cumference, The Cube Root Of Infinity and Anonymous: 29
- **Cold dark matter** *Source:* https://en.wikipedia.org/wiki/Cold_dark_matter?oldid=786983568 *Contributors:* Bryan Derksen, Roadrunner, Tim Starling, Llywrch, Looxix~enwiki, Schneelocke, Doradus, Rho~enwiki, Eequor, Christopherlin, Keith Edkins, Balcer, Karl Dickman, Rich Farmbrough, StephanKetz, 9SGjOSfyHJaQVsEmy9NS, Fwb22, Lysdexia, Uris, Rjwilmsi, Chobot, 2over0, Kungfuadam, KnightRider~enwiki, Nickst, Nightbat, MalafayaBot, Cmanser, Farseer, OhioFred, John, Robofish, Eridani, Joeyfox10, Alaibot, Mbell, Headbomb, CosineKitty, Magioladitis, Rod57, Ryan WMD, Idioma-bot, 0-Jenny-0, Michael H 34, BotKung, Mazarin07, Paradoctor, Niceguyedc, Auntof6, Addbot, Lightbot, Sebas310, Yobot, Systemizer, AnomieBOT, Materialscientist, GrouchoBot, Waleswatcher, IAP Astro, Erik9bot, Kikuyu3, Sae1962, Citation bot 4, Pinethicket, Jonesey95, Tom.Reding, Trappist the monk, Gwyneth99, RA0808, Slightsmile, Zurich Astro, AvicAWB, SporkBot, AThinkingScientist, ClueBot NG, CocuBot, BBCDM, Helpful Pixie Bot, Bibcode Bot, BG19bot, Dualus, Winston Trechane, Layzeeboi, Willoakley, Prokaryotes, Sjzaslaw, Monkbot, Franz Sciortino, Sofia Koutsouveli, Astronome de Meudon, Stefania.deluca, Galaxy Mantis, Jmesser5, TychosElk, John "Hannibal" Smith, Bear-rings, Jerry Rivera Mendrez, DbDotNet1 and Anonymous: 50
- **Baryon** *Source:* https://en.wikipedia.org/wiki/Baryon?oldid=774901288 *Contributors:* AxelBoldt, Tobias Hoevekamp, Bryan Derksen, Ben-Zin~enwiki, Heron, Tim Starling, Alan Peakall, Paul A, Salsa Shark, Glenn, Mxn, Charles Matthews, The Anomebot, ElusiveByte, Phys, Bevo, Traroth, Donarreiskoffer, Robbot, Korath, Kristof vt, Merovingian, Ojigiri~enwiki, Sunray, Wikibot, Giftlite, DocWatson42, ShaunMacPherson, Herbee, Xerxes314, Dratman, DÅ‚ugosz, Kaldari, OwenBlacker, Icairns, JohnArmagh, Rich Farmbrough, Guanabot, Mani1, E2m, Tompw, El C, Bobo192, 9SGjOSfyHJaQVsEmy9NS, Giraffedata, Physicistjedi, Jumbuck, Gary, ABCD, Oleg Alexandrov, Woohookitty, Tevatron~enwiki, BD2412, Kbdank71, Nightscream, Ae77, MZMcBride, Chekaz, R.e.b., Erkcan, Ttwaring, Maxim Razin, Oo64eva, Chobot, Roboto de Ajvol, YurikBot, Bambaiah, Jimp, Salsb, Ergzay, DragonHawk, SCZenz, E2mb0t~enwiki, Bota47, Simen, Sbyrnes321, Lainagier, Timotheus Canens, Bluebot, Colonies Chris, Kingdon, Shadow1, Bigmantonyd, Drphilharmonic, Kseferovic, Wierdw123, Physicsdog, Torrazzo, Verdy p, Michael C Price, Thijs!bot, Headbomb, Marek69, Hcobb, Orionus, QuiteUnusual, Spartaz, Plantsurfer, Amaterial1121, Diamond2, Swpb, BatteryIncluded, David Eppstein, Hveziris, Saxophlute, Gwern, Ben MacDui, R'n'B, Ash, Thirdright, Maurice Carbonaro, STBotD, VolkovBot, GimmeBot, NoiseEHC, Tearmeapart, BotKung, BrianADesmond, Antixt, AlleborgoBot, Lou427, SieBot, VVVBot, Gerakibot, LeadSongDog, Keilana, Paolo.dL, Doctorfluffy, TrufflesTheLamb, OKBot, Hamiltondaniel, TubularWorld, ClueBot, Artichoker, ChandlerMapBot, CalumH93, Addbot, LaaknorBot, CarsracBot, Ehrenkater, Jonhstone12, Legobot, Luckas-bot, Yobot, Bugbrain 04, AnomieBOT, JackieBot, Materialscientist, Citation bot, ArthurBot, Xqbot, Omnipaedista, SassoBot, Spellage, WaysToEscape, FrescoBot, Citation bot 1, Tom.Reding, FoxBot, Noommos, EmausBot, John of Reading, Dcirovic, JSquish, ZéroBot, StringTheory11, Stibu, Ethaniel, Markinvancouver, ClueBot NG, Koornti, Kasirbot, Rezabot, Bibcode Bot, Atomician, CitationCleanerBot, Zedshort, Marioedesouza, David.moreno72, ChrisGualtieri, WorldWideJuan, CoolHandLouis, Monkbot, KasparBot, Srednuas Lenoroc, LixinZheng, Bear-rings, Dukwon and Anonymous: 115
- **Equation of state (cosmology)** *Source:* https://en.wikipedia.org/wiki/Equation_of_state_(cosmology)?oldid=760711151 *Contributors:* AxelBoldt, Zandperl, Spiko-carpediem~enwiki, 9SGjOSfyHJaQVsEmy9NS, Joke137, Quale, Netrapt, Caco de vidro, EncMstr, Colonies Chris, E4mmacro, JRSpriggs, Kjknohw, Dr.enh, Michael C Price, Peter Gulutzan, STBot, Jorfer, Ruffwiki, VolkovBot, SwordSmurf, Neparis, Alexbot, Kentgen1, Scog, Addbot, Mathieu Perrin, Yobot, Maximilian Reininghaus, FrescoBot, ZéroBot, Bibcode Bot, Tetra quark, Sir Cumference, Csumstudent, TerrainAhead and Anonymous: 19
- **Lambda-CDM model** *Source:* https://en.wikipedia.org/wiki/Lambda-CDM_model?oldid=782277039 *Contributors:* Bryan Derksen, The Anome, Roadrunner, Boud, Michael Hardy, Dcljr, Charles Matthews, Timwi, Dragons flight, Forseti, Gandalf61, Wjhonson, Giftlite, Andycjp, Pjacobi, Vsmith, Jonathanischoice, AdamSolomon, Art LaPella, 9SGjOSfyHJaQVsEmy9NS, Diego Moya, Plumbago, Ceyockey, Joke137, Rnt20, Drbogdan, Rjwilmsi, Zbxgscqf, Mike Peel, Bubba73, Srleffler, Acefitt, Karch, YurikBot, Vuvar1, Gadget850, CharlesHBennett, Caco de vidro, McGeddon, Bluebot, Jdthood, Hve, Yannick Copin, Titus III, JorisvS, Beetstra, Hypnosifl, Spebudmak, Petr Matas, CmdrObot, Dr.enh, Thijs!bot, Headbomb, Vertium, Peter Gulutzan, Escarbot, Rico402, Huttarl, Drollere, Yobol, Iakane49, DAID, KylieTastic, STBotD, Sheliak, VolkovBot, RedAndr, MariAlexan, BotKung, SwordSmurf, Catdogqq, SieBot, Hertz1888, Droog Andrey, BartekChom, IlkkaP, Sunrise, Coldcreation, Duae Quartunciae, Astrohou, DragonBot, Copyeditor42, Telekenesis, Brews ohare, Cenarium, Scog, Panos84, Ich42, Parejkoj, Addbot, LaaknorBot, Yobot, Ptbotgourou, Amirobot, Amble, Azcolvin429, AnomieBOT, Citation bot, Xqbot, Gap9551, Dendropithecus, Omnipaedista, Mnmngb, FrescoBot, Paine Ellsworth, Craig Pemberton, SF88, Jonesey95, Tom.Reding, Pmokeefe, Puzl bustr, Sehatfield, Dr. Salvia, Earthandmoon, RjwilmsiBot, Ripchip Bot, Newty23125, John of Reading, Italia2006, A2soup, Midas02, Suslindisambiguator, Timetraveler3.14, Brandmeister, One.Ouch.Zero, Senator2029, Milk Coffee, Fire Vortex, ClueBot NG, Jj1236, Bibcode Bot, Technical 13, BG19bot, Flekkie, Harizotoh9, SoylentPurple, Darylgolden, Khazar2, Wjs64, Junjunone, Thewarriltonsiegedoc, Mcgaugh, Prokaryotes, Christophe1946, Mfb, Orrerysky, Sjzaslaw, Monkbot, Unatnas1986, Sofia Koutsouveli, Verdana Bold, Mof-tan, Tetra quark, Crito10, Jeffgr9, Duganc525, Jmc76, ????, TychosElk, Youknowwhatimsayin, P.Shiladitya, InternetArchiveBot, JGS952, Riki71144, Vive la Chimie, Saurabh astro, WikiMasterGhibif, Tosci, MinervasOwl and Anonymous: 78
- **Cosmological constant** *Source:* https://en.wikipedia.org/wiki/Cosmological_constant?oldid=786749618 *Contributors:* AxelBoldt, Magnus Manske, Vicki Rosenzweig, Bryan Derksen, The Anome, Ed Poor, Enchanter, William Avery, Roadrunner, Schewek, Hephaestos, Boud, Bcrowell, Lquilter, TakuyaMurata, Minesweeper, Stevenj, Kimiko, Samw, Timwi, Reddi, Asar~enwiki, Dogface, Bevo, Anupamsr, Johnleemk, BenRG, Phil Boswell, Robbot, Goethean, Wjhonson, Wereon, Giftlite, Bobblewik, Jonel, Rjpetti, Icairns, Rgrg, Burschik, JimJast, 4pq1injbok, Pjacobi, Vsmith, StephanKetz, Dbachmann, Pavel Vozenilek, Dmr2, Bender235, RJHall, Pt, El C, Frankenschulz, RoyBoy, Rbj, 9SGjOSfyHJaQVsEmy9NS, Knucmo2, Jumbuck, Falcorian, Angr, OwenX, Linas, StradivariusTV, Kzollman, Mpatel, Joke137, Wisq, Christopher Thomas, Rnt20, Ashmoo, Coneslayer, Rjwilmsi, Coemgenus, Nightscream, RE, Itinerant1, Srleffler, Chobot, Amaurea, PointedEars, YurikBot, Hillman, RussBot, Ytrottier, SpuriousQ, Gaius Cornelius, Salsb, Sir48, Muu-karhu, DeadEyeArrow, Helge Rosé, Petri Krohn, KasugaHuang, SmackBot, Incnis Mrsi, WilyD, Nickst, Cush, Colonies Chris, Avb, Cybercobra, Ligulembot, Yevgeny Kats, Lambiam, Matt489, Paladinwannabe2, Ckatz, Onionmon, Basicdesign, Newone, Sirwhiteout, Chetvorno, CmdrObot, Orannis, Hardrada, Mlsmith10, MaxEnt, Phatom87, Forthommel, Frostlion, Dr.enh, Michael C Price, Tawkerbot4, Clovis Sangrail, Christian75, Thijs!bot, Mathmoclaire, Headbomb, Peter Gulutzan, Gnarlyocelot, Escarbot, AntiVandalBot, Tim Shuba, JAnDbot, LinkinPark, .anacondabot, Magioladitis, WolfmanSF, SHCarter, Ling.Nut, Jlerner,

DAGwyn, Nikopopl, MartinBot, Mschel, Morris729, Lantonov, BobEnyart, Jorfer, Blckavnger, Fylwind, Atheuz, TXiKiBoT, Rei-bot, Mathwhiz 29, Thrawn562, Venny85, SieBot, El Wray, Puzhok, Gerakibot, BartekChom, OKBot, Coldcreation, ClueBot, The Thing That Should Not Be, Frdayeen, Niceguyedc, Excirial, Bender2k14, Brews ohare, Kentgen1, Scog, Panos84, Louis925, Alphatronic, XLinkBot, DCCougar, Sesquihypercerebral, Torchflame, Addbot, DOI bot, Zahd, Delaszk, Legobot, Luckas-bot, Yobot, Aldebaran66, Amble, Perusnarpk, AnomieBOT, Materialscientist, Citation bot, Louelle, Srich32977, Omnipaedista, Waleswatcher, A. di M., ??, Paine Ellsworth, Citation bot 1, Newt Scamander, Gil987, Tom.Reding, BlackHades, Jordgette, Michael9422, Earthandmoon, Vekov, UpdateNerd, RjwilmsiBot, Racerx11, Dcirovic, Solomonfromfinland, Italia2006, Hhhippo, ZéroBot, Liquidmetalrob, Arbnos, Quondum, Ewa5050, Iiar, Zueignung, Khestwol, ClueBot NG, Astrocog, Frietjes, Jhmmok, Rezabot, Const.S, Helpful Pixie Bot, Bibcode Bot, Rascal Sage, Jeffloiselle, Hippokrateszholdacskai, RiseUpAgain, Makecat-bot, Kryomaxim, Wjs64, Andyhowlett, Jp4gs, Blackbombchu, Prokaryotes, Inanygivenhole, Kogge, Paspaspas, Christophe1946, RandomAgentNation, Monkbot, Tetra quark, KasparBot, Jmc76, Maha Abdelmoneim, RandomEditor99, Lemesb, Ted.tem.parker, Bobbie73, Bender the Bot, Magic links bot and Anonymous: 150

- **Friedmann–Lemaître–Robertson–Walker metric** *Source:* https://en.wikipedia.org/wiki/Friedmann%E2%80%93Lema%C3%AEtre%E2%80%93Robertson%E2%80%93Walker_metric?oldid=787592934 *Contributors:* The Anome, XJaM, Roadrunner, Boud, Michael Hardy, CesarB, Looxix~enwiki, Cyan, Silvonen, BenRG, Robbot, Sho Uemura, Giftlite, Fropuff, Varlaam, Fimbulvetr, AmarChandra, JDoolin, Urhixidur, Burschik, Rich Farmbrough, Ascánder, Bender235, Ben Standeven, El C, Art LaPella, Cje~enwiki, 9SGjOSfyHJaQVsEmy9NS, Slambo, JHG, Eddie Dealtry, Sburke, Jeff3000, Mpatel, Joke137, BD2412, Ketiltrout, Rjwilmsi, Mike Peel, Ligulem, Itinerant1, RexNL, Hillman, KSmrq, Gaius Cornelius, Cryptic, Salsb, SEWilcoBot, Anotherwikipedian, Tony1, ColdFusion650, Netrapt, Caco de vidro, KasugaHuang, That Guy, From That Show!, Sardanaphalus, KnightRider~enwiki, SmackBot, Nickst, Hbackman, Bluebot, Hve, Xiner, Jmnbatista, Dan Gluck, JRSpriggs, Thijs!bot, Markus Pössel, Headbomb, Marek69, Peter Gulutzan, R'n'B, Eifelgeist~enwiki, Astro-davis, Lseixas, Sheliak, VolkovBot, CosmicAl, BotKung, Senemmar, Synthebot, N.Thorpe, AlleborgoBot, Adavis3, PaddyLeahy, SieBot, Bobathon71, Pomona17, Frdayeen, Richerman, Cldv71, Brews ohare, Kentgen1, MelonBot, DumZiBoT, TimothyRias, InMemoriamLuangPu, Sesquihypercerebral, Addbot, Lightbot, OlEnglish, Meisam, Legobot, Luckas-bot, Yobot, AnomieBOT, Citation bot, ArthurBot, Dendropithecus, False vacuum, Omnipaedista, RibotBOT, Wiklol, Tom.Reding, IVAN3MAN, RockSolidCosmo, Earthandmoon, McSaks, AmigoCgn, Brazmyth, JFB80, مهدی بردبار, Law of Entropy, X-men2011, Helpful Pixie Bot, Bibcode Bot, BG19bot, Layzeeboi, Al'Beroya, DPolez, Soll, Frinthruit, AHusain3141, Tetra quark, Absolutelypuremilk, AWK1947, Magic links bot and Anonymous: 43
- **Friedmann equations** *Source:* https://en.wikipedia.org/wiki/Friedmann_equations?oldid=785154165 *Contributors:* XJaM, Michael Hardy, Ahoerstemeier, Ciphergoth, Charles Matthews, David Shay, Topbanana, BenRG, Gandalf61, Ancheta Wis, Giftlite, Lethe, SWAdair, Spoirier~enwiki, Constantine, 9SGjOSfyHJaQVsEmy9NS, Haham hanuka, Keflavich, JHG, Cmapm, Japanese Searobin, FeanorStar7, Mpatel, GregorB, Zzyzx11, Joke137, Tevatron~enwiki, Canderson7, Dennis Estenson II, Chobot, McGinnis, Wavelength, Hillman, Doctorsundar, Salsb, JocK, Sir48, Alain r, Caco de vidro, Sardanaphalus, SmackBot, Colonies Chris, Hve, OrphanBot, E4mmacro, Jbergquist, Vina-iwbot~enwiki, Mets501, Brienanni, Dan Gluck, HelloAnnyong, Spebudmak, LethargicParasite, JRSpriggs, Forthommel, Meznaric, Michael C Price, Coccoinomane, Thijs!bot, Headbomb, Proximo.xv~enwiki, Gmarsden, Colin MacLaurin, MSBOT, Kerotan, R'n'B, Maurice Carbonaro, Lantonov, Sheliak, Sylar~enwiki, VolkovBot, TXiKiBoT, AntonioBigazzi, SwordSmurf, PaddyLeahy, SieBot, Andrewjlockley, Ranzan, Anchor Link Bot, Ottoshmidt, ChandlerMapBot, Rememberlands, Estirabot, Kentgen1, Panos84, MystBot, Dazza79, Addbot, DOI bot, MrOllie, E.pajer, AndersBot, SpBot, Quaristice, Joe9320, Lightbot, Luckas-bot, Yobot, Aldebaran66, Perusnarpk, AnomieBOT, Ciphers, Citation bot, Gap9551, Omnipaedista, FrescoBot, GaryAKent, Pinethicket, RedBot, IVAN3MAN, Albiero, Earthandmoon, EmausBot, Thesecretaryofwar, Maschen, Llightex, ClueBot NG, Itai33, Biju-rp, Bibcode Bot, MadameBruxelles, B wik, Yours.NDV, Vkpd11, Glacialfox, Bakkedal, Andyhowlett, Frinthruit, Monkbot, Waters.Justin, Boehm, Magic links bot and Anonymous: 66
- **Hubble's law** *Source:* https://en.wikipedia.org/wiki/Hubble'{}s_law?oldid=781341973 *Contributors:* AxelBoldt, Mav, Bryan Derksen, Tarquin, AstroNomer, Andre Engels, XJaM, Diatarn iv~enwiki, Spiff~enwiki, Patrick, Boud, Liftarn, Cyde, Natbat, Minesweeper, 168..., Looxix~enwiki, Mark Foskey, Jeff Relf, Andres, Hashar, Timwi, Random832, Evgeni Sergeev, Tpbradbury, Hyacinth, Phoebe, Xaven, Anupamsr, Kulnor, BenRG, Carbuncle, Donarreiskoffer, Sverdrup, Rorro, Borislav, Xanzzibar, Enochlau, Giftlite, Jao, Karn, Noone~enwiki, Waltpohl, Alexander.stohr, Hugh2414, Just Another Dan, Bobblewik, Tagishsimon, Dinojerm, Csmiller, Johnflux, H Padleckas, Iantresman, Urhixidur, Eisnel, Ta bu shi da yu, JimJast, Discospinster, Oliver Lineham, Vsmith, Jpk, Chub~enwiki, Dbachmann, Dmr2, Bender235, ESkog, Elwikipedista~enwiki, RJHall, Ylee, Pt, Laurascudder, Art LaPella, Vervin, .:Ajvol:., Jjk, 9SGjOSfyHJaQVsEmy9NS, Sjmcd, Kjkolb, Nk, Nsaa, Tom Yates, Jumbuck, Rsholmes, Eric Kvaalen, Andrewpmk, Wtmitchell, Velella, Cmapm, Gmelfi, Gene Nygaard, Falcorian, Oleg Alexandrov, WilliamKF, Nuno Tavares, OwenX, FeanorStar7, RHaworth, Logomancer, StradivariusTV, ?, Steinbach, Sengkang, Isnow, Joke137, Christopher Thomas, Yegorm, Graham87, Drbogdan, Rjwilmsi, Zbxgscqf, Kevinwparker, Bubba73, The wub, Mohawkjohn, Fivemack, Nickpowerz, Hashproduct, Nimur, RobyWayne, Fresheneesz, Alphachimp, ScottAlanHill, DVdm, Amaurea, PointedEars, YurikBot, Wavelength, RobotE, Wester, Jimp, Hillman, Durand101, JabberWok, Chris Mid, Archelon, Sir48, Raskolnikov The Penguin, WAS 4.250, 2over0, Acctorp, Vicarious, Caco de vidro, Argo Navis, RG2, John Broughton, GrinBot~enwiki, SmackBot, Tarret, Delldot, Yamaguchi??, Onsly, Gilliam, Hmains, Chris the speller, MK8, OrangeDog, Dabigkid, Sadads, Farry, Yurigerhard, Darth Panda, Gracenotes, Scwlong, Trekphiler, Can't sleep, clown will eat me, Allemannster, Hve, LouScheffer, Aldaron, John D. Croft, Mr Minchin, SpaceTiger, DantheCowMan, Wirbelwind, Kleuske, Just plain Bill, Kendrick7, ALK, Ligulembot, Vina-iwbot~enwiki, Lambiam, Silvem, J 1982, Mr. Vernon, Imi2, Hypnosifl, Mets501, Iridescent, Astrobayes, UncleDouggie, Richard75, Phoenixrod, Az1568, Encyclopediarocketman, Tem2, Tommysun, Harold f, J Milburn, CRGreathouse, CmdrObot, David s graff, Olaf Davis, MrFizyx, NE Ent, Adrianinos, Vectro, Cydebot, Abeg92, ZioX, Agne27, Michael C Price, Christian75, Kozuch, Abtract, Malleus Fatuorum, Epbr123, Headbomb, Marek69, Peter Gulutzan, Iviney, Greg L, X96lee15, AntiVandalBot, Luna Santin, Danger, JAnDbot, Spacehippy, Brendand, 100110100, Greensburger, GermanSoccer3, Christopher Cooper, Rothorpe, Gekedo, Michele123, Catgut, Vanished user ty12kl89jq10, Charliet, Dravick, Shijualex, Connor Behan, R'n'B, CommonsDelinker, Uncle Dick, Maurice Carbonaro, Ian.thomson, TomyDuby, M-le-mot-dit, NewEnglandYankee, Joshmt, Dorftrottel, Sheliak, Xnuala, Orthologist, TXiKiBoT, Ssri1983, GcSwRhIc, Qxz, LeaveSleaves, BotKung, SwordSmurf, PaddyLeahy, SieBot, Jim77742, CircafuciX, Hertz1888, Dawn Bard, Wing gundam, Droog Andrey, Tombomp, OKBot, CharlesGillingham, Cosmo0, Sfan00 IMG, ClueBot, Ohmygodeven, EoGuy, Wendy.krieger, Agge1000, Sun Creator, Brews ohare, NuclearWarfare, Tnxman307, Kentgen1, Dj manton, Wnt, Darkicebot, QYV, Tarl N., Ariconte, Madkasse, Addbot, Mortense, DOI bot, Jojhutton, DougsTech, CarsracBot, Glane23, Bob K31416, Lzkelley, Lightbot, Finley08, Hartz, Legobot, Yobot, TaBOTzerem, Aldebaran66, Henriettaleavitt, Tempodivalse, AnomieBOT, Rubinbot, The High Fin Sperm Whale, ArdWar, Citation bot, ArthurBot, Xqbot, Lordelicht, Nickkid5, Xmrbearx, Gap9551, Srich32977, Frosted14, QMarion II, SassoBot, A. di M., CES1596, FrescoBot, AllCluesKey, Paine Ellsworth, Jc3s5h, D'ohBot, S0fakingdie, Citation bot 1, Citation bot 4, Careful With That Axe, Eugene, Gaba p, A412, Tom.Reding,

Rushbugled13, Naturehead, Serols, Liweitianux, Coraifeartaigh, RockSolidCosmo, Heurisko, Vrenator, Chicodroid, Stroppolo, Minimac, Bongdentoiac, Newty23125, EmausBot, Energy Dome, WikitanvirBot, Primefac, Fotoni, Minimac's Clone, Mcguyver2, Tolly4bolly, Brandmeister, HCPotter, AndyTheGrump, Esilence, Llightex, Sven Manguard, Mattanel, Gestrid, ClueBot NG, Anagogist, MelbourneStar, Gilderien, Juanreza, Braincricket, Synethos, ساجد امجد ساجد, Helpful Pixie Bot, Bibcode Bot, BG19bot, Ymblanter, ElphiBot, ChronHigherEdReader, CitationCleanerBot, Glevum, Jason from nyc, Stigmatella aurantiaca, Javiramos, Cerabot~enwiki, Twhitguy14, Wjs64, Tony Mach, Renerpho, Epicgenius, Al'Beroya, Joe LHC Portal, Beavertron, Spartacus99, Tokrabelgium, Ashorocetus, The Herald, PirtleShell, JeanLucMargot, Aabrucadubraa, Monkbot, SlavaRodionov, Jwalker444, Cyrej, Tetra quark, Gcarpenter83, Rutzrutz, Halexus, Huritisho, Jpiquette, The Cube Root Of Infinity, Entranced98, Bender the Bot, Φιλοσ and Anonymous: 279

- **Matter** *Source:* https://en.wikipedia.org/wiki/Matter?oldid=785330546 *Contributors:* Carey Evans, CYD, The Anome, Tarquin, Ted Longstaffe, XJaM, Roadrunner, Ben-Zin~enwiki, Heron, Camembert, Ryguasu, Isis~enwiki, Stevertigo, Patrick, D, Tim Starling, Gabbe, Menchi, Ixfd64, Tomos, Minesweeper, Looxix~enwiki, Ahoerstemeier, Suisui, Angela, Glenn, Cyan, Mxn, Charles Matthews, Reddi, Hyacinth, Rm, Xevi~enwiki, Gakrivas, Jerzy, BenRG, RadicalBender, Rashack~enwiki, Chuunen Baka, Robbot, Kizor, Gandalf61, Ashley Y, Auric, Hadal, Papadopc, Lupo, HaeB, Jan Lapère, Alan Liefting, Enochlau, Giftlite, Djinn112, Art Carlson, FeloniousMonk, Bensaccount, Solipsist, Alexf, Quadell, Antandrus, OverlordQ, Lesgles, Karol Langner, Mikko Paananen, JimWae, DragonflySixtyseven, Bumm13, Kevin B12, Okapi~enwiki, Int19h, Nike, Trevor MacInnis, ELApro, Brianjd, EugeneZelenko, Discospinster, Rich Farmbrough, Vsmith, Jpk, Autiger, Martpol, Bender235, Kbh3rd, Ghitis, JoeSmack, RJHall, MisterSheik, Mr. Billion, Zegoma beach, Remember, Jashiin, CDN99, Adambro, Bobo192, Army1987, Whosyourjudas, Rrh02, Smalljim, Orbst, .:Ajvol:., Dungodung, Maurreen, 9SGjOSfyHJaQVsEmy9NS, Giraffedata, PaRaLyZeDHoRSe, Jojit fb, Kjkolb, Nk, Charonn0, MPerel, Sam Korn, Haham hanuka, Jayakar, Alansohn, Arthena, Atlant, Paleorthid, AzaToth, Kocio, Walkerma, Jaw959, Wdfarmer, Avenue, Bart133, Metron4, Snowolf, Wtmitchell, Oking83, Ott, Angr, Woohookitty, Madchester, Canaen, Sandrapalaje, CharlesC, RuM, Graham87, Dpr, Sjö, Rjwilmsi, Jake Wartenberg, Strait, MarSch, Quiddity, Tangotango, Nneonneo, Vav11, Yamamoto Ichiro, Exeunt, FayssalF, FlaBot, Lorkki, Naraht, Nihiltres, Ectoraige, Crazycomputers, Andy85719, RexNL, TimSE, TheDJ, BabyNuke, TheMightyGrecian, Cloudo, King of Hearts, Chobot, DaGizza, Sharkface217, DVdm, Bgwhite, Gwernol, The Rambling Man, Wavelength, RobotE, Sceptre, Jimp, Bhny, Juansmith, Stephenb, Wimt, Ugur Basak, NawlinWiki, Injinera, Wiki alf, Grafen, Janarius, SCZenz, Retired username, Anetode, Dhollm, Jpbowen, Chichui, Alex43223, Natkeeran, DeadEyeArrow, Dna-webmaster, Wknight94, FF2010, Vadept, Enormousdude, Zzuuzz, Leliathomas, E Wing, Sean Whitton, JuJube, Aeon1006, JoanneB, Chez37, CWenger, Katieh5584, Junglecat, Banus, Mejor Los Indios, DVD R W, SmackBot, Unschool, KnowledgeOfSelf, Hydrogen Iodide, Unyoyega, C.Fred, Bomac, Gilliam, Skizzik, Fogster, Dauto, Andy M. Wang, Grokmoo, Rmosler2100, Master of Puppets, SchfiftyThree, Sbharris, Colonies Chris, Hallenrm, Darth Panda, Can't sleep, clown will eat me, OrphanBot, Rrburke, Addshore, Nakon, Jiddisch~enwiki, John D. Croft, Dreadstar, SpiderJon, DMacks, Ultraexactzz, Kotjze, Kalathalan, Where, BobbyPeru, Yevgeny Kats, Byelf2007, Dbtfz, LarchOye, Ocanter, Gobonobo, Breno, AstroChemist, Edwy, IronGargoyle, Vidit1, Simonalexander2005, Loadmaster, Munita Prasad, Noah Salzman, Hiiiiiiiiiiiiiiiiiiiii, Kyoko, Dicklyon, Waggers, Dr.K., RichardF, MHWiki, Supaman89, CrazedEwok, Levineps, Lord Anubis, Joseph Solis in Australia, Newone, Cyon, Casull, Tony Fox, Esurnir, Courcelles, JRSpriggs, JForget, DSatYVR, Ale jrb, Dycedarg, Van helsing, Vyznev Xnebara, MargyL, Flapping Fish, McVities, MarsRover, Penbat, Gregbard, Fl, Bvcrist, Peterdjones, Gogo Dodo, 01011000, Anonymi, Rracecarr, Studerby, Dr.Kane, Abtract, RickDC, Epbr123, Barticus88, Mbell, O, TonyTheTiger, Giorgio51, N5iln, Headbomb, Trevyn, Marek69, Grayshi, Nick Number, MichaelMaggs, Troy392004, Dualactionblend, Escarbot, Oreo Priest, Dantheman531, AntiVandalBot, Majorly, JHFTC, Cpkondas, Quintote, Jayron32, CHollman82, Jj137, Madbehemoth, Danger, Astavats, Canadian-Bacon, Curlingpro47, ClassicSC, Ioeth, JAnDbot, Jimothytrotter, Elias Enoc, The Transhumanist, Sanchom, Blood Red Sandman, Smiddle, Db099221, Andonic, Tergadare, Kerotan, Acroterion, Magioladitis, Connormah, Bongwarrior, VoABot II, AuburnPilot, JamesBWatson, Kinston eagle, Kajasudhakarababu, Think outside the box, Nyttend, Cic, Avicennasis, Indon, Animum, Dirac66, Cpl Syx, MindReality, Vssun, DerHexer, Hbent, Arnesh, Tojo940, Greenguy1090, S3000, FisherQueen, MartinBot, Rettetast, Anaxial, Sm8900, Dan.g, Dogatdog, R'n'B, AlexiusHoratius, Qwertuy, Ash, LedgendGamer, J.delanoy, Trusilver, Bogey97, Maurice Carbonaro, Ginsengbomb, Eliz81, It Is Me Here, DarkFalls, Ben robbins, Gurchzilla, Vanished User 4517, Tcisco, NewEnglandYankee, In Transit, Newtman, Minesweeper.007, Ionescuac, Juliancolton, Cometstyles, Tiggerjay, DH85868993, DorganBot, Natl1, Squids and Chips, Funandtrvl, Wikieditor06, Vranak, HamatoKameko, Deor, CWii, Christophenstein, JohnBlackburne, Bry9000, Haade, Philip Trueman, Eastgate, TXiKiBoT, Kww, SCriBu, Anonymous Dissident, Qxz, Seraphim, Melsaran, DennyColt, Abdullais4u, Noformation, UnitedStatesian, Vgranucci, Whammes2, Shadowlapis, CaptColon, Isis4563, Iluso, Ilkali, Wykypydya, Brainmuncher, Lamro, Synthebot, Lova Falk, MCTales, Spinningspark, Kchiles, Conostrov, The Strange Kid, Insanity Incarnate, Alcmaeonid, Fireglowe, AlleborgoBot, Logan, TheXenocide, BriEnBest, J. Naven, SieBot, Vijai Singh, Portalian, Nihil novi, Gerakibot, YourEyesOnly, Nathan, Triwbe, Wing gundam, Gravitan, Flyer22 Reborn, Tiptoety, Oda Mari, Prestonmag, Frank.hedlund, Granf, Oxymoron83, Harry-, Steven Crossin, Iain99, Techman224, The-G-Unit-Boss, Pantuflas, Kudret abi, OKBot, Jonlandrum, Asperal, Ascidian, Denisarona, ClueBot, GorillaWarfare, Snigbrook, Fox, The Thing That Should Not Be, Rjd0060, Wwheaton, Drmies, TheOldJacobite, Goshwak, CounterVandalismBot, Classified as matter, Rotational, Superguy342, Puchiko, Excirial, Jusdafax, Afoxtrotn00b, Editorman12342, MorrisRob, Bassoonboy, Rhododendrites, Brews ohare, NuclearWarfare, Cenarium, World, Jotterbot, Mr.24SevenCrashHolly, Tnxman307, Razorflame, SchreiberBike, Oswald07, Saebjorn, Polly, La Pianista, Calor, Taranet, Thingg, Aitias, Versus22, Phynicen, MelonBot, SoxBot III, Bibibita, Goodvac, Hattiel, Nishu pcp, DumZiBoT, Templarion, TimothyRias, Jmanigold, JKeck, AlexGWU, Mattermatters, -: Wik3d Playful:-, Lord pain377, Rreagan007, Zkunz1, Facts707, WikHead, SilvonenBot, Mifter, Vianello, ZooFari, ElMeBot, Alexcs123, RyanCross, Thatguyflint, HexaChord, Gimie the beat boys, Gatorsalldawa, Addbot, Willking1979, Some jerk on the Internet, DougsTech, Ronhjones, Fieldday-sunday, Laurinavicius, Leszek Jańczuk, Ashanda, Protonk, Brentdeezee, Chamal N, Chzz, XRK, Favonian, 5 albert square, Kisbesbot, Numbo3-bot, Ehrenkater, Tide rolls, Luckas Blade, Lrrasd, Zorrobot, MuZemike, Grandpsykick, Yobot, Ht686rg90, Senator Palpatine, Legobot II, THEN WHO WAS PHONE?, Brougham96, Azcolvin429, AnomieBOT, Somecrazydude, Joule36e5, Eminem69041, Killiondude, Galoubet, 9258fahsflkh917fas, Piano non troppo, Icalanise, Ipatrol, Kingpin13, Yachtsman1, Abshirdheere, Ulric1313, Bluerasberry, Materialscientist, Magixdx, ShikyoSays, The High Fin Sperm Whale, Citation bot, Vuerqex, Neurolysis, Bagumba, Ianwestapleton, Xqbot, Lloydsd, Bihco, YakbutterT, Pvkeller, Jsharpminor, Grim23, The Evil IP address, Crzer07, Tricko20, J04n, Paul Sinclair, Omnipaedista, Amaury, Reflections of Memory, Doulos Christos, Codyfitz8, Tjsnuff, Bigger digger, Nisamayarg, A. di M., Peter470, Griffinofwales, Joel grover, CES1596, FrescoBot, Singhking97, Alaphent, Sebastiangarth, Machine Elf 1735, Citation bot 1, Redrose64, Pinethicket, I dream of horses, Vicenarian, The Arbiter, Tom.Reding, A8UDI, SpaceFlight89, Merlion444, Heiji hattori is LOVE, Irbisgreif, IVAN3MAN, PhilOak, Gamewizard71, FoxBot, Calle Cool, Trappist the monk, Kimtiger12345, Daniel G J, Vrenator, Darsie42, Reaper Eternal, Wst Nam, Devildude666, Suffusion of Yellow, Tbhotch, Oualidi13, Keegscee, DARTH SIDIOUS 2, Regancy42, Devcas, DASHBot, John of Reading, Ajraddatz, Washout4, Fotoni, Klbrain, Solarra, Turnurban, Wikipelli, Dcirovic, P. S. F. Freitas, Anirudh Emani, Tiniet, Thecheesykid, Jpfairweather, Jmanvball, JSquish, ZéroBot, Anir1uph, Quiqui1, ElvisPresley1, Hazard-SJ, Aeonx, Quondum, L1A1 FAL, Arman Cagle, YvonneM, Inka 888, Damirgraffiti, Carmichael,

Negovori, RockMagnetist, TYelliot, DASHBotAV, Wussification, Morgis, Efiiamagus, BigBarrelRollingHoss, Khestwol, Petrb, Xanchester, ClueBot NG, Mifapetrenko, Satellizer, Piast93, Waroe, Movses-bot, Erichardson2626, Wallawaller, Gamerstuff75, Dittymathew, Goober6521, Dream of Nyx, CaroleHenson, 336, Be94ware, Widr, Helpful Pixie Bot, Ramaksoud2000, Bibcode Bot, Trunks ishida, BG19bot, Sanjeevgen, Soptx, Vagobot, Moustacheluvr69, Aozf05, MusikAnimal, Thumani Mabwe, Mark Arsten, Lolbading, Soukhoi, Altaïr, CitationCleanerBot, Jaloqin, Rs2360, MehFooL, GSMOL, Chiles Malesters, Glacialfox, Cimorcus, Pratyya Ghosh, Mdann52, Khazar2, EuroCarGT, Southwing17, Maxgaineducation, EagerToddler39, Brittany taylorr, Djdjordje12345, CoreyM1232, Lugia2453, Frosty, Graphium, Weather228, Wywin, NievesO, Reatlas, Tuckah parent, Hacker8484, Faizan, Asharia283, Austinben1011, Sosaalhabab, Bumoa, Professor Caramel Revenge, Guymandudeme, Leompersi, Trololsir, Ginsuloft, DavRosen, Jackmcbarn, Frank12123, Demigod mcg1, Victortran2000, JaconaFrere, AspaasBekkelund, Kiran cb, IiKkEe, Kaltakus, Gvdjcfhhdfvuxhhrfbj, Havj, Harrydafitz, FourViolas, Fartser, Steven ngsteven, Epic2500, Vicallsssssshdggsj, CV9933, Jayster101, Uwharrie MilSim, NateBillingsley, KasparBot, Kafishabbir, InternetArchiveBot, Zhakhan9er, Riki71144, Emir of Wikipedia, 72, SuperCat and Anonymous: 1142

- **Radiation** *Source:* https://en.wikipedia.org/wiki/Radiation?oldid=787817717 *Contributors:* Trelvis, Olof, Zundark, Szopen, AstroNomer, Rjstott, RK, Alex.tan, AdamW, PierreAbbat, William Avery, Heron, Hephaestos, Olivier, MimirZero, Edward, Patrick, RTC, Michael Hardy, Tim Starling, Tompagenet, Wapcaplet, Ixfd64, Eric119, Ahoerstemeier, William M. Connolley, Jebba, Aarchiba, Glenn, Andres, Smack, Malbi, Reddi, Dysprosia, Zoicon5, SEWilco, Omegatron, Raul654, David.Monniaux, Rogper~enwiki, Chuunen Baka, Robbot, Zandperl, Altenmann, Academic Challenger, Tsavage, Diberri, Xanzzibar, Dmn, Giftlite, Harp, Mintleaf~enwiki, Kmote, Niteowlneils, Leonard G., Bovlb, Rchandra, Tweenk, Bobblewik, Utcursch, Alexf, Antandrus, Karol Langner, H Padleckas, Thincat, Gscshoyru, Usrnme h8er, Deglr6328, Ehamberg, Iwilcox, ELApro, Mike Rosoft, Discospinster, Rich Farmbrough, Banditoitaliano, Vsmith, Jpk, Xezbeth, Arthur Holland, Bender235, RJHall, El C, Art LaPella, Dalf, Bobo192, Whosyourjudas, Smalljim, Keron Cyst, Zetawoof, MPerel, Haham hanuka, Merope, Msh210, Alansohn, Duffman~enwiki, Neonumbers, Ricky81682, Riana, Fritzpoll, Redfarmer, Malo, Snowolf, Velella, Dabbler, Danhash, RainbowOfLight, VMajander, Bsadowski1, Versageek, DV8 2XL, Kazvorpal, Zntrip, Woohookitty, Camw, TomTheHand, Rtdrury, Pdn~enwiki, Eras-mus, Christopher Thomas, Mandarax, Tslocum, RichardWeiss, Graham87, Kbdank71, Sjö, Rjwilmsi, Strait, Seraphimblade, SMC, Yamamoto Ichiro, Lcolson, Alejo2083, Nihiltres, Nivix, Gurch, Nwatson, TheDJ, AnthonyA7, Srleffler, MGSpiller, Zotel, Antikon, Lpsiphi, Chobot, Evilphoenix, DVdm, Gwernol, The Rambling Man, Wavelength, Stephenb, Okedem, Shell Kinney, Ksyrie, Salsb, Anomalocaris, NawlinWiki, Injinera, Wiki alf, Aeusoes1, Trovatore, Tjarrett, Stijn Calle, Scottfisher, User27091, 2over0, Closedmouth, Zerodamage, Arthur Rubin, CharlesHBennett, CWenger, Anclation~enwiki, Jedwards05, Shadowin, Friman~enwiki, SmackBot, Unschool, InverseHypercube, KnowledgeOfSelf, Hydrogen Iodide, Goldfishbutt, Rokfaith, AndreasJS, Clpo13, Pennywisdom2099, Matthuxtable, Renesis, Frymaster, Canthusus, Onebravemonkey, BillyBleach, Commander Keane bot, Yamaguchi□□, Gilliam, Chaojoker, BirdValiant, ERcheck, Andy M. Wang, Chris the speller, Bluebot, Asclepius, Persian Poet Gal, Jprg1966, SchfiftyThree, Complexica, RayAYang, Sbharris, Gracenotes, DethmeOw, Can't sleep, clown will eat me, Милан Јелисавчић, Rrburke, Addshore, Mr.Z-man, Flyguy649, Jaimie Henry, Khukri, Kntrabssi, Dreadstar, Polonium, Hammer1980, WoodyWerm, Jóna Þórunn, Richard0612, Kukini, Evbneto, Robomaeyhem, DO11.10, Bakic, John, Scientizzle, Crazyviolinist, Gobonobo, Aleenf1, Paulrach, IronGargoyle, Muadd, Dicklyon, Ζεύς, Inquisitus, NinjaCharlie, Iridescent, Dragix, Igoldste, RekishiEJ, Cotinis, CapitalR, Domitori, Ewulp, Origin415, Tawkerbot2, Dlohcierekim, Jh12, SkyWalker, Wikipeedio, Dycedarg, Ninetyone, KyraVixen, El aprendelenguas, FlyingToaster, Fortran~enwiki, Cydebot, Michaelas10, MarS, Mrstonky, Rracecarr, Raomap, Kotiwalo, Tawkerbot4, Aajaja, Doug Weller, DumbBOT, Btharper1221, Dark Samus, NotQuiteEXPComplete, Thijs!bot, Epbr123, Vegarith, Qwyrxian, LeeG, Pathbinder, Mojo Hand, Dougsim, Marek69, RickinBaltimore, Dfrg.msc, Dgies, Big Bird, MichaelMaggs, Natalie Erin, Maximz2005, AntiVandalBot, Seaphoto, QuiteUnusual, Quintote, Billscottbob, Yoosq, SkoreKeep, Altamel, Elaragirl, The Utterly Annoying Pedant, Dreaded Walrus, Caper13, JAnDbot, Husond, MER-C, Inks.LWC, FaerieInGrey, Penubag, Gekedo, Hroðulf, Bongwarrior, VoABot II, Wikidudeman, JamesBWatson, Kinston eagle, Swpb, Sojourner001, Wormcast, Dlempa, Oroso, 0612, NatureA16, Hdt83, MartinBot, Chris lawton, Vigyani, Rettetast, Jonathan Hall, R'n'B, CommonsDelinker, AlexiusHoratius, Pekaje, PrestonH, Thirdright, J.delanoy, Trusilver, Hans Dunkelberg, Brynblob, Lantonov, Tannkremen, McSly, Pyrospirit, NewEnglandYankee, SJP, ThatDeadDude, Touch Of Light, Paul19321, Cmichael, Geekdiva, Treisijs, Collisionman5000, Zelky, SoCalSuperEagle, Idioma-bot, FeralDruid, ACSE, MBlue2020, VolkovBot, CWii, Johnfos, Jeff G., Amikake3, Lear's Fool, CART fan, Ryan032, Philip Trueman, DoorsAjar, Zidonuke, Marskuzz, Alan Rockefeller, Weareqpr, Hank Ramsey, Qxz, Oxfordwang, Kirkpthompson, Gingerscot, JhsBot, Cremepuff222, Ilyushka88, Guest9999, Slash 195, Madhero88, Pamputt, Andy Dingley, Yk Yk Yk, Lamro, SmileToday, Falcon8765, Whiteandnerdy1234, Burntsauce, Dorkules, Rafaor7r~enwiki, Insanity Incarnate, Doc James, Jesseknowsall, Faisalhafsamaria, Burgercat, Starkrm, Thw1309, Joe A K, SieBot, Moonriddengirl, Scarian, WereSpielChequers, Malcolmxl5, Jauerback, Josh the Nerd, Plinkit, Caltas, Billygettothechopper, Yintan, Coreycotterill, Flyer22 Reborn, Tiptoety, Radon210, Qst, Oda Mari, Doctorfluffy, Oxymoron83, Brokendata, Faradayplank, Harry-, Steven Crossin, RW Marloe, Techman224, Hobartimus, Svick, HAVINGASHITINMANHATTEN, Josh2000200, Sean.hoyland, Dabomb87, Denisarona, Escape Orbit, XDanielx, WikipedianMarlith, Sokari, Atif.t2, ClueBot, LAX, Trojancowboy, Hockeysteve54, Bubbletruble, The Thing That Should Not Be, Plastikspork, Jan1nad, Taroaldo, Arakunem, Drmies, Mild Bill Hiccup, Polyamorph, CounterVandalismBot, Wonderland42, Blanchardb, Fortheemeperor, Dupdater, Marselan, Eadthem, Excirial, Anonymous101, Kjramesh, Jusdafax, Theryguy512, Sourabhchopra, Wikitumnus, 12 Noon, Abrech, Robert the dee dee dee, Rhododendrites, Bob the dee dee dee, NuclearWarfare, Rbe008, PhySusie, StopblockingmyIPaddress, Morel, ChrisHodgesUK, Kakofonous, Thingg, Aitias, Jonverve, ForestDim, Versus22, Introductory adverb clause, SoxBot III, Egmontaz, Peachcream, Oore, Rishi.bedi, Lukeozoid2k8, XLinkBot, Spitfire, Nathan Johnson, Tconroy13, FellGleaming, Liltomboy25, Sakura Cartelet, Avoided, Mavoss1007, Lemchesvej, BlownSugar, Kodster, Jts5039, Addbot, Xp54321, Chrissweet, Shawn Savaya, Willking1979, Some jerk on the Internet, Tcncv, Magus732, Ketchup176, Ronhjones, CanadianLinuxUser, Fluffernutter, Mentisock, Download, Glane23, Chzz, Dillweed155, Tomanator111, Goddaz, Shosho6969, Nvurdien, Nolelover, Whammyhappy, Tide rolls, Mjquinn id, Gail, Hartz, Eliword, MissAlyx, Legobot, Luckas-bot, Fraggle81, Dfe6543, II MusLiM HyBRiD II, Yngvadottir, Mmxx, THEN WHO WAS PHONE?, Sarrus, KamikazeBot, Mydoctor93, Azcolvin429, Eric-Wester, Tempodivalse, Synchronism, Backslash Forwardslash, AnomieBOT, HairyPerry, AUG, Kristen Eriksen, Jim1138, Dwayne, Piano non troppo, Tuanglen, Cyanidethistles, Materialscientist, ImperatorExercitus, Bigsmallmatt, ArthurBot, Alkalk, Gsmgm, Szxd, Xqbot, Sketchmoose, Ionactive, Melmann, Addihockey10, Capricorn42, Research Method, POWPOWPOWER, Sadlerjackmn, AbigailAbernathy, Monkey0101, Coretheapple, GrouchoBot, Wikicatti, Wizardist, حامد میرزاحسینی, Bellerophon, Eugene-elgato, SD5, FrescoBot, Angel katt, HJ Mitchell, JassmineBreeanna, Zachbz, Ysyoon, Alipson, CarstenHermann, Pinethicket, I dream of horses, 10metreh, Calmer Waters, Jusses2, Serols, SpaceFlight89, Piandcompany, PRONIZ, Wabul, RandomStringOfCharacters, Robo Cop, Thecurran91, Shanmugamp7, Reconsider the static, TehRabbitt, Double sharp, TobeBot, کاشف عقیل, Johntaylor312, Kalaiarasy, ItsZippy, Lotje, Vrenator, LilyKitty, Clarkcj12, Daleman112290, Lang1877, Cowlibob, Neisan52, Brian the Editor, Tbhotch, Minimac, DARTH SIDIOUS 2, Onel5969, Mean as custard, Hajatvrc, Niffirg1, EmausBot, Orphan Wiki, Immunize, Blingaman123, ScottyBerg, Sadalsuud, Akjar13, Racerx11, Yt95, RA0808, RenamedUser01302013, Solarra, Tommy2010,

Winner 42, Wikipelli, K6ka, Hhhippo, Mjaked, Jwgealt, Fæ, Josve05a, Klavierspieler, Jamjr7895, Chris141496, Lateg, A930913, Vividadmit, Zap Rowsdower, Wayne Slam, Ocaasi, Tolly4bolly, Rcsprinter123, Moeali123, GeorgeBarnick, IGeMiNix, L Kensington, Sparkey989, FilthyLampost, RockMagnetist, Folgabil, Herk1955, Newtrend19, TheAgamer, Jeff1111111111, ClueBot NG, Anderson-b001, Ulflund, Martin Dluhos, MelbourneStar, Gilderien, Srothen84, O.Koslowski, Kasirbot, Widr, Youmnasalah, Miah781227, Mtking, MerlIwBot, Jonfool, TrollingTrollingson, Rage308, More3118, Regulov, Lowercase sigmabot, J991, Shikhars24, Silvrous, Altaïr, Glevum, TheLeprosy, DARIO SEVERI, M.a.Padmanabha Rao, Snow Blizzard, Zedshort, Robcw, Klilidiplomus, Tutelary, Pratyya Ghosh, Ajaxfiore, Cyberbot II, ToBeFree, FoCuSandLeArN, Mogism, G.Buttersnaps1, Lugia2453, Frosty, Jamesx12345, JustAMuggle, Reatlas, Katilyn russell2014, Sevıntі faıv, Lounajus, Everymorning, Ontilu, Comp.arch, Dynoman 555, Noobfailhack, Jmcmail, Ginsuloft, Limnalid, Maryann monteza, Anon 013189, ThatRusskiiGuy, DFVV92, 321emikiw, Bipav Aoxke, Riddleh, Sofia Koutsouveli, Trackteur, Kite In Tree, ChamithN, Crystallizedcarbon, Beanstash, Speedygarcias, Joshie0067, Mypowerpuff, Editing656, Pussy6969696969, GamerBoy613, McDonald of Kindness, Alen drummer, KasparBot, Fantommanman, Pandg2, Wywyit, CAPTAIN RAJU, Surripere36, Serridd, CLCStudent, Allthefoxes, ThEePiCtRoLeR, Xx Salty12V xX, Bear-rings and Anonymous: 1197

- **Standard ruler** *Source:* https://en.wikipedia.org/wiki/Standard_ruler?oldid=727547157 *Contributors:* Florian Blaschke, RJHall, Physicistjedi, Philbull, SmackBot, Alaibot, Epbr123, Chris goulet, Patar knight, Addbot, Luckas-bot, ZX81, Arashdeli, NOrbeck, Erik9bot, Acky69, FrescoBot, Tom.Reding and Anonymous: 6

- **Chandrasekhar limit** *Source:* https://en.wikipedia.org/wiki/Chandrasekhar_limit?oldid=784156849 *Contributors:* Wesley, Bryan Derksen, The Anome, Tarquin, AstroNomer, Ap, Verloren, Wayne Hardman, Andre Engels, Josh Grosse, Montrealais, Stevertigo, Alan Peakall, Alfio, Suisui, Marco Krohn, Lenaic, Phr, The Anomebot, Jnc, Mathus~enwiki, Robbot, Arkuat, Mirv, Xanzzibar, Davidcannon, Xyzzyva, ShaunMacPherson, Harp, Wolfkeeper, Wyss, Duncharris, Sundar, DefLog~enwiki, Eranb, Sam Hocevar, B.d.mills, Jayjg, Rich Farmbrough, Vsmith, Chowells, MuDavid, Bender235, ZeroOne, RJHall, El C, Olve Utne, La goutte de pluie, Tezeti, Burn, Allen McC.~enwiki, Adrian.benko, David Haslam, Miaow Miaow, Joke137, Christopher Thomas, Rnt20, Search4Lancer, Avia, Pradeepsinghhbti, Jehochman, Nihiltres, Gparker, Lithiyum, Lynxara, Physchim62, Chobot, DVdm, Spacepotato, RobotE, Hairy Dude, Gaius Cornelius, David R. Ingham, Leutha, The Yeti, SmackBot, Incnis Mrsi, Unyoyega, Richard B, Slaniel, Ohnoitsjamie, Kevin Ryde, DHN-bot~enwiki, Colonies Chris, Salmar, SundarBot, T-borg, Alcuin, Andrei Stroe, SashatoBot, Malixsys, Fimus, JorisvS, Kyoko, Antonio Prates, PhoenixSeraph, Friendly Neighbour, CRGreathouse, WeggeBot, Myasuda, Cydebot, Asymptote, Thijs!bot, CharlotteWebb, Deflective, Quentar~enwiki, Epinheiro, Coolhandscot, WolfmanSF, JustinGreen, MartinBot, STBot, Gah4, Fconaway, BigrTex, Numbo3, Peter Chastain, IdLoveOne, DorganBot, Idioma-bot, JCMP, John Darrow, Philip Trueman, TXiKiBoT, Ask123, Markp93, SieBot, Sahilm, OsamaBinLogin, Arthur Smart, OKBot, Martarius, ClueBot, Garyzx, Aggel1000, Jotterbot, BOTarate, Techroach, Daffydd, MystBot, Mortense, Darko.veberic, DOI bot, Colinb007, Lightbot, Mira7, Zorrobot, Legobot, Yobot, Donfbreed, Amirobot, PianoDan, Jim1138, Materialscientist, Citation bot, Teleprinter Sleuth, ArthurBot, Xqbot, Nfr-Maat, Tomwsulcer, Gap9551, Andrestand, Jonesey95, A412, Rosa67, Σ, Puzl bustr, DARTH SIDIOUS 2, Andrea105, Ripchip Bot, EmausBot, John of Reading, WikitanvirBot, Solomonfromfinland, Hhhippo, L Kensington, Llightex, Whoop whoop pull up, ClueBot NG, Braincricket, Helpful Pixie Bot, Bibcode Bot, Majee chinmay, Mynameisnoted, Dodshe, RickV88, 786b6364, Dexbot, Frosty, Poipoise, Monkbot, Lince meladath, RKhehehe, William Yang-62300, Pulsarwind, Vinayak123321, Prajaman, Magic links bot and Anonymous: 119

- **Luminosity** *Source:* https://en.wikipedia.org/wiki/Luminosity?oldid=785923078 *Contributors:* AstroNomer, Malcolm Farmer, XJaM, Heron, JohnOwens, Michael Hardy, Ixfd64, Tango, Looxix~enwiki, Pizza Puzzle, Robbot, Rursus, Giftlite, Jyril, Lethe, Xerxes314, Bensaccount, Guanaco, Karol Langner, Tomruen, Icairns, JohnRDaily, Ylai, Bender235, Zaslav, RJHall, Tom, LutzL, Jumbuck, Dhanak~enwiki, Gene Nygaard, Alai, Arent, Linas, Palica, Magister Mathematicae, Qorkfiend, Josh Parris, Coneslayer, Lzz, Kolbasz, Srleffler, Smithbrenon, Kjlewis, YurikBot, NTBot~enwiki, Gaius Cornelius, SCZenz, Sir48, E2mb0t~enwiki, Thnidu, Ásgeir IV.~enwiki, GrinBot~enwiki, SmackBot, Unyoyega, Hardyplants, PeterSymonds, Skizzik, DHN-bot~enwiki, Darth Panda, DMacks, J.smith, SashatoBot, JorisvS, Bjankuloski06en~enwiki, 16@r, Dicklyon, Melicans, Cydebot, Thijs!bot, Headbomb, Glennchan, Orionus, JAnDbot, Magioladitis, Nyq, Catgut, Cgingold, Schmloof, Andre.holzner, Kometsuga, Leyo, J.delanoy, Thucydides411, DorganBot, VolkovBot, Larryisgood, Elphion, Jackfork, Waycool27, RayNorris, Enviroboy, SieBot, Meldor, DivineBurner, Maelgwnbot, ClueBot, Niceguyedc, Lartoven, Thingg, Arianewiki1, Emoloisirc, SilvonenBot, Addbot, Darko.veberic, Elishabet, Bte99, CarsracBot, ChenzwBot, Luckas-bot, Yobot, AlexPenson, AnomieBOT, DemocraticLuntz, Jim1138, Lgm432, Materialscientist, Yunus.sendag, Citation bot, Quebec99, MauritsBot, Xqbot, Tomdo08, Lithopsian, Rainald62, Astronomyinertia, Gaba p, Pinethicket, I dream of horses, Tom.Reding, عباد مجاهد ديارزة, SciCorrector, Wikiborg4711, Onel5969, Ripchip Bot, Alzarian16, EmausBot, Sadalsuud, Primefac, Japs 88, Traveller1234, Lamb99, Wackywace, AvicAWB, DacodaNelson, Bamyers99, Wikfr, MichiganY, One.Ouch.Zero, BR84, ClueBot NG, Phidica, Braincricket, Doctore, Bibcode Bot, DBigXray, Boesingm, KitchiRUs, Astros4477, SoylentPurple, Menelaost., StarryGrandma, Khazar2, PeerRevision, Yamaha5, Doraemon325, Finnusertop, Mfb, Marxide, Monkbot, Stenbocken92, Quarticle, Wkfnsodjeoajdj manfjv, Ruchikarajpal, Sir Cumference, CyberWarfare, CheChe and Anonymous: 135

- **Distance measures (cosmology)** *Source:* https://en.wikipedia.org/wiki/Distance_measures_(cosmology)?oldid=778867796 *Contributors:* Michael Hardy, BenRG, Ctachme, Ben Standeven, Huntster, 9SGjOSfyHJaQVsEmy9NS, Aarghdvaark, Physchim62, DVdm, Bgwhite, Caco de vidro, Mejor Los Indios, SmackBot, Hftf, JorisvS, Phuzion, Amalas, Banedon, Myasuda, Meznaric, Alaibot, Publicola, Peter Gulutzan, Michael Wood-Vasey, Brianpschmidt, R'n'B, Migran, Wesino, DoorsAjar, Matthias Buchmeier, Anoko moonlight, Aggel1000, PixelBot, Gushaa1977, Panos84, Parejkoj, Addbot, Yobot, Gap9551, FrescoBot, DrilBot, Tom.Reding, RockSolidCosmo, TobeBot, Octaazacubane, Wikipelli, ZéroBot, Bkocsis, ClueBot NG, Quantum Doughnut, CasualVisitor, Mn-imhotep, CuriousMind01, Vouliskp10, Zhermes, Tetra quark, Fmadd and Anonymous: 18

- **Baryon acoustic oscillations** *Source:* https://en.wikipedia.org/wiki/Baryon_acoustic_oscillations?oldid=778397061 *Contributors:* Reddi, RJHall, Wtmitchell, Drbogdan, Welsh, Malcolma, SmackBot, Modest Genius, Sambot, JorisvS, Merryjman, CmdrObot, Banedon, Martin Hogbin, Headbomb, Peter Gulutzan, Venturesum, Fellwalker57, J.delanoy, Spiral5800, James McBride, ImageRemovalBot, Martarius, Gaia Octavia Agrippa, Iohannes Animosus, SoxBot, Mastertek, DCCougar, MidwestGeek, Parejkoj, Addbot, Skippy le Grand Gourou, AnomieBOT, Citation bot, Anna Frodesiak, Gap9551, Erik9bot, FrescoBot, Berkeleyjess, HRoestBot, Plucas58, Tom.Reding, Trappist the monk, WildBot, Cogiati, Owenmann, ClueBot NG, Braincricket, Timestheyare, Bibcode Bot, BG19bot, Trevayne08, Zedshort, Mtysan, Dough34, Monkbot, Jiale8331, Sofia Koutsouveli, Tetra quark, Esadri21 and Anonymous: 28

- **Recombination (cosmology)** *Source:* https://en.wikipedia.org/wiki/Recombination_(cosmology)?oldid=786607037 *Contributors:* Boud, Julesd, Pierre Boreal, Dbachmann, Bender235, Firsfron, K Lepo, GünniX, Fresheneesz, Tom Morris, Rcbutcher, Peter17, Wwhat, Ntsimp, Michael C Price, Headbomb, Peter Gulutzan, GeorgeDishman, PStrait, Maurice Carbonaro, Kevinecahill, Pleroma, UnitedStatesian, James

McBride, Addbot, KleinGordon, Yobot, Sdcoonce, Br77rino, FrescoBot, Citation bot 1, Tom.Reding, Michael9422, John of Reading, OldRoute66, ZéroBot, Ulflund, CocuBot, Mariguld, Helpful Pixie Bot, FiveColourMap, I am One of Many, Jwratner1, Wikipedyah, IiKkEe, Tetra quark and Anonymous: 11

- **Decoupling (cosmology)** *Source:* https://en.wikipedia.org/wiki/Decoupling_(cosmology)?oldid=769961503 *Contributors:* Jerzy, Rjwilmsi, Ashill, Headbomb, Yobot, Tinton5, Heurisko, Ulflund, Bibcode Bot, PDiracDelta, Mark viking, HannahFord428, Tetra quark, Bender the Bot and Anonymous: 2
- **Structure formation** *Source:* https://en.wikipedia.org/wiki/Structure_formation?oldid=778758228 *Contributors:* Zundark, Edward, Rtfisher, Everyking, Grm wnr, FT2, Vsmith, Nabla, El C, Cmdrjameson, 9SGjOSfyHJaQVsEmy9NS, RHaworth, Mu301, Jeff3000, Joke137, Drbogdan, Rjwilmsi, R Lee E, Wavelength, Gaius Cornelius, Redgolpe, That Guy, From That Show!, SmackBot, Ashill, MalafayaBot, Droll, Colonies Chris, J 1982, Myasuda, BobQQ, Corpx, Thijs!bot, Headbomb, Lfstevens, Magioladitis, R'n'B, Warut, Sheliak, AlnoktaBOT, SwordSmurf, Lamro, SieBot, Shahidur Rahman, Jonmtkisco, Panos84, Dana boomer, Tailedkupo, DumZiBoT, RexxS, DCCougar, Hess88, Addbot, DOI bot, AkhtaBot, Samiswicked, Glane23, Yobot, Amirobot, AnomieBOT, Christopher.Gordon3, Citation bot, Icosmology, FrescoBot, Paine Ellsworth, Kikuyu3, Citation bot 1, Berkeleyjess, Tom.Reding, RockSolidCosmo, Puzl bustr, JLincoln, RjwilmsiBot, TjBot, John of Reading, Italia2006, RaptureBot, Brownie Charles, EdoBot, Machina Lucis, Lincoln Josh, Bibcode Bot, Khazar2, Tzymne, Penitence, PirtleShell, Jwratner1, Kogge, Xibalban Alchemist, Cosmic connection, Monkbot, Tetra quark, KasparBot, Mboerwink, TychosElk, Youknowwhatimsayin and Anonymous: 16
- **Cosmic microwave background** *Source:* https://en.wikipedia.org/wiki/Cosmic_microwave_background?oldid=787684837 *Contributors:* AxelBoldt, Bryan Derksen, The Anome, Tarquin, AstroNomer, AdamW, XJaM, Roadrunner, SimonP, Jaknouse, Youandme, Tedernst, Boud, Michael Hardy, EddEdmondson, Modster, Dominus, Loisel, Alfio, Egil, Stevenj, Ec5618, Timwi, Reddi, Cmbant, Chuunen Baka, Donarreiskoffer, Robbot, Jredmond, Peak, Mirv, Rursus, Mark Krueger, Carlj7, JerryFriedman, Kevin Saff, Graeme Bartlett, Harp, Art Carlson, Guanaco, Gzornenplatz, SWAdair, Bobblewik, TerokNor, Quadell, Beland, Karol Langner, Oneiros, MFNickster, Infradig, Sam Hocevar, Lumidek, Iantresman, Tsemii, Burschik, Mschlindwein, Deglr6328, Flyhighplato, Safety Cap, Moverton, Rich Farmbrough, Guanabot, Pjacobi, Wrp103, Vsmith, MuDavid, Bender235, Jonathanischoice, RJHall, JustinWick, Livajo, Pt, Edward Z. Yang, Art LaPella, Army1987, Mtruch, MITalum, Svenax, 9SGjOSfyHJaQVsEmy9NS, Physicistjedi, (aeropagitica), Haham hanuka, Alansohn, Anthony Appleyard, 119, Free Bear, Eric Kvaalen, Andrew Gray, Proxide, JHG, Denniss, GeorgeStepanek, Kdau, EAi, Count Iblis, Cmapm, DV8 2XL, Gene Nygaard, Ceyockey, Falcorian, Oleg Alexandrov, Ian Moody, Kelly Martin, TheNightFly, Pol098, Yukawa~enwiki, Jok2000, Uris, Wdanwatts, Joke137, Tevatron~enwiki, Rnt20, Graham87, Grammarbot, Edison, Josh Parris, Sjö, Drbogdan, Rjwilmsi, Zbxgscqf, Strangethingintheland, Marasama, Hjb26, Mike s, Mike Peel, Ttwaring, Fragglet, Acefitt, DVdm, Bgwhite, Ahpook, YurikBot, Wavelength, Hairy Dude, Jimp, RussBot, Gaius Cornelius, Eleassar, David R. Ingham, NawlinWiki, DragonHawk, Grafen, Chrisbrl88, Deckiller, FF2010, Smkolins, Orioane, Jules.LT, CWenger, QmunkE, ArielGold, Kungfuadam, Profero, Teply, GrinBot~enwiki, BenWandelt, SmackBot, Esradekan, Ashill, Saravask, Onsly, Gilliam, Benjaminevans82, Hmains, RobertKennedy, Lindosland, Andyzweb, Bluebot, Kurykh, H2ppyme, Myxsoma, Silly rabbit, Sangrolu, DHN-bot~enwiki, Scwlong, Modest Genius, Wikipedia brown, Rrburke, LouScheffer, Aldaron, Wen D House, Cybercobra, Bowlhover, A.R., Clean Copy, DenisRS, Zadignose, Ligulembot, Rossp, SashatoBot, Robomaeyhem, JzG, UberCryxic, Hypnosifl, Wwagner, Johnny 0, KJS77, Iridescent, Pathosbot, Mssgill, Chetvorno, Friendly Neighbour, CWY2190, Makuabob, Cydebot, Peripitus, Hsxavier~enwiki, Tbird1965, Alaibot, ZombieLoffe, הסרפד, Astrophysics Kid, Headbomb, Peter Gulutzan, Davidhorman, Raphiq, Gioto, Widefox, Orionus, TexMurphy, Rico402, Arch dude, Hut 8.5, .anacondabot, Antelan, VoABot II, Catslash, Pcp071098, Bubba hotep, First Harmonic, Allstarecho, LorenzoB, Kornfan71, Davidjcmorris, Keith D, R'n'B, Rrostrom, Yonidebot, Tgebbie, Jpwest, Migran, Александр Сигачёв, Austin512, Novis-M, Tarotcards, Rominandreu, Wesino, DadaNeem, DorganBot, Epistemenical, Sheliak, VolkovBot, Svmich, Sporti, Craigheinke, TXiKiBoT, MusicScience, Anonymous Dissident, Michael H 34, Broadbot, SwordSmurf, James McBride, Kbrose, Biscuittin, SieBot, Paradoctor, Hertz1888, Csmart287, Wing gundam, Zbvhs, Csblack, Mimihitam, Jdaloner, RMB1987, Duae Quartunciae, Anchor Link Bot, Wyattmj, Wjmummert, Martarius, GarbagEcol, ClueBot, The Thing That Should Not Be, Niceguyedc, Aggel000, ChandlerMapBot, I am a violinist, Excirial, Homonihilis, Nymf, Alexbot, Jefflayman, SolomonFreer, PixelBot, Bob108, Telekenesis, Tnxman307, Mastertek, Natty sci~enwiki, BOTarate, Panos84, Aitias, Nakomaru, Jonverve, DumZiBoT, BarretB, XLinkBot, DCCougar, BodhisattvaBot, Gwark, Dthomsen8, ErgoSum88, Ich42, Addbot, Dryphi, DOI bot, Ronhjones, Chotabe, Ka Faraq Gatri, Proxima Centauri, Ehrenkater, Astro-norte, Lightbot, OlEnglish, Zorrobot, Ben Ben, Luckas-bot, Yobot, Ptbotgourou, Legobot II, Aldebaran66, Amble, Wireader, Azcolvin429, AnomieBOT, Tryptofish, Stuffed cat, Captain Quirk, Jim1138, Hunnjazal, Citation bot, Palitzsch250, Xqbot, Plastadity, Seb.mag, Nnivi, Cydelin, Srich32977, Lithopsian, J04n, GrouchoBot, EqualMusic, Omnipaedista, Frankie0607, Kyng, Amaury, Mnmngb, Bigger digger, Fotaun, CES1596, GliderMaven, Nagualdesign, FrescoBot, LucienBOT, Paine Ellsworth, Binrdow, Citation bot 2, HamburgerRadio, Citation bot 1, HRoestBot, MoonGirl78, Jonesey95, Tom.Reding, Lithium cyanide, Pmokeefe, RedBot, Brian Everlasting, IVAN3MAN, RockSolidCosmo, TobeBot, Trappist the monk, Comet Tuttle, Michael9422, LI995, Earthandmoon, Tbhotch, Marie Poise, Wikiborg4711, Siranmichel, DexDor, Cwsavage78, Mathewsyriac, EmausBot, WikitanvirBot, Immunize, Quantanew, GoingBatty, Klbrain, Snorgway, Italia2006, Grondilu, ZéroBot, Medeis, Quondum, AManWithNoPlan, Miguelzuma, Iiar, Pumpkinking0192, Tbgriswold, Hang Li Po, ChiZeroOne, DASHBotAV, Mikhail Ryazanov, ClueBot NG, Ulflund, Factorial8, Helpful Pixie Bot, Bibcode Bot, BG19bot, Omegafold, Mdak06, AvocatoBot, Socal212, Ninney, Altaïr, Natalia.missana, Sparkie82, Fivemusketeers, U-95, ChrisGualtieri, JYBot, Dexbot, Neicdk, Manjolis, LightandDark2000, Antunesi, Reatlas, Rfassbind, User74~enwiki, Qmgsobserver, Praemonitus, Zlelik2000, OxygenBlueDansk, Friedlicherkoenig, AbiLtoCen, Johndric Valdez, Exoplanetaryscience, Mfb, Jlarks73, MSheshera, Monkbot, Filedelinkerbot, Falcon9v1.1, Unatnas1986, Trackteur, Werzaz, Anthul, SkyFlubbler, Samoniel1, Tullyojr, SwagYolo420ilovethis, Tetra quark, Isambard Kingdom, Anand2202, GeneralizationsAreBad, Freakcrane870, Sir Cumference, Feelthhis, Outedexits, Yohoyo, InternetArchiveBot, Aoziwe, Barkana, RubidiumZ, Mohd Tabish916, Bear-rings, Redhat101, Bender the Bot, MinusBot, Magic links bot and Anonymous: 316
- **Angular diameter distance** *Source:* https://en.wikipedia.org/wiki/Angular_diameter_distance?oldid=770374791 *Contributors:* SimonP, Patrick, Hjb26, Physchim62, Caco de vidro, SmackBot, Ekrenor, OrphanBot, Radagast83, Kurtan~enwiki, Hermitage17, Cydebot, Casliber, R'n'B, CommonsDelinker, Migran, Wesino, Squids and Chips, Oshwah, Cosmo0, Djr32, Addbot, Fgnievinski, Chzz, LilHelpa, Locos epraix, NOrbeck, Louperibot, OgreBot, RedBot, Newty23125, John of Reading, Milad pourrahmani, ???, Bender the Bot and Anonymous: 14
- **Galaxy cluster** *Source:* https://en.wikipedia.org/wiki/Galaxy_cluster?oldid=781729381 *Contributors:* DavidLevinson, Rossumcapek, RichardWeiss, Drbogdan, Tomtheman5, Fresheneesz, DVdm, Gadget850, Rrburke, J 1982, Gobonobo, Mgiganteus1, George100, Hermitage17, Infophile, Magioladitis, Oroso, CommonsDelinker, Oshwah, Peter Erwin, UnitedStatesian, James McBride, Mverleg, Ost316, SilvonenBot, Addbot, HerculeBot, AnomieBOT, Materialscientist, Anna Frodesiak, Gap9551, SassoBot, Paine Ellsworth, Originalwana, Sae1962, Q'piraq,

Gaba p, Tom.Reding, Dude1818, EmausBot, Jmencisom, Italia2006, Röhmöfantti, ChuispastonBot, ClueBot NG, BattyBot, Dexbot, Rfassbind, Spencer.mccormick, Tetra quark, Geniousankittripathi, Shezi110, Jin.cantab, Zhakhan9er, Greenelastic, Cepheidcatnip, ParshallMice and Anonymous: 28

- **Number density** *Source:* https://en.wikipedia.org/wiki/Number_density?oldid=765886457 *Contributors:* Michael Hardy, Giftlite, Curps, Karol Langner, Icairns, Jeff3000, BD2412, Physchim62, Hairy Dude, Silverchemist, Enormousdude, SmackBot, Melchoir, Eskimbot, DinosaursLoveExistence, Jan.Kamenicek, JForget, Kehrli, N2e, Cydebot, Christian75, Thijs!bot, Epbr123, Mbell, RolfSander, Grimlock, R'n'B, Stan J Klimas, Xenonice, VolkovBot, Swind, PixelBot, Harlock81, Addbot, Fgnievinski, Lightbot, Luckas-bot, AnomieBOT, Anne Bauval, GrouchoBot, قلی زادگان, Maggyero, TobeBot, EmausBot, Primefac, ZéroBot, ChuispastonBot, ClueBot NG, CocuBot, Helpful Pixie Bot, BZTMPS, Trackteur, Fmadd and Anonymous: 17
- **Phantom energy** *Source:* https://en.wikipedia.org/wiki/Phantom_energy?oldid=773373336 *Contributors:* Alfio, JWSchmidt, HangingCurve, Nickptar, Mike Rosoft, Brim, 9SGjOSfyHJaQVsEmy9NS, Axeman89, Joke137, Drbogdan, Rjwilmsi, Adoniscik, Cryptic, Wiki alf, SmackBot, P0rkup, Vyznev Xnebara, Mudd1, Peter Gulutzan, LordAnubisBOT, BrettAllen, HowardFrampton, Addbot, DOI bot, Luckas-bot, JohnHarold, Dreamer08, AnomieBOT, Farin12, Citation bot 1, ZéroBot, ClueBot NG, Leejoe Schar, Bibcode Bot, Guanghuilin, Kevin12xd, BerFinelli, Ceosad, Tetra quark, Multiverse Guy, Sir Cumference, Cyberpupk, MAdmiKe and Anonymous: 17
- **Quintessence (physics)** *Source:* https://en.wikipedia.org/wiki/Quintessence_(physics)?oldid=786423895 *Contributors:* AxelBoldt, Bryan Derksen, Ted Longstaffe, Bth, Montrealais, Michael Hardy, Nixdorf, Muriel Gottrop~enwiki, Loren Rosen, Emperorbma, Alex S, Timwi, Reddi, Markhurd, DW40, Jeffq, Barbara Shack, Snowdog, Beland, Bbbl67, Thorwald, Rich Farmbrough, Pjacobi, Gianluigi, Dbachmann, Eric Forste, El C, 9SGjOSfyHJaQVsEmy9NS, Dirac1933, Richard Weil, GregorB, Joke137, Rjwilmsi, RE, Diza, Hairy Dude, NawlinWiki, Dbfirs, 2over0, Thnidu, Petri Krohn, KasugaHuang, SmackBot, Rentier, Cybercobra, LoveEncounterFlow, Ossipewsk, Kurtan~enwiki, Jerald Frazier (aka DJ Adrenaline), Penbat, Thijs!bot, N5iln, Headbomb, Peter Gulutzan, Widefox, Magioladitis, Jpod2, Robin S, Iakane49, Tarotcards, Izno, VolkovBot, Don4of4, Lamro, EoGuy, Andwor, AbJ32, Agentxyz, Rreagan007, Addbot, Legobot, Yobot, Baxxterr, AnomieBOT, Nighthawk008, Citation bot, Finncarey, Lsj, Tom.Reding, RjwilmsiBot, Mmpcq, Klbrain, Hhhippo, Pas:7131, Khestwol, ClueBot NG, P0lise, Mr Gearloose, MerlIwBot, Bibcode Bot, BG19bot, Mediran, Andyhowlett, Prokaryotes, Blanclar, Monkbot, Diloshwan, Tetra quark, The Rocky Road and Anonymous: 63
- **Gravitational interaction of antimatter** *Source:* https://en.wikipedia.org/wiki/Gravitational_interaction_of_antimatter?oldid=777279254 *Contributors:* Bryan Derksen, Bkell, Intangir, WhiteDragon, Rjolly, Pjacobi, Vsmith, Pearle, Anthony Appleyard, Keenan Pepper, RJFJR, Count Iblis, Linas, Christopher Thomas, Mandarax, Ketiltrout, Rjwilmsi, Strait, Ian Pitchford, Srleffler, DVdm, Bgwhite, Hillman, Praetonia, Ripper234, Reyk, Petri Krohn, Jeff Silvers, SmackBot, Jcarroll, JorisvS, Bobamnertiopsis, Lynch82, Davidryan168, Dycedarg, Vyznev Xnebara, Banedon, Boardhead, Bm gub, DJ Rubbie, Tyco.skinner, RogueNinja, Darklilac, Cardamon, David Eppstein, Cadwaladr, Usp, Barraki, Squids and Chips, Objectivist, Agesworth, Flyer22 Reborn, Malawi craig, Arjen Dijksman, CultureDrone, Arunsingh16, ResidueOfDesign, Kbdankbot, Addbot, Roentgenium111, Download, Proxima Centauri, Munkel Davidson, Yobot, Aldebaran66, AnomieBOT, VanishedUser sdu9aya9fasdsopa, LilHelpa, N0rbeck, GrouchoBot, A. di M., Carlog3, Kikuyu3, Jonesey95, Full-date unlinking bot, Wellsmax, Teaser47401, JfLowell, Bibcode Bot, Massimozanardi, BG19bot, Acharneski, Tiscando, Reader505, Blaspie55, Pikachu Bros., Mdann52, Dexbot, 331dot, Exacerangutan, TwoTwoHello, Reatlas, Marco.bs, Varkman, AgentLym, Signoredexter, Conundrum84720, Epigogue, Looper12312321, MRMOMMYMAN, MrAntigravity, Dr.R.C.Gupta and Anonymous: 70
- **Future of an expanding universe** *Source:* https://en.wikipedia.org/wiki/Future_of_an_expanding_universe?oldid=786977091 *Contributors:* Kragen, Dmytro, Jni, Giftlite, Alison, Gracefool, Kainaw, Jackol, Thincat, RJHall, El C, Lycurgus, Hadlock, Christopher Thomas, Drbogdan, Rjwilmsi, Koavf, Ronocdh, Jehochman, ScottJ, DoubleBlue, Bgwhite, Spacepotato, RussBot, JocK, Gulliveig, Allens, Katieh5584, Serendipodous, SmackBot, Ashill, Incnis Mrsi, Jab843, Onebravemonkey, Gilliam, Izzynn, Thumperward, Colonies Chris, GrahameS, Bowlhover, Pulu, Kendrick7, Acdx, J 1982, Muadd, AdultSwim, Joseph Solis in Australia, Banedon, Keraunos, Headbomb, Najro, Peter Gulutzan, Magioladitis, Email4mobile, Cgingold, Captain panda, NewEnglandYankee, KylieTastic, Squids and Chips, Michaelpremsrirat, Oshwah, PlanetStar, BartekChom, Lightmouse, DragonZero, ClueBot, Plastikspork, CooPs89, Jusdafax, Larphenflorp, Millionsandbillions, Arjayay, Panos84, Johnuniq, Antti29, Maldek, Addbot, Eric Drexler, Lightbot, Yobot, Knownot, Suntag, Maldek2, Robert Treat, AnomieBOT, Letuño, Jim1138, Piano non troppo, Materialscientist, Draco de Mos Fulmen, Cyphoidbomb, Gap9551, Volvo B9TL, Jennli, Mnmngb, Unideanet, FreeKnowledgeCreator, George585, Paine Ellsworth, Eronel189, Chard513, Pinethicket, Tom.Reding, Serols, Jim37hike, BlackHades, Jeroen De Dauw, Doublebackslash, VEO15, Double sharp, Trappist the monk, Nickyus, Reach Out to the Truth, Abcsrfun123, Tesseract2, Acather96, Bt8257, K6ka, Thecheesykid, Hhhippo, Jetman508, Yiosie2356, Donner60, Surajt88, Whoop whoop pull up, ClueBot NG, Lord Roem, SenseiAC, 497glbig, Widr, Bibcode Bot, Regulov, BG19bot, ISTB351, MusikAnimal, Alvin Lee, Pikachu Bros., U-95, Dexbot, Frosty, SteenthIWbot, Sauropodomorph, Eyesnore, Madreterra, Strangenight, DavidLeighEllis, Jwratner1, Leroybrown2000, Noyster, Patbdwll, Dough34, Fixuture, MarioProtIV, Thundergodz, SkyFlubbler, Wikipedian 2, MGChecker, Aberlamps, Tetra quark, Isambard Kingdom, GarStazi, GSS-1987, Christopherwiki3!, Naymyo1971, JJMC89 bot, NickTheRipper, Chrissymad, Megaskizzen, Cyrus noto3at bulaga, Here2help, Labjt21, Magic links bot and Anonymous: 155
- **De Sitter space** *Source:* https://en.wikipedia.org/wiki/De_Sitter_space?oldid=786638462 *Contributors:* Michael Hardy, Looxix~enwiki, Timwi, Phys, Mporter, ShaunMacPherson, Lethe, Fropuff, Eequor, Serenus~enwiki, Cacycle, Mytg8, Rgdboer, John Vandenberg, Kevin Lamoreau, Ceyockey, Linas, Rjwilmsi, R.e.b., Chobot, YurikBot, Hillman, Msikma, Bota47, SDS, SmackBot, Tonyr68uk, Silly rabbit, Nbarth, Ulner, Torrazzo, Vanisaac, Thijs!bot, CZmarlin, JAnDbot, Lantonov, Mrmahdiarnt, Schucker, Speaker to wolves, YohanN7, SieBot, Likebox, Alexbot, 1ForTheMoney, Addbot, Oberflaechenelement, Delaszk, Luckas-bot, Ptbotgourou, TaBOT-zerem, AnomieBOT, Citation bot, Dendropithecus, FrescoBot, Dogbert66, Argumzio, Citation bot 1, Seattle Jörg, WikitanvirBot, Chasrob, Quondum, Afjvanraan, Mgvongoeden, ChrisGualtieri, Spray787, Enyokoyama, Keith the Koala, Nerd1a4i and Anonymous: 34
- **Ultimate fate of the universe** *Source:* https://en.wikipedia.org/wiki/Ultimate_fate_of_the_universe?oldid=787586419 *Contributors:* Derek Ross, Vicki Rosenzweig, Bryan Derksen, The Anome, RK, Mintguy, Edward, JohnOwens, Michael Hardy, Gabbe, Ixfd64, IZAK, Alfio, Goatasaur, Looxix~enwiki, Theresa knott, Schneelocke, Feedmecereal, Reddi, Dysprosia, Tempshill, VeryVerily, Bevo, Wiwaxia, TravelingDude, BenRG, UninvitedCompany, Meelar, Robinh, Carlj7, Xanzzibar, Mattflaschen, Tea2min, David Gerard, DocWatson42, Barbara Shack, Misterkillboy, Ferkelparade, Herbee, Average Earthman, Everyking, Curps, DJSupreme23, Eequor, SWAdair, SonicAD, Gadfium, Hoxu, Quadell, Antandrus, Beland, Karol Langner, Zantolak, Nickptar, B.d.mills, Burschik, Grunt, Aponar Kestrel, Lacrimosus, Perey, Freakofnurture, Poccil, GaidinBDJ, Discospinster, Rich Farmbrough, Leibniz, Vsmith, Freestylefrappe, StephanKetz, Silence, SocratesJedi, Dbachmann,

Mani1, JoeSmack, Konstantin~enwiki, Mr. Billion, El C, C1k3, Art LaPella, Cigarette, Marathoner, Func, BrokenSegue, L33tminion, 9SGjOSfyHJaQVsEmy9NS, La goutte de pluie, Apostrophe, (aeropagitica), Pearle, Cyrillic, Jakew, Mareino, Jason One, Alansohn, Tablizer, Polarscribe, Hipocrite, Andrewpmk, Plumbago, Samahnub, Echuck215, Kocio, Snowolf, TaintedMustard, Rafti Institute, Harej, GabrielF, Versageek, DV8 2XL, Falcorian, Oleg Alexandrov, WilliamKF, Richard Arthur Norton (1958-), Roboshed, Mindmatrix, FeanorStar7, Uncle G, Plek, Aaron McDaid, Robert K S, Ruud Koot, Jok2000, Christopher Thomas, Lazareth, Mandarax, RichardWeiss, Chupon, FreplySpang, Ketiltrout, Sjö, Drbogdan, Rjwilmsi, Ckoenigsberg, Mikehambleton, Urbane Legend, Tawker, Mike Peel, The wub, Remurmur, AySz88, Yamamoto Ichiro, JohnDBuell, Fuzzyeric, Naraht, Nihiltres, Harmil, ElfQrin, BitterMan, Diza, Imnotminkus, DVdm, Sasoriza, Cactus.man, Gwernol, Amaurea, Kummi, The Rambling Man, Spacepotato, Posxt, Jimp, Hillman, Mark Ironie, Durand101, SpuriousQ, CambridgeBayWeather, Pseudomonas, Daveswagon, Thechosenone021, Msikma, Astral, Długosz, SingularityZero, MarkBrooks, Wknight94, WAS 4.250, Mfryc, Deville, Glome83, El benito, Arthur Rubin, E Wing, Xaxafrad, CWenger, Geoffrey.landis, HereToHelp, Ilmari Karonen, Caco de vidro, Katieh5584, RG2, Ryūkotsusei, SmackBot, Haymaker, Davepape, KnowledgeOfSelf, GalaxyGirl, McGeddon, Dieboybun, Crigaux, Felix Dance, Onebravemonkey, Half-Shadow, Commander Keane bot, IW4, Gilliam, Portillo, DividedByNegativeZero, Skizzik, Carl.bunderson, RobertM525, Stuart P. Bentley, Valley2city, Anachronist, Bluebot, JONJONAUG, Gabrinae, WikiPedant, Kotra, Can't sleep, clown will eat me, Voyajer, Mihxil, TheKMan, XQ fan, LouScheffer, Kntrabssi, Danhilton, PsychoJosh, NickPenguin, EdGl, Kendrick7, RossF18, SashatoBot, Lambiam, Valfontis, Dark Formal, J 1982, Stefan2, IronGargoyle, Ckatz, Bless sins, Noah Salzman, Martinp23, Larrymcp, Dr.K., C17H19NO3, D Hill, LeyteWolfer, Happy-melon, Donder~enwiki, Tdmg, Tawkerbot2, RookZERO, JRSpriggs, Perryathlete09, SkyWalker, CRGreathouse, CmdrObot, Naterock, Wafulz, David s graff, Van helsing, Ilikefood, Banedon, Anih, ShelfSkewed, Cracker017, Gregbard, Gmusser, Travelbird, Sa.vakilian, Acs4b, Nick Ottery, Michael C Price, Tawkerbot4, Torc2, DumbBOT, DarkLink, Theirishpianist~enwiki, BetacommandBot, Epbr123, Coelacan, Paktype, Mbell, Interested2, Keraunos, Manyana, Second Quantization, Peter Gulutzan, Gnurkel, Nick Number, Dawnseeker2000, Mmortal03, Alessandriana, Mentifisto, AntiVandalBot, Seaphoto, StantheGarbageMan, Pollira, Scepia, Itistoday, Zero g, Tim Shuba, MECU, Oddity-, Legare, Daggoth, Gökhan, JDitto, LinkinPark, Thundernlightning, Dcooper, MegX, SiobhanHansa, Acroterion, Chaxes, ILSS, Celithemis, Bongwarrior, VoABot II, Carlwev, Brandt Luke Zorn, JamesBWatson, Steven Walling, Theroadislong, Cgingold, Kevinwiatrowski, Lyonscc, Canyouhearmenow, Der-Hexer, Gyakusetsu, Skylights76, Thompson.matthew, Jonomacdrones, El Zoof, Hdt83, MartinBot, Philomelius, Twigletmac, Octopug, Roasty-toast, Mschel, CommonsDelinker, MapleTree, LedgendGamer, RaccoonFox, J.delanoy, Alasdair22, Pharaoh of the Wizards, Mindgames11, Trusilver, Ginsengbomb, Tokyogirl79, Realelite, Shawn in Montreal, Univisions~enwiki, Mikael Häggström, ElectricValkyrie, Jedimaster9459, Richard D. LeCour, Ferahgo the Assassin, Orion2k4, Xiahou, Lights, Qwesa7, VolkovBot, TreasuryTag, RingtailedFox, Mrlion~enwiki, Seattle Skier, Xdt, STLocutus, Speaker to wolves, Lonnie randall, Oshwah, Dojarca, Pwnage8, TJJH, Zanardm, Anonymous Dissident, Someguy1221, Terence7, Jubajube, Bentley4, Cremepuff222, Antman27, SwordSmurf, Dirkbb, Bryonss, Lamro, Wasted Sapience, Halfthought, Enviroboy, Vector Potential, Gstarkman, Rod1234, Burntsauce, ObiWanBillKenobi, Rileyboughton, Christopher546, Northfox, Bluedenim, Gogogosox, PaddyLeahy, Subh83, SieBot, Sonicology, Hertz1888, Caltas, Sounnga, Kylemew, Vinaymangal, WildWildBil, Fluck, Qst, Faisalzainabdullah, Dominik92, Wmpearl, Oxymoron83, Nuttycoconut, Zharradan.angelfire, KathrynLybarger, Manway, Ranzan, Dravecky, Iknowyourider, Nick1824, Hamiltondaniel, Wahrmund, Pinkadelica, Ghetsmith, Kanonkas, Explicit, Soporaeternus, Martarius, Tanvir Ahmmed, Elassint, ClueBot, Kumagoro-42, Fyyer, The Thing That Should Not Be, HairyFotr, Av99, Drmies, DanielDeibler, 10euro, XxCoNdEmNeDxX, Rotational, Fatladi, Gboy2325, Malinaccier Public, Kitsunegami, Jusdafax, Blickmaestro, SeeYouNextTuesday999, NuclearWarfare, CMW275, Jason-AQuest, Olpcgga, Usingthewavequation, Youssef4342, VatoFirme, Thingg, Silenceisgolden123, Kate Phaye, ApostleJoe, Versus22, Mythdon, Johnuniq, Arianewiki1, XLinkBot, Tuxlie, Rob 188, Tarheel95, Tarl N., KriticKill, Avoided, NellieBly, Aunt Entropy, D.M. from Ukraine, Tre2, Chimpstop, Addbot, Xp54321, Willking1979, Feel the mike bob, Atethnekos, Earth is over, Devrana, Protonk, Morning277, Prankster-Turtle, Glane23, Bassbonerocks, Favonian, Sappho'd, Bouncingball2, E.R.UT, Gail, Greyhood, Luckas-bot, Yobot, Systemizer, TaBOT-zerem, Amble, AnomieBOT, Dirjarmocksorz, Jamihiller, Forrestnoble, Flewis, Jaymeethepirate, Citation bot, Maxis ftw, Tutterz, Phlembowper99, Cureden, Renaissancee, Smk65536, DSisyphBot, Yoyodigman, Gap9551, Measles, Buckelz, Omnipaedista, Poorya13, The Wiki ghost, Erik9, Puffino, PortiaDarby, NonHypocriteType, HRIN, IamRandom, FrescoBot, Paine Ellsworth, Io Herodotus, Pepper, TimonyCrickets, Wiki4ata, Thorenn, Goodbye Galaxy, Sae1962, Olithorpe1234, Tom.Reding, Efficiency1101e, BRUTE, BlackHades, Angie Maria, Reconsider the static, IJBall, Crusoe8181, Elmoro, IVAN3MAN, RockSolidCosmo, Proffsl, Diannaa, DARTH SIDIOUS 2, K-ceron, Whisky drinker, EmausBot, Stevefrise, TageMA, Mlazear, GoingBatty, Rowan casey, Bt8257, Slightsmile, Wikipelli, TeleComNasSprVen, Gretzky hbk, KampferXeon, Solomonfromfinland, Permenent, 1234r00t, EWikist, Prvák, Pbjr23, Donner60, Blueberry321, Rock'n'Wraith, FL0per, Weltherr, Dermiel, ClueBot NG, GeekX, Ypnypn, A520, Vacation9, Jesspiper, Shmuj, Widr, ScienceDawns, Mightymights, Helpful Pixie Bot, DmitryKsWikis, Titodutta, Wbm1058, Bibcode Bot, Northamerica1000, CoatesField, MusikAnimal, The Almightey Drill, Harizotoh9, Darkso l8, Fuchsiavelvet, Sourabh.khot, Uyrki8, Wannabemodel, Mdann52, Nathanielfirst, Raymond1922A, Flygon's friend, Armanschwarz, Alexander1257, Alexlyoko13, McBenjamin, Jxmorris, I am One of Many, Howicus, BerFinelli, Mcdonaldsworkerdude12, Madreterra, CipherNemo, CensoredScribe, Babitaarora, Jwratner1, Jupitus Smart, Anrnusna, Entirbloos, Monkbot, JoshLemer, WanderingLost, Joebob is awesome, Amortias, Thundergodz, A Great Catholic Person, Wikipedian 2, SA 13 Bro, The Page Fixer, Eteethan, ScrapIronIV, Tetra quark, DrooodleCreeper, Isambard Kingdom, Pippamagnolia, MB298, Just a Rube, Buckbill10, Edsolis90, Dysmys, Legatusamu, HRI$HIKE$H, Yohoyo, Dawangy, InternetArchiveBot, KGirlTrucker81, Tevlev3, MassChaos23, Imminent77, המכבת, Lollypodjebwjdd, MyHypothesisOntheUniverse, XenonNSMB, Edbe 01 and Anonymous: 736

- **Big Rip** *Source:* https://en.wikipedia.org/wiki/Big_Rip?oldid=780605508 *Contributors:* AxelBoldt, Vicki Rosenzweig, Bryan Derksen, Tarquin, Dcljr, Cyp, Glenn, Nikai, Jeandré du Toit, Schneelocke, Reddi, Maximus Rex, BenRG, Lumos3, Academic Challenger, Rursus, Tea2min, Smjg, Wolfkeeper, Herbee, Henry Flower, Matthäus Wander, Nickptar, Burschik, Discospinster, Guanabot, StephanKetz, Dbachmann, Nard the Bard, RJHall, Art LaPella, Rpresser, 9SGjOSfyHJaQVsEmy9NS, La goutte de pluie, HasharBot~enwiki, Free Bear, Arthena, Super-Magician, Versageek, Thryduulf, Scjessey, Aristotle Pagaltzis, Joke137, Banpei~enwiki, Drbogdan, Rjwilmsi, Pjetter, Koavf, Mike Peel, Maxim Razin, Crazycomputers, Wars, Angus Lepper, Hairy Dude, Bhny, Trovatore, Omega 13, Roman Soldier, ColinMcMillen, SmackBot, AaronM, C.Fred, Bluebot, Hibernian, Darthgriz98, Richard001, J 1982, Don't give an Ameriflag, Kirbytime, Danilot, Dan Gluck, Iridescent, Tawkerbot2, Henrickson, JForget, Kjknohw, Gregbard, Cydebot, Gogo Dodo, Peter Gulutzan, MarkGyver, AgentPeppermint, Noclevername, Northumbrian, EmRunTonRespNin, WikiSlasher, Spartaz, LinkinPark, RadicalPi, Swpb, Nyttend, DGG, Sm8900, Monkeybait99, R'n'B, Mbweissman, J.delanoy, Foober, Framhein, Juliancolton, Lights, VolkovBot, RingtailedFox, Kinneytj, HowardFrampton, TXiKiBoT, Oshwah, Bentley4, W.a.tas, CorusTV, Wasted Sapience, Seresin, Bobo The Ninja, SieBot, Lalala98, Lalala95, Wing gundam, Flyer22 Reborn, Format43, Duae Quartunciae, Hamiltondaniel, Atilla98, Michael Petersen, Ratemonth, ClueBot, Estevoaei, AndyFielding, Editor2020, Staticshakedown, Addbot, Jncraton, OlEnglish, MattDredd, Luckas-bot, Yobot, Legobot II, Amble, KamikazeBot, AnomieBOT, LED65, Obersachsebot, Capricorn42, RibotBOT, CaZeRillo, LucienBOT, Io Herodotus, Citation bot 1, MusBeaumont, 70virginousc, Trappist the monk, NerdyScienceDude, EmausBot, WikitanvirBot,

ZéroBot, Pas:7131, Teapeat, Rememberway, ClueBot NG, Pennykohl, Bibcode Bot, Kvark92, Ikjyotsingh, Epicgenius, Madreterra, Amrellithy, Jwratner1, Kogge, Urmomisachode, Phlegl, Christolav, Wikipedian 2, Tonathan100, AlphaBetaGamma01, Tetra quark, Isambard Kingdom, Pulkitmidha, Sir Cumference, CLCStudent, Bender the Bot and Anonymous: 133

- **Big Bounce** *Source:* https://en.wikipedia.org/wiki/Big_Bounce?oldid=782800323 *Contributors:* Bryan Derksen, Edward, Furrykef, Omegatron, Korath, Peak, Academic Challenger, HaeB, GreatWhiteNortherner, Jason Quinn, Junuxx, LucasVB, Antandrus, FT2, Bender235, El C, 9SGjOSfyHJaQVsEmy9NS, Pearle, Knucmo2, Sade, Velella, Gpvos, Vashti, Sin-man, Drbogdan, Rjwilmsi, Linuxbeak, Acefitt, Bgwhite, Stephenb, Gaius Cornelius, JonathanD, CWenger, SmackBot, Armeria, Dane Sorensen, J 1982, Zzzzzzzzzzz, TPIRFanSteve, Robertinventor, Thijs!bot, Headbomb, Peter Gulutzan, BlytheG, Mentifisto, DagosNavy, VoABot II, Skylights76, Rob Lindsey, Ildus58, MooresLaw, Fullmetal2887, KylieTastic, Mike V, CardinalDan, Andyvphil, Knightshield, PaddyLeahy, MarcelloBarnaba, Filos96, Hatster301, Atif.t2, ClueBot, Eetvartti, Unbuttered Parsnip, CohesionBot, Antiquary, Simon Villeneuve, Nonunitary, NonvocalScream, Addbot, Luckas-bot, Fraggle81, Backslash Forwardslash, AnomieBOT, Ipatrol, Rtyq2, Citation bot, StrontiumDogs, Adrianilias, Theone567hunter, Finncarey, Citation bot 1, I dream of horses, Seryo93, Trappist the monk, Aiurdin, Nick Moyes, Slightsmile, Dpieski, SporkBot, Crux007, Terraflorin, ClueBot NG, Helpful Pixie Bot, Bibcode Bot, BG19bot, NUMB3RN7NE, Layzeeboi, Dannie996, Lianatajo, Anthony Felizardo, Monkbot, Loraof, Cyrej, Tetra quark, Safderg, Equinox, A shell, Dilbilen, Magic links bot and Anonymous: 97

- **Big Crunch** *Source:* https://en.wikipedia.org/wiki/Big_Crunch?oldid=778718101 *Contributors:* The Epopt, Bryan Derksen, The Anome, Roadrunner, Patrick, Modster, Cyde, Delirium, Looxix~enwiki, Ahoerstemeier, LittleDan, Glenn, Poor Yorick, Schneelocke, Timwi, Joseaperez, BenRG, Sanders muc, Ianb, Blainster, CdaMVvWgS, Centrx, Giftlite, DocWatson42, Curps, Alison, Eequor, SWAdair, Isidore, Piotrus, Oddball990, CesarFelipe, Sam Hocevar, Burschik, Mschlindwein, Freakofnurture, Rich Farmbrough, Dbachmann, Scumbag, JoeSmack, AdamSolomon, El C, Cigarette, Chessphoon, 9SGjOSfyHJaQVsEmy9NS, Slicky, La goutte de pluie, Deryck Chan, HasharBot~enwiki, Knucmo2, Jumbuck, Danski14, Nurban, Wtmitchell, Rick Sidwell, Ceyockey, Dismas, WilliamKF, 2004-12-29T22:45Z, Behonkiss, Christopher Thomas, Chun-hian, GBoehm, Ketiltrout, Drbogdan, TheIncredibleEdibleOompaLoompa, Wragge, FlaBot, Robmods, Wars, Acefitt, E Pluribus Anthony, YurikBot, Spacepotato, Retodon8, Gaius Cornelius, Nick, Silverhill, Scott Adler, Deville, LeonardoRob0t, Katieh5584, SmackBot, Anastrophe, Septegram, Gilliam, Portillo, Betacommand, WikiPedant, Can't sleep, clown will eat me, MadameArsenic, Frap, Drsmoo, JonHarder, Mmathu, Richard001, Eynar, Ironcito, Dane Sorensen, Eliyak, Soap, J 1982, Disavian, Shadowcaster187, Laogeodritt, Dan Gluck, Megawattbulbman, Cxat, Joseph Solis in Australia, Newone, Tophtucker, JHP, Fairyhairycarpetfluff, Sax Russell, Cumulus Clouds, Lurlock, Gregbard, Sopoforic, Cydebot, Eu.stefan, Bookgrrl, Underpants, JayW, Crum375, Thijs!bot, Mathias315, Peter Gulutzan, JustAGal, AgentPeppermint, Dawnseeker2000, Escarbot, Sanchom, Jenattiyeh, Doctorhawkes, Thedoorhinge, JamesBWatson, Nyttend, Orochi nem0, Kevinwiatrowski, Skylights76, Tonicthebrown, Francis Tyers, J.delanoy, FrummerThanThou, KylieTastic, Heliogabalus227, Steel1943, VolkovBot, CWii, AlnoktaBOT, Anonymous Dissident, Qxz, UnitedStatesian, Rex Imperator, Wasted Sapience, Vector Potential, AlleborgoBot, Theowningone, Proxmire, SieBot, Sonicology, Caltas, GlassCobra, Byrialbot, Trucizna, OKBot, Cosmo0, Doom2099, ClueBot, LAX, Fyyer, Wfraga, Niceguyedc, Rotational, Chantal Marguerite, Azinmagehero, Mkativerata, Excirial, Ember of Light, Youssef4342, The Wicked Twisted Road, Johnuniq, XLinkBot, Orgthingy~enwiki, BIlbo The Hobbit, Nepenthes, SilvonenBot, Addbot, Nohomers48, Amosyjos2, Tide rolls, Luckas-bot, Yobot, Granpuff, TaBOT-zerem, DarqueSoul, KamikazeBot, AnomieBOT, Zhieaanm, VX, Materialscientist, ArthurBot, Hi878, Permethius, Jezhotwells, Shadowjams, Io Herodotus, Krj373, Machine Elf 1735, Dvartian, MastiBot, MondalorBot, Meaghan, Diannaa, Khan197khan, DASHBot, EmausBot, ImprovingWiki, John of Reading, WikitanvirBot, Dcirovic, Hhhippo, Ὁ οἶστρος, Hashiq, SporkBot, Carmichael, Intraceptor, ClueBot NG, Jack Greenmaven, Pennykohl, BG19bot, Mark Arsten, U-95, SNAAAAKE!!, Raymond1922A, Hmainsbot1, Frosty, Ikjyotsingh, Bingbangcosmology, Madreterra, Jwratner1, Tacobelll3, Jodetert, Wikipedian 2, AHusain3141, Loraof, Smokingbull, Johanna, Tetra quark, Isambard Kingdom, Anand2202, GeneralizationsAreBad, ExperiencedArticleFixer, KasparBot, Siveini, Sir Cumference, Doug8796, Daynoel436, Gmcsween, Lukejosiah03 and Anonymous: 224

- **Proton decay** *Source:* https://en.wikipedia.org/wiki/Proton_decay?oldid=779061499 *Contributors:* Uriyan, Bryan Derksen, Tarquin, Taw, Alex.tan, Maury Markowitz, Stevertigo, Edward, Michael Hardy, Bcrowell, TakuyaMurata, Alfio, Looxix~enwiki, J'raxis, Bluelion, Maximus Rex, Phys, TravelingDude, BenRG, Securiger, Henrygb, Bkell, Wereon, Giftlite, Herbee, Xerxes314, DJSupreme23, Jason Quinn, CryptoDerk, Cglassey, B.d.mills, Deglr6328, Rich Farmbrough, FT2, Pjacobi, Ben Standeven, El C, 9SGjOSfyHJaQVsEmy9NS, La goutte de pluie, Jason One, Jérôme, Miranche, Zyqqh, Reaverdrop, DV8 2XL, Feezo, MattJakel, Joke137, Christopher Thomas, Rjwilmsi, RexNL, Jimp, Limulus, Jengelh, Lucinos~enwiki, GeeJo, Długosz, Tachs, Nikkimaria, Smurrayinchester, RodVance, Groyolo, MacsBug, SmackBot, Melchoir, Jrockley, Winterheart, Tigerhawkvok, Colonies Chris, Scwlong, Audriusa, V1adis1av, QFT, LouScheffer, Wikiwikiwiki3~enwiki, Doug Bell, John, JorisvS, Groggy Dice, Simkiott, Happy-melon, CRGreathouse, Sahrin, Michael C Price, Headbomb, Stannered, WinBot, Maliz, DerHexer, Momojeng, Jotempe, Dzogchenpa, TXiKiBoT, Hqb, Andysoh, Someguy1221, Bentley4, Ptrslv72, Northfox, SieBot, Chandrahas9, BlueAzure, Martarius, Mild Bill Hiccup, Chieron, Arjayay, SkyLined, Addbot, Barak Sh, Yobot, Legobot II, Robert Treat, AnomieBOT, Rubinbot, Materialscientist, Citation bot, W.stanovsky, Ender's Shadow Snr, MeDrewNotYou, Jan Krieg, GreenRoot, Citation bot 1, Tom.Reding, Double sharp, ZéroBot, Vbrun237, Timetraveler3.14, Brandmeister, Surajt88, Nerdok, Fbrugmans, Teaktl17, ClueBot NG, Widr, Helpful Pixie Bot, Bibcode Bot, RageOfGod, BG19bot, CitationCleanerBot, GregorDS, EnzaiBot, Matherforthewin, Osteologia, Cjean42, 22merlin, Monkbot, Aberlamps, Nidj123, Isambard Kingdom, Philip.Ratcliffe, NoToleranceForIntolerance, KAP03 and Anonymous: 81

43.11.2 Images

- **File:080998_Universe_Content_240_after_Planck.jpg** *Source:* https://upload.wikimedia.org/wikipedia/commons/b/b6/080998_Universe_Content_240_after_Planck.jpg *License:* Public domain *Contributors:* http://map.gsfc.nasa.gov/media/080998/index.html updated data from http://www.nasa.gov/mission_pages/planck/news/planck20130321.html *Original artist:* NASA, Modified by User:⍰⍰

- **File:14-283-Abell2744-DistantGalaxies-20141016.jpg** *Source:* https://upload.wikimedia.org/wikipedia/commons/d/d2/14-283-Abell2744-DistantGalaxies-20141016.jpg *License:* Public domain *Contributors:* http://www.nasa.gov/sites/default/files/14-283_0.jpg *Original artist:* NASA, J. Lotz, (STScI)

- **File:14-296-GalaxyClusters-PerseusVirgo-ChandraXRay-20141027.jpg** *Source:* https://upload.wikimedia.org/wikipedia/commons/8/8d/14-296-GalaxyClusters-PerseusVirgo-ChandraXRay-20141027.jpg *License:* Public domain *Contributors:* http://www.nasa.gov/sites/default/files/thumbnails/image/14-296.jpg *Original artist:* NASA/CXC/Stanford/I. Zhuravleva et al

- **File:1e0657_scale.jpg** *Source:* https://upload.wikimedia.org/wikipedia/commons/a/a8/1e0657_scale.jpg *License:* Public domain *Contributors:* Chandra X-Ray Observatory: 1E 0657-56 *Original artist:* NASA/CXC/M. Weiss
- **File:Accretion_Disk_Binary_System.jpg** *Source:* https://upload.wikimedia.org/wikipedia/commons/0/0c/Accretion_Disk_Binary_System.jpg *License:* Public domain *Contributors:* ? *Original artist:* ?
- **File:Aleksandr_Fridman.png** *Source:* https://upload.wikimedia.org/wikipedia/commons/6/62/Aleksandr_Fridman.png *License:* Public domain *Contributors:* Cropped From *Original artist:* Unknown<a href='https://www.wikidata.org/wiki/Q4233718' title='wikidata:Q4233718'><img alt='wikidata:Q4233718' src='https://upload.wikimedia.org/wikipedia/commons/thumb/f/ff/Wikidata-logo.svg/20px-Wikidata-logo.svg.png' width='20' height='11' srcset='https://upload.wikimedia.org/wikipedia/commons/thumb/f/ff/Wikidata-logo.svg/30px-Wikidata-logo.svg.png 1.5x, https://upload.wikimedia.org/wikipedia/commons/thumb/f/ff/Wikidata-logo.svg/40px-Wikidata-logo.svg.png 2x' data-file-width='1050' data-file-height='590' /></a>
- **File:Alfa_beta_gamma_radiation_penetration.svg** *Source:* https://upload.wikimedia.org/wikipedia/commons/6/61/Alfa_beta_gamma_radiation_penetration.svg *License:* CC BY 2.5 *Contributors:*
- Alfa_beta_gamma_radiation.svg *Original artist:* Alfa_beta_gamma_radiation.svg: User:Stannered
- **File:Ambox_current_red.svg** *Source:* https://upload.wikimedia.org/wikipedia/commons/9/98/Ambox_current_red.svg *License:* CC0 *Contributors:* self-made, inspired by Gnome globe current event.svg, using Information icon3.svg and Earth clip art.svg *Original artist:* Vipersnake151, penubag, Tkgd2007 (clock)
- **File:Ambox_important.svg** *Source:* https://upload.wikimedia.org/wikipedia/commons/b/b4/Ambox_important.svg *License:* Public domain *Contributors:* Own work based on: Ambox scales.svg *Original artist:* Dsmurat
- **File:Ambox_question.svg** *Source:* https://upload.wikimedia.org/wikipedia/commons/1/1b/Ambox_question.svg *License:* Public domain *Contributors:* Based on Image:Ambox important.svg *Original artist:* Mysid, Dsmurat, penubag
- **File:Angular-size-redshift-relation.png** *Source:* https://upload.wikimedia.org/wikipedia/commons/d/d0/Angular-size-redshift-relation.png *License:* Public domain *Contributors:* No machine-readable source provided. Own work assumed (based on copyright claims). *Original artist:* No machine-readable author provided. Hjb26 assumed (based on copyright claims).
- **File:Arrow_Blue_Down_001.svg** *Source:* https://upload.wikimedia.org/wikipedia/commons/a/a5/Arrow_Blue_Down_001.svg *License:* Public domain *Contributors:* Own work *Original artist:* User:Sameboat
- **File:Asymmetric_Ashes_(artist'{}s_impression).jpg** *Source:* https://upload.wikimedia.org/wikipedia/commons/d/db/Asymmetric_Ashes_%28artist%27s_impression%29.jpg *License:* CC BY 4.0 *Contributors:* http://www.eso.org/public/images/eso0644a/ *Original artist:* ESO
- **File:BSicon_BHF.svg** *Source:* https://upload.wikimedia.org/wikipedia/commons/7/76/BSicon_BHF.svg *License:* Public domain *Contributors:* Own work *Original artist:* Bernina & axpde
- **File:BSicon_KBHFa.svg** *Source:* https://upload.wikimedia.org/wikipedia/commons/1/19/BSicon_KBHFa.svg *License:* Public domain *Contributors:* Own work *Original artist:* user:axpde
- **File:BSicon_STR.svg** *Source:* https://upload.wikimedia.org/wikipedia/commons/3/3c/BSicon_STR.svg *License:* Public domain *Contributors:* Own work *Original artist:* de:User:Bernina & de:User:axpde
- **File:BSicon_eBHF.svg** *Source:* https://upload.wikimedia.org/wikipedia/commons/1/19/BSicon_eBHF.svg *License:* Public domain *Contributors:* Own work *Original artist:* Bernina & axpde
- **File:BSicon_eKBHFe.svg** *Source:* https://upload.wikimedia.org/wikipedia/commons/6/6a/BSicon_eKBHFe.svg *License:* Public domain *Contributors:* Own work *Original artist:* user:axpde
- **File:BSicon_hWSTR.svg** *Source:* https://upload.wikimedia.org/wikipedia/commons/d/de/BSicon_hWSTR.svg *License:* Public domain *Contributors:* Own work *Original artist:* de:User:axpde
- **File:BSicon_tBHF.svg** *Source:* https://upload.wikimedia.org/wikipedia/commons/2/27/BSicon_tBHF.svg *License:* Public domain *Contributors:* Own work *Original artist:* axpde
- **File:BSicon_tSTRa.svg** *Source:* https://upload.wikimedia.org/wikipedia/commons/d/da/BSicon_tSTRa.svg *License:* Public domain *Contributors:* Icons von Bernina *Original artist:* T.h.
- **File:BSicon_tSTRe.svg** *Source:* https://upload.wikimedia.org/wikipedia/commons/d/d0/BSicon_tSTRe.svg *License:* Public domain *Contributors:* Icons von Bernina *Original artist:* T.h.
- **File:Baryon-decuplet-small.svg** *Source:* https://upload.wikimedia.org/wikipedia/commons/7/78/Baryon-decuplet-small.svg *License:* Public domain *Contributors:* Own work *Original artist:* Trassiorf
- **File:Baryon-octet-small.svg** *Source:* https://upload.wikimedia.org/wikipedia/commons/b/b5/Baryon-octet-small.svg *License:* Public domain *Contributors:* Own work *Original artist:* Trassiorf
- **File:Big_Crunch.gif** *Source:* https://upload.wikimedia.org/wikipedia/commons/f/f4/Big_Crunch.gif *License:* Public domain *Contributors:* No machine-readable source provided. Own work assumed (based on copyright claims). *Original artist:* No machine-readable author provided. Rogilbert~commonswiki assumed (based on copyright claims).
- **File:Big_crunch.png** *Source:* https://upload.wikimedia.org/wikipedia/commons/5/52/Big_crunch.png *License:* CC-BY-SA-3.0 *Contributors:* ? *Original artist:* ?
- **File:BlackHole.jpg** *Source:* https://upload.wikimedia.org/wikipedia/commons/d/d4/BlackHole.jpg *License:* Public domain *Contributors:* http://web.archive.org/web/20100416132936/http://www.gsfc.nasa.gov/topstory/20011015blackhole.html (direct link) http://earthsky.org/space/comparing-theory-to-observation-in-eating-habits-of-giant-black-holes (direct link) *Original artist:* XMM-Newton, ESA, NASA

- **File:BoRG-58.jpg** *Source:* https://upload.wikimedia.org/wikipedia/commons/b/be/BoRG-58.jpg *License:* CC BY 3.0 *Contributors:* http://www.spacetelescope.org/images/heic1201a/ *Original artist:* NASA, ESA, M. Trenti (University of Cambridge, UK and University of Colorado, Boulder, USA), L. Bradley (STScI), and the BoRG team
- **File:CMB_Timeline300_no_WMAP.jpg** *Source:* https://upload.wikimedia.org/wikipedia/commons/6/6f/CMB_Timeline300_no_WMAP.jpg *License:* Public domain *Contributors:* Original version: NASA; modified by Ryan Kaldari *Original artist:* NASA/WMAP Science Team
- **File:COBE_cmb_fluctuations.png** *Source:* https://upload.wikimedia.org/wikipedia/commons/a/a3/COBE_cmb_fluctuations.png *License:* Public domain *Contributors:*
- http://lambda.gsfc.nasa.gov/product/cobe/dmr_image.cfm *Original artist:* The COBE datasets were developed by the NASA Goddard Space Flight Center under the guidance of the COBE Science Working Group.
- **File:COSMOS_3D_dark_matter_map.png** *Source:* https://upload.wikimedia.org/wikipedia/commons/3/31/COSMOS_3D_dark_matter_map.png *License:* Public domain *Contributors:* http://spacetelescope.org/images/heic0701b/
Original artist: NASA/ESA/Richard Massey (California Institute of Technology)
- **File:Cmbr.svg** *Source:* https://upload.wikimedia.org/wikipedia/commons/c/cd/Cmbr.svg *License:* Public domain *Contributors:* Own work *Original artist:* Quantum Doughnut
- **File:Collage_of_six_cluster_collisions_with_dark_matter_maps.jpg** *Source:* https://upload.wikimedia.org/wikipedia/commons/0/03/Collage_of_six_cluster_collisions_with_dark_matter_maps.jpg *License:* CC BY 4.0 *Contributors:* http://www.spacetelescope.org/images/heic1506a/ *Original artist:* NASA, ESA, D. Harvey (École Polytechnique Fédérale de Lausanne, Switzerland), R. Massey (Durham University, UK), the Hubble SM4 ERO Team, ST-ECF, ESO, D. Coe (STScI), J. Merten (Heidelberg/Bologna), HST Frontier Fields, Harald Ebeling(University of Hawaii at Manoa), Jean-Paul Kneib (LAM)and Johan Richard (Caltech, USA)
- **File:Color_image_of_galaxy_cluster_MCS_J0416.1–2403.jpg** *Source:* https://upload.wikimedia.org/wikipedia/commons/a/a2/Color_image_of_galaxy_cluster_MCS_J0416.1%E2%80%932403.jpg *License:* CC BY 4.0 *Contributors:* http://www.spacetelescope.org/images/heic1416a/ (image link) *Original artist:* ESA/Hubble, NASA, HST Frontier Fields
Acknowledgement: Mathilde Jauzac (Durham University, UK and Astrophysics & Cosmology Research Unit, South Africa) and Jean-Paul Kneib (École Polytechnique Fédérale de Lausanne, Switzerland)
- **File:Commons-logo.svg** *Source:* https://upload.wikimedia.org/wikipedia/en/4/4a/Commons-logo.svg *License:* PD *Contributors:* ? *Original artist:* ?
- **File:Cosmic_distance_ladder.jpg** *Source:* https://upload.wikimedia.org/wikipedia/commons/2/23/Cosmic_distance_ladder.jpg *License:* CC BY 4.0 *Contributors:* http://www.spacetelescope.org/images/heic1611a/ *Original artist:* NASA,ESA, A. Feild (STScI), and A. Riess (STScI/JHU)
- **File:CosmoDistanceMeasures_z_to_1e4.png** *Source:* https://upload.wikimedia.org/wikipedia/commons/5/5d/CosmoDistanceMeasures_z_to_1e4.png *License:* Public domain *Contributors:* Transferred from en.wikipedia to Commons by Drongou using CommonsHelper. *Original artist:* Wesino at English Wikipedia
- **File:CosmoDistanceMeasures_z_to_onehalf.png** *Source:* https://upload.wikimedia.org/wikipedia/commons/8/8a/CosmoDistanceMeasures_z_to_onehalf.png *License:* Public domain *Contributors:* Transferred from en.wikipedia to Commons by Drongou using CommonsHelper. *Original artist:* Wesino at English Wikipedia
- **File:Crab_Nebula.jpg** *Source:* https://upload.wikimedia.org/wikipedia/commons/0/00/Crab_Nebula.jpg *License:* Public domain *Contributors:* HubbleSite: gallery, release. *Original artist:* NASA, ESA, J. Hester and A. Loll (Arizona State University)
- **File:DMPie_2013.svg** *Source:* https://upload.wikimedia.org/wikipedia/commons/1/1f/DMPie_2013.svg *License:* CC BY-SA 3.0 *Contributors:* Own work *Original artist:* Szczureq
- **File:Dark_Energy.jpg** *Source:* https://upload.wikimedia.org/wikipedia/commons/c/ce/Dark_Energy.jpg *License:* Public domain *Contributors:* http://hubblesite.org/newscenter/archive/releases/2001/09/image/g/ OR http://science.nasa.gov/astrophysics/focus-areas/what-is-dark-energy/ *Original artist:* Ann Feild (STScI)
- **File:Dark_matter_map_of_KiDS_survey_region_(region_G12).jpg** *Source:* https://upload.wikimedia.org/wikipedia/commons/4/44/Dark_matter_map_of_KiDS_survey_region_%28region_G12%29.jpg *License:* CC BY 4.0 *Contributors:* https://www.eso.org/public/images/eso1642a/ *Original artist:* Kilo-Degree Survey Collaboration/H. Hildebrandt & B. Giblin/ESO
- **File:Distanza_di_diametro_angolare.png** *Source:* https://upload.wikimedia.org/wikipedia/commons/c/c8/Distanza_di_diametro_angolare.png *License:* CC BY 3.0 *Contributors:* Transferred from en.wikipedia to Commons. *Original artist:* Migran at English Wikipedia
- **File:Drop_closeup.jpg** *Source:* https://upload.wikimedia.org/wikipedia/commons/7/74/Drop_closeup.jpg *License:* CC BY 2.0 *Contributors:* File:Flickr_-_NewsPhoto!_-_drup.jpg - original: http://www.flickr.com/photos/13088710@N02/3581239516/ *Original artist:* Jos van Zetten
- **File:Earth-moon.jpg** *Source:* https://upload.wikimedia.org/wikipedia/commons/5/5c/Earth-moon.jpg *License:* Public domain *Contributors:* NASA: old link, new link *Original artist:* Apollo 8 crewmember Bill Anders
- **File:Edit-clear.svg** *Source:* https://upload.wikimedia.org/wikipedia/en/f/f2/Edit-clear.svg *License:* Public domain *Contributors:* The *Tango! Desktop Project*. *Original artist:*

  The people from the Tango! project. And according to the meta-data in the file, specifically: "Andreas Nilsson, and Jakub Steiner (although minimally)."
- **File:Electromagnetic-Spectrum.png** *Source:* https://upload.wikimedia.org/wikipedia/commons/8/8a/Electromagnetic-Spectrum.png *License:* CC BY-SA 2.5 *Contributors:* taken from en.wikipedia en:Image:Electromagnetic-Spectrum.svg and en:Image:Electromagnetic-Spectrum.png (deleted) *Original artist:* transferred by Penubag (talk · contribs) on 05:04, 15 May 2008
- **File:End_of_universe.jpg** *Source:* https://upload.wikimedia.org/wikipedia/commons/9/98/End_of_universe.jpg *License:* Public domain *Contributors:* ? *Original artist:* ?

- **File:Extragalactic_distance_ladder.JPG** *Source:* https://upload.wikimedia.org/wikipedia/commons/1/13/Extragalactic_distance_ladder.JPG *License:* CC BY-SA 3.0 *Contributors:* This file was derived from: Extragalactic Distance Ladder.svg
Original artist: brews_ohare
- **File:FarNearFields-USP-4998112-1.svg** *Source:* https://upload.wikimedia.org/wikipedia/commons/5/5d/FarNearFields-USP-4998112-1.svg *License:* Public domain *Contributors:* US Patent 6657596 *Original artist:* Goran M Djuknic
- **File:Folder_Hexagonal_Icon.svg** *Source:* https://upload.wikimedia.org/wikipedia/en/4/48/Folder_Hexagonal_Icon.svg *License:* Cc-by-sa-3.0 *Contributors:* ? *Original artist:* ?
- **File:Friedmann_universes.svg** *Source:* https://upload.wikimedia.org/wikipedia/commons/d/dc/Friedmann_universes.svg *License:* Public domain *Contributors:* Own work *Original artist:* BenRG
- **File:G299-Remnants-SuperNova-Type1a-20150218.jpg** *Source:* https://upload.wikimedia.org/wikipedia/commons/b/bd/G299-Remnants-SuperNova-Type1a-20150218.jpg *License:* Public domain *Contributors:* http://www.nasa.gov/sites/default/files/thumbnails/image/g299.jpg *Original artist:* NASA/CXC/U.Texas
- **File:GalacticRotation2.svg** *Source:* https://upload.wikimedia.org/wikipedia/commons/b/b9/GalacticRotation2.svg *License:* CC-BY-SA-3.0 *Contributors:* Own work in Inkscape 0.42 *Original artist:* PhilHibbs
- **File:Galaxy_cluster_Abell_2218_gravitaitonal_lens.jpg** *Source:* https://upload.wikimedia.org/wikipedia/en/4/42/Galaxy_cluster_Abell_2218_gravitaitonal_lens.jpg *License:* PD *Contributors:*
NASA
Original artist: WFPC2, HST, NASA, STScI, Andrew Fruchter [#cite_note-2 [2]]
- **File:Galaxy_cluster_IDCS_J1426.jpg** *Source:* https://upload.wikimedia.org/wikipedia/commons/0/08/Galaxy_cluster_IDCS_J1426.jpg *License:* CC BY 4.0 *Contributors:* http://www.spacetelescope.org/images/opo1602a/ *Original artist:* NASA, ESA, and M. Brodwin (University of Missouri)
- **File:Gravitational_lens-full.jpg** *Source:* https://upload.wikimedia.org/wikipedia/commons/0/02/Gravitational_lens-full.jpg *License:* Public domain *Contributors:* ? *Original artist:* ?
- **File:Gravitationell-lins-4.jpg** *Source:* https://upload.wikimedia.org/wikipedia/commons/0/0b/Gravitationell-lins-4.jpg *License:* Public domain *Contributors:* http://hubblesite.org/newscenter/newsdesk/archive/releases/2003/01/image/a *Original artist:* NASA, N. Benitez (JHU), T. Broadhurst (Racah Institute of Physics/The Hebrew University), H. Ford (JHU), M. Clampin (STScI),G. Hartig (STScI), G. Illingworth (UCO/Lick Observatory), the ACS Science Team and ESA
- **File:HST-Smiling-GalaxyClusterSDSS-J1038+4849-20150210.jpg** *Source:* https://upload.wikimedia.org/wikipedia/commons/b/b9/HST-Smiling-GalaxyClusterSDSS-J1038%2B4849-20150210.jpg *License:* Public domain *Contributors:* http://www.nasa.gov/sites/default/files/thumbnails/image/15861603283_3579db3fc6_o.jpg *Original artist:* NASA/ESA
- **File:Heic0612b_H.jpg** *Source:* https://upload.wikimedia.org/wikipedia/commons/2/22/Heic0612b_H.jpg *License:* Public domain *Contributors:* ? *Original artist:* ?
- **File:Heic1401a-Abell2744-20140107.jpg** *Source:* https://upload.wikimedia.org/wikipedia/commons/4/4d/Heic1401a-Abell2744-20140107.jpg *License:* Public domain *Contributors:* HubbleSite: NewsCenter *Original artist:* NASA, ESA, and J. Lotz, M. Mountain, A. Koekemoer, and the HFF Team (STScI). "Image/Video Use Policy Space Telescope imagery is free for anyone to use." => http://www.spacetelescope.org/images/heic1401a/ and http://www.spacetelescope.org/copyright/
- **File:Hertzsprung-Russel_StarData.png** *Source:* https://upload.wikimedia.org/wikipedia/commons/1/17/Hertzsprung-Russel_StarData.png *License:* CC BY 4.0 *Contributors:* https://www.eso.org/public/images/eso0728c/ *Original artist:* ESO
- **File:Horizonte_inflacionario.svg** *Source:* https://upload.wikimedia.org/wikipedia/commons/b/b4/Horizonte_inflacionario.svg *License:* CC-BY-SA-3.0 *Contributors:* Transferred from en.wikipedia to Commons.; original: *I created this work in Adobe Illustrator.* *Original artist:* Joke137 at English Wikipedia
- **File:Horn_Antenna-in_Holmdel,_New_Jersey.jpeg** *Source:* https://upload.wikimedia.org/wikipedia/commons/f/f7/Horn_Antenna-in_Holmdel%2C_New_Jersey.jpeg *License:* Public domain *Contributors:* Great Images in NASA Description *Original artist:* NASA
- **File:Hubble_constant.JPG** *Source:* https://upload.wikimedia.org/wikipedia/commons/2/2c/Hubble_constant.JPG *License:* CC BY-SA 3.0 *Contributors:* Own work *Original artist:* Brews ohare
- **File:Hubble_stretches_the_stellar_tape_measure_ten_times_further.jpg** *Source:* https://upload.wikimedia.org/wikipedia/commons/0/00/Hubble_stretches_the_stellar_tape_measure_ten_times_further.jpg *License:* CC BY 4.0 *Contributors:* http://www.spacetelescope.org/images/opo1423a/ *Original artist:* Illustration Credit: NASA, ESA, and A. Feild (STScI)Science Credit: NASA, ESA, A. Riess (JHU/STScI), S. Casertano (STScI/JHU), J. Anderson and J. MacKenty (STScI), and A. Filippenko (University of California, Berkeley)
- **File:Ilc_9yr_moll4096.png** *Source:* https://upload.wikimedia.org/wikipedia/commons/3/3c/Ilc_9yr_moll4096.png *License:* Public domain *Contributors:* http://map.gsfc.nasa.gov/media/121238/ilc_9yr_moll4096.png *Original artist:* NASA / WMAP Science Team
- **File:Image_of_the_galaxy_cluster_SpARCS1049.jpg** *Source:* https://upload.wikimedia.org/wikipedia/commons/2/23/Image_of_the_galaxy_cluster_SpARCS1049.jpg *License:* CC BY 4.0 *Contributors:* http://www.spacetelescope.org/images/heic1519a/ *Original artist:* NASA/STScI/ESA/JPL-Caltech/McGill
- **File:Inverse_square_law.svg** *Source:* https://upload.wikimedia.org/wikipedia/commons/2/28/Inverse_square_law.svg *License:* CC BY-SA 3.0 *Contributors:* Own work *Original artist:* Borb
- **File:LCDM.jpg** *Source:* https://upload.wikimedia.org/wikipedia/commons/7/7d/LCDM.jpg *License:* CC BY-SA 3.0 *Contributors:* Own work *Original artist:* Michael L. Umbricht

- **File:Lambda-Cold_Dark_Matter,_Accelerated_Expansion_of_the_Universe,_Big_Bang-Inflation.jpg** *Source:* https://upload.wikimedia.org/wikipedia/commons/c/c2/Lambda-Cold_Dark_Matter%2C_Accelerated_Expansion_of_the_Universe%2C_Big_Bang-Inflation.jpg *License:* CC BY-SA 3.0 *Contributors:* Own work *Original artist:* User:Coldcreation
- **File:Lemaitre.jpg** *Source:* https://upload.wikimedia.org/wikipedia/commons/5/52/Lemaitre.jpg *License:* Public domain *Contributors:* ? *Original artist:* ?
- **File:Lock-green.svg** *Source:* https://upload.wikimedia.org/wikipedia/commons/6/65/Lock-green.svg *License:* CC0 *Contributors:* en:File:Free-to-read_lock_75.svg *Original artist:* User:Trappist the monk
- **File:Magnifying_the_distant_Universe.jpg** *Source:* https://upload.wikimedia.org/wikipedia/commons/b/b9/Magnifying_the_distant_Universe.jpg *License:* CC BY 4.0 *Contributors:* http://www.spacetelescope.org/images/potw1412a/ *Original artist:* ESA/Hubble & NASA Acknowledgement: Nick Rose
- **File:Matter_Distribution.JPG** *Source:* https://upload.wikimedia.org/wikipedia/commons/5/50/Matter_Distribution.JPG *License:* CC BY-SA 3.0 *Contributors:* Own work *Original artist:* Brews ohare
- **File:Merge-arrows.svg** *Source:* https://upload.wikimedia.org/wikipedia/commons/5/52/Merge-arrows.svg *License:* Public domain *Contributors:* ? *Original artist:* ?
- **File:NO2-N2O4.jpg** *Source:* https://upload.wikimedia.org/wikipedia/commons/8/8b/NO2-N2O4.jpg *License:* Public domain *Contributors:* en:Image:N02-N2O4.jpg *Original artist:* en:User:Greenhorn1
- **File:Office-book.svg** *Source:* https://upload.wikimedia.org/wikipedia/commons/a/a8/Office-book.svg *License:* Public domain *Contributors:* This and myself. *Original artist:* Chris Down/Tango project
- **File:Open_Access_logo_PLoS_transparent.svg** *Source:* https://upload.wikimedia.org/wikipedia/commons/7/77/Open_Access_logo_PLoS_transparent.svg *License:* CC0 *Contributors:* http://www.plos.org/ *Original artist:* art designer at PLoS, modified by Wikipedia users Nina, Beao, and JakobVoss
- **File:PIA16874-CobeWmapPlanckComparison-20130321.jpg** *Source:* https://upload.wikimedia.org/wikipedia/commons/6/64/PIA16874-CobeWmapPlanckComparison-20130321.jpg *License:* Public domain *Contributors:* http://photojournal.jpl.nasa.gov/catalog/PIA16874 (direct link) *Original artist:* NASA/JPL-Caltech/ESA
- **File:PIA20052-GalaxyCluster-MOO-J1142+1527-20151103.jpg** *Source:* https://upload.wikimedia.org/wikipedia/commons/5/53/PIA20052-GalaxyCluster-MOO-J1142%2B1527-20151103.jpg *License:* Public domain *Contributors:* http://photojournal.jpl.nasa.gov/jpeg/PIA20052.jpg *Original artist:* NASA/JPL-Caltech/Gemini/CARMA
- **File:Pages_from_Prop1994.jpg** *Source:* https://upload.wikimedia.org/wikipedia/commons/b/bb/Pages_from_Prop1994.jpg *License:* CC BY-SA 3.0 *Contributors:* Own work *Original artist:* HiddenValley123
- **File:ParallaxeV2.png** *Source:* https://upload.wikimedia.org/wikipedia/commons/d/df/ParallaxeV2.png *License:* CC-BY-SA-3.0 *Contributors:* ? *Original artist:* ?
- **File:Phase_change_-_en.svg** *Source:* https://upload.wikimedia.org/wikipedia/commons/0/0b/Phase_change_-_en.svg *License:* Public domain *Contributors:* Own work *Original artist:* F l a n k e r, penubag
- **File:Phase_diagram_for_pure_substance.JPG** *Source:* https://upload.wikimedia.org/wikipedia/commons/e/e6/Phase_diagram_for_pure_substance.JPG *License:* CC BY-SA 3.0 *Contributors:* Own work *Original artist:* Brews ohare
- **File:Phot-27e-07.jpg** *Source:* https://upload.wikimedia.org/wikipedia/commons/4/40/Phot-27e-07.jpg *License:* CC BY 4.0 *Contributors:* http://www.eso.org/gallery/v/ESOPIA/Galaxies/phot-27e-07.jpg.html *Original artist:* ESO: observations by Susana Randall, Claudio Melo, Swetlana Hubrig; day astronomer Dominique Naef; Henri Boffin (ESO) processed the data and made the colour-composite, and Haennes Heyer (ESO) made the final adjustments.
- **File:Photon_waves.png** *Source:* https://upload.wikimedia.org/wikipedia/commons/f/ff/Photon_waves.png *License:* CC-BY-SA-3.0 *Contributors:* ? *Original artist:* ?
- **File:Plasma-lamp_2.jpg** *Source:* https://upload.wikimedia.org/wikipedia/commons/2/26/Plasma-lamp_2.jpg *License:* CC-BY-SA-3.0 *Contributors:*
- own work www.lucnix.be *Original artist:* Luc Viatour
- **File:Portal-puzzle.svg** *Source:* https://upload.wikimedia.org/wikipedia/en/f/fd/Portal-puzzle.svg *License:* Public domain *Contributors:* ? *Original artist:* ?
- **File:PowerSpectrumExt.svg** *Source:* https://upload.wikimedia.org/wikipedia/commons/1/16/PowerSpectrumExt.svg *License:* Public domain *Contributors:* lambda.gsfc.nasa.gov PowerSpectrumExt.pdf in a collection of WMAP Data Product Images. Converted to svg with pdf2svg. *Original artist:* NASA/WMAP Science Team
- **File:Progenitor_IA_supernova.svg** *Source:* https://upload.wikimedia.org/wikipedia/commons/c/ce/Progenitor_IA_supernova.svg *License:* CC BY 3.0 *Contributors:* http://hubblesite.org/newscenter/archive/releases/star/supernova/2004/34/image/d/ *Original artist:* NASA, ESA and A. Feild (STScI); vectorisation by chris ?
- **File:Proton_decay.svg** *Source:* https://upload.wikimedia.org/wikipedia/commons/8/85/Proton_decay.svg *License:* CC BY-SA 3.0 *Contributors:* Own work *Original artist:* Cjean42
- **File:Proton_decay2.svg** *Source:* https://upload.wikimedia.org/wikipedia/en/6/65/Proton_decay2.svg *License:* PD *Contributors:*

  I drew this diagram using Inkscape.

  Original artist:

  GreenRoot (talk)

- **File:Proton_decay3.svg** *Source:* https://upload.wikimedia.org/wikipedia/commons/b/bc/Proton_decay3.svg *License:* Public domain *Contributors:* I drew this diagram using Inkscape. *Original artist:* GreenRoot (talk)
- **File:Proton_decay4.svg** *Source:* https://upload.wikimedia.org/wikipedia/en/5/51/Proton_decay4.svg *License:* PD *Contributors:*

  I drew this diagram using Inkscape.

  Original artist:

  GreenRoot (talk)
- **File:Quark_structure_proton.svg** *Source:* https://upload.wikimedia.org/wikipedia/commons/9/92/Quark_structure_proton.svg *License:* CC BY-SA 2.5 *Contributors:* Own work *Original artist:* Arpad Horvath
- **File:Quartz_oisan.jpg** *Source:* https://upload.wikimedia.org/wikipedia/commons/4/4a/Quartz_oisan.jpg *License:* CC BY-SA 4.0 *Contributors:* Own work *Original artist:* Didier Descouens
- **File:Question_book-new.svg** *Source:* https://upload.wikimedia.org/wikipedia/en/9/99/Question_book-new.svg *License:* Cc-by-sa-3.0 *Contributors:*

  Created from scratch in Adobe Illustrator. Based on Image:Question book.png created by User:Equazcion *Original artist:*

  Tkgd2007
- **File:Question_dropshade.png** *Source:* https://upload.wikimedia.org/wikipedia/commons/d/dd/Question_dropshade.png *License:* Public domain *Contributors:* Image created by JRM *Original artist:* JRM
- **File:R-parity_violating_decay.svg** *Source:* https://upload.wikimedia.org/wikipedia/commons/7/71/R-parity_violating_decay.svg *License:* Public domain *Contributors:* en:Image:RPD.png *Original artist:* en:User:Maliz, User:Stannered
- **File:Radioactive.svg** *Source:* https://upload.wikimedia.org/wikipedia/commons/b/b5/Radioactive.svg *License:* Public domain *Contributors:* Created by Cary Bass using Adobe Illustrator on January 19, 2006. *Original artist:* Cary Bass
- **File:Radioactivity_and_radiation.png** *Source:* https://upload.wikimedia.org/wikipedia/commons/6/6e/Radioactivity_and_radiation.png *License:* CC BY-SA 3.0 *Contributors:* Own work *Original artist:* Doug Sim
- **File:Recent_Hubble'{}s_Constant_Values.png** *Source:* https://upload.wikimedia.org/wikipedia/commons/b/b7/Recent_Hubble%27s_Constant_Values.png *License:* CC BY-SA 3.0 *Contributors:* Own work *Original artist:* Primefac
- **File:Rotation_curve_(Milky_Way).JPG** *Source:* https://upload.wikimedia.org/wikipedia/commons/7/77/Rotation_curve_%28Milky_Way%29.JPG *License:* CC BY-SA 3.0 *Contributors:* Own work *Original artist:* Brews ohare
- **File:SN1994D.jpg** *Source:* https://upload.wikimedia.org/wikipedia/commons/a/a2/SN1994D.jpg *License:* CC BY 3.0 *Contributors:* http://www.spacetelescope.org/images/opo9919i/ *Original artist:* NASA/ESA, The Hubble Key Project Team and The High-Z Supernova Search Team
- **File:SN1998aq_max_spectra.svg** *Source:* https://upload.wikimedia.org/wikipedia/commons/2/25/SN1998aq_max_spectra.svg *License:* CC BY-SA 3.0 *Contributors:* self-made, data from Matheson et al. 2008. *Original artist:* Falcorian
- **File:SNIacurva.png** *Source:* https://upload.wikimedia.org/wikipedia/commons/8/88/SNIacurva.png *License:* CC-BY-SA-3.0 *Contributors:* No machine-readable source provided. Own work assumed (based on copyright claims). *Original artist:* No machine-readable author provided. Xenoforme~commonswiki assumed (based on copyright claims).
- **File:Science.jpg** *Source:* https://upload.wikimedia.org/wikipedia/commons/5/54/Science.jpg *License:* Public domain *Contributors:* ? *Original artist:* ?
- **File:Size_IK_Peg.svg** *Source:* https://upload.wikimedia.org/wikipedia/commons/4/4a/Size_IK_Peg.svg *License:* CC-BY-SA-3.0 *Contributors:* Own work *Original artist:* RJHall, chris ? (vector)
- **File:Solar_system.jpg** *Source:* https://upload.wikimedia.org/wikipedia/commons/8/83/Solar_system.jpg *License:* Public domain *Contributors:* http://photojournal.jpl.nasa.gov/catalog/PIA03153 *Original artist:* NASA/JPL
- **File:Spacetime_curvature.png** *Source:* https://upload.wikimedia.org/wikipedia/commons/2/22/Spacetime_curvature.png *License:* CC-BY-SA-3.0 *Contributors:* ? *Original artist:* ?
- **File:Spacetime_lattice_analogy.svg** *Source:* https://upload.wikimedia.org/wikipedia/commons/6/63/Spacetime_lattice_analogy.svg *License:* CC BY-SA 3.0 *Contributors:* Own work. Self -made in Blender & Inkscape. *Original artist:* Mysid
- **File:Standard_Model_of_Elementary_Particles.svg** *Source:* https://upload.wikimedia.org/wikipedia/commons/0/00/Standard_Model_of_Elementary_Particles.svg *License:* Public domain *Contributors:* Own work by uploader, PBS NOVA [1], Fermilab, Office of Science, United States Department of Energy, Particle Data Group *Original artist:* MissMJ
- **File:Structure_mode_history.svg** *Source:* https://upload.wikimedia.org/wikipedia/commons/e/ef/Structure_mode_history.svg *License:* CC-BY-SA-3.0 *Contributors:* This is my work, calculated using a code I wrote myself, plotted in Gnuplot and edited in Adobe Illustrator. *Original artist:* Joke137 at English Wikipedia
- **File:Stylised_Lithium_Atom.svg** *Source:* https://upload.wikimedia.org/wikipedia/commons/6/6f/Stylised_atom_with_three_Bohr_model_orbits_and_stylised_nucleus.svg *License:* CC-BY-SA-3.0 *Contributors:* based off of Image:Stylised Lithium Atom.png by Halfdan. *Original artist:* SVG by Indolences. Recoloring and ironing out some glitches done by Rainer Klute.
- **File:Supernova&galaxia.png** *Source:* https://upload.wikimedia.org/wikipedia/commons/c/c5/Supernova%26galaxia.png *License:* CC-BY-SA-3.0 *Contributors:* No machine-readable source provided. Own work assumed (based on copyright claims). *Original artist:* No machine-readable author provided. Xenoforme~commonswiki assumed (based on copyright claims).
- **File:Supernova_remnant_N103B.jpg** *Source:* https://upload.wikimedia.org/wikipedia/commons/1/12/Supernova_remnant_N103B.jpg *License:* CC BY 4.0 *Contributors:* https://www.spacetelescope.org/images/heic1707a/ *Original artist:* ESA/Hubble, NASA

- **File:Symbol_book_class2.svg** *Source:* https://upload.wikimedia.org/wikipedia/commons/8/89/Symbol_book_class2.svg *License:* CC BY-SA 2.5 *Contributors:* Mad by Lokal_Profil by combining: *Original artist:* Lokal_Profil
- **File:Symbol_list_class.svg** *Source:* https://upload.wikimedia.org/wikipedia/en/d/db/Symbol_list_class.svg *License:* Public domain *Contributors:* ? *Original artist:* ?
- **File:System-search.svg** *Source:* https://upload.wikimedia.org/wikipedia/commons/1/17/System-search.svg *License:* Public domain *Contributors:* The Tango! Desktop Project *Original artist:* The people from the Tango! project
- **File:The_Astonomer.jpg** *Source:* https://upload.wikimedia.org/wikipedia/commons/e/e7/The_Astonomer.jpg *License:* CC BY-SA 3.0 *Contributors:* Own work *Original artist:* Bidgee
- **File:Type_Ia_supernova_simulation_-_Argonne_National_Laboratory.jpg** *Source:* https://upload.wikimedia.org/wikipedia/commons/b/b7/Type_Ia_supernova_simulation_-_Argonne_National_Laboratory.jpg *License:* CC BY 2.0 *Contributors:* http://www.flickr.com/photos/argonne/5352540236/ *Original artist:* Argonne National Laboratory / U.S. Department of Energy
- **File:UniverseComposition.svg** *Source:* https://upload.wikimedia.org/wikipedia/commons/9/9d/UniverseComposition.svg *License:* CC0 *Contributors:* Own work - Derived from jpg version *Original artist:* Aflafla1
- **File:Universe_expansion2.png** *Source:* https://upload.wikimedia.org/wikipedia/commons/3/37/Universe_expansion2.png *License:* Public domain *Contributors:* Created by uploader from public domain source *Original artist:* Gnixon at English Wikipedia
- **File:Universecolor.jpg** *Source:* https://upload.wikimedia.org/wikipedia/commons/7/78/Universecolor.jpg *License:* CC BY 3.0 *Contributors:* Own work *Original artist:* Drschawrz
- **File:Velocity-redshift.JPG** *Source:* https://upload.wikimedia.org/wikipedia/commons/b/bd/Velocity-redshift.JPG *License:* CC BY-SA 3.0 *Contributors:* Own work *Original artist:* Brews ohare
- **File:WhiteDwarf_mass-radius_en.svg** *Source:* https://upload.wikimedia.org/wikipedia/commons/1/14/WhiteDwarf_mass-radius_en.svg *License:* CC BY 3.0 *Contributors:* File:WhiteDwarf mass-radius fr.svg *Original artist:*
- user:AllenMcC. (original jpg file)
- **File:Wiki_letter_w.svg** *Source:* https://upload.wikimedia.org/wikipedia/en/6/6c/Wiki_letter_w.svg *License:* Cc-by-sa-3.0 *Contributors:* ? *Original artist:* ?
- **File:Wikibooks-logo.svg** *Source:* https://upload.wikimedia.org/wikipedia/commons/f/fa/Wikibooks-logo.svg *License:* CC BY-SA 3.0 *Contributors:* Own work *Original artist:* User:Bastique, User:Ramac et al.
- **File:Wikinews-logo.svg** *Source:* https://upload.wikimedia.org/wikipedia/commons/2/24/Wikinews-logo.svg *License:* CC BY-SA 3.0 *Contributors:* This is a cropped version of Image:Wikinews-logo-en.png. *Original artist:* Vectorized by Simon 01:05, 2 August 2006 (UTC) Updated by Time3000 17 April 2007 to use official Wikinews colours and appear correctly on dark backgrounds. Originally uploaded by Simon.
- **File:Wikiquote-logo.svg** *Source:* https://upload.wikimedia.org/wikipedia/commons/f/fa/Wikiquote-logo.svg *License:* Public domain *Contributors:* Own work *Original artist:* Rei-artur
- **File:Wikisource-logo.svg** *Source:* https://upload.wikimedia.org/wikipedia/commons/4/4c/Wikisource-logo.svg *License:* CC BY-SA 3.0 *Contributors:* Rei-artur *Original artist:* Nicholas Moreau
- **File:Wikiversity-logo-Snorky.svg** *Source:* https://upload.wikimedia.org/wikipedia/commons/1/1b/Wikiversity-logo-en.svg *License:* CC BY-SA 3.0 *Contributors:* Own work *Original artist:* Snorky
- **File:Wiktionary-logo-en-v2.svg** *Source:* https://upload.wikimedia.org/wikipedia/commons/9/99/Wiktionary-logo-en-v2.svg *License:* CC-BY-SA-3.0 *Contributors:* ? *Original artist:* ?
- **File:Wiktionary-logo-v2.svg** *Source:* https://upload.wikimedia.org/wikipedia/en/0/06/Wiktionary-logo-v2.svg *License:* CC-BY-SA-3.0 *Contributors:* ? *Original artist:* ?
- **File:Wz-z.jpg** *Source:* https://upload.wikimedia.org/wikipedia/commons/f/f0/Wz-z.jpg *License:* CC BY-SA 4.0 *Contributors:* Own work *Original artist:* Esadri21

43.11.3 Content license

- Creative Commons Attribution-Share Alike 3.0

www.ingramcontent.com/pod-product-compliance
Lightning Source LLC
Chambersburg PA
CBHW081717220526
45468CB00008B/1874

* 9 7 8 1 9 7 4 0 1 3 3 9 5 *